D0183612

Communications
Electronics
Circuits

Communications Electronics Circuits

SECOND EDITION

J. J. DeFrance

Professor, Department of Electrical Technology
New York City Community College

RINEHART PRESS
SAN FRANCISCO

5643 Paradise Drive
Corte Madera, Calif. 94925

A division of Holt, Rinehart and Winston, Inc.

Library of Congress Catalog Card Number 71-187116

ISBN 0-03-083139-3

9 038 14131211109

Preface

PREFACE TO SECOND EDITION

Since the writing of the first edition, major advances have been made in the semiconductor industry. Consequently, these devices have replaced vacuum tubes in low-power applications. New receiver designs do not use tubes, and so semiconductors—particularly field-effect transistors (FETs)—will be given priority treatment in such applications. Yet, even in receivers, vacuum-tube treatment cannot be eliminated completely because vacuum-tube equipment is still on the market. In fact, equipment is still being manufactured with tubes. On the other hand, in the high-power field (transmitters), the picture has not changed; vacuum tubes are still the only suitable active devices. Whereas one tube can deliver power outputs as much as 400 kilowatts, transistors, at best, are limited to below the 100-watt level.

Because students of communications electronics have shown great interest in obtaining FCC licenses, another change has been made in this second edition. A list of FCC questions, pertinent to each chapter, has been added at the end of each chapter. This is explained more fully in the next sections.

PREFACE TO THE FIRST EDITION

This text is intended for use as a second course in electronics circuitry. It is aimed primarily at the level of the engineering technician. Whereas the previous volume (*General Electronics Circuits*) discussed circuits that might be found in *any* area of the electronics field, this follow-up volume deals specifically with circuits used in the broad area of electronic communications.

Up until about 1940, the "electronics" industry was concerned only with the transmission and reception of voice and music through the air. In fact, the term radio (rather than electronics) was used to describe this field. Today, with industrial and computer applications, electronics has spread until it touches almost every aspect of our lives—not only in the entertainment area, but into medicine, commerce, transportation, and industry. Yet, in spite of these rapid advances, the field of production, transmission, and reception of radio-frequency waves is still—by far—the most important and largest application of electronics, and it is still growing. The intelligence transmitted via radio waves (in addition to voice and music) is now used for aircraft and vessel navigational systems; depth, range, and altitude finders; missile guidance; detection and tracking of moving targets; anticollision devices; and telemetry systems.

The many varieties of radar devices are functionally the same as the radio transmitter at a broadcasting station and the radio receiver at a listener's home. The entire space industry would collapes were it not for the ability to "communicate" with satellite or space vehicle. Control signals are transmitted to the vehicle to change its trajectory, or to start (or stop) some device such as a camera or recorder. Pictures or scientific data gathered by a space probe are sent back to earth stations. All of this requires radio communications.

Prerequisite to understanding the subject matter in this volume is a good foundation in direct-current fundamentals, alternating-current fundamentals, vacuum-tube and semiconductor characteristics, and basic electronic circuitry, such as power supplies and untuned amplifiers. A background in algebra, vector algebra, and basic trigonometry is also required. A knowledge of calculus, although not essential, would enhance an understanding of the mathematical aspects of this book.

The author uses the same direct, personal approach and conversational style that has proved to be successful in previous texts. The emphasis is on practical considerations without sacrificing technical depth or accuracy. Extensive use is made of circuit diagrams, illustrative problems, and problems at the end of each chapter, to illustrate the practical applications of theory. Numerous review questions are also given at the end of each chapter. These can be used for self-evaluation by the student, or for classroom discussion, using the *programmed machine question* technique discussed in the *Instructional Notes*.

Garden City, New York J. J. DEFRANCE

ACKNOWLEDGMENTS

The author is indebted to the following industrial organizations for their cooperation in furnishing photographs, tables, characteristics, schematics, and technical information on their products and on the latest advances in the field:

Andrew Corporation
Bogen Communications Division
Lear Siegler, Inc.
Communications Associates, Inc.
Delco Radio Division
General Motors Corporation
Gates Radio Company
General Electric Company
General Radio Company
Hewlett Packard
P. R. Mallory & Company
McCoy Electronics Company
J. W. Miller Company
Motorola Consumer, Industrial & Military Electronics
Motorola Semiconductor Products, Inc.
Pacific Semiconductors, Inc.
Phelps-Dodge Copper Products Corporation
Radio Corporation of America
H. H. Scott, Inc.
Sylvania Electric Products, Inc.
Texas Instruments, Inc.
Zenith Radio Corporation

The author further wishes to express his deep appreciation to the persons listed below, for their generous assistance in answering his many questions.

Messrs. W. Brack and J. Provenzano of Communications Associates, Inc.
Messrs. B. Trout and E. Mueller of Motorola, Inc.
Messrs. E. O. Johnson and J. L. Preston of R.C.A.
Mr. D. R. Von Recklinghausen of H. H. Scott, Inc.
Mr. S. Weaver of Texas Instruments, Inc.

FCC License

Anyone employed in the operation or maintenance of transmitting equipment in the United States must have an appropriate commercial license from the Federal Communications Commission (FCC). However, even when such a license is not legally required, it is recommended that the license be obtained for its prestige value. It is proof that the holder is highly qualified in his field—in much the same way as a "license" is given to doctors and lawyers who have passed their qualifying examinations. Furthermore, employers often give preference and/or higher salaries to licensed applicants.

There are several FCC licenses. The best one (and the one recommended) is the first-class radiotelephone license. The written examinations for FCC licenses are divided into eight *elements*:

Element 1—Twenty questions on basic law.

Element 2—Fifty questions on basic operating practice.

Element 3—One hundred questions on basic radiotelephone theory.

Element 4—Fifty questions on advanced radiotelephone theory.

Element 5—Fifty questions on radiotelegraph operating practice.

Element 6—One hundred questions on advanced radiotelegraph.

Element 7—One hundred questions on aircraft radiotelegraph (regulation and treaty). (This is for an *Aircraft Endorsement.*)

Element 8—Fifty questions on ship radar techniques (*Radar Endorsement*).

An applicant for the radiotelephone first-class license must pass Elements 1, 2, 3, and 4. To prepare an applicant for these examinations, the FCC provides a manual, *Study Guide and Reference Material for Commercial Radio Operator Examinations*, and a *Supplement* to the manual. These give typical questions for each element. (No answers are given.) Although question-and-answer books are available, it is preferable—in this author's opinion—to study the pertinent theory and then to develop your own answers.

Toward this end, the FCC Study Guide questions have been reshuffled, and those questions pertinent to each chapter have been added at the end of each chapter. A seemingly peculiar numbering system is used to identify each question from the original study guide. The first digit represents the element from which the question was taken; the decimal portion represents the actual number of the question within the element. For example, the text question 4.019 is the original Question 19 in Element 4 of the Study Guide. In similar fashion, the text question S-3.126 is Question 126 in Element 3 of the Supplement.

Note: Since the various licenses do not require passing the same elements, a number of basic questions are repeated from one element to another.

Instructional Notes

Many studies of the psychology of learning have shown that effective learning must involve active participation by the learner, and his correct responses must be "rewarded." Teaching machines developed in keeping with these basic principles have been very successful. In general, these machines give factual information, ask questions (in small steps) based on this information, elicit some form of response, and finally give or confirm the correct answer.

However, machines have serious limitations. One type, although it allows the student to make any answer, merely states the correct response and continues. Another type restricts the student to a choice of one out of only four answers. If a wrong response is selected, the machine indicates why it is wrong and then allows the student to make another response. When he selects the correct response, the program advances to the next step.

In a classroom situation, a live instructor can combine the better features of each type of machine. Not only can he allow a student complete freedom of response, but he can also modify his teaching "program" instantaneously to fit any response.

This text was written with such a teaching technique in mind. In the body of the text, factual information is given and circuit operation is described in more detail than usual, so that the instructor will not have to spend valuable class time lecturing at length to supplement missing items or skimpy treatment. Instead, the lesson time can be spent using a question-discussion-guidance technique, with heavy emphasis on student participation. To help implement this type of lesson, the author has incorporated many review questions at the end of each chapter. These questions follow the text sequence and represent small "bits" of each topic, much in the manner of a teaching machine. Sufficient class time should also be allowed for a satisfying analysis of all problems assigned for homework. If additional time is available, it can be well spent in enriching and motivating each lesson from the instructor's own practical experience.

This system has been tested by the author with several classes, with very gratifying results. Not only were class averages raised, but even more important, the students actually enjoyed these lessons and came to class better prepared to join in the general discussion. An interesting development is that some students began applying this learning technique to their other subjects, making up their own "small-bit" questions for self-study, or bringing the questions into class for discussion.

The letter-symbol nomenclature used with semiconductors and vacuum tubes can at times be quite confusing. To avoid further confusion, the symbols used in this text are in accord with IEEE standards. For ready reference, an explanation of these symbols is given in the Appendix.

Contents

1

Resonant Circuits

In the early 1900s, radio communications—the transmission and reception of voice and music through the air—was probably the only application of electronics of any significance. In radio broadcasting, the intelligence signal, voice or music, is "raised" to some high radio-frequency (RF) level. The specific frequency selected is called the *carrier frequency,* and the process of elevation is called *modulation.* The end result is a *modulated wave,* which contains the original intelligence as a relatively narrow band of component frequencies to either side of the carrier. These components are known as *sidebands.* The total frequency spread of carrier and *sidebands* is often referred to as a *channel.* By using a variety of channels within the radio-frequency spectrum, it is possible to transmit many programs simultaneously. At the receiving locations, the voltage developed by these radio waves is extremely low, in the order of a few microvolts. It is necessary first to amplify these signals by means of *RF voltage or small-signal amplifiers,* and then to extract the intelligence from the amplified RF wave. This latter process, which brings the voice or music back to its own natural frequency range, is called *demodulation* or *detection.* Meanwhile, back at the transmitting location, before modulation can take place, the carrier frequency itself must be generated. This is the function of *oscillator* circuits. Furthermore, if the transmitted signals are to be picked up at remote locations, high transmitting power levels are needed. For this purpose, *RF power amplifiers* are used. Finally, to get these signals into space, *transmission lines* feed the RF energy up to the *antenna,* and the antenna radiates this energy into space.

These terms are presented here merely by way of introduction. They will be discussed in detail in the chapters to follow.

Electronics, since those early days, has spread until it touches almost every aspect of our lives—not only in the entertainment area, but into medicine, commerce, transportation, and industry. Yet, with all this expansion, the field of production, transmission, and reception of radio-frequency waves is still by far the most important and largest application of electronics. The intelligence transmitted via radio waves (in addition to voice and music) now includes aircraft and vessel navigational systems; depth, range, and altitude finders; missile guidance; detection and tracking of moving targets; anticollision devices; and telemetering systems. The same basic functions, mentioned above with regard to radio communications, apply equally well to these newer uses. In addition, high power RF waves are also used for many industrial applications of induction and dielectric heating, including cooking. These industrial applications have increased the importance of radio-frequency circuits in the study of electronics.

Circuits designed to operate at radio frequencies are generally limited to a specific frequency, or to a relatively narrow band of frequencies, within the RF spectrum. The specific frequency is the *resonant* frequency of the circuit, and the narrow band is the *bandwidth* of the circuit. In this chapter these factors are discussed as they apply to series and parallel resonant circuits. The characteristics of these circuits are analyzed in detail, in preparation for the circuit applications to follow in subsequent chapters.

SERIES-RESONANT CIRCUITS

In order for resonance to occur, a circuit must contain inductance and capacitance; it may also (and generally does) have some resistance. This resistance may be the effective resistance of the coil itself (all practical coils have some resistance), or it may be a resistor deliberately introduced to create some desired effect. Although the value of resistance greatly affects current and voltage values of a circuit that is in resonance, it does not determine *when* resonance will occur. Before we tackle the condition known as resonance, let us review briefly the mathematical relationships pertinent to *any* series circuit containing R, L, and C:

1. $X_L = 2\pi fL$
2. $X_C = 1/(2\pi fC)$
3. $X_0 = X_L - X_C$
   where X_0 is the *net* reactance, and is a positive value (inductive)

if X_L is greater than X_C, or a negative value (capacitive) if X_C is greater than X_L.

*4. $Z = \sqrt{R^2 + X_0^2}$

5. $I = E/Z$

Resonance occurs in an R-L-C circuit when the inductive reactance equals the capacitive reactance, $X_L = X_C$. Regardless of the value of inductance and capacitance used, there is always a frequency at which these two reactances are equal. This resonant frequency (f_0) can be found by equating the two reactance values:

$$X_L = X_C \qquad \text{or} \qquad 2\pi f_0 L = \frac{1}{2\pi f_0 C}$$

Solving for frequency, we get $(2\pi f_0)^2 = 1/LC$ and

$$f_0 = \frac{1}{2\pi\sqrt{LC}} \tag{1-1}$$

At this resonant frequency (since $X_L = X_C$), the net reactance (X_0) is zero. Obviously, then, the circuit impedance (Z) must be a minimum and equal to the circuit resistance R. Since the impedance is a minimum, the line current will be a maximum, and since the impedance is purely resistive, the line current is in phase with the line voltage. The circuit phase angle (θ_c) is zero. Summarizing these key points concerning a series-resonant circuit, we get

1. $X_L = X_C$
2. $f_0 = \dfrac{1}{2\pi\sqrt{LC}}$
3. $Z = \text{minimum} = R$
4. $I = \text{maximum}$
5. $\theta_c = 0$ deg

EXAMPLE 1–1

A series circuit consists of $L = 15.8$ millihenries (mH), $C = 0.1$ microfarad (μF), $R = 10$ ohms (Ω), and a line voltage $E_T = 10$ volts (V). Find the resonant frequency, the current at this frequency, and the voltage across each component at this frequency.

*As a simplification, it should be noted that $Z = X_0$ if the reactance is equal to or greater than ten times the resistance value. (The error from such an assumption is less than 1 percent even for $X = 8R$.) Similarly, the impedance Z can be considered to be equal to the resistance value, for $X = 0.1R$ or smaller.

Solution

1. $f_0 = \dfrac{1}{2\pi\sqrt{LC}} = \dfrac{0.159}{\sqrt{15.8 \times 10^{-3} \times 0.1 \times 10^{-6}}} = \mathbf{4000\ Hz}$

2. Since the circuit is resonant ($X_L = X_C$), $Z = R$, and

$$I = \frac{E}{R} = \frac{10}{10} = \mathbf{1.0\ A}$$

3. To find the voltage across L and across C we must first find X_L and X_C.

(*a*) $X_L = 2\pi f L = 2\pi \times 4000 \times 15.8 \times 10^{-3} = 398\ \Omega$

(*b*) $X_C = \dfrac{1}{2\pi f C} = \dfrac{0.159 \times 10^6}{4000 \times 0.1} = 398\ \Omega$

(Notice that $X_L = X_C$. This checks our calculations so far.)

(*c*) $E_L = IX_L = 1.0 \times 398 = 398\ \text{V}$

$E_C = IX_C = 1.0 \times 398 = 398\ \text{V}$

$E_R = IR = 1.0 \times 10 = 10\ \text{V}$

Q Rise in Voltage

Notice the value of the voltage across the reactive components in the above problem. With a line voltage of only 10 V, the voltages across the inductor and capacitor are each 398 V. Such a rise will occur in any resonant circuit—if the resistance is low compared to the inductive and capacitive reactances. But the ratio of inductive reactance to resistance—with reference to a coil—is a measure of the quality of the coil and is known as the Q of the coil. Similarly, in a complete circuit, the ratio of reactance to circuit resistance is known as the *Q of the circuit.* It does not matter here whether the circuit resistance is due to the coil, the capacitor, or to a separate resistive component. The resistance is *total resistance R_T*. In calculating the circuit Q at resonance, either reactance (X_L or X_C) can be used, since they are equal values.

To find the Q of the circuit in the above problem, we proceed as follows: the reactances X_L and X_C are each 398 Ω and the circuit resistance is 10 Ω; therefore,

$$Q_c = \frac{X_L}{R_T} = \frac{398}{10} = 39.8$$

Now compare the coil or capacitor voltage to the line voltage. Notice that these reactive voltages are 39.8 times greater than the line voltage, or exactly Q times the line voltage. From this we can draw the general conclusion that *in a series-resonant circuit there is a Q rise in voltage.* This effect can also be shown mathematically as follows:

$$E_L = IX_L$$

At resonance

$$I = \frac{E_T}{R_T}$$

Replacing I by its algebraic equivalent, we get

$$E_L = \frac{E_T X_L}{R_T}$$

But the ratio of reactance to resistance is Q_c. Therefore

$$E_L = Q_c E_T \qquad (1\text{–}2)$$

Similarly it can be shown that

$$E_C = Q_c E_T \qquad (1\text{–}3)$$

To the five key points for a series-resonant circuit we can now add a sixth:

6. $E_L = E_C = Q_c E_T$

Effect of Frequency

In the above problem, we quite readily calculated the circuit values at the resonant frequency. However, if the frequency of the supply source were changed, the problem would become more complex. It would be necessary to calculate the reactance, impedance, current, and voltage values for various input frequencies. Table 1–1 shows the effect of frequency on the circuit used in Example 1–1. To simplify the numerical aspects, we have taken the resonant value of X_L and X_C as 400 Ω (instead of 398).

At a frequency of 1000 Hz (as compared to the resonant frequency of 4000 Hz), the inductive reactance must decrease in proportion: $X_L = 100$ Ω. X_C must increase by the same proportion to 1600 Ω. The net reactance $X_0 = X_L - X_C$ will rise to -1500 Ω, the negative sign indicating that it is capacitive. The circuit phase angle (arctan X/R = arctan 150) is practically 90°. Since the reactance is much greater than $10R$, the circuit impedance Z can be considered to be 1500 Ω and purely capacitive. The line current is $I = E/Z$ or 6.70 mA. The component voltages E_C and E_L are respectively IX_C and IX_L, or 10.7 V and 0.67 V. Now, starting from these values at 1000 Hz, we can, by proportion, readily find the circuit values for several other frequencies. These are given in Table 1–1.

Table 1–1
EFFECT OF FREQUENCY ON CIRCUIT VALUES
($L = 15.8$ mH $C = 0.1\ \mu$F $R = 10\ \Omega$)

FREQUENCY (Hz)	X_L (Ω)	X_C (Ω)	X_0 (Ω)	θ_c	Z (Ω)	Z (CHARACTER)	I (mA)	E_C (V)	E_L (V)
1000	100	1600	−1500	90°	1500	Capacitive	6.70	10.7	0.67
2000	200	800	−600	89°	600	Capacitive	16.70	13.3	3.34
3000	300	533	−233	88°	233	Capacitive	43.00	25.1	12.9
4000	**400**	**400**	**0**	**0°**	**10**	**Resistive**	**1 A**	**400**	**400**
5000	500	320	+180	87°	180	Inductive	55.00	18.1	27.5
8000	800	200	+600	89°	600	Inductive	16.70	3.34	13.3
16000	1600	100	+1500	90°	1500	Inductive	6.70	0.67	10.7

Analyzing and summarizing these results, we get:

1. *At resonance*
   (*a*) The impedance is minimum and resistive (equal to R).
   (*b*) The current is at a maximum and is in phase with the line voltage.
   (*c*) There is a resonant rise in voltage: $E_C = E_L = Q_c E_T$
2. *Decreasing frequencies—below resonance*
   (*a*) The impedance increases rapidly and becomes capacitive.
   (*b*) The line current drops rapidly, then tends to level off.
   (*c*) E_L and E_C also drop rapidly and then tend to level off.
3. *Increasing frequencies—above resonance*
   (*a*) The impedance increases rapidly but this time becomes inductive.
   (*b*) The line current again drops rapidly, then tends to level off.
   (*c*) E_L and E_C also drop rapidly and then tend to level off.

This variation of impedance with frequency is shown in Fig. 1–1.

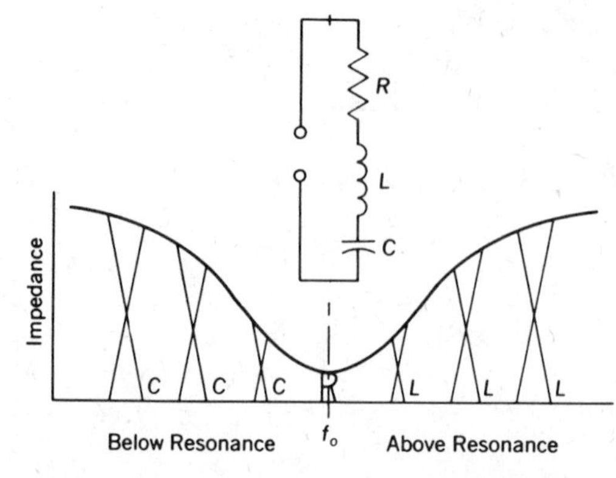

Figure 1-1 The effect of frequency on the impedance of a series-resonant circuit.

Effect of Circuit Resistance

It has already been mentioned in this chapter that the value of circuit resistance has no effect on resonant frequency, but that it does have a great effect on the current and voltage values at resonance. Let us analyze this more carefully and see also what happens at frequencies above and below resonance. For a numerical analysis we can use the *L-C* circuit values given in Table 1–1, raising the circuit resistance from 10 Ω to 20 and 60 Ω (see Table 1–2). It should be obvious that the net reactance X_0 will be the same as before.

Table 1–2

EFFECT OF RESISTANCE ON CIRCUIT VALUES

($L = 15.8$ mH $C = 0.1$ μF)

FREQUENCY (Hz)	X_0 (Ω)	$R = 20$ Ω			$R = 60$ Ω		
		Z (Ω)	I (mA)	E_c (V)	Z (Ω)	I (mA)	E_c (V)
1000	−1500	1500	6.70	10.7	1500	6.70	10.7
2000	−600	600	16.70	13.3	600	16.70	13.3
4000	**0**	**20**	**500**	**200**	**60**	**167.00**	**66**
8000	+600	600	16.70	3.34	600	16.70	3.34
16000	+1500	1500	6.70	0.67	1500	6.70	0.67

Notice from Table 1–2 that even the 60-Ω circuit resistance is negligible in comparison to the circuit net reactance, when off resonance. Consequently, except at resonance—or for a very small frequency change to each side—the total circuit impedance is unaffected by the change in resistance. Compare the line current values and the E_C values for $R = 10$, 20, and 60 Ω. Here again you can see that, at off resonance, there is no effect. At resonance, however, notice how much lower the voltage across the capacitor is when $R = 60$, as compared to the 400-V value when $R = 10$ Ω. Because of the higher resistance, the circuit Q is much lower, and the resonant rise is not as great. Figure 1–2 summarizes, in a qualitative way, the effect of resistance on the line current and on the voltage

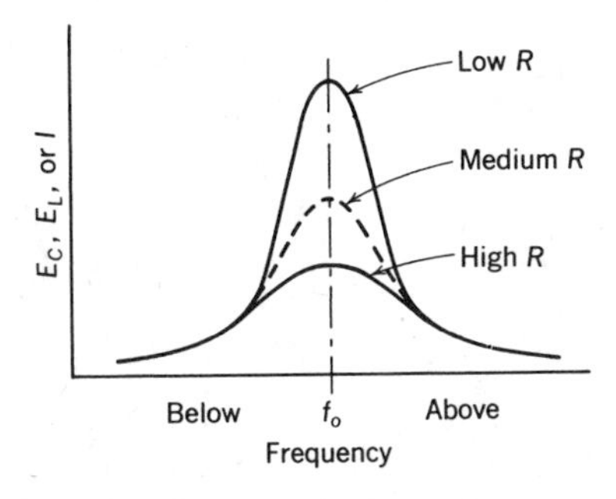

Figure 1-2 The effect of resistance on the frequency response curve of a series-resonant circuit.

across the capacitor or coil in a series-resonant circuit. These curves are known as *response curves* or *resonance curves*. The sides or *skirts* of the curve are much steeper when the circuit resistance is low. The low-resistance curve is a sharp curve, whereas a high-resistance value produces a broad resonance curve. The significance (or application) of this feature is explained in more detail further on in this chapter.

Effect of L/C Ratio

In Example 1–1, a 15.8-mH coil and a 0.1-μF capacitor were used to resonate the circuit at 4000 Hz. (The circuit resistance was 10 Ω.) If the inductance value is increased by ten times (to 158 mH), we can still obtain resonance at 4000 Hz if we simultaneously *reduce* the capacitance value to one-tenth of its previous value, or 0.01 μF. Since the LC product is the same, the resonant frequency is unchanged. However, these changes increase the *L/C ratio* to 100 times its previous value. Let us see what effect, if any, this change will produce. To simplify the discussion, we will assume that the change in total circuit resistance is negligible.* Also, instead of employing a mathematical solution of the problem as previously done, we will this time use a qualitative analysis.

1. *At resonance,* X_L will still equal X_C, and the net reactance will be zero. The impedance is purely resistive and equal to R. Since the circuit resistance has not changed, Z is unchanged, and the line current value—at resonance—is not affected. However, X_L and X_C each increases tenfold. Therefore the circuit Q and the voltages across L and C increase in proportion to the increase in reactance. Consequently, a higher L/C ratio produces a higher Q rise in voltage.
2. *Below resonance,* because of the higher L/C ratio, the capacitor is smaller, and X_C increases more rapidly as the frequency is reduced. Meanwhile, even though the inductance value and the inductive reactance are also larger than in Example 1–1, the value of X_L is still negligible compared to the increased X_C value. Therefore, the net reactance increases, and the circuit impedance Z increases with an increase in the L/C ratio. The line current will therefore *decrease* by the same factor. The voltages across the capacitor and coil are the product of line current and the respective reactance. The decrease in current is offset by the increase in reactance, and therefore *below resonance* these voltages are not affected by the change in L/C ratio.

*Although the ohmic resistance of the coil may increase, the effective resistance of the coil may not increase much. Furthermore the coil resistance is only part of the total circuit effective resistance. This assumption is therefore quite valid.

3. *Above resonance,* because of the larger inductance value, X_L rises rapidly. The capacitive reactance (even though larger because of the smaller capacitor) is negligible in comparison to this higher X_L value. The net reactance and total circuit impedance again increase with an increase in the L/C ratio. As above, the line current is lower, but E_C and E_L—at *off resonance*—will not be affected by the change in L/C ratio.

Conversely, if the L/C ratio is reduced (but still maintaining the same resonant frequency and circuit resistance), the circuit Q is reduced; the Q rise at resonance is reduced, but the skirts of the response curve—at off resonance—are not affected. Figure 1–3 shows the effect of L/C ratio on the response curve. An example will verify these conclusions.

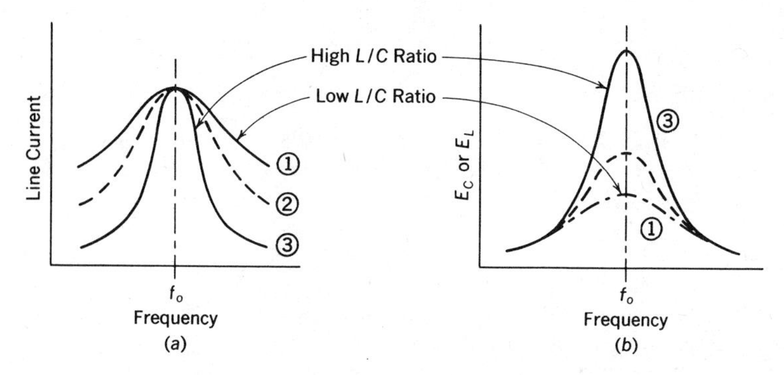

Figure 1-3 The effect of the L/C ratio on the response curve of a series-resonant circuit.

EXAMPLE 1–2

Compare the line current and the voltage across the capacitor of the circuit in Example 1–1 ($L = 15.8$ mH, $C = 0.1\ \mu$F, $R = 10\ \Omega$, and $E_T = 10$ V) at frequencies of 1000, 4000, and 16,000 Hz for (1) L increased to 158 mH and $C = 0.01\ \mu$F and (2) L decreased to 1.58 mH and $C = 1.0\ \mu$F.

Solution

In each of the above cases, the LC product remains the same; therefore, the resonant frequency is still 4000 Hz. The reactance, impedance, line current, and capacitor voltage for three key frequencies from Table 1–2 are repeated in Table 1–3, and for each of these frequencies the values for the new L/C ratios are added. Reactance and impedance values can be obtained by proportions, and current and voltage values by Ohm's law.

Notice that the results from this problem agree with the qualitative analysis. At resonance (4000 Hz), the circuit impedance and line current are not affected by the L/C ratio, but the Q rise in voltage varies directly

Table 1–3
COMPARATIVE EFFECTS OF L/C RATIO (R = 10 Ω)

FREQUENCY (Hz)	CIRCUIT CONDITIONS L	C	X_L (Ω)	X_C (Ω)	Z (Ω)	I (mA)	E_C (V)
	15.8 mH	0.1 μF	100	1600	−1500	6.70	10.7
1000	158 mH	0.01 μF	1000	16,000	−15,000	0.67	10.7
	1.58 mH	1.0 μF	10	160	−150	67.00	10.7
	15.8 mH	0.1 μF	400	400	10	1 A	400
4000	158 mH	0.01 μF	4000	4000	10	1 A	4000
	1.58 mH	1.0 μF	40	40	10	1 A	40
	15.8 mH	0.1 μF	1600	100	+1500	6.70	0.67
16,000	158 mH	0.01 μF	16,000	1000	+15,000	0.67	0.67
	1.58 mH	1.0 μF	160	10	+150	67.00	0.67

with the inductance value. On the other hand, at off resonance—above or below—the voltage E_C is not affected by the L/C ratio, but the circuit impedance varies directly with the inductance value, and the line current varies inversely with this value.

Gain of a Resonant Circuit

At radio frequencies, a series-resonant circuit is often used to obtain a voltage gain, in a manner somewhat similar to a step-up, untuned, iron-core transformer. However, whereas the iron-core transformer is intended to step up *all* input signal voltages by the same ratio, the series-resonant circuit provides this step-up or voltage gain only at or near its resonant frequency. To obtain this voltage gain, the input signal voltage is fed to the total series circuit and the output signal voltage is obtained from across the capacitor only (or across the coil). This is shown in Fig. 1-4. Since the output voltage is the capacitor voltage E_C, the amount of

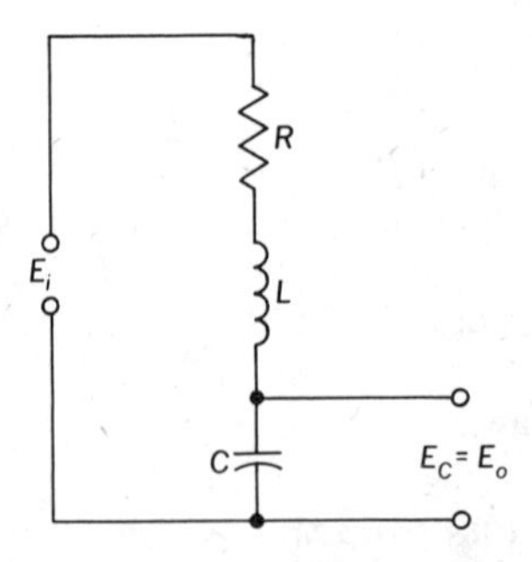

Figure 1-4 A series-resonant circuit used for voltage gain.

voltage gain is determined by the Q of the circuit, and this, in turn, is dependent on the circuit resistance and on the L/C ratio. Notice how this is substantiated by Figs. 1–2 and 1–3. A higher resonant rise, and therefore a higher gain, is obtained when the circuit resistance is low (Fig. 1–2) and when the L/C ratio is high (Fig. 1–3(b)).

Sharpness of Resonance—Bandwidth

The major application of resonant circuits is in transmitters and receivers. Regardless of whether these transmitters and receivers are used for television entertainment, or whether they are used to transmit control signals or receive information from an earth satellite, the resonant circuits have two closely related functions: (1) to produce, or pass through, a specific frequency (the carrier) or a band of frequencies (the channel) and (2) to reject, or stop passage of, any other frequency not within the channel band. The success of this "mission" of a resonant circuit is dependent upon the *selectivity* or *sharpness of resonance* of the response curve. A circuit has good selectivity if the sides of the response curve drop off steeply. Compare the curves of Fig. 1–2. Obviously, the curve for the low-resistance circuit has the best selectivity. Now examine the curves of Fig. 1–3(b). Curve 3 has the highest L/C ratio, the steepest sides, and the highest selectivity. This leads us to a general conclusion. If high selectivity is required, we should select or design our resonant circuit so as to have a minimum of series resistance and as high an L/C ratio as possible.

High selectivity (a steep-sided curve) is necessary if we wish to amplify or pass one specific frequency while stopping or rejecting another frequency close by. In this respect, selectivity could be defined as the ratio of the response at resonance (the desired signal) to the response at some frequency off resonance (the undesired signal). Figure 1–5 shows this more clearly. Curve *a* has moderately high selectivity. The undesired signal, at frequency f_1, has only twenty percent of the amplitude of the desired signal. On the other hand, curve *b* has rather poor selectivity. For the same two signals at frequencies f_1 and f_0, the undesired signal (f_1) develops almost as strong an output level as the desired signal.

Unfortunately, selectivity is not the only desirable characteristic wanted in a resonant circuit. It was mentioned that such a circuit may have to pass a *band of frequencies*. The ideal response curve should therefore be flat (horizontal) at the top, and the span or *bandwidth* of this flat top should cover the frequency band that it is desired to pass. The sides or skirts of this curve should then drop straight down, vertically. The response to any frequency outside of the desired band would be zero. Such

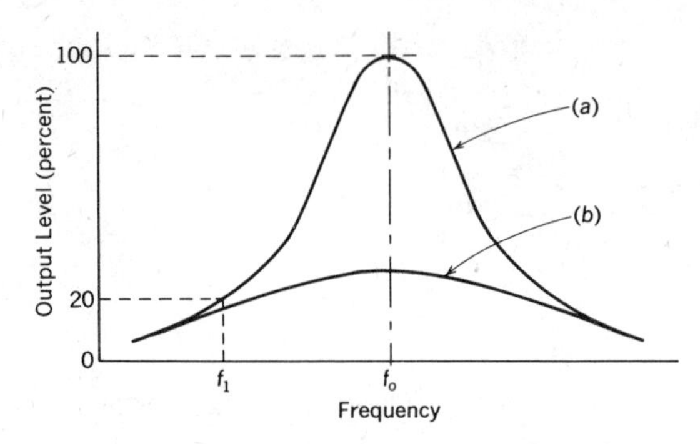

Figure 1-5 The effect of sharpness of resonance on adjacent channel selectivity.

an ideal response curve is shown in Fig. 1–6 as curve *a*. The bandwidth of this response curve is 70 kilohertz (565–635 kHz).

The simple series-resonant circuit cannot produce this ideal response. Instead, it produces more of a "bell-shaped" response, shown by curve *b* of Fig. 1–6. Notice that this curve has the same total amplitude and, in general, spans the same frequency range as the ideal curve. But, because of the sloping sides, how do we evaluate the bandwidth of curve *b*? If we measure it near the peak, it will be very narrow. If we measure it near the base of the curve, it will be much wider. To remove this ambiguity, the measuring points have been standardized at a level corresponding to an output (voltage) equal to 70.7% or 0.707 of the maximum value. This corresponds to the half-power point. Expressed in decibels, the output level at the extremes of the frequency band is down 3 dB.

In the discussion of sharpness of resonance, it was concluded that the lower the circuit resistance and the higher the inductance (for the same frequency), the sharper the response curve. But these two factors

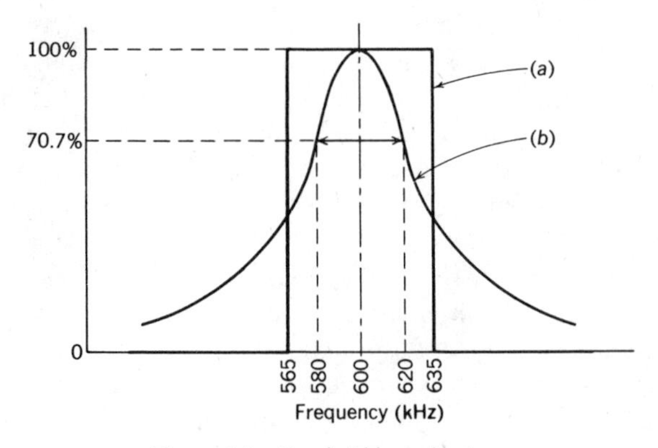

Figure 1-6 Bandwidth evaluation.

determine the Q of the circuit. Therefore a high-Q circuit will have a sharp response curve. Since bandwidth is the opposite of sharpness, low-Q circuits will provide wider bandwidth, that is, the bandwidth of any circuit varies inversely with the circuit Q. Actually, for a given Q, the bandwidth also depends on the resonant frequency, or

$$\text{BW} = \frac{f_0}{Q} \tag{1-4}$$

EXAMPLE 1–3

A series R-L-C circuit is resonant at 1200 kHz and has a Q of 80 at this frequency. Find the bandwidth.

Solution

$$\text{BW} = \frac{f_0}{Q} = \frac{1200\ (\text{kHz})}{80} = \mathbf{15\ kHz}$$

EXAMPLE 1–4

A series circuit is to be designed that will be resonant at 300 kHz and have a bandwidth of 60 kHz. What value of Q should be used?

Solution

$$Q = \frac{f_0}{\text{BW}} = \frac{300}{60} = \mathbf{5}$$

Example 1–4 was introduced to emphasize the fact that bandwidth and Q are inversely related, and that if a wide bandwidth is desired, the resulting Q may be very low. This is bad. Not only will the adjacent-channel selectivity be very poor, but the gain of the resonant circuit will be very low. We can avoid this effect and still get a wide bandwidth if we operate the circuit at a higher resonant frequency. For example, if in the above problem the operating frequency is raised from 300 to 6000 kHz, the desired 60-kHz bandwidth can be obtained with a Q of 100. This is a factor in the selection of suitable intermediate-frequency values for superheterodyne receivers.

General Resonance Curve

If we wanted to know the frequency-response (E_o versus f) characteristic of a series R-L-C circuit, we could obtain it from a point-by-point calculation. Using conventional ac circuit theory, we would have to calculate X_L, X_C, X_0, Z, I, and E_C for a number of frequencies to either side of resonance, and then plot these E_C values versus frequency. This is a slow, laborious procedure. There is a much simpler technique using resonant circuit theory. This method makes one assumption—that for a small frequency change, to either side of resonance, the Q of the circuit remains constant at the resonant value. (The error so introduced is less than 1%.) With this assumption, it can be proven mathematically* that the relative output drops off as follows:

1. At f_0, output = 100%
2. At a frequency deviation of $\pm\frac{1}{2}$ BW, output = 70.7%
3. At a frequency deviation of ± 1 BW, output = 44.7%
4. At a frequency deviation of $\pm 1\frac{1}{2}$ BW, output = 32%
5. At a frequency deviation of ± 2 BW, output = 24%
6. At a frequency deviation of ± 4 BW, output = 13%

EXAMPLE 1-5

Plot the frequency response curve for a series circuit wherein $C = 160$ pF, $L = 250\ \mu$H, and the circuit resistance is 12.57 Ω.

Solution

1. $f_0 = \dfrac{1}{2\pi\sqrt{LC}} = \dfrac{0.159}{\sqrt{160 \times 10^{-12} \times 250 \times 10^{-6}}} = 795 \text{ kHz}$
2. $X_L = 2\pi f L = 2\pi \times 795 \times 10^3 \times 250 \times 10^{-6} = 1250\ \Omega$
3. $Q = \dfrac{X_L}{R} = \dfrac{1250}{12.57} = 99.4$
4. $\text{BW} = \dfrac{f_0}{Q} = \dfrac{795}{99.4} = 8.0 \text{ kHz}$
5. At 795 kHz, the output level = 100%

   At 791 and 799 kHz ($\pm\frac{1}{2}$ BW), output = 70%

   At 787 and 803 kHz (± 1 BW), output = 45%

   At 783 and 807 kHz ($\pm 1\frac{1}{2}$ BW), output = 32%

   At 779 and 811 kHz (± 2 BW), output = 24%

   At 763 and 827 kHz (± 4 BW), output = 13%

These points are plotted, and the response curve is shown, in Fig. 1-7.

* F. E. Terman, *Electronic and Radio Engineering* (New York, McGraw-Hill, Inc., 1955), Chap. 3.

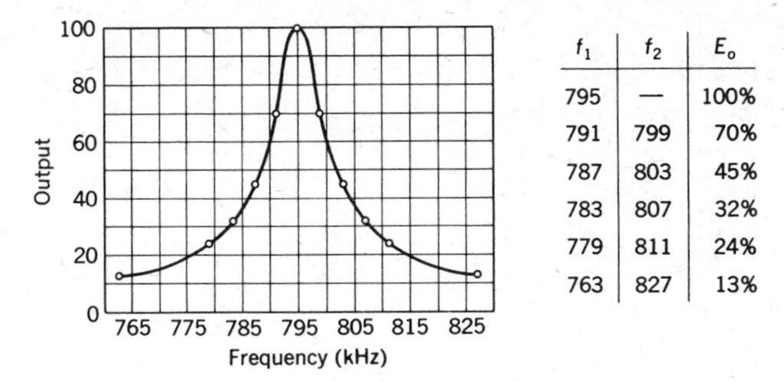

f_1	f_2	E_o
795	—	100%
791	799	70%
787	803	45%
783	807	32%
779	811	24%
763	827	13%

Figure 1-7 Response curve for Example 1-5.

PARALLEL-RESONANT CIRCUITS

When a coil and capacitor are connected in parallel to each other—*with respect to the supply source*—a condition known as parallel resonance can occur. Before discussing this condition, let us review the relationships that apply to a two-branch *L-C* circuit at any frequency. Such a circuit is shown in Fig. 1–8. This *L-C* parallel circuit is often referred to as a *tank* circuit. Branch 1 contains the capacitor. In general, the losses in a capacitor are negligible. The wiring resistances are also negligible; so this branch is shown as a *pure* capacitive branch. Branch 2 contains the coil. The coil will have some effective resistance. Furthermore, as is

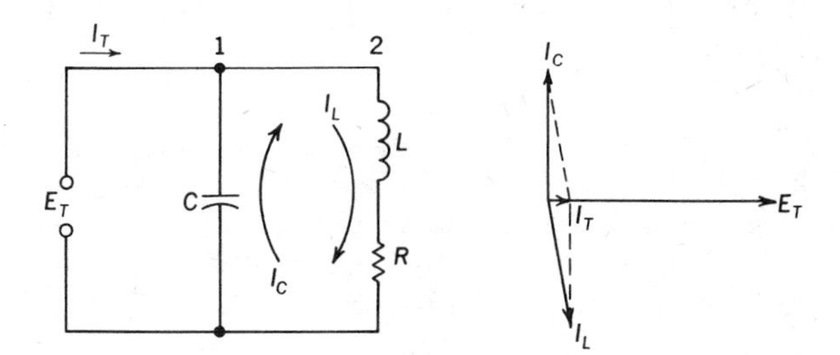

Figure 1-8 Parallel circuit.

seen later, any additional circuit resistance is also generally associated with the inductance. Therefore this branch is shown as containing *L* and *R*. However, in any well-designed tank circuit, the reactance of the inductive branch is at least ten times the resistance value. In other words, the *Q* of this branch is at least 10. In such a circuit—*at any frequency*—we find that:

1. $I_1 = I_C = \dfrac{E_T}{Z_1} = \dfrac{E_T}{X_C}$
2. I_1 (or I_C) will lead the line voltage by 90°
3. $I_2 = I_L = \dfrac{E_T}{Z_2} = \dfrac{E_T}{X_L}$

(Since the Q of the inductive branch is 10 or higher, we can consider that Z_2 is equal to X_L.)

4. I_2 (or I_L) will lag the line voltage by *approximately* 90°
5. $I_T = \dot{I}_1 + \dot{I}_2$ (a phasor addition)
6. $Z_T = \dfrac{E_T}{I_T}$

Resonant Relations

PERFECT COIL. In the ideal tank circuit, both branches are perfect reactances, the inductive branch *resistance* being zero. Resonance occurs (as in the series circuit) when $X_L = X_C$. Therefore the frequency at resonance is

$$f_0 = \frac{1}{2\pi\sqrt{LC}}$$

Since $X_L = X_C$, then $I_L = I_C$, and since the inductive branch resistance is zero, these two currents are not only equal but also opposite in phase. The total line current must be zero. The impedance of this parallel circuit at resonance is infinite.

PRACTICAL CASE. The above case is, of course, a theoretical one, since the resistance of the inductive branch is not zero. However, in a practical case, the Q of the inductive branch is generally 10 or higher, and the resonant frequency is still the frequency at which $X_L = X_C$. Also, the two branch currents I_L and I_C are equal in *magnitude*. However, since the inductive branch current is not quite at 90° lagging, the phasor sum of the branch currents is not zero—but it is a minimum value. This resultant current I_T is in phase with the line voltage. Finally, since the line current is a minimum, the impedance of the parallel circuit ($Z_{\parallel}$) is a maximum; and since this current is in phase with the voltage, this impedance is resistive. Summarizing these key points for a parallel-resonant circuit:

1. $X_L = X_C$ and $I_L = I_C$
2. $f_0 = \dfrac{1}{2\pi\sqrt{LC}}$

3. I_T = minimum
4. $Z_{\|}$ = maximum and resistive
5. $\theta_c = 0°$

Parallel Circuit Impedance

Qualitatively, the impedance of a parallel-resonant circuit is a maximum and is resistive. Furthermore, the higher the Q of the circuit, the closer this impedance approaches infinity. The actual value can be determined as follows. In any two-branch tank circuit, the parallel impedance ($Z_{\|}$) can be found from product over sum, or

$$Z_{\|} = \frac{\dot{Z}_1 \times \dot{Z}_2}{\dot{Z}_1 + \dot{Z}_2} \tag{1-5}$$

but $Z_1 = X_C$, and if $Q = 10$ or higher, $Z_2 = X_L$. Also $\dot{Z}_1 + \dot{Z}_2$ is actually the series impedance (Z_s) of the three components C, L, and R. Therefore

$$Z_{\|} = \frac{X_L \times X_C}{Z_s}$$

But, at resonance, the series impedance Z_s is equal to the resistance R of the inductive branch. Also $X_L = X_C$. Therefore *at resonance*

$$Z_{\|} = \frac{X_L \times X_C}{R} = QX_C \tag{1-6}$$

or

$$Z_{\|} = \frac{X_L^2}{R} = QX_L \tag{1-7}$$

Furthermore, replacing X_L by $2\pi fL$ and X_C by $1/(2\pi fC)$, we get

$$Z_{\|} = \frac{2\pi fL}{2\pi fC \times R} = \frac{L}{CR} \tag{1-8}$$

There we have a variety of equations for calculating the impedance of a parallel-resonant circuit. (Actually they are merely variations of the same equation.) Sometimes one form is more convenient than another, but any one of these provides a much more direct solution than the step-by-step general ac circuit method.

EXAMPLE 1–6

A parallel circuit has a capacitor of 100 pF in one branch and an inductance of 100 μH plus a resistance of 10 Ω in the second branch. The line voltage is 100 V. Find: f_0, I_L, I_C, I_T, and $Z_{\|}$.

Solution

1. $f_0 = \dfrac{1}{2\pi\sqrt{LC}} = \dfrac{0.159}{\sqrt{100 \times 10^{-6} \times 100 \times 10^{-12}}} = \mathbf{1590\ kHz}$

2. $Z_{\|} = \dfrac{L}{CR} = \dfrac{100 \times 10^{-6}}{100 \times 10^{-12} \times 10} = \mathbf{100{,}000\ \Omega}$

3. $I_T = \dfrac{E_T}{Z_{\|}} = \dfrac{100}{100{,}000} = \mathbf{1\ mA}$

4. (*a*) $X_L = 2\pi fL = 2\pi \times 1590 \times 10^3 \times 100 \times 10^{-6} = 1000\ \Omega$

   (*b*) $Z_L = X_L$ (since X_L is greater than $10R$)

   (*c*) $I_L = \dfrac{E_T}{Z_L} = \dfrac{100}{1000} = \mathbf{100\ mA}$

5. $I_C = I_L = \mathbf{100\ mA}$ (resonance)

Notice the directness with which this problem can be handled using resonant-circuit theory. For comparison, this same problem is assigned at the end of this chapter for solution using general ac circuit theory (see Problem 16).

Circulating Current

The circuit Q of a parallel circuit is found (as in series circuits) from the ratio of X_L to R. In the above problem it would be

$$Q_c = \frac{X_L}{R} = \frac{1000}{10} = 100$$

Now also notice the ratio of capacitor and coil branch currents to the line current. Again, we have a ratio of 100. The conclusion is obvious. *In a parallel-resonant circuit, there is a Q rise in current,* that is, the branch currents (I_L and I_C) are equal and are Q times greater than the line current I_T. These currents circulate round and round in the tank circuit and are therefore referred to as the *circulating current.* However, although equal *in magnitude,* these currents are not exactly in phase-opposition because of the inductive branch resistance (see Fig. 1–8(b)). In Example 1–6, I_c leads the voltage by 90°, but the inductive branch current I_L lags by 89.45°.* The in-phase component of $I_L = I_L \cos\theta = 100 \times 0.01 = 1.0$ mA. This is the line current, and since it is the in-phase component of I_L, it is in phase with the line voltage.

*$Q = 100$; therefore, $\tan\theta = 100$, $\theta = 89.45°$, and $\cos\theta = 0.0100$.

Effect of Frequency

We have seen that the impedance of the parallel circuit is at a maximum and is resistive at resonance. Now let us analyze what happens to this impedance for frequencies above and below resonance. (For this discussion, refer to Fig. 1–8.) Since at resonance $X_L = X_C$, $I_L = I_C$ and the line current $(\dot{I}_L + \dot{I}_C)$ is a very low value and is in phase with the line voltage. If the input signal frequency is increased above resonance, then:

1. X_L increases and I_L decreases;
2. X_C decreases and I_C increases;
3. the resultant line current will be capacitive and will lead the line voltage; and
4. the circuit acts as a *capacitive* circuit.
5. Furthermore, the greater the frequency swing above resonance, the greater the imbalance of I_C over I_L and the higher the resultant line current. Effectively, then, the impedance decreases rapidly as the frequency rises above the resonant value.

By similar reasoning, it should be readily obvious that, for frequencies below resonance, the inductive branch current increases, the capacitive branch current decreases, and the line current lags the line voltage. The impedance of the tank circuit is inductive and decreases as the frequency is reduced below resonance. This impedance variation with frequency is shown in Fig. 1–9.

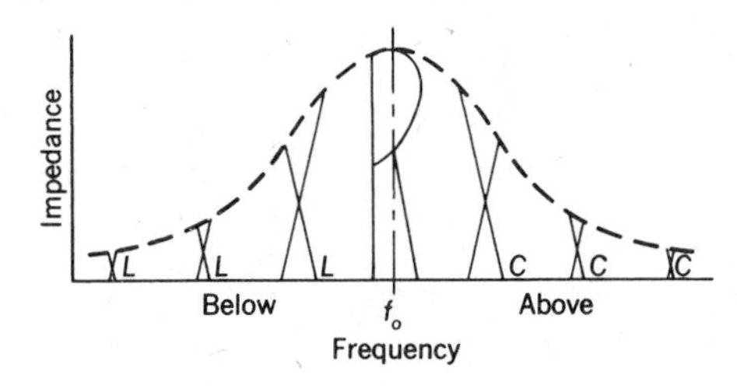

Figure 1-9 Variation of the impedance of a parallel (tank) circuit with frequency.

Now let us consider the same problem from a quantitative approach. The impedance of any two-branch tank circuit is given by:

$$Z_{\|} = \frac{\dot{Z}_1 \times \dot{Z}_2}{\dot{Z}_1 + \dot{Z}_2}$$

If the Q of the tank circuit is 10 or more, Z_1 and Z_2 can be replaced by their equivalent values of X_C and X_L. Furthermore, *for small frequency deviations* to either side of resonance, the increase (or decrease) in the X_L

value is offset by an opposite change in the X_C value, and the numerator may be considered as remaining constant at the *resonant value* of X_L^2. Since the denominator is the phasor sum of Z_1 and Z_2, it is actually the series impedance Z_s of the two branches. The equation for the impedance of the tank circuit—for small deviations off resonance—simplifies to

$$Z_{\|} = \frac{X_L^2}{Z_s} \tag{1-9}$$

where X_L is the inductive reactance *at resonance* and Z_s is the series impedance *at the desired frequency*.

EXAMPLE 1–7

Find the impedance of the tank circuit of Example 1–6 ($C = 100$ pF, $L = 100$ μH, and $R = 10\ \Omega$) at frequencies of 10 kHz below and 80 kHz above resonance.

Solution

1. From Example 1–6:
   (*a*) $f_0 = 1590$ kHz
   (*b*) $X_L = 1000\ \Omega$; $Z_{\|} = 100{,}000\ \Omega$ at resonance.
2. At 1580 kHz (10 kHz below resonance):
   (*a*) $X_L = 1000 \times \frac{1580}{1590} = 994\ \Omega$
   (*b*) $X_C = 1000 \times \frac{1590}{1580} = 1005\ \Omega$
   (*c*) $X_0 = X_L - X_C = 1005 - 994 = 11\ \Omega$
   (*d*) $Z_s = \sqrt{R^2 + X_0^2} = \sqrt{10^2 + 11^2} = 14.9\ \Omega$
   (*e*) $Z_{\|} = \frac{X_L^2}{Z_s} = \frac{1000 \times 1000}{14.9} = \mathbf{67{,}200\ \Omega}$
3. At 1670 kHz (80 kHz above resonance):
   (*a*) $X_L = 1000 \times \frac{1670}{1590} = 1050\ \Omega$
   (*b*) $X_C = 1000 \times \frac{1590}{1670} = 952\ \Omega$
   (*c*) $X_0 = X_L - X_C = 1050 - 952 = 98\ \Omega$
   (*d*) $Z_s = \sqrt{R^2 + X_0^2} = \sqrt{10^2 + 98^2} = 98.6\ \Omega$

Notice that, at this frequency off resonance, the series resistance becomes negligible compared to the reactance.

(e) $Z_{\|} = \dfrac{X_L^2}{Z_s} = \dfrac{1000 \times 1000}{98.6} = \mathbf{10{,}150\ \Omega}$

Effect of Series Resistance

If we plot a curve of impedance versus frequency for a parallel-resonant circuit, we get the same type of bell-shaped response curve as we got for the series-resonant circuit (see Fig. 1–2). In the series circuit, however, we plotted current (or E_C or E_L) versus frequency. Resistance in the series circuit reduced the height (gain) of the curve and broadened or increased the bandwidth. Now let us see what effect a change in the *series* resistance (in the inductive branch) will have on a parallel-resonant circuit. At resonance, the impedance is given by L/CR. Obviously any increase in the series resistance will have an *inverse* effect on the circuit impedance. More specifically, if the series resistance is doubled, the circuit impedance is cut in half. Also, since the Q of the circuit is given by X_L/R, an increase in resistance will lower the circuit Q and reduce the Q rise of the branch currents.

On the other hand, as we go off resonance, the circuit impedance is given by X_L^2/Z_s. Compared to the net reactance, the circuit series resistance can become negligible (see Example 1–7 at 1670 kHz). Therefore, a change in the series resistance has little or no effect on the skirts of the response curve. The overall action is similar to that in the series circuit. A low series resistance will result in high Q, high gain, high impedance, and high sharpness of resonance or selectivity. On the other hand, increasing the series resistance reduces the above characteristics but produces a wider bandwidth. These effects are shown in Fig. 1–10(a). Notice the similarity with Fig. 1–2.

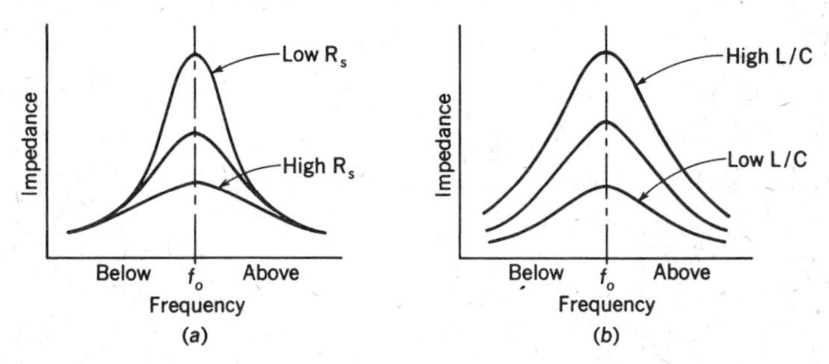

Figure 1-10 The effect on the response curve of a parallel resonant circuit of (a) inductive branch resistance (b) *L/C* ratio.

Effect of L/C Ratio

A larger inductance value can be used in a parallel-resonant circuit, and the same resonant frequency can be maintained, if the capacitance value is reduced in proportion. The combined change will, however, raise the L/C ratio. Since *at resonance* the impedance of the tank circuit is L/CR, an increase in the L/C ratio will cause a proportionate increase in the circuit impedance. (As in the series circuit discussion, we are assuming negligible change in the series resistance of the inductive branch.) It should also be obvious that a higher L/C ratio, because of the higher inductance value, will result in a higher X_L, a higher Q, and a higher resonant rise in the branch currents, as compared to the line current and sharper selectivity. This effect is shown in Fig. 1–10(b).

Gain and Bandwidth—General Resonance Curve

From the preceding discussion, you can see that the gain and bandwidth of a parallel-resonant circuit are affected by the same factors—and in the same manner—as is the series-resonant circuit. For high gain, the circuit should have a high L/C ratio and/or a low resistance in the inductive branch—in other words, a high Q. On the other hand, for wide bandwidth, a lower Q is necessary, and this can be obtained by using a lower L/C ratio and/or a higher series resistance in the inductive branch. In a parallel-resonant circuit, as in the series-resonant circuit, the bandwidth is again

$$\mathrm{BW} = \frac{f_0}{Q}$$

The full response curve for a parallel-resonant circuit can therefore also be obtained using the general resonance-curve technique. Since this was covered earlier under series resonance, it will not be repeated here.

Effect of Shunt Resistance

In practical applications, it sometimes happens that a resistive loading effect is shunted across a parallel-resonant circuit. At other times, an actual resistor may be deliberately connected across a tank circuit in order to obtain some desired effect. In either case the circuit now contains three branches—an inductive branch with series resistance, a pure capacitive branch, and a pure resistive branch. Such a circuit is shown in Fig. 1–11. The resistor R_2 acts as a load across the tank circuit. Let us analyze the effect of this loading resistor for three factors: impedance; circuit Q and gain; and sharpness of resonance.

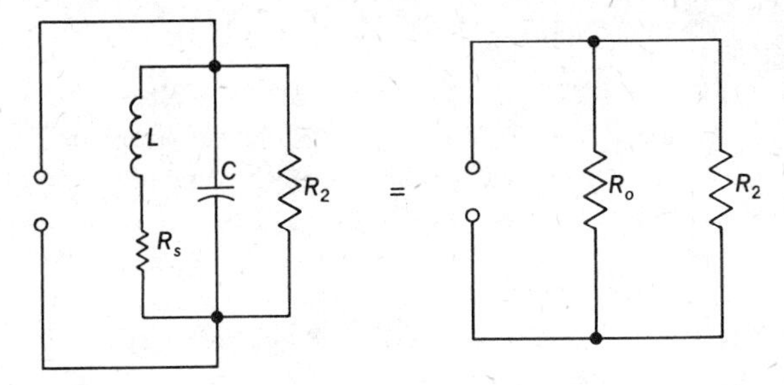

Figure 1-11 A loaded tank circuit.

If we consider the shunt resistor R_2 to be temporarily disconnected, the impedance of the tank circuit at resonance would be $Z_{\|} = L/CR_s$ (R_s is the *series* resistance in the inductive branch). Since this impedance is resistive, it can be represented by the symbol R_0. Now, if we reconnect the loading resistor, we have R_2 in parallel with R_0. Obviously, the equivalent impedance of the *loaded* tank circuit, Z_e, will decrease. This value can be found using the two-branch parallel resistance formula. An example should clarify this analysis:

EXAMPLE 1–8

A tank circuit has a capacitor of 100 pF and an inductor of 150 μH. The series resistance is 15 Ω. Find the impedance and Q of this tank circuit (1) not loaded, (2) shunted with a 2-megohm resistor, (3) shunted with a 0.1-megohm resistor, and (4) with a loading resistor of 15,000 Ω.

Solution

1. *Unloaded*

(*a*) $$Z_{\|} = \frac{L}{CR} = \frac{150 \times 10^{-6}}{100 \times 10^{-12} \times 15} = \mathbf{100{,}000\ \Omega}$$

(*b*) $$Z_{\|} = \frac{X_L^2}{R_s},$$ and solving for X_L

$$X_L = \sqrt{Z_{\|} \times R_s} = \sqrt{100{,}000 \times 15} = 1225\ \Omega$$

(*c*) But from Eq. (1–7), $Z_{\|} = X_L Q$, and solving for Q

$$Q = \frac{Z_{\|}}{X_L} = \frac{100{,}000}{1225} = \mathbf{81.6}$$

2. *With a 2-megohm load*

(*a*) The unloaded impedance $Z_{\|} = 100{,}000\ \Omega$

(*b*) $$Z_e = \frac{Z_{\|} \times R_2}{Z_{\|} + R_2} = \frac{0.1 \times 2}{2.1} = \mathbf{95{,}300\ \Omega}$$

(*c*) $Q_e = \frac{Z_e}{X_L} = \frac{95{,}300}{1225} = \mathbf{78.8}$

where Q_e is the Q of the loaded circuit. Notice that since the shunt resistance value is fairly high compared to the unloaded impedance, it has very little effect on the tank circuit.

3. *With a 0.1-megohm load*

(*a*) Since the shunting resistance value is equal to the unloaded tank circuit impedance, the impedance of the loaded circuit will be cut in half, or $Z_e = \mathbf{50{,}000\ \Omega}$

(*b*) Since the impedance is halved, Q is also halved, or

$$Q_e = \mathbf{40.8}$$

4. *With a 15,000-ohm load*

(*a*) $Z_e = \frac{Z_{\|} \times R_2}{Z_{\|} + R_2} = \frac{100\text{k} \times 15\text{k}}{115\text{k}} = \mathbf{13{,}050\ \Omega}$

(*b*) $Q_e = \frac{Z_e}{X_L} = \frac{13{,}050}{1225} = \mathbf{10.65}$

From this example, several general conclusions can be made as to the effects of shunting a resistive load across a tank circuit:

1. The impedance of the parallel circuit is decreased.
2. The Q of the circuit is decreased.
3. Since the Q is reduced, the gain is decreased, but the bandwidth is increased.
4. The larger the value of shunt resistance (compared to the unloaded impedance), the less the effect.

If we compare the effect of this shunting resistance with the earlier discussion on the effect of series resistance, we will notice that the actions are similar but *reversed.* Adding a shunt resistor is equivalent to *increasing* the series resistance. This effect is shown in Fig. 1–12. The original circuit is given in diagram *a*, where R_{su} is the actual (or unloaded) series resistance, and R_{sh} is the shunting resistance. This circuit can be replaced by diagram *b*, where R_{sa} is the "added equivalent series resistance," and R_{sl} is the loaded series resistance. Notice that the effect of the shunt resistor is to increase the total effective series resistance.

A further examination of Example 1–8 will also show that the *lower* the value of the shunt resistor, the *larger the added equivalent series resistance.* For any given shunt resistance, we can evaluate the magnitude of the equivalent series resistance by comparing the loaded and unloaded impedances. Since $Z_{\|} = X_L^2/R_s$, and the X_L value is unchanged, any change

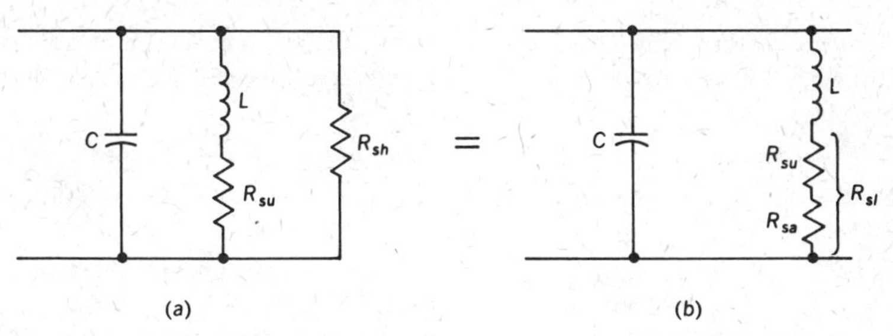

Figure 1-12 Effect of shunt resistance on the equivalent series resistance of a tank circuit.

in impedance must be caused by a change in the total series resistance. As a quick illustration, in Example 1–8, step 3, a 0.1-megohm shunt resistor was added to a tank circuit having a 15-Ω *series* resistance. This shunt load cut the circuit impedance and Q in half. The same effect can be achieved by doubling the series resistance. Therefore, the addition of the 0.1-megohm shunt resistor was equivalent to increasing the total series resistance to 30 Ω, or increasing the original series resistance by another 15 Ω. Expressed mathematically in general terms,

$$\frac{R_{s\,(\text{loaded})}}{R_{s\,(\text{unloaded})}} = \frac{Z_{(\text{unloaded})}}{Z_{(\text{loaded})}} = \frac{Q_{(\text{unloaded})}}{Q_{(\text{loaded})}}$$

where the R_s values represent the *total* series resistance in the inductive branch of the tank circuit. Let us use this relation to find the value of the equivalent series resistance of the 15,000-Ω shunt load in Example 1–8, step 4.

EXAMPLE 1–9

In Example 1–8, a shunt resistance value of 15,000 Ω lowered the circuit impedance from 100,000 Ω to 13,050 Ω. The actual series resistance is 15 Ω. Find the equivalent added series resistance value due to the 15,000-Ω load.

Solution

1. $R_{s\,(\text{loaded})} = R_{s\,(\text{unloaded})} \dfrac{Z_{(\text{unloaded})}}{Z_{(\text{loaded})}} = \dfrac{15 \times 100{,}000}{13{,}050} = 115\ \Omega$

2. Equivalent added $R_s = 115 - 15 = 100\ \Omega$

REVIEW QUESTIONS

1. (*a*) What circuit components are necessary to produce resonance in a series circuit? (*b*) If the resistance in a series-resonant circuit is doubled, what happens to the resonant frequency? (*c*) If the inductance of the circuit is doubled, what happens to the resonant frequency? (*d*) If the capacitance is reduced to one-fourth, what happens to the resonant frequency? (*e*) Give the equation for determining the resonant frequency of a series *L-C-R* circuit.

2. What circuit condition is responsible for causing resonance in a series circuit?

3. State three circuit conditions other than the one indicated in Question 2 that apply to a series circuit at resonance.

4. A series circuit containing inductance, capacitance, and resistance is connected to an ac supply source. All circuit values are variable. At present, the circuit is operating below its resonant frequency. State three ways by which resonance can be produced.

5. (*a*) In a series circuit at resonance, what is the relationship between the applied voltage and the voltage across *L*? across *C*? (*b*) What determines this factor?

6. Refer to Table 1–1. (*a*) Compared to the 4000-Hz value, why does $X_L = 200\ \Omega$ at 2000 Hz? (*b*) Why does $X_C = 800\ \Omega$ at 2000 Hz? (*c*) What does X_0 represent? (*d*) How is its value obtained? (*e*) What does θ_c represent? (*f*) How is its value obtained? (*g*) What determines whether *Z* is capacitive or inductive? (*h*) Why does $Z = 10\ \Omega$ and why is it resistive at 4000 Hz? (*i*) Why are *Z* and X_0 the same value at 3000 Hz? (*j*) How are the values in column *I* obtained? (*k*) How are the values in column E_c obtained?

7. A series circuit is resonant to some given frequency. If the input signal frequency is reduced, what is the effect on the magnitude and character of the circuit impedance? Explain.

8. Repeat Question 7 for an increase in the signal frequency to above resonance.

9. If the resistance of a series *L-C-R* circuit is increased, what happens to the impedance at (*a*) frequencies appreciably off resonance? Explain. (*b*) resonance? Explain.

10. With the circuit change of Question 9, what happens to the line current at (*a*) frequencies appreciably off resonance? Explain. (*b*) resonance? Explain.

11. With the circuit change of Question 9, what happens to the voltage across the capacitor at (*a*) frequencies appreciably off resonance? Explain. (*b*) resonance? Explain.

12. Refer to Fig. 1–3(a). (*a*) What kind of circuit produces these response curves? (*b*) How does the circuit for curve 1 differ from the circuit for curve 2? How does it differ from the circuit for curve 3? (*c*) How does the difference in circuit constants affect the resonant frequency of each curve? Explain. (*d*) Since the circuit constants are different, why is the line current at frequency f_0 the same for each case? (*e*) Why does circuit 1 have a higher current than circuit 2 at frequencies above f_0? below f_0? (*f*) Why does circuit 3 have a lower current than circuit 2 at frequencies above f_0? below f_0?

13. Refer to Fig. 1–3(b). (*a*) Why is the voltage at f_0 higher for circuit 3 as compared to circuit 1? (*b*) Why are the voltages for all three cases identical at frequencies far off resonance?

14. (*a*) Under what condition will the circuit of Fig. 1–4 provide a voltage gain?

Explain. (*b*) How much power could this "stepped-up" output voltage supply? Explain.

15. If higher gain is desired from a series-resonant circuit, should we use (*a*) a circuit with high or low resistance? Explain. (*b*) a circuit with high or low L/C ratio? Explain.

16. (*a*) What is meant by the sharpness of resonance of a response curve? (*b*) In Fig. 1–3(a), which curve has the "sharper" resonance curve? (*c*) Which is the "sharper" curve in Fig. 1–3(b)?

17. How is the sharpness of resonance affected by (*a*) resistance? (*b*) L/C ratio?

18. (*a*) What is meant by the selectivity of a resonant circuit? (*b*) State two design features that would produce high selectivity. (*c*) Why is selectivity desirable? (*d*) Why may "razor-sharp" selectivity be undesirable?

19. Receivers *a* and *b* have response curves as indicated in Fig. 1–5(a) and (b), respectively. Compare the sound output from each of these two receivers if there are local transmitting stations at frequencies f_0 and f_1.

20. (*a*) Describe the ideal receiver response curve. (*b*) Why is such a response curve desirable?

21. Refer to Fig. 1–6. What is the bandwidth for (*a*) curve *a*? (*b*) curve *b*? (*c*) How is bandwidth evaluated?

22. What is the bandwidth of the response curve in Fig. 1–7?

23. Give the equation for calculating the bandwidth of a resonant circuit.

24. What is the relationship between bandwidth and selectivity?

25. If a high degree of selectivity is desired, should circuit Q be high or low? Explain.

26. If the choice of operating frequency is optional, should a high or low frequency be selected to obtain (*a*) a high degree of selectivity? Explain. (*b*) a wide bandwidth? Explain.

27. A resonant circuit was designed to operate at a given frequency. On test, it was found to have insufficient bandwidth. State two ways to redesign it for increased bandwidth without affecting the resonant frequency.

28. What is the relative output from any resonant circuit at frequencies off resonance by (*a*) $+\frac{1}{2}$ BW, (*b*) -1 BW, (*c*) $+4$ BW, (*d*) $-1\frac{1}{2}$ BW, and (*e*) $+2$ BW.

29. Refer to the tabulated data in Fig. 1–7. (*a*) What is the bandwidth? (*b*) How are the frequency values corresponding to 45% of E_o obtained? (*c*) How are the frequency values for 24% of E_o obtained? (*d*) What is the technique used to obtain these values called? (*e*) Is there any advantage to this method?

30. What name is commonly used when referring to a two-branch parallel-tuned circuit?

31. Given a two-branch parallel-resonant circuit: (*a*) What is the main component in each branch? (*b*) In which of these branches is the circuit resistance shown? Why? (*c*) For RF circuit applications, what is the minimum value of Q for the branch of part (*b*)?

32. (*a*) Give the equation for the resonant frequency of a parallel *L-C* circuit when Q is 10 or higher. (*b*) State four other circuit conditions that apply at this frequency.

33. (*a*) What is the impedance of an ideal tank circuit? (*b*) How does series resistance in the inductive leg affect this impedance?

34. (*a*) Give the *general* equation for the parallel impedance of any two-branch circuit. (*b*) Give three versions of this equation applicable at resonance.

35. How does Q rise in a parallel-resonant circuit differ from the Q rise in a series-resonant circuit?

36. (*a*) To what type of circuit does the term *circulating* current apply? (*b*) Where does this current exist? (*c*) Does it differ from line current? Explain.

37. In a parallel-resonant circuit, since $I_L = I_C$, why isn't the line current zero?

38. Refer to Fig. 1–8. (*a*) At what frequency do these phasor relations apply? (*b*) How does the magnitude of I_L compare with I_C? (*c*) How does the magnitude of I_L compare with I_T?

39. Given a parallel-tuned circuit: (*a*) At what frequency will the circuit impedance be resistive? (*b*) Is this impedance value high or low? Explain. (*c*) What happens to the impedance—its magnitude and character (L, C, or R)—as the frequency is reduced? Why?

40. Under what condition will the impedance of a parallel-tuned circuit appear as a low, capacitive reactance? Explain.

41. Refer to the equation $Z_{\parallel} = X_L^2/Z_s$. (*a*) Does it apply to a tank circuit at resonance? (*b*) If so, what would Z_s become? (*c*) Does the equation apply to frequencies removed but close to resonance? (*d*) If so, what do X_L and Z_s stand for?

42. In a parallel-tuned circuit, if the resistance in the inductive branch is increased, what effect will it have on (*a*) the resonant frequency? (*b*) the impedance at resonance? (*c*) the impedance off resonance?

43. Refer to Fig. 1–10(a). (*a*) What basic type of circuit would produce the curves shown? (*b*) What does R_s stand for? (*c*) Why do lower R_s values produce higher impedances at resonance? (*d*) Why are the impedances well above resonance not affected by the change in R_s?

44. Refer to Fig. 1–10(b). (*a*) To what basic type of circuit do these curves apply? (*b*) What does L/C stand for? (*c*) Why does a low L/C ratio produce low impedance at resonance?

45. In a parallel-resonant circuit, if wide bandwidth is the prime consideration, should the circuit be designed with a high or low value of (*a*) inductive branch resistance? (*b*) L/C ratio? (*c*) circuit Q?

46. Repeat Question 45*a*, *b*, and *c*, if the prime requirement is high gain.

47. Repeat Question 45*a*, *b*, and *c*, if the prime requirement is selectivity.

48. If a resistor is shunted across a resonant tank circuit, what is its effect on (*a*) circuit impedance? (*b*) circuit Q? (*c*) sharpness of resonance? (*d*) bandwidth? (*e*) gain? (*f*) circulating tank current?

49. Explain the effect of each of the following on the impedance and Q of a parallel-resonant circuit: (*a*) Adding a high value of shunt resistance. (*b*) Adding a high value of resistance in series with the inductive branch. (*c*) Adding a low value of resistance in series with the inductive branch. (*d*) Adding a low value of shunt resistance.

50. With reference to a parallel-resonant circuit, compare the value of a shunt resistance with its equivalent series resistance value.

51. Refer to Fig. 1–2. (*a*) In diagram *a*, what does R_{su} represent? (*b*) What does R_{sh} represent? (*c*) What is the significance of the symbol = between diagrams *a* and *b*? (*d*) In diagram *b*, how does R_{su} compare with the R_{su} of diagram *a*? (*e*) What is R_{sa}? (*f*) What is R_{sl}?

PROBLEMS

1. A series circuit consists of $L = 80\ \mu H$, $C = 400$ pF, and $R = 15\ \Omega$. Find the resonant frequency.

2. In Problem 1, we wish to change the resonant frequency to 750 kHz. (*a*) Can this be done by changing the value of *L*? If so, to what value? (*b*) Can this be done by changing the value of *C*? If so, to what value? (*c*) Can this be done by changing the value of *R*? If so, to what value?

3. A series circuit has the following circuit constants: $C = 300$ pF, $L = 220\ \mu H$, and $R = 20\ \Omega$. The supply voltage is 60 V. Find (*a*) the resonant frequency, (*b*) the impedance at resonance, (*c*) the line current at resonance, (*d*) the circuit *Q*, and (*e*) the voltage across each component.

4. A series circuit contains $L = 159\ \mu H$, $C = 159$ pF, and $R = 20\ \Omega$. The supply voltage is 100 V. (*a*) Calculate the resonant frequency. (*b*) Calculate the circuit impedance *Z* at resonance and for frequencies of 25, 50, 100, and 200 kHz above and below resonance. Indicate whether these impedances are resistive, inductive, or capacitive. Plot a curve of *Z* versus *f*. (*c*) Calculate the line current for the above frequencies. Plot a curve of *I* versus *f* on a second sheet. (*d*) Calculate the voltage across the capacitor E_C for the same frequencies. Plot this curve on a third sheet. (*Note:* Save these curves; they are used again in Problems 5, 6, 7.)

5. Repeat Problem 4*b*, *c*, and *d* for $R = 50\ \Omega$. Plot the curves of *Z*, *I*, and E_C on the same sheets as the corresponding curves of Problem 4.

6. Repeat Problem 4*b*, *c*, and *d*, using $L = 15.9\ \mu H$, $C = 1590$ pF, and $R = 20\ \Omega$, for frequencies of 100, 200, and 400 kHz each side of resonance. Plot curves of *Z*, *I*, and E_C on the same sheets as the corresponding curves of Problem 4.

7. (*a*) Calculate the bandwidth for each of the circuits of Problems 4, 5, and 6, using the equation and circuit constants. (*b*) Measure the bandwidth of each of these circuits from the E_C versus frequency curves.

8. An RF transmission circuit requires a bandwidth of 200 kHz. How can this bandwidth be obtained without sacrificing the gain available with a *Q* of 50?

9. Using the general resonance curve technique, calculate and plot current versus frequency for the circuit of Problem 4 ($L = 159\ \mu H$, $C = 159$ pF, $R = 20\ \Omega$, $E_T = 100$ V). Compare this plot with the corresponding curve of Problem 4.

10. Using the general resonance curve method, calculate and plot current versus frequency for the circuit of Problem 6 ($L = 15.9\ \mu H$, $C = 1590$ pF, $R = 20\ \Omega$, $E_T = 100$ V). Compare this plot with the corresponding curve of Problem 6.

11. A coil of 300 μH and a capacitor of 300 pF are connected in series across a supply source of 5.0 V and variable frequency. The circuit resistance is 19 Ω. (*a*) Find the resonant frequency and bandwidth. (*b*) Tabulate data for a resonance curve of current versus frequency.

12. A capacitor of 100 pF is connected in parallel with a coil of 253 μH and 8.2 Ω resistance. Find the resonant frequency. (*Note:* Save these results for use in Problems 16, 18, 20, 22, 23.)

13. A parallel circuit resonant at 1000 Hz is desired. A capacitor of 0.2 μF is available. What value of inductance will be needed?

14. What value of capacitance will be needed with a coil of 180 μH to resonate at a frequency of 1200 kHz?

15. A variable capacitor has a range 40–400 pF. The highest resonant frequency desired is 1700 kHz. (*a*) What value of inductance is needed? (*b*) What is the frequency range of this tank circuit?

16. A parallel circuit has a capacitor of 100 pF in one branch and an inductance of 100 μH and 10 Ω resistance in the second branch. The line voltage is 100 V. Find: f_0, I_L, I_C, I_T, and $Z_{\|}$. *Use conventional ac circuit theory.* Compare these answers with the answers from Example 6.

17. The tank circuit of Problem 12 ($L = 253\ \mu H$, $C = 100$ pF, $R = 8.2\ \Omega$) is connected across a 3-V signal source at the resonant frequency. Using resonant-circuit theory, find the (*a*) circuit impedance, (*b*) line current, (*c*) current in each branch, and (*d*) circuit Q.

18. A parallel-resonant circuit has a series resistance of 10 Ω. At resonance, the inductive reactance is 1000 Ω and the line current is 0.5 mA. What is the value of the circulating current?

19. Find the impedance of the tank circuit of Problem 12 ($L = 253\ \mu H$, $C = 100$ pF, $R = 8.2\ \Omega$) for frequencies of 50 and 100 kHz above and below resonance.

20. In Problem 19, if the line current remains constant at 1.0 mA, find the voltage developed across the tank circuit at (*a*) resonance, (*b*) 50 and 100 kHz above resonance, (*c*) 50 and 100 kHz below resonance.

21. In Problem 12, if the resistance in the inductive branch is increased to 82 Ω, calculate the effect on the (*a*) resonant frequency, (*b*) impedance at resonance, and (*c*) impedance at 100 kHz off resonance.

22. A parallel-tuned circuit is resonant at 1200 kHz. At 1192 kHz and at 1208 kHz the resonance curve falls to 0.707 of its maximum amplitude. (*a*) What is the bandwidth of this circuit? (*b*) What is the Q of this circuit? (*c*) Tabulate data for the full response curve. Plot this curve.

23. The tank circuit of Problem 12 ($L = 253\ \mu H$, $C = 100$ pF, $R_s = 8.2\ \Omega$) is shunted with a load resistance R_L of 0.5 MΩ. Find the new (loaded) value of (*a*) circuit impedance, (*b*) circuit Q, and (*c*) bandwidth. (*d*) What resistance in series with the coil would have the same effect as this shunt resistance?

24. Repeat Problem 23*a*, *b*, and *c* for a shunt resistance value of 150,000 Ω.

25. The output circuit of a transmitter has an inductance of 4.2 mH and a resistive effect of 10 Ω in one branch, and a capacitor of 120 pF in the other branch. When this tank circuit is properly loaded, the effective resistance in the inductive branch rises to 500 Ω. Find the (*a*) resonant frequency, (*b*) impedance of the loaded tank circuit, and (*c*) impedance, unloaded.

26. In Problem 25, with the coil inductance decreased to 1.4 mH and the capacitance increased to 360 pF, find (*a*) the resonant frequency and (*b*) the impedance of the loaded tank circuit. (*c*) What is the effect of this L/C change?

TYPICAL FCC QUESTIONS*

3.063 State the formula for determining the resonant frequency of a circuit when the inductance and capacitance are known.

3.067 What is the formula for determining the wavelength when the frequency, in kilohertz, is known?

3.322 What is the value of total reactance in a series resonant circuit at the resonant frequency?

3.323 What is the value of reactance across the terminals of the capacitor of a parallel resonant circuit, at the resonant frequency and assuming zero resistance in both legs of the circuit?

3.324 Given a series resonant circuit consisting of a resistance of 6.5 ohms, and equal inductive and capacitive reactances of 175 ohms, what is the voltage drop across the resistance, assuming the applied circuit potential is 260 volts?

3.325 Given a series resonant circuit consisting of a resistance of 6.5 ohms, and equal inductive and capacitive reactances of 175 ohms, what is the voltage drop across the inductance when the applied circuit potential is 260 volts?

3.326 Under what conditions will the voltage drop across a parallel tuned circuit be a maximum?

3.520 What is meant by carrier frequency?

3.522 What are the frequency ranges included in the following frequency subdivisions: MF (medium frequency), HF (high frequency), VHF (very high frequency), UHF (ultra high frequency) and SHF (super high frequency)?

4.004 If, in a given ac series circuit, the resistance and capacitive reactance are of equal magnitude of 11 ohms, and the frequency is reduced to 0.411 of its value at resonance, what is the resultant impedance of the circuit at the new frequency?

4.019 If an ac series circuit has a resistance of 12 ohms, an inductive reactance of 7 ohms and capacitive reactance of 7 ohms, at the resonant frequency, what will be the total impedance at twice the resonant frequency?

4.020 In a parallel circuit composed of an inductance of 150 microhenrys and a capacitance of 160 micromicrofarads, what is the resonant frequency?

* Original FCC wording has in general been retained in all Typical FCC Questions in this book.

4.021 What value of capacitance must be shunted across a coil having an inductance of 56 microhenrys in order that the circuit resonate at 5000 kilohertz?

4.103 If a frequency-doubler stage has an input frequency of 1000 kilohertz and the plate inductance is 60 microhenrys, what value of plate capacitance is necessary for resonance, neglecting stray capacitances?

4.169 What effect does a loading resistance have on a tuned radio-frequency circuit?

6.083 State the formula for determining the resonant frequency of a circuit when the inductance and capacitance are known.

6.103 What changes in circuit constants will double the resonant frequency of a resonant circuit?

6.104 How may the Q of a parallel resonant circuit be increased?

6.105 If a parallel circuit, resonant at 1000 kilohertz, has its values of inductance halved and capacitance doubled, what will be the resultant resonant frequency?

6.106 What is the resonant frequency of a tuned circuit consisting of a capacitor of 500 micromicrofarads, a tuning coil of 150 microhenrys and a resistance of 10 ohms?

S-3.034 What is the Q of a circuit? How is it affected by the circuit resistance? How does the Q of a circuit affect bandwidth?

2

Coupled Circuits

Whenever energy is to be transferred from a source to a load, the circuits must be interconnected by a suitable coupling network. If maximum voltage output is desired from the source, the coupling network should make the load look like a high resistance value, compared to the internal resistance of the source. On the other hand, for maximum power transfer, the coupling circuit must effect a match between the source resistance and the load resistance. The same basic principles are used in coupling nonresonant amplifier stages. The discussion in this chapter centers on applications to RF circuits.

TYPES OF COUPLING

Many varieties of coupling networks can be used. However, regardless of the type, coupling is effected by having a circuit element common to both the input (or source side) and the output (or load side). The degree of coupling, or interaction, is measured by a *coupling coefficient, k,* and is evaluated as the ratio of the impedance Z_m of the common element to the square root of the product of the *total* impedance of the two circuits. Expressed mathematically, this is

$$k = \frac{Z_m}{\sqrt{Z_1 Z_2}} \quad \text{(2-1)}$$

The impedances Z_1 and Z_2 must be of the same type, for example, both inductors or capacitors. Now, let us see how this applies to some commonly used coupling circuits.

Common-Resistance Coupling

The voltage-divider circuit of Fig. 2–1 can be considered as a common-resistance coupling network. The common circuit element is resistor R_m. The voltage developed across this resistor by the action of the supply source is effectively applied to the output side or load circuit. The larger the value of this mutual resistor, the greater the voltage across the load, or the higher the degree of coupling. This circuit, however, would not normally be used in a practical application because of the losses introduced by the resistors.* It is presented here merely as an introduction to the remaining circuits.

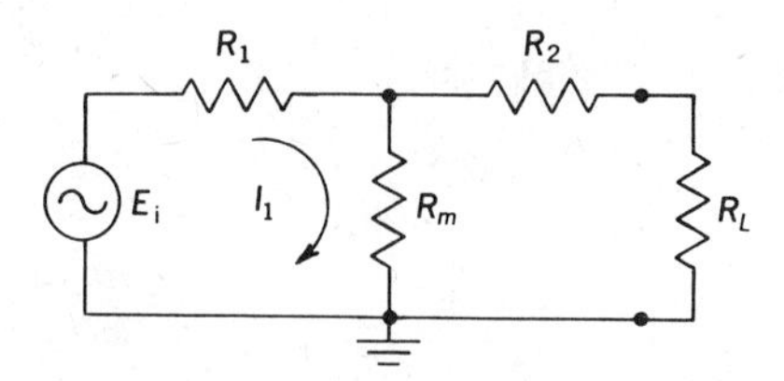

Figure 2-1 Common-resistance coupling.

Common-Inductance Coupling

To eliminate the losses of the above circuit, the resistors can be replaced with inductors. This produces the common-inductance coupling network of Fig. 2–2(a). The voltage developed across the common element L_m by the source (or input) voltage is effectively applied to the load (or output). The larger this inductance value, the greater the load voltage. In other words, the coupling coefficient increases with the value of L_m. The exact relation is given by

$$k = \frac{L_m}{\sqrt{(L_1 + L_m)(L_2 + L_m)}} \tag{2–2}$$

Notice that this equation is a direct application of the general expression

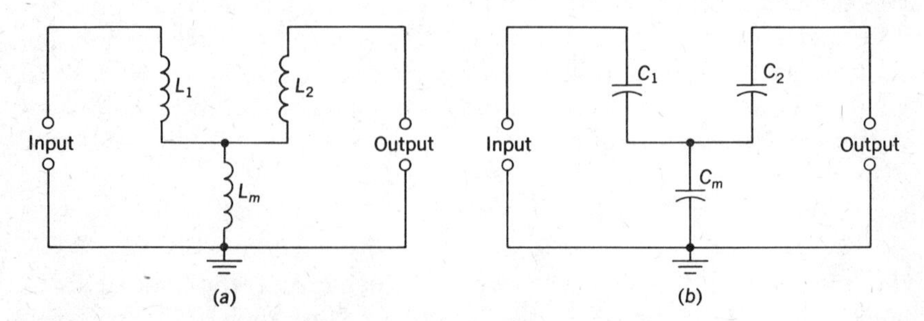

Figure 2-2 Common-reactance coupling.

* It would be used only if attenuation of the input signal were desired.

given earlier in this discussion. The degree of coupling is unaffected by any change in the signal frequency. Since XL_m, XL_1, and XL_2 all vary in the same proportion, the ratio remains constant.

Common-Capacitance Coupling

A coupling network, quite akin to the common-inductance network uses capacitive elements in place of inductors. This produces the common-capacitance circuit of Fig. 2-2(b). Since capacitors have negligible resistance, this network does not introduce any losses. The degree of coupling varies directly with the reactance of the common element, or *inversely* with the capacitance of C_m. The exact relationship is given by

$$k = \sqrt{\frac{C_1 C_2}{(C_1 + C_m)(C_2 + C_m)}} \tag{2-3}$$

In this equation, since all the circuit elements are capacitive, a change in frequency affects the numerator and the denominator in the same manner. Therefore, the coefficient of coupling remains constant regardless of a change in frequency.

Mutual-Inductance Coupling

If two coils are close enough to each other, some of the flux of one coil will link or cut the second coil. A voltage will be *induced* in this second coil because of the flux changes *produced* by the first coil. The inductance of any coil signifies the ability or property of that coil to induce a voltage within its own winding. More specifically, this should be referred to as *self-inductance* (L), but when we have one coil inducing a voltage in a second coil, this "transfer" ability or property is called *mutual inductance* (M). The two coils are said to be *coupled.* In the previous types of coupling, the common or mutual element was a real component—R, L, or C; in this case, coupling is achieved via an intangible mutual inductance M created by flux linkage. Yet it has all the properties of a real component, similar to L_m in the common-inductance network of Fig. 2–2(a).

The degree of coupling between mutually coupled coils depends on what portion of the flux of coil 1 reaches and cuts the turns of coil 2 (or vice versa: what portion of the flux produced by coil 2 links coil 1). Obviously, the closer the coils, the greater the interaction and the greater the coupling coefficient. When all of the flux of one coil links the other, we have unity coupling ($k = 1$). This is achieved, for all practical purposes, only when the coils are mounted on a properly designed common iron core. In RF circuits using air cores (or powdered iron cores), the coupling coefficients are much lower—generally around 0.05 or less. Since mutual inductance depends on flux linkages, it is directly affected by the value of coupling coefficient. Mathematically the relation of these two terms is given by*

$$M = k\sqrt{L_1 L_2}$$

EXAMPLE 2–1

The primary and secondary of an RF air-core transformer each have an inductance of 300 μH. If the mutual inductance between these windings is 6 μH, find the coefficient of coupling.

Solution

$$k = \frac{M}{\sqrt{L_1 L_2}} = \frac{6}{\sqrt{300 \times 300}} = \mathbf{0.02}$$

A schematic diagram showing mutually coupled coils in an air-core transformer is shown in Fig. 2–3(a).

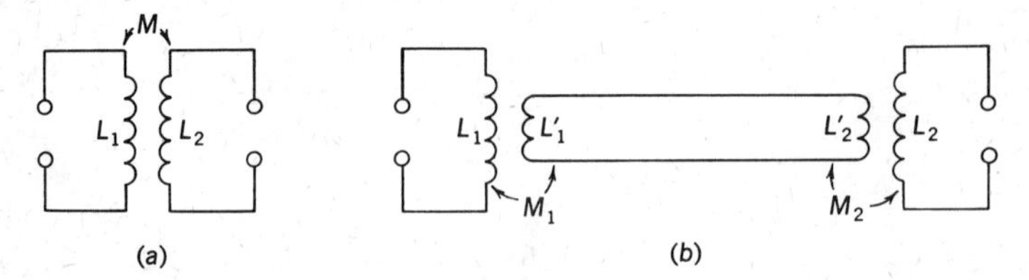

Figure 2-3 Mutually coupled coils.

Link Coupling

In some cases, where it is desirable to use inductive (mutual) coupling, it is impossible for physical reasons to bring coils L_1 and L_2 close together. One such situation would arise when the coils (L_1 and L_2) are part of the output and input circuits, respectively, of two separate assemblies, such as a preamplifier and main amplifier, or a final power amplifier and an antenna coupling unit. The solution is to use a *link coupling* system, as shown in Fig. 2–3(b). The coils L_1 and L_1' are in one unit and closely coupled, while coils L_2 and L_2' are in the second unit and also closely coupled. The length of the line depends, of course, on the separation between the two units. Energy is transferred from L_1 to L_1' by the mutual inductance M_1, fed by direct connection to L_2', and finally coupled to L_2 by the mutual inductance M_2.† The overall coupling coefficient is

*J. J. DeFrance, *Electrical Fundamentals* (Englewood Cliffs, N.J., Prentice-Hall, Inc., 1969), Chap. 19, Part 1.

†Notice the similarity between this circuit and the audio lines used when audio components are separated by excessive distances.

given by

$$k = \frac{M_1 M_2}{L_m \sqrt{L_1 L_2}} \tag{2-4}$$

where $L_m = L_1' + L_2'$. The degree of coupling can be varied by changing L_1', L_2' or by varying their relation to the main coils.

Combination Type Coupling

In all of the above circuits, the degree of coupling remains constant regardless of frequency. This is so because the impedance of the common (or mutual) element varies with frequency, in the same manner as the "main" elements. Sometimes, however, it is advantageous to have the coupling increase (or decrease) automatically for different frequencies. To obtain such an effect, combination types of coupling must be used. Three such networks are shown in Fig. 2–4. In the circuit of Fig. 2–4(a), common-impedance coupling is used, but the common impedance is $(X_L - X_C)$. At their resonant frequency the common impedance is zero and the degree of coupling drops to zero. The circuit of Fig. 2–4(b) employs mutual-inductance coupling and also common-capacitance coupling. This circuit has a higher degree of coupling at the lower frequencies because the common impedance (X_C) increases. The reverse effect is obtained with the circuit of Fig. 2–4(c). Here we have mutual-inductance and direct-capacitive coupling. The capacitance value used is very small, resulting in a high X_C value in series with X_{L2}. This voltage-divider action limits the portion of the L_1 voltage that is applied directly across L_2. However, at the higher frequencies, X_C decreases, X_L increases, and a greater proportion of the primary voltage is applied to the secondary coil.

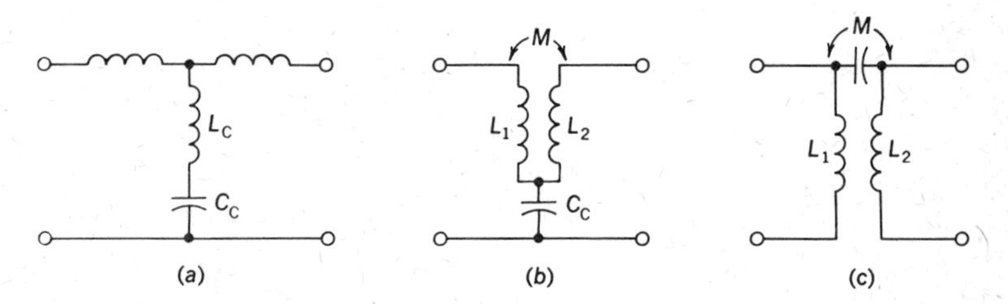

Figure 2-4 Combination coupling circuits.

INDUCTIVE COUPLING

The most common type of coupling used in RF circuits is the mutual-inductance coupling. It is more commonly referred to as simply *inductive coupling* or *transformer coupling*. These coupling circuits are further classified as *single-tuned* circuits, if either the primary or the secondary circuit is tuned to resonance, or as *double-tuned* circuits when both the primary and secondary sides are tuned. We will analyze the relations that exist in these circuits with respect to induced voltage, interaction between windings, and the types of response curves obtainable.

Secondary Induced Voltage

If an RF voltage is applied to a primary winding, a current I_p will flow in this winding. This, in turn, produces a flux ϕ_1. The magnitude of this flux will depend on the current value I_p and on the number of turns N_1 in the primary winding. Since the inductance of any winding is proportional to the square of the number of turns,

$$N_1 \propto \sqrt{L_1} \qquad \text{and} \qquad \phi_1 \propto I_p\sqrt{L_1}$$

Only part of this flux will link the secondary, depending on the coefficient of coupling k. The flux that cuts the secondary turns will induce a voltage E_s in the secondary winding. The magnitude of this induced voltage will depend not only on the amount of flux linkage, but also on the number of turns in the secondary winding N_2. But N_2 can be replaced by its equivalent $\sqrt{L_2}$. Also, the flux linking the secondary is $k\phi_1$ or $kI_p\sqrt{L_1}$. Combining these factors, we get

$$E_s \propto (kI_p\sqrt{L_1})(\sqrt{L_2}) \propto k\sqrt{L_1L_2}\, I_p$$

This relationship shows that the induced voltage is *proportional* to the flux and to the inductance values. There is one other factor to be considered—the speed at which the flux changes. This, in turn, depends on the frequency of the applied voltage or, more specifically, on the angular velocity $2\pi f$ or ω. Also note the term $k\sqrt{L_1L_2}$ in the proportionality above. It is the mutual inductance M. Making this replacement and adding the speed-of-cutting factor, we get

$$E_s = \omega M I_p \tag{2–5}$$

This equation gives the *magnitude* of the induced voltage with respect to the mutual inductance, frequency of operation, and primary current. With regard to phase relations, we know that flux is in phase with current, and that an induced voltage lags the flux that caused it by 90°. To show this phase relation, the induced voltage equation is often shown as

$$E_s = -j\omega M I_p \tag{2–6}$$

The $-j$ term indicates that the voltage *lags* the current by 90°.

EXAMPLE 2–2

An RF transformer has a primary inductance of 400 μH, a secondary inductance of 220 μH, and a coefficient of coupling of 0.025. The primary input signal has frequency of 800 kHz and is of such magnitude as to produce a primary current of 5.0 mA. Find the induced secondary voltage.

Solution

1. $M = k\sqrt{L_1 L_2} = 0.025\sqrt{400 \times 220 \times 10^{-12}} = 7.42\ \mu\text{H}$
2. $E_s = \omega M I_p = (2\pi \times 800 \times 10^3)(7.42 \times 10^{-6})(5 \times 10^{-3}) = \mathbf{0.186\ V}$

Reflected Impedance

If a load (R, L, C, or combination) is connected across the secondary winding of a coupled circuit, current will flow when a voltage is induced in this winding. The value of this secondary current I_s will depend on the magnitude of the induced voltage and on the total *series* impedance (Z_{s2}) (winding plus load) of the secondary circuit.

$$I_s = \frac{E_s}{Z_{s2}} = \frac{-j\omega M I_p}{Z_{s2}}$$

This secondary current causes a secondary flux, and part of this secondary flux—because of the mutual inductance—will link with the primary turns, inducing a voltage $E_{1,2}$ in the primary winding. Since it is an induced voltage, its magnitude and phase must be given by the induced voltage equation, or

$$E_{1,2} = -j\omega M I_s$$

Replacing I_s by its equivalent value

$$E_{1,2} = -j\omega M\left(\frac{-j\omega M I_p}{Z_{s2}}\right) = j^2\frac{(\omega M)^2}{Z_{s2}} I_p = -\frac{(\omega M)^2}{Z_{s2}} I_p$$

This induced voltage is in opposition to the applied primary voltage, so that now the applied voltage must overcome not only the primary impedance voltage drop ($I_p Z_p$), but also this counter voltage. Expressed mathematically,

$$E_T = I_p Z_p + E_{1,2} = I_p Z_p - \frac{(\omega M)^2}{Z_s} I_p = I_p\left[Z_p - \frac{(\omega M)^2}{Z_s}\right]$$

This equation is of the form $E = IZ$, where I is the primary current I_p, and Z is the bracketed quantity $Z_p - (\omega M)^2/Z_s$. Obviously, then, the effect of the voltage induced in the primary is the same as if the primary impedance were changed from its original value of Z_p to $Z_p - (\omega M)^2/Z_s$. In other words, because of the coupling, we now have an equivalent primary impedance

$$Z'_p = Z_p + \frac{(\omega M)^2}{Z_s} \qquad \textbf{(2–7)}$$

Solving for the equivalent primary inductance by dividing both sides by ω and then simplifying, we get

$$L_p' = L_p(1 - k^2) \tag{2–8}$$

where k is the coefficient of coupling.

Referring once more to Eq. (2–7), we see that the equivalent primary impedance consists of the "uncoupled" primary impedance Z_p and an impedance factor equal to $(\omega M)^2/Z_s$. Since this effect is caused by the action of, or the *reflection* from the secondary circuit, it is known as the *reflected impedance,* Z_r. The voltage induced in the primary due to the current flowing in the secondary can mathematically be replaced by this reflected impedance. The magnitude of this reflected impedance is given by

$$Z_r = \frac{(\omega M)^2}{Z_s} \tag{2–9}$$

However, when evaluating the equivalent primary impedance, notice from Eq. (2–7) that the reflected impedance is subtracted from Z_p. *This is a phasor equation.* All quantities must be expressed in phasor notation and handled using vector algebra methods.

The Character of Reflected Impedance

In the above derivations, the reflected impedance term is preceded by a negative sign. This sign is ignored when we specify the magnitude of Z_r, but it is of significance when we consider the character (L, C, or R) of the reflected quantity. Let us consider the *phasor* meaning of this sign from four practical aspects.

OPEN SECONDARY CIRCUIT. If the secondary circuit is open, there is no secondary current, no voltage induced back into the primary, and no reflected impedance. $Z_r = 0$. This situation arises when the secondary circuit is untuned and feeds a very high impedance load, such as the gate-source input of a class-A FET circuit. (See Fig. 2–5(a).) Since gate current does not flow and the input capacitance of the FET is very low, the input impedance of this type of transistor is so high that we can consider the secondary circuit to be open. (This also applies to class-A vacuum-tube circuits.)

RESISTIVE SECONDARY CIRCUIT. When the secondary circuit is tuned and resonant (series or parallel), it acts as a resistive circuit. Whatever the load connected to such a circuit—vacuum tube, transistor, resistive, or otherwise—as long as it remains tuned to resonance, the series impedance Z_s of such a circuit is pure resistance and is equal to the series resistance R_s of the circuit. Since the denominator of the reflected impedance equation is pure number (no j factor), the total quantity is also a pure number. In other words, the reflected impedance Z_r becomes *reflected resistance* R_r. *When the secondary circuit is resonant, the reflected quantity is pure resistance,* or

$$R_r = \frac{(\omega M)^2}{R_s} \tag{2-10}$$

where R_s is the *series* resistance of the secondary circuit. Two such circuit applications are shown in Fig. 2–5(b) and (c).

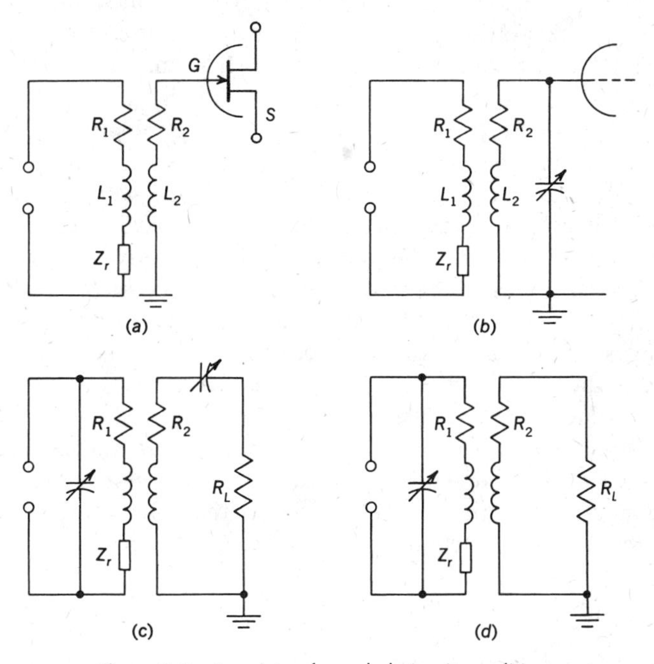

Figure 2-5 A variety of coupled circuit conditions.

CAPACITIVE SECONDARY CIRCUIT. When the impedance of the secondary circuit is capacitive, the general term Z_s becomes $0 - jX_C$. Substituting this in the equation for reflected impedance and rationalizing, we get

$$Z_r = \frac{(\omega M)^2}{-jX_C} \cdot \frac{+jX_C}{+jX_C} = +j\frac{(\omega M)^2}{X_C}$$

The $+j$ term indicates that the character of the reflected impedance is *inductive.* Notice that *a capacitive secondary circuit reflects into the primary as an added inductive reactance.* To see this condition in a practical application, refer to Fig. 2–5(c). The secondary is a series *L-C-R* circuit. If the applied signal frequency is lower than the resonant frequency of the circuit, X_C is greater than X_L, and the secondary is capacitive. Stated in another way, the circuit is tuned above the frequency of the incoming signal.

This same effect can also apply to the parallel-resonant tank circuit of Fig. 2–5(b). With respect to its load, tube or transistor, this *L-C-R* circuit is a parallel-resonant circuit, *but with respect to the induced voltage* E_s, *it is a series-resonant circuit.* (Also remember that term Z_s refers to the *series* impedance of the secondary circuit.) Therefore, as before, if the secondary is tuned above the frequency of the incoming signal, it is capacitive and reflects an inductive reactance into the primary.

INDUCTIVE SECONDARY CIRCUIT. It should be obvious without further proof that, if the series impedance of the secondary circuit is inductive, it will reflect a capacitive reactance into the primary. This condition could occur when the secondary is tuned (resonates) below the frequency of the incoming signal. Since the signal frequency is higher than the resonant value, X_L exceeds X_C, and the circuit reactance is inductive. The same effect could also be produced with a relatively low-resistance load. (See Fig. 2–5(d).) If the combined resistance of R_L and the effective resistance of the coil is low compared to the reactance X_L, the secondary is essentially an inductive circuit.

When the resistance in the secondary circuit is not negligible, the series secondary impedance becomes $Z_s = R + jX_L$. The reflected impedance is now a complex quantity, but of opposite nature (i.e., $Z_r = R - jX_C$). This value can be found from the general equation $Z_r = (\omega M)^2/Z_s$. It is interesting to note that the X/R ratio for the reflected impedance will be the same as for the secondary impedance.

EXAMPLE 2–3

The circuit of Fig. 2–5(b) has the following values: $L_1 = 400\ \mu\text{H}$; $L_2 = 220\ \mu\text{H}$; $k = 0.05$; $R_1 = 10\ \Omega$; $R_2 = 4.1\ \Omega$; signal frequency = 10 V at 800 kHz. Find the capacitance needed to tune the secondary current to resonance, and find the value of the primary current at resonance.

Solution

1. For resonance at 800 kHz:

(*a*) $\omega = 2\pi f = 6.28 \times 800 \times 10^3 = 5.02 \times 10^6$

(*b*) $X_{L2} = \omega L_2 = (5.02 \times 10^6)(220 \times 10^{-6}) = 1105\ \Omega$

(*c*) $X_C = X_L = 1105\ \Omega$

$$C = \frac{1}{\omega X_C} = \frac{10^{12}}{5.02 \times 10^6 \times 1105} = \mathbf{180\ pF}$$

2. To find I_p at resonance, we must find the resistance reflected into the primary and the effective primary impedance, Z'_p:

   (*a*) $M = k\sqrt{L_1L_2} = 0.05\sqrt{400 \times 220 \times 10^{-12}} = 14.8\ \mu\text{H}$

   (*b*) $R_r = \dfrac{(\omega M)^2}{R_s} = \dfrac{(5.02 \times 10^6 \times 14.8 \times 10^{-6})^2}{4.1} = 1350\ \Omega$

   (*c*) $X_{L1} = \omega L_1 = 5.02 \times 10^6 \times 400 \times 10^{-6} = 2008\ \Omega$

   (*d*) $Z'_p = \sqrt{(X_{L1})^2 + (R_1 + R_r)^2} = \sqrt{(2008)^2 + (1360)^2} = 2425\ \Omega$

   (*e*) $I_p = \dfrac{E_p}{Z'_p} = \dfrac{10}{2425} = \mathbf{4.13\ mA}$

EXAMPLE 2–4

Find the primary current in Example 2–3 with the capacitance value increased to 190 pF.

Solution

1. $X_C = \dfrac{1}{\omega C} = \dfrac{10^{12}}{5.02 \times 10^6 \times 190} = 1050\ \Omega$

2. Net secondary reactance:

$$X_0 = X_{L2} - X_C = 1105 - 1050 = 55\ \Omega$$

3. $Z_s = X_0 = 55\ \Omega \qquad (X_0 > 10R_2;\ R_2 \text{ is negligible})$

4. $Z_r = \dfrac{(\omega M)^2}{Z_s} = \dfrac{(5.02 \times 10^6 \times 14.8 \times 10^{-6})^2}{55}$
   $= 100.5\ \Omega$ (*capacitive reactance*)

5. $Z'_p = X_{L1} - Z_r = 2008 - 100 = 1908\ \Omega$

6. $I_p = \dfrac{E_p}{Z'_p} = \dfrac{10}{1908} = \mathbf{5.24\ mA}$

Untuned-Primary Tuned-Secondary Circuits

A common type of coupled circuit found in RF amplifiers uses an untuned primary winding and a tuned secondary winding. Such a circuit is shown in Fig. 2–5(b) and is shown again in Fig. 2–6(a) coupling the drain of one FET to the gate of the next FET.* Let us analyze the effect of frequency on the output voltage E_o applied to the gate of the second stage. Figure 2–6(b) shows the Thévenin (or constant voltage) equivalent circuit. Because of the high value of drain resistance, r_d, and because these RF coils are generally made with high-impedance primaries,

*A similar circuit can be used between collector and emitter of bipolar transistors, or between plate and grid of vacuum tubes.

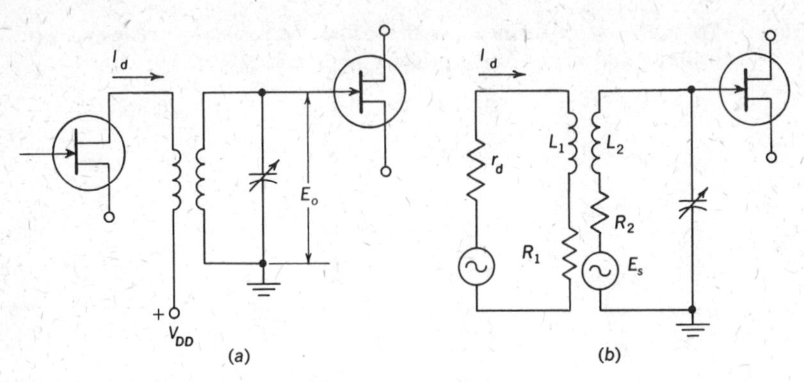

Figure 2-6 A single-tuned inductive-coupled circuit and its equivalent diagram.

the effect of reflected impedance from the secondary into the primary is negligible in comparison to the primary impedance. Furthermore, for small changes in frequency to either side of resonance, the primary impedance Z_p can be considered to be approximately constant. Therefore, for a fixed input signal level, the current I_p flowing through the primary winding is essentially constant. The induced voltage E_s is proportional to this primary current. ($E_s = \omega M I_p$.) For a small frequency swing to either side of resonance, it also is essentially constant. Notice from the equivalent circuit diagram that this induced voltage feeds a *series* secondary circuit containing L, C, and R. At resonance, the series impedance is a minimum, the secondary current is a maximum, and there is a Q rise in the voltage across the capacitor E_C. This is also the voltage E_o applied to the gate of the next stage. Therefore E_o is a maximum at resonance. The curve of gate voltage versus frequency is identical to the response curve for a resonant circuit as discussed in Chapter 1. The bandwidth is given by f_0/Q_c* and the full response curve can be obtained by the general resonance curve method.

EXAMPLE 2–5

The circuit of Fig. 2–6(a) has the following values: $L_1 = 5$ mH, 60 Ω; $L_2 = 220$ μH, 9.1 Ω; the coefficient of coupling between the coils is 0.05; $C = 180$ pF. Find the output voltage applied to the gate of the second FET, at resonance, if the ac component of drain current is 88 μA, and the drain resistance of the FET is 50 kΩ.

* For *bandwidth* consideration, it is first necessary to find the resistance reflected into the secondary by the primary circuit. Then Q_e is $X_L/(R_2 + R_r)$.

Solution

1. $f_0 = \dfrac{1}{2\pi\sqrt{LC}} = \dfrac{0.159}{\sqrt{220 \times 10^{-6} \times 180 \times 10^{-12}}} = 800\ kHz$

2. $\omega = 2\pi f = 6.28 \times 800 \times 10^3 = 5.02 \times 10^6$

3. $\omega M = \omega k\sqrt{L_1 L_2} = 5.02 \times 10^6 \times 0.05\sqrt{5 \times 10^{-3} \times 220 \times 10^{-6}} = 263\ \Omega$

4. The induced secondary voltage is

$$E_s = \omega M I_p = 263 \times 88 \times 10^{-6} = 23.2 \text{ mV}$$

5. To find the Q of this secondary circuit (in order to get the Q rise in voltage):

   (*a*) $X_{L2} = \omega L_2 = 5.02 \times 10^6 \times 220 \times 10^{-6} = 1105\ \Omega$

   (*b*) $Q_2 = \dfrac{X_{L2}}{R_2} = \dfrac{1105}{9.1} = 121.5$

   (*c*) $E_C = E_o = Q_2 E_s = 121.5 \times 23.2 \times 10^{-3} = 2.82 \text{ V}$

Effect of Coupling — Single-Tuned Circuits

If the coupling between the primary and secondary of the above circuit is increased, the mutual inductance increases; ωM increases and the voltage induced in the secondary winding increases. The current I_s in the secondary circuit will also increase. This in turn produces a higher value for E_C (since $E_C = I_s X_C$) and for the gate voltage E_g. This is true, not only at resonance, but also for frequencies to either side of resonance. Obviously, the tighter the coupling, the higher the level of the response curve. In addition, as the coupling coefficient is increased and ωM increases, the reflected resistance also increases. On the primary side, this still has negligible effect, compared to the r_d and X_L values already present in the

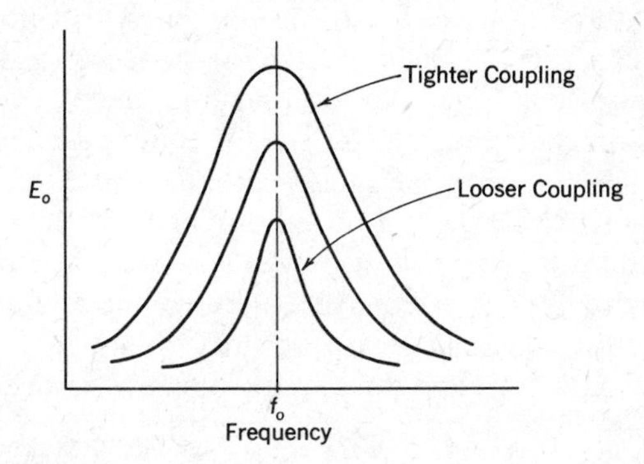

Figure 2-7 The effect of coupling on the response curve of an untuned-primary, tuned-secondary transformer.

primary circuit. On the other hand, since the secondary circuit is a high-Q resonant circuit, the increased reflected resistance lowers the effective Q of the secondary. Meanwhile, because of the reduced Q_e, the bandwidth of the response curve widens. This effect can be seen from the curves of Fig. 2–7. From these curves it should also be obvious that, if a high degree of selectivity is desired, the coupling should be loose and that gain or output voltage level must be sacrificed.

Tuned-Primary Tuned-Secondary Circuits

Another type of inductively coupled circuit that is commonly used—particularly for operation at a fixed frequency—is the *double-tuned circuit,* wherein both the primary and the secondary side of the coupling transformer are tuned. Such a circuit, and its equivalent diagram, are shown in Fig. 2–8.* Since both the primary and secondary are high-Q resonant circuits, reflected resistance and impedance have a much more pronounced effect in this circuit than in the single-tuned transformer. Let us analyze the resonance characteristics of a double-tuned circuit for various degrees of coupling.

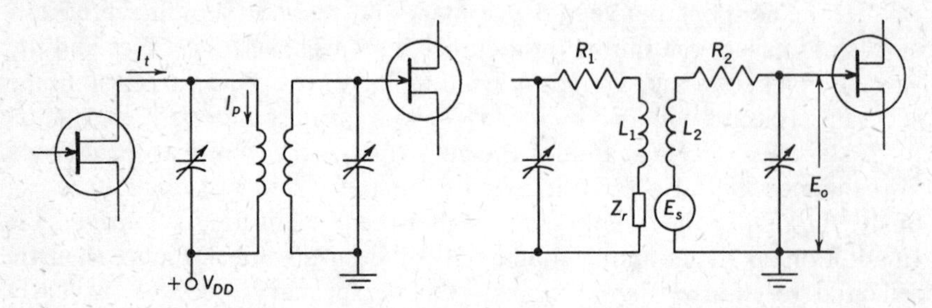

Figure 2-8 A double-tuned inductive-coupled circuit and its equivalent diagram.

LOOSE COUPLING. We will start our analysis with a coupling coefficient low enough so that the reflected resistance at resonance is negligible. With reference to Fig. 2–8, since the impedance of the primary tank circuit is approximately constant for small frequency deviations to either side of resonance, the current I_t will also be approximately constant. However, there will be a Q rise in the primary current I_p at resonance, and the induced voltage E_s will also be a maximum at resonance. Since the secondary circuit is also resonant, there will be another Q rise in E_o as compared to E_s. It should be realized that, because of the loose coupling ωM is low, E_s is low, and the output voltage E_o is also relatively low.

* It is not always necessary to use parallel-tuned circuits. For example, when feeding into a low-resistance load (such as a bipolar common-emitter input circuit), the secondary circuit could be series tuned as shown in Fig. 2–5(c). These details are discussed later.

However, because of the multiple Q rise, the output voltage for this circuit at resonance is appreciably higher than for a single-tuned circuit. Meanwhile, for frequencies off resonance, the combined effect of a drop in I_p compared to I_t, and a drop in E_o compared to E_s, produces a sharper resonance curve than would be obtained with a single-tuned circuit. This response curve is shown in Fig. 2–9.

INCREASED COUPLING. If the coupling is increased, ωM increases and the reflected impedance increases. At resonance, the impedance reflected into the primary is resistive and causes the Q of the primary to decrease. This lowers the primary current I_p and broadens the primary-current resonance curve. Meanwhile, the increase in ωM more than offsets the decrease in the primary current so that the product $\omega M I_p$ increases. The induced voltage E_s increases and E_o also increases. Furthermore, the broader primary-current resonance curve maintains E_s at higher values over a wider range of frequency swing. Consequently, the response curve of the secondary circuit—E_o versus frequency—is also broader. Increasing the coupling has raised the output voltage and broadened the response curve.

CRITICAL COUPLING. If the coupling is increased still further, a point is reached wherein the resistance reflected into the primary, at resonance, is equal to the resistance of the primary circuit itself, or $R_r = R_p$. (The "coupled" Q of the primary circuit is cut in half.) This condition is known as *critical coupling,* and the coefficient of coupling is designated as k_c. Because of the lowered Q, the tank circuit primary current drops. At this coupling level, the decrease in I_p is just offset by the increase in ωM and the *induced voltage E_s reaches its maximum possible value.* The output voltage E_o also reaches a maximum. Meanwhile, because of the lower primary Q, and the broader primary-current resonance curve, the secondary response curve also becomes broader. This curve is shown in Fig. 2–9 as curve 2.

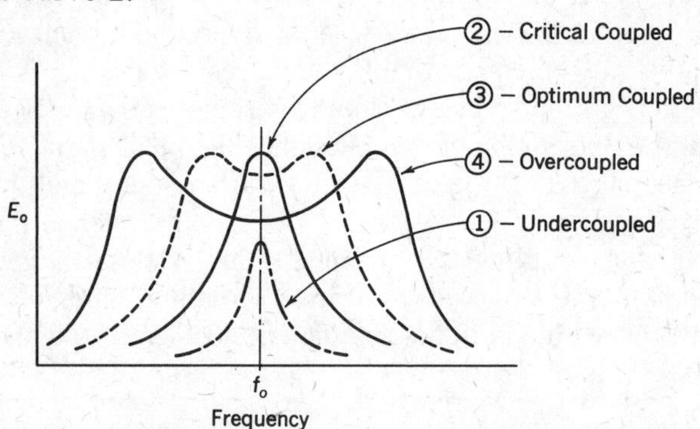

Figure 2-9 The effect of coupling on the response curve of a double-tuned inductive-coupled RF transformer.

OVERCOUPLING. When coupling is increased beyond the critical coupling value, the circuit is said to be *overcoupled.* Let us analyze the effect of a small increase beyond the critical value. At resonance, the reflected resistance is greater than the original primary resistance. The primary Q drops and the tank current I_p drops. The drop in I_p is no longer offset by the increase in ωM, so that the induced secondary voltage also drops, and with it the voltage E_o falls below the resonant peak reached with critical coupling. The loss, however, is not too great.

At some input frequency slightly below the resonant frequency of the secondary tank circuit, the series impedance Z_s of the secondary circuit becomes capacitive. Now, the impedance reflected into the primary is no longer resistive, but inductive. The *added inductance retunes the primary circuit into resonance at this lower input frequency.* The full Q rise in primary current is again obtained; E_s rises to its maximum value, and E_o also reaches its maximum value. However, if the input signal frequency is reduced further, the amount of reflected inductance cannot keep the primary circuit in resonance. The primary current drops off sharply and E_o, in the secondary circuit, falls rapidly.

A similar effect occurs for an input frequency slightly above the resonant frequency of the secondary tank circuit. The reflected impedance, this time a capacitance, is in series with the inductive branch of the primary tank circuit. This reduces the net inductance value and tunes the primary tank circuit to resonance at this slightly higher input frequency. Again we have a Q rise in primary current, and E_o reaches its maximum value. For frequencies further off resonance, the reflected capacitance value can no longer produce primary resonance. The primary current drops off sharply and the secondary output voltage decreases rapidly. The response curve for an overcoupled condition has a characteristic double-humped shape, with a dip in the center and sharply dropping sides. Two such curves are shown in Fig. 2–9 as curves 3 and 4.

Notice the narrower of these two overcoupled curves, marked *optimum coupled.* Remember that the ideal response curve is a rectangular shape, with a flat top and sides sloping vertically down. (See Fig. 1–6.) This particular overcoupled curve resembles the ideal shape fairly well—the top is reasonably flat, the center dip is rather slight, and the sides are fairly steep. Since such a curve most closely approaches the ideal response curve, the coupling that produces this response curve is called *optimum coupling*. Compared to critical coupling, the best response curve is obtained with a coupling coefficient 50% greater than critical, or

$$k_{\text{opt}} = 1.5k_c \qquad \textbf{(2–11)}$$

If the coupling is made still tighter than the optimum value, the reflected resistance at resonance increases further, producing a deeper center dip. The reflected impedance off resonance also increases, so that the primary resonant frequencies occur at greater deviations from the under-coupled resonant value. In other words, the output voltage peaks are further apart. When the coupling is considerably greater than critical, the actual frequencies at which these peaks occur can be found from

$$f_{\text{peaks}} = \frac{f_0}{\sqrt{1 \pm k}} \tag{2–12}$$

Bandwidth – Double-Tuned Circuits

Examine the curves of Fig. 2–9 again. Notice that the bandwidth increases with the *degree* of coupling. Yet no mention has been made of the Q of these circuits. Obviously then, the determining factor in a double-tuned circuit is not Q but the coupling. For a given frequency, the tighter the coupling – or the higher the coupling coefficient – the greater the bandwidth. Expressed mathematically, this is

$$\text{BW}_{dt} = kf_0 \tag{2-13}$$

where the subscript dt is used to indicate a double-tuned circuit.

EXAMPLE 2–6

It is desired to obtain a bandwidth of 200 kHz at an operating frequency of 10 MHz, using a double-tuned circuit. What value of coupling coefficient should be used?

Solution

$$k = \frac{\text{BW}_{dt}}{f_0} = \frac{200}{10{,}000} = \mathbf{0.02}$$

Effect of Q on Degree of Coupling

In the above problem, a bandwidth of 200 kHz is obtained from a double-tuned circuit operating at 10 MHz because the coefficient of coupling is 0.02. However, we cannot tell whether this will produce an undercoupled, critical-coupled, optimum-coupled, or an excessively over-coupled circuit. The actual *degree* of coupling depends on the Q of the two circuits, as can be shown from the following analysis. The critical coupling condition is obtained when the reflected resistance is equal to the primary resistance, or, at critical coupling

$$R_r = R_p$$

and, since

$$R_r = \frac{(\omega M)^2}{R_s}$$

then

$$\omega M = \sqrt{R_p R_s} \qquad \text{(for critical coupling)}$$

But, considering each circuit by itself (no reflected resistance)

$$R_s = \frac{X_{Ls}}{Q_s} \qquad \text{and} \qquad R_p = \frac{X_{Lp}}{Q_p}$$

Substituting gives

$$\omega M = \sqrt{\frac{X_{Lp}}{Q_p} \times \frac{X_{Ls}}{Q_s}}$$

Dividing both sides by ω, we get

$$M = \sqrt{\frac{L_p}{Q_p} \times \frac{L_s}{Q_s}} \qquad \text{(for critical coupling)}$$

By definition

$$M = k\sqrt{L_p L_s}$$

Equating and dividing both sides by $\sqrt{L_p L_s}$, we get

$$k = k_c = \frac{1}{\sqrt{Q_p Q_s}} \qquad \text{(for critical coupling)} \qquad \textbf{(2–14)}$$

If the two tuned circuits have equal Q values

$$k_c = \frac{1}{Q} \qquad \textbf{(2–15)}$$

From the *uncoupled* Q of each circuit, we can now find what value of k corresponds to critical coupling. Then by comparing the *actual* k value with this critical value we can determine if a circuit is under, critical, or overcoupled. Even more important, we can use this relation to adjust the circuit Q values so as to obtain the desired *degree* of coupling.

EXAMPLE 2–7

The primary and secondary circuits of Example 2–6 each (separately) have a Q of 100. The coupling coefficient is 0.02 and the operating frequency is 10 MHz.

1. What is the degree of coupling (under, critical, optimum, or further overcoupled)?
2. What change in the value of secondary circuit Q would produce critical coupling?
3. What change in the secondary Q value would produce optimum coupling?
4. How can (2) or (3) be effected?

Solution

1. (*a*) For critical coupling, with Q_1 and Q_2 each equal to 100, the coupling coefficient should be

$$k_c = \frac{1}{Q} = \frac{1}{100} = 0.01$$

(*b*) For optimum coupling, the coefficient should be

$$k_{opt} = 1.5k_c = 0.01 \times 1.5 = 0.015$$

(*c*) Since this circuit uses $k = 0.02$, it is overcoupled beyond optimum.

2. (*a*) In order for 0.02 to correspond to the critical coupling value,

$$k_c = 0.02 = \frac{1}{\sqrt{Q_1 Q_2}} \qquad \text{and} \qquad \sqrt{Q_1 Q_2} = 50$$

(*b*) Since Q_1 is being maintained at 100,

$$Q_2 = \frac{(50)^2}{Q_1} = \frac{(50)^2}{100} = \mathbf{25}$$

3. (*a*) For optimum coupling, the k value of 0.02 must be 50% greater than the critical coupling value. Therefore

$$k_c = \frac{k_{opt}}{1.5} = \frac{0.02}{1.5} = 0.0133$$

(*b*) Using the same technique as in step 2 above,

$$\sqrt{Q_1 Q_2} = \frac{1}{0.0133} = 75$$

$$Q_2 = \frac{(75)^2}{100} = \mathbf{56.3}$$

4. Since the required Q in steps 2 or 3 is lower than the actual Q of 100, the Q of the secondary circuit can be reduced by shunting a suitable resistance value across the tank circuit.

From this example we can conclude that, for a given coupling coefficient, the lower the circuit Q, the lower the *degree* of coupling. Let us apply this idea to another situation.

EXAMPLE 2–8

A double-tuned inductively coupled circuit has a primary and secondary Q of 80. The coupling coefficient is 0.015.

1. Find the degree of coupling.
2. Does the response curve have a single- or double-peak shape?
3. If the response is not a single-peak curve, how can such a curve be produced?

Solution

1. (*a*) $k_c = \frac{1}{Q} = \frac{1}{80} = 0.0125$

   (*b*) Since the given k is greater than k_c, the circuit is overcoupled.
2. The response curve is double-peaked (overcoupled).
3. To convert the response to single-peak, the Q of the primary and/or secondary should be lowered. (Shunt with a resistor.)

For example, if Q_2 is lowered to 45, critical coupling would require

$$k_c = \frac{1}{\sqrt{Q_1 Q_2}} = \frac{1}{\sqrt{80 \times 45}} = 0.0167$$

Since the given k is less (0.015), the circuit is now undercoupled, or single-peak.

This example bears out the conclusion that, for a given coupling coefficient, a lower Q produces a lower degree of coupling. An important application of this basic principle involves the alignment of receivers with wideband IF amplifiers (see Chapter 8).

RF FILTER CIRCUITS

In some RF circuit applications, it may be desirable to pass through, or to stop the passage of, a group of frequencies from a relatively wider frequency spectrum. This function is accomplished by the use of filter circuits. Depending on the specific portion of the spectrum to be passed, or stopped, filter circuits can be divided into four classifications, as follows.

1. *Low-pass filters.* This type of circuit will pass all frequency components of the input signal, with little or no attenuation, up to a certain critical frequency. This critical value, the *cutoff* fre-

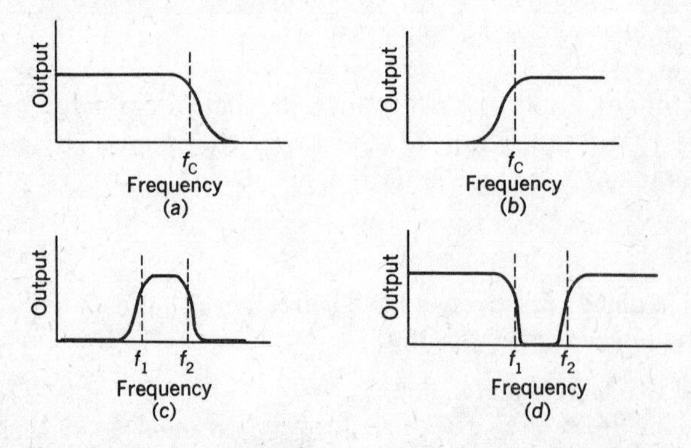

Figure 2-10 Frequency characteristics of various filter circuits.

quency, is the frequency at which the output level is down 3 decibels. For frequencies higher than cutoff, the output level drops rapidly toward zero. The response curve for a low-pass filter is shown in Fig. 2–10(a).

2. *High-pass filters.* The action of this type of filter is opposite to that of the previous circuit. All high frequencies are passed with little or no attenuation, down to the cutoff value. Again, at the cutoff frequency, the attenuation is 3 decibels. For frequencies below this cutoff value, the output is attenuated rapidly. (See Fig. 2–10(b).)
3. *Band-pass filters.* This type of filter passes only a relatively narrow band of frequencies. Frequencies above or below the desired channel are attenuated rapidly, as shown in Fig. 2–10(c).
4. *Band-rejection filters.* This filter is also known as a *band-stop* or *band-elimination* filter, or a *wave trap.* As these names indicate, this circuit will prevent the passage of a relatively narrow band of frequencies, whereas all frequencies above or below this band will be passed through with little or no attenuation. This response curve is shown in Fig. 2–10(d).

Two techniques can be used to attenuate or stop the passage of undesired frequency components. One method is to put a high impedance in series with the signal path to oppose the flow of current at these frequencies. The actual circuit element(s) used to accomplish this vary, depending on the specific frequency range to be blocked. For example:

1. To attenuate high frequencies—use an inductor.
2. To attenuate low frequencies—use a capacitor.
3. To attenuate a specific band of frequencies—use a parallel *L-C* circuit resonant over this band of frequencies.
4. To oppose all frequencies—except some narrow band—use a series *L-C* circuit, resonant over this narrow band of frequencies.

A second method for attenuating undesired frequency components is to shunt a low impedance across the signal path so as to bypass these frequencies to ground. Again the circuit components used will depend on the specific frequencies to be eliminated. However, it should be noted that this requirement is opposite to the series effect, so the circuit components are used in reverse manner:

1. To bypass high frequencies—use a capacitor.
2. To bypass low frequencies—use an inductor.
3. To bypass a specific band of frequencies—use a series-resonant circuit.
4. To bypass all frequencies—except some narrow band—use a parallel-resonant circuit.

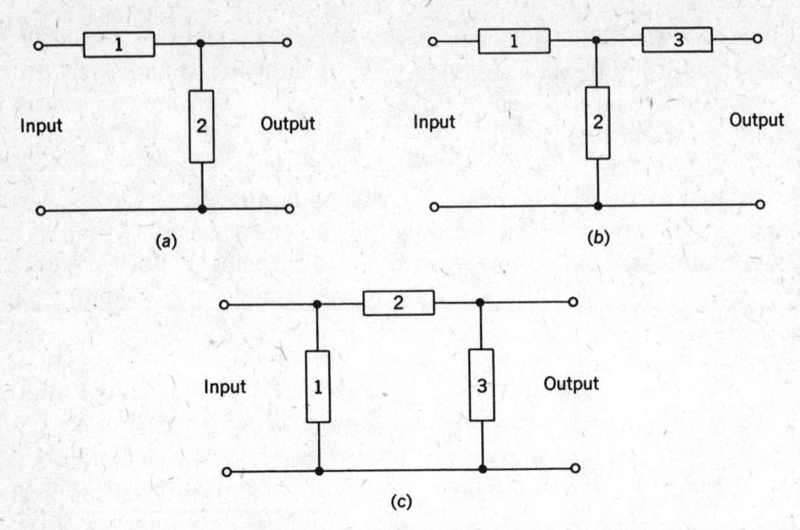

Figure 2-11 Three basic filter circuit configurations: (a) L type, (b) T type, (c) π type.

The final point to be considered is the interconnection of these components. Three basic circuits are in use. They are shown in block-diagram form in Fig. 2–11:

1. The L circuit, Fig. 2–11(a).
2. The T circuit, Fig. 2–11(b).
3. The π circuit, Fig. 2–11(c).

Notice that the name of the circuit tends to match the circuit configuration. Of course, the L circuit is really an inverted L. The circuit components going into each block depend on the specific frequency requirements. For example, if a high-pass filter (to attenuate low frequencies) is required, the series elements in any one of the three circuits would be capacitors, and the shunt elements would be coils. The selection of specific circuit components will be left as an exercise for the student. (See Review Questions 36, 37, 38.)

The block diagrams shown in Fig. 2–11 are for single filter sections. If the attenuation characteristic as obtained from a single section is not sharp enough, multiple sections can be used, with the output of one section being fed to the next.

Filter Circuit Values

When a filter is used in an electronic circuit, it is necessary for optimum results that the input and output impedances of the filter circuit match the impedance of the circuit or line into which it is inserted. The values of the inductors and capacitors used will determine this impedance and the cutoff frequency. Regardless of the type of filter, the impedance "looking into the filter" is known as the *characteristic impedance,* Z_0, and is given by

$$Z_0 = \sqrt{\frac{L}{C}} \tag{2–16}$$

The values of L and C for a particular cutoff frequency vary, depending on the type of filter. For low-pass filters the following equations* apply:

1. *L-type, low-pass:*

$$L = \frac{Z_0}{\pi f_c}, \qquad C = \frac{1}{\pi f_c Z_0} \tag{2–17}$$

2. *T-type, low-pass:* Since it uses two inductors in series, each L value is one-half the above.
3. *π-type, low-pass:* Since it uses two shunt capacitors, each C value is one-half the above.

Similar equations can be given for high-pass circuits:

1. *L-type, high-pass:*

$$L = \frac{Z_0}{4\pi f_c}, \qquad C = \frac{1}{4\pi f_c Z_0} \tag{2–18}$$

2. *T-type, high-pass:* Since it uses two capacitors in series, each C value is twice the above.
3. *π-type, high-pass:* Since it uses two shunt inductors, each L value is twice the above.

Equations for the band-pass and band-stop filters have not been given. These circuits are actually combinations of low-pass and high-pass circuits. For example, examine the response curve of the band-stop filter in Fig. 2–10(d). This effect can be obtained by paralleling a low-pass and a high-pass filter, with the low-pass circuit having a lower cutoff frequency than the high-pass circuit. Similarly, the response of Fig. 2–10(c) can be obtained from a low-pass and a high-pass circuit whose cutoff frequencies overlap.

IMPEDANCE MATCHING BY COUPLED CIRCUITS

When power is to be transferred from one circuit to another, proper impedance matching becomes important. This applies not only to vacuum-tube power amplifiers, but also to any transistor amplifier, small-signal or

* F. Langford-Smith, *Radiotron Designers Handbook,* 4th ed. (Sydney, Australia, Radio Corporation of America, 1952).

large-signal. At radio frequencies, tuned circuits can be used not only to provide the desired selectivity, but also to effect proper impedance matching. A number of circuit variations are possible. Each has its advantages, depending on the source and load to be matched.

Direct Coupling

The impedance of a parallel-resonant tank circuit is resistive and generally quite high. In fact, if the Q of the coil is excellent, this impedance approaches infinity. Obviously, such a tank circuit would not be suitable for matching sources or loads having finite resistance values. On the other hand, although increasing the *actual* resistance of the tank circuit would reduce the tank circuit impedance ($Z_{\parallel} = L/CR$), power would be wasted in this tank circuit resistance, rather than transferred from source to load. This is no solution, but there are two techniques that work quite well. One or the other is used, depending on whether the load resistance is lower or higher than the needed value.

Let us consider a case where the load impedance is low (for example, $R_L = 100\ \Omega$); the impedance required to match the source is moderately high (such as 7000 Ω); and the tank circuit consists of $C = 300$ pF, $L = 300\ \mu$H, and $R = 20\ \Omega$. This circuit is shown in Fig. 2-12(a). The impedance of this tank circuit (L/CR) is 50,000 Ω. If we connected the load directly across the tank circuit, the loading would be excessive; the Q of the tank circuit would be destroyed. Furthermore, the impedance of

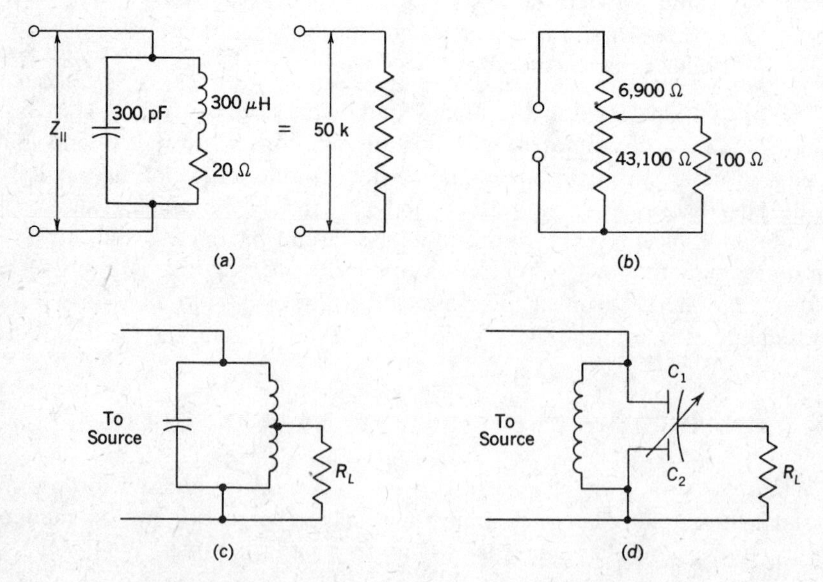

Figure 2-12 Impedance matching—direct-coupled load.

this loaded circuit would be approximately 100 Ω. Certainly, this would not match the source, and very little power would be transferred to the load. However, consider the resistor of Fig. 2–12(b) as representing the impedance of the tank circuit at resonance, and let us assume that we can tap this equivalent resistor at any convenient portion of its total value. Such a tap point is used in Fig. 2–12(b), and the load is connected between this tap and the lower end of the resistor. The parallel network of 43,100 and 100 Ω has a combination resistance of approximately 100 Ω. Add to this the resistance of the upper series portion (6900 Ω), and we get a total circuit resistance of 7000 Ω. This is the correct value for a perfect impedance match.

Figures 2–12(c) and (d) show two methods for applying this principle to an actual tank circuit. When a tapped coil is used, a higher value of load resistance requires a higher tap point for impedance matching. The same tap effect is achieved in the circuit of Fig. 2–12(d) by using a special differential capacitor. The total capacitance remains constant, regardless of the position of the rotor. Only the ratio of C_1 to C_2 changes. Making C_2 large compared to C_1 (X_{C2} small compared to X_{C1}) is similar to moving the tap point down. This would match a relatively low-impedance load. For higher-impedance loads, the effective tap point is moved up by decreasing the capacitance value of C_2 and increasing the value of capacitor C_1.

When the impedance of the load is higher than the value required for matching the source, the above process is reversed and the *source* is connected to a tap on the tank circuit. Such a circuit is shown in Fig. 2–13(a). As discussed before, the lower the source resistance, the lower is the tap point for a proper impedance match. An example will bring out the quantitative aspects of this matching technique.

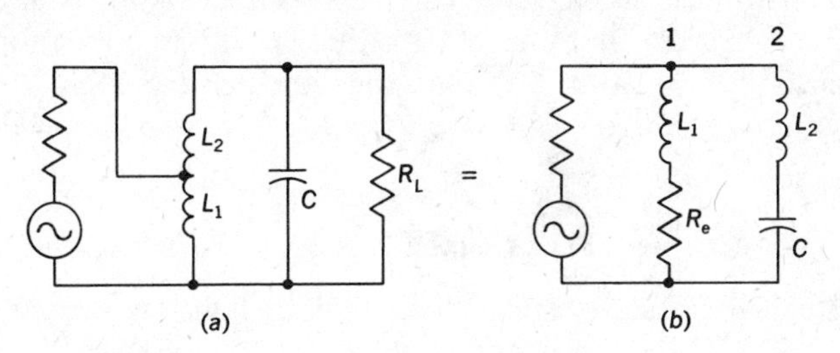

Figure 2-13 Direct coupling to a low-impedance source.

EXAMPLE 2–9

A tank circuit has $L = 300\ \mu\text{H}$, $C = 300$ pF, and a *total effective* series resistance (including the shunt effect of the load) of 125 Ω. This loaded tank circuit is to be matched, at its resonant frequency, to a transistor requiring a 2000-Ω load.

Find the inductance value at which the coil should be tapped.

Solution

1. The impedance of the loaded tank circuit is

$$Z_{\|} = \frac{L}{CR} = \frac{300 \times 10^{-6}}{300 \times 10^{-12} \times 125} = 8000\ \Omega$$

2. Since the source impedance is less than the loaded tank circuit value, a circuit as in Fig. 2–13(a) is required. This circuit is redrawn for easier analysis in Fig. 2–13(b). Notice in branch 2 that a portion L_2 of the coil is in series with the capacitor, reducing the reactance of that branch to $X_C - X_{L2}$. Simultaneously, the impedance of branch 1 is reduced to approximately X_{L1} (if we neglect R_e compared to X_{L1}).
3. Now to find the value of L_1:

   (*a*) The desired $Z_{\|} = 2000\ \Omega$

   (*b*) $Z_{\|} = \dfrac{X_{L1}^2}{R_e}$ or $X_{L1} = \sqrt{Z_{\|} \times R_e}$

   $$X_{L1} = \sqrt{2000 \times 125} = 500\ \Omega$$

   (*c*) $$f_0 = \frac{1}{2\pi\sqrt{LC}} = \frac{0.159}{\sqrt{300 \times 10^{-6} \times 300 \times 10^{-12}}} = 530\ \text{kHz}$$

   (*d*) $$L_1 = \frac{X_{L1}}{2\pi f} = \frac{500}{2\pi \times 530 \times 10^3} = \mathbf{150\ \mu H}$$

   The coil should be tapped such that $L_1 = 150\ \mu$H.

The direct-coupled circuit is capable of matching any impedance load to any source by feeding either the source or the load (*or both*) to a suitable tap on the tank circuit. Yet the very fact that it is adaptable to such a wide range of values makes it rather critical to adjust for any specific value. There are other circuits that are better suited to matching loads for a limited range.

Inductive-Coupled, Series-Tuned Loads

A circuit particularly suited for matching into low-impedance loads is shown in Fig. 2–14. The load is inductively coupled to the source or generator, and it forms part of a series-tuned *L-C-R* circuit. The load resistance value is generally large compared to the resistance of the secondary circuit itself and can therefore be considered as the total series resistance.* At resonance, the secondary circuit reflects a resistance R_r

* If this were not so, too much of the power delivered to the secondary would be wasted as I^2R losses in the *actual* circuit resistance.

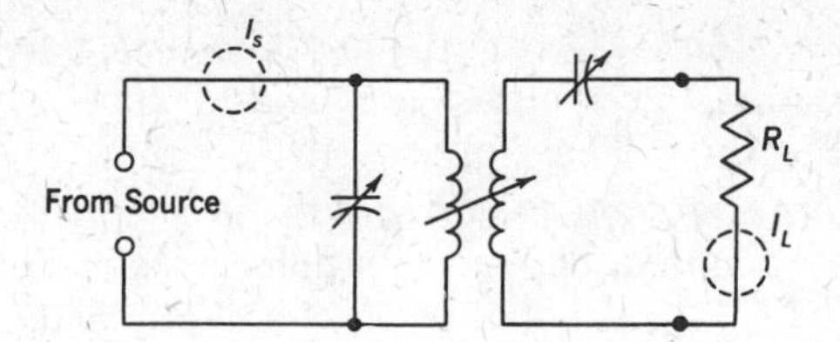

Figure 2-14 Impedance matching to low-impedance loads.

into the inductive branch of the primary tank circuit. The value of reflected resistance depends on the mutual inductance and the load resistance, or

$$R_r = \frac{(\omega M)^2}{R_L}$$

In turn, the impedance $Z_{\|}$ of the primary tank circuit will depend on this reflected resistance. ($Z_{\|} = L/CR$, where R is the total of the primary circuit resistance and the reflected resistance. Again, for efficiency, the *load effect* R_r should be large compared to the *actual* primary circuit resistance.)

To obtain an impedance match for a given load value, the coupling between the primary and secondary is adjusted so that the primary tank circuit impedance (because of R_r) equals the source impedance. In this respect, notice that as the coupling is increased ωM increases, R_r increases, and the primary circuit impedance decreases.

If a circuit has been adjusted for a good impedance match, any increase in the load resistance will reduce the reflected resistance value, destroying the impedance match. This effect can be countered by tightening the coupling until the increase in $(\omega M)^2$ offsets the increase in R_L. This brings out a limitation of this circuit. Once the two coils are coupled as close as possible *physically,* no further increase in ωM can be obtained. It is therefore impossible to produce the required value of reflected resistance if the load resistance is too high. This circuit will produce matching only for relatively low-impedance loads, in the approximate range of $0 - 250\ \Omega$.

With reference to Fig. 2–14, ammeters I_s and I_L (broken circles) can be inserted to measure the generator (or source) current and the output load current, respectively. Starting with the circuits at minimum coupling, let us analyze the effect of increasing the coupling on each current value. On the primary side at loose coupling the reflected resistance is low, the tank circuit impedance is high, and the primary or source current I_s is low. As the coupling is increased R_r increases, $Z_{\|}$ decreases, and the current I_s will increase. The tighter the coupling, the higher is this current value.

On the secondary side the picture is not quite the same. The power delivered to the load is $I_L^2R_L$. This power is a maximum when the load is matched to the source. Therefore, as the coupling is tightened, I_L increases *until the coupling condition for maximum power transfer is reached.* Tighter coupling beyond this point will result in a decrease in the load current I_L. Proper coupling can therefore be determined by noting when the current in the load circuit reaches a maximum.

Inductive-Coupled, Shunt-Tuned Loads

When it is necessary to transfer power to high-impedance loads, the secondary circuit is reconnected into a parallel-resonant circuit and the load is shunted across this matching network. Such a circuit is shown in Fig. 2–15. As discussed in Chapter 1, a resistor shunted across a tank circuit has the same effect as a resistor in series with L and C—except that the larger the shunt resistance value, the lower is the equivalent series resistance value. Therefore, the load resistor R_L in Fig. 2–15 can be replaced by an equivalent resistance R_e (shown by dashed lines). Now the circuit (L, C, and R_e) is identical to the series-tuned circuit of Fig. 2–14. The following conclusions therefore apply:

1. A high-resistance load (equivalent to a low value of R_e) tends to produce a high value of reflected resistance and therefore requires looser coupling to produce a given $Z_{\|}$ value for impedance matching.
2. The lower the load resistance value R_L, the higher is the equivalent R_e, the smaller is the reflected resistance, and the higher $Z_{\|}$. If the impedance match is to be maintained, the coupling must be tightened.

Here again we are faced with a physical limitation. If the shunt load resistance value is too low, it may not be physically possible to obtain a tight

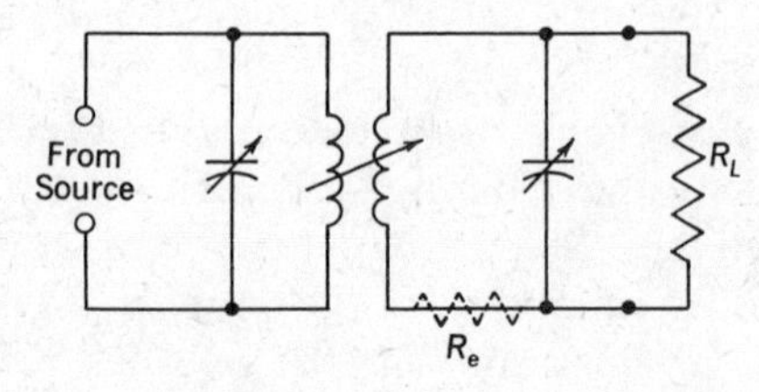

Figure 2-15 Impedance matching to high-impedance loads.

enough coupling. Impedance match thus becomes impossible. This circuit is suitable for relatively high-impedance loads from approximately 250 Ω and up. As in the previous circuit, proper matching results in a maximum current in the load, whereas the primary current increases as the coupling is tightened, even beyond the matching point.

Pi Network Impedance Matching

Notice that the two previous circuits are more or less well suited for the opposite ends of the range scale, and each is sort of at the "ragged edge" for intermediate values. A circuit particularly suited to matching intermediate values of loads is the π *network* shown in Fig. 2–16(a). The capacitors C_1 and C_2 could be a differential unit, as described for Fig. 2–12(d). The total capacitance is constant, but the ratio of C_1 to C_2 can be varied. The source "looks into" an impedance as shown by the equivalent diagram in Fig. 2–16(b), where R_e represents the equivalent series resistance of the load. The impedance of this equivalent tank circuit is reduced as capacitor C_1 is made larger and C_2 smaller. This technique will match a given load into low-impedance sources. In a similar analysis, the load "looks into" an impedance as shown by the equivalent circuit of Fig. 2–16(c), where R_e is the equivalent series resistance of the source. The impedance of this equivalent circuit is increased as C_2 is made smaller and C_1 larger. This technique will match a given source into higher-impedance loads. In a practical situation, since the source impedance is fixed, the ratio of C_1 to C_2 is varied to match the load to the source. For low-impedance loads, the capacitance of C_2 is increased (C_1 decreasing); to match high-impedance loads, the capacitance of C_2 is decreased (C_1 increasing). This circuit is best suited for load values approximately 5–2000 Ω.

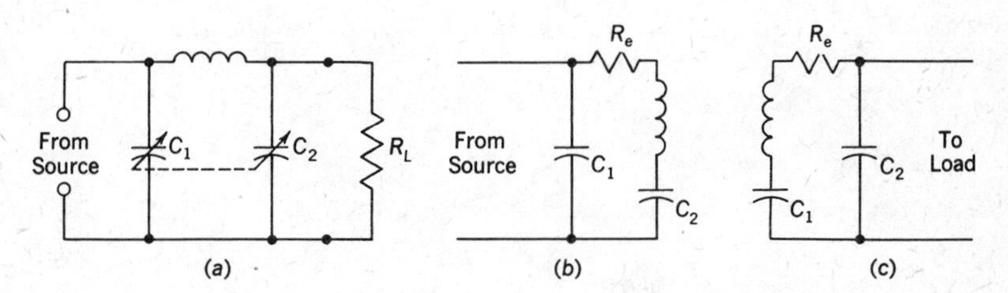

Figure 2-16 Impedance matching using a π network.

EXAMPLE 2–10

A vacuum tube requires a load of 8000 Ω for optimum power transfer. It is to be inductively coupled to a 5500-Ω load using an air-core transformer L_1 and L_2, each 60 μH. The operating frequency is to be 1200 kHz. The effective resistance of each winding (R_1 and R_2) at this frequency is 10 Ω. Find the capacitance needed to tune each coil to this operating frequency, and find the coefficient of coupling needed to obtain a proper impedance match. Try a series-tuned load and a shunt-tuned load, and determine which is preferable.

Solution

1. Find the capacitance value needed:

   (a) $X_C = X_L = 2\pi fL = 6.28 \times 1200 \times 10^3 \times 60 \times 10^{-6} = 452\ \Omega$

   (b) $C = \dfrac{1}{2\pi f X_C} = \dfrac{0.159 \times 10^{12}}{1200 \times 10^3 \times 452} = \mathbf{295\ pF}$

2. Find the primary reflected resistance (R_r) needed to match the primary tank impedance $Z_\parallel$ to 8000 Ω (see Fig. 2–17(a)):

   (a) From $Z_\parallel = \dfrac{L}{CR_{s1}}$, where $R_{s1} = R_1 + R_r$,

   $$R_{s1} = \frac{L}{CZ_\parallel} = \frac{60 \times 10^{-6}}{295 \times 10^{-12} \times 8000} = 25.4\ \Omega$$

   $$R_r = 25.4 - 10 = 15.4\ \Omega$$

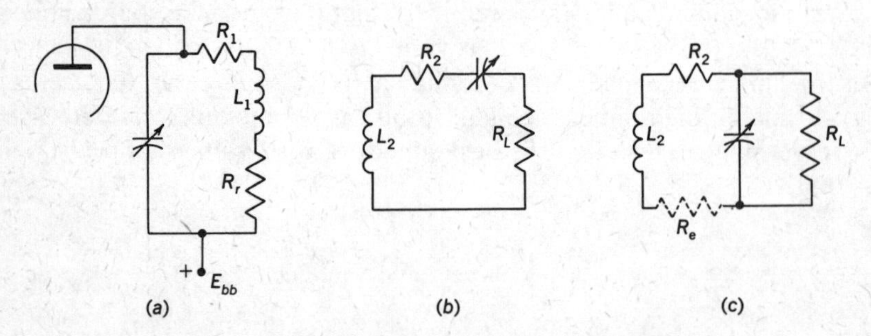

Figure 2-17 Impedance matching—equivalent circuits.

3. Using a series-tuned secondary circuit, we will get $R_r = 15.4\ \Omega$ when:

   (a) $\omega M = \sqrt{R_r R_{s2}} = \sqrt{15.4 \times (5500 + 10)} = 291\ \Omega$

   (b) $\omega = 2\pi f = 6.28 \times 1200 \times 10^3 = 7.53 \times 10^6$

   (c) $M = \dfrac{\omega M}{\omega} = \dfrac{291}{7.53} \times 10^{-6} = 38.7\ \mu\text{H}$

   (d) $k = \dfrac{M}{\sqrt{L_1 L_2}} = \dfrac{38.7}{60} = \mathbf{0.645}$

4. Using a shunt-tuned secondary (Fig. 2–17(c)), we get:

(*a*) $Z_{\|} \text{ (unloaded)} = \dfrac{L}{CR_2} = \dfrac{60 \times 10^{-6}}{295 \times 10^{-12} \times 10} = 20{,}400\ \Omega$

(*b*) $Z_{\|} \text{ (loaded)} = \dfrac{Z_{\|} \times R_L}{Z_{\|} + R_L} = \dfrac{2.04 \times 10^4 \times 0.55 \times 10^4}{2.95 \times 10^4} = 4330\ \Omega$

(*c*) $R_{s2} = R_2 + R_e = \dfrac{L}{CZ_{\|}} = \dfrac{60 \times 10^{-6}}{295 \times 10^{-12} \times 4330} = 47.0\ \Omega$

(*d*) $\omega M = \sqrt{R_r \times R_{s2}} = \sqrt{15.4 \times 47.0} = 26.9\ \Omega$

(*e*) $M = \dfrac{\omega M}{\omega} = \dfrac{26.9}{7.53 \times 10^6} = 3.57\ \mu\text{H}$

(*f*) $k = \dfrac{M}{\sqrt{L_1 L_2}} = \dfrac{3.57}{60} = 0.0595$

5. The shunt-tuned load is preferable by far. A coupling coefficient of 0.0595 can be readily attained, whereas the coefficient of 0.645 is much too tight and cannot be realized with air-core transformers.

REVIEW QUESTIONS

1. (*a*) What is the basic purpose of any coupling circuit? (*b*) What is the specific function of such a circuit when voltage output is the prime consideration? (*c*) How does this differ if power transfer is involved? (*d*) State a feature common to all coupling circuits.

2. (*a*) What is meant by the term *coupling coefficient*? (*b*) Give the symbol used to represent this term. (*c*) Give the general equation for this term.

3. Refer to Fig. 2–1. (*a*) What component is responsible for the coupling? (*b*) How can the coupling be increased? (*c*) Would this circuit be useful for maximum power transfer? Explain.

4. Refer to Fig. 2–2(a). (*a*) What type of coupling is this? (*b*) How can the degree of coupling be increased? (*c*) What is the effect of frequency on the degree of coupling? Explain.

5. Refer to Fig. 2–2(b). (*a*) What type of coupling is this? (*b*) How can the degree of coupling be decreased? (*c*) At what frequency would the degree of coupling be a maximum? Explain.

6. Refer to Fig. 2–3(a). (*a*) What type of coupling is this? (*b*) How is coupling achieved? (*c*) What is the *common element* called? (*d*) By what symbol is it represented? (*e*) How is the degree of coupling increased? (*f*) Is this type of coupling frequency-sensitive? Explain.

7. Refer to Fig. 2–3(b). (*a*) What is this type of coupling called? (*b*) What basic type of coupling is it? (*c*) When is this type of coupling necessary? (*d*) How is the coupling varied?

8. Under what condition is some form of combination coupling preferred to the above simple coupling circuits?

9. Refer to Fig. 2–4(a). (*a*) What basic types of coupling does this circuit employ? (*b*) At what frequency will the coupling be a minimum? Why?

10. Refer to Fig. 2–4(b). (*a*) What basic types of coupling are used? (*b*) As the frequency increases, what happens to the degree of coupling? Explain.

11. Refer to Fig. 2–4(c). (*a*) What basic types of coupling are used? (*b*) How is the coupling affected by operation at higher frequencies?

12. (*a*) State four factors that determine the magnitude of the induced voltage in the secondary of an inductively coupled circuit. (*b*) Give the equation for this voltage in terms of the mutual inductance. (*c*) What is the phase relation between this voltage and the primary current?

13. (*a*) What is the actual significance of the term *reflected impedance*? (*b*) What causes this effect? (*c*) What letter symbol is used to represent it? (*d*) Give the equation for evaluating its magnitude.

14. Refer to Fig. 2–5. (*a*) Which of these diagrams uses inductive coupling? (*b*) Which uses single-tuned inductive coupling? (*c*) Which uses double-tuned inductive coupling? (*d*) Could a double-tuned circuit be drawn in any other way? Explain. (*e*) What does R_1 represent? (*f*) What does R_2 represent? (*g*) What does Z_r represent?

15. In Fig. 2–5(a), what is the character and magnitude of the reflected impedance? Why?

16. Refer to Fig. 2–5(b). (*a*) What is the character of the reflected impedance when the secondary is resonant to the signal frequency? (*b*) Why? (*c*) Give the equation for the reflected quantity for this specific condition. (*d*) Can this effect be obtained with any of the other circuits in Fig. 2–5? Explain.

17. (*a*) In Fig. 2–5(c), if the secondary is tuned to resonance, what is the character of the reflected impedance? (*b*) What is the relative magnitude of the reflected quantity if R_L is large? (*c*) If R_L is small?

18. In Fig. 2–5(b), if the secondary is tuned above the frequency of the incoming signal, what is the character of the reflected impedance? Explain.

19. In Fig. 2–5(c), what is the character of the reflected impedance if the secondary is tuned (*a*) above the frequency of the incoming signal? Explain. (*b*) below the frequency of the incoming signal? Explain.

20. In Fig. 2–5(d), what is the character of the reflected impedance for low values of R_L? Explain.

21. With reference to Fig. 2–6(b), what do each of the following symbols represent: r_d, R_1, L_1, L_2, R_2, E_s?

22. In Fig. 2–6(b): (*a*) Why isn't the reflected impedance from secondary to primary shown? (*b*) With respect to E_s, is the secondary a series or a parallel circuit? Explain.

23. Refer to Fig. 2–7. (*a*) To what type of circuit do these curves apply? (*b*) What does E_o represent? (*c*) What component voltage in Fig. 2–6 corresponds to E_o? (*d*) What does f_0 represent? (*e*) Why is E_o a maximum at this frequency? (*f*)

Why does tighter coupling produce a higher output voltage? (*g*) Why does the bandwidth increase with tighter coupling?

24. (*a*) What type of circuit is shown in Fig. 2–8(a)? (*b*) Can a double-tuned circuit take any other form? Explain. (*c*) When would this variation be of advantage?

25. For similar circuit values, which circuit (single-tuned or double-tuned inductively coupled) would have the higher output voltage? Why?

26. (*a*) In what type of circuit can the condition of critical coupling be achieved? (*b*) What circuit values are necessary to produce this condition?

27. Refer to curve 3 of Fig. 2–9. (*a*) Why is this response curve considered as the optimum curve? (*b*) What degree of coupling produces this effect? (*c*) Why is the E_o value at f_0 lower than for curve 2? (*d*) Why does E_o rise for a small frequency change below f_0? (*e*) Why does it also rise for a small change above f_0? (*f*) Why does E_o drop at frequencies beyond these peak values?

28. In curve 4 of Fig. 2–9, why are the peaks further apart?

29. In Fig. 2–9, does curve 2, 3, or 4 have the highest peak value? Explain.

30. In a double-tuned RF transformer (all other factors being equal), how does the bandwidth obtained vary with (*a*) the Q of the coils? (*b*) the resonant frequency? (*c*) the coupling coefficient? (*d*) Give the equation for bandwidth.

31. In a double-tuned circuit, what is the relation between the Q of the coils and (*a*) the coupling coefficient? Explain. (*b*) the degree of coupling? Explain. (*c*) Give the equation showing the relationship of part *b*.

32. A double-tuned transformer produces a double-peaked response curve. State two ways by which a single-peak response curve can be obtained.

33. What is meant by the *cutoff* frequency of a filter?

34. With reference to Fig. 2–10, what type of filter produces the output shown in (*a*) diagram (a)? (*b*) diagram (b)? (*c*) diagram (c)? (*d*) diagram (d)?

35. Refer to Fig. 2–11(a). (*a*) Based on connections, what type of filter is this? (*b*) Based on frequency, what type is it? Explain. (*c*) Which is the *series* element?

36. In Fig. 2–11(*a*), what circuit component would be used for elements 1 and 2 to obtain (*a*) low-pass action? Why? (*b*) high-pass action? Why? (*c*) band-pass action? Why? (*d*) band-rejection action? Why?

37. In the π filter of Fig. 2–11, what circuit component would be used for elements 1, 2, and 3 to obtain (*a*) low-pass action? Why? (*b*) high-pass action? Why? (*c*) band-rejection action? Why? (*d*) band-pass action? Why?

38. Repeat parts *a, b, c,* and *d* of Question 37 for the T filter of Fig. 2–11.

39. When is it important to match the impedance of a source to its load? Why?

40. Refer to Fig. 2–12. (*a*) What is the meaning of the 50k resistor alongside circuit (a)? (*b*) For what frequency condition does this apply? (*c*) What is the total resistance of circuit (b)? (*d*) If the potentiometer arm is moved down, what happens to this total resistance? (*e*) What type of coupling is used between source and load of Fig. 2–12(c)? (*f*) In diagram (c), if a given load is to be matched to a higher source impedance, how is this effected? Explain.

41. Refer to Fig. 2–12(d). (*a*) How does the total capacitance vary as the shaft is turned clockwise? (*b*) What is the effect of rotation? (*c*) What is the purpose of this type of capacitor? (*d*) For a low-impedance load, how should the values of C_1 and C_2 compare? Explain.

42. Refer to Fig. 2–12(c). What change should be made if the load impedance is higher than the desired matching value?

43. The circuit of Fig. 2–13(a) is to be used with a source having a somewhat higher internal impedance. How is an impedance match maintained?

44. Refer to Fig. 2–13(b). (*a*) At the resonant frequency of this circuit, which is the higher reactance: X_{L2} or X_C? Explain. (*b*) How is the total reactance of branch 2 determined? (*c*) Is this reactance inductive or capacitive? (*d*) Compared to a typical tank circuit, what does this branch represent? (*e*) What does R_e represent? (*f*) Knowing X_{L1}, X_{L2}, X_C, and R_e, what is the most direct way of solving for the impedance of this circuit at resonance? (*g*) Would this impedance be higher, lower, or the same as with the source connected at the top (Fig. 2–13(a)) of the tank?

45. Refer to Fig. 2–14. (*a*) What type of coupling is this? (*b*) For what range of load values is this circuit suitable? (*d*) How does the load resistance compare with the resistance of the secondary winding? (*d*) What is the significance of the arrow through the coils? (*e*) What is this feature used for? (*f*) Assuming that the circuit had been matched, if the load resistance is decreased, how is proper matching restored? (*g*) Explain why this is so.

46. In Fig. 2–14, as the coupling is tightened from too loose a value through the proper value and beyond, what will happen to the current registered by (*a*) I_s? (*b*) I_L?

47. Refer to Fig. 2–15. (*a*) How does this circuit differ from Fig. 2–14? (*b*) What does the dotted resistor R_e represent? (*c*) If the R_L value were increased, how would this affect the value of R_e? (*d*) For what range of load values is this circuit suitable? (*e*) Assuming that the circuit had been matched, if the load resistance is increased, how is proper matching restored? (*f*) Explain why this is so.

48. Refer to Fig. 2–16(a). (*a*) How many capacitors are used in this circuit? (*b*) What is this unit called? (*c*) What is the effect on the capacitance values (C_1, C_2, and C_{tot}) as the rotor is turned? (*d*) For what range of values is this circuit suitable? (*d*) How is impedance matching obtained?

49. Refer to Fig. 2–16(b). (*a*) What does this schematic represent? (*b*) What does R_e represent? (*c*) What happens to the impedance of this circuit as C_1 is decreased? Explain. (*d*) What matching would this achieve?

50. Refer to Fig. 2–16(c). (*a*) What does this schematic represent? (*b*) What does R_e represent? (*c*) What happens to the impedance of this circuit as C_1 is decreased? Explain. (*d*) What matching would this achieve?

51. Refer to Fig. 2–17. (*a*) What do each of the following quantities represent: R_1, R_2, R_r, and R_L? (*b*) Which of these is (are) unknown? (*c*) Do diagrams (b) and (c) represent primary or secondary circuits? (*d*) What is the essential difference between these diagrams? (*e*) What does R_e in diagram (c) represent?

PROBLEMS

1. Two coils have inductances of 2.0 mH and 4.5 mH, respectively. What value of coupling is needed in order to create a mutual inductance of 0.12 mH?

2. Find the mutual inductance between a 200-μH coil and a 350-μH coil, if the coefficient of coupling is 0.022.

3. What coefficient of coupling would be needed in an RF transformer to produce an induced voltage of 0.3 V, if the winding inductances are 300 μH primary and 380 μH secondary, the primary current is 3.5 mA, and the input signal frequency is 1100 kHz?

4. An RF transformer has the following conditions: $L_1 = L_2 = 260$ μH and $k = 0.055$. What value of primary current is needed to induce 0.15 V in the secondary winding at a frequency of 750 kHz?

5. An inductive-coupled circuit using an untuned primary and tuned secondary has the following values: $L_1 = 250$ μH; $L_2 = 200$ μH; effective resistances of the windings = 10 and 6 Ω, respectively; the secondary capacitor = 200 pF; the coefficient of coupling = 0.042. Calculate the value of the reflected impedance and the total primary impedance at (*a*) the resonant frequency of the secondary, (*b*) 770 kHz, (*c*) 880 kHz.

6. A FET ($r_d = 50$ kΩ) is coupled to a second FET via an untuned-primary, tuned-secondary transformer. The ac component of drain current in the first FET is 2.0 μA. The coupling values are $L_1 = 10$ mH, $L_2 = 150$ μH and 9.42 Ω, $k = 0.02$, $C_2 = 169$ pF. Find (*a*) the voltage at the gate of the second FET at the resonant frequency, (*b*) the bandwidth of the resonance curve. (*c*) Repeat part *a* for 20 kHz below resonance.

7. A double-tuned circuit is resonant at 1700 kHz. Find (*a*) the bandwidth, if $k = 0.013$, and (*b*) the coupling coefficient required for a bandwidth of 10 kHz.

8. A double-tuned circuit has a primary Q of 83 and a secondary Q of 120. Both windings are tuned to 1700 kHz, and their coupling coefficient is 0.013. Is the circuit over, critical, or undercoupled?

9. In Problem 8, what value of Q would make the coupling correspond to optimum? (Use equal Q values for the two coils.)

10. A double-tuned transformer is to operate at 800 kHz, have a bandwidth of 12 kHz, and give best band-pass response. (*a*) What value of coupling coefficient is needed? (*b*) Assuming equal values of primary and secondary Q, what value of Q would be required?

11. An RF double-tuned transformer is designed for use at an intermediate frequency of 450 kHz. Primary and secondary are alike—$L = 2$ mH and $Q = 100$. (*a*) What value of coupling should be used for 18-kHz bandwidth? (*b*) What should the Q of the coils be for best response? (*c*) What should the effective resistance of the coils be for this value of Q? (*d*) What are the resistances of the given coils? (*e*) What value of series resistance should be added? (*f*) What value of shunt resistance would produce this same effect?

12. A tuned-primary, tuned-secondary circuit (see Fig. 2–8) has the following values: $L_1 = 159\ \mu H$; $R_1 = 10\ \Omega$; $C_1 = 159$ pF; $L_2 = 150\ \mu H$; $R_2 = 9.42\ \Omega$; $C_2 = 169$ pF; $k = 0.01$. The ac component of the total tank current (I_t) is 0.05 mA. Find (*a*) the impedance of the primary tank circuit, (*b*) the output voltage E_o at resonance, (*c*) the output voltage at 10 kHz above resonance, and (*d*) the degree of coupling (under, critical, or over).

13. A double-tuned transformer has the following values: $f_0 = 600$ kHz, $L_1 = 265\ \mu H$, $R_{L1} = 10\Omega$, $X_{L2} = 1000\ \Omega$, $R_{L2} = 10\ \Omega$, $k = 0.01$, and applied primary voltage $= 5.0$ V. Find (*a*) the Q of the primary circuit (consider the action of the secondary), (*b*) the total primary tank current, (*c*) the induced secondary voltage, (*d*) the output voltage across C_2, (*e*) the degree of coupling (under, critical, or over).

14. Determine the proper values for a low-pass, π-type filter having a characteristic impedance of 600 Ω and a cutoff frequency of 1000 kHz.

15. Repeat Problem 14 for a T-type high-pass filter.

16. A load of 20,000 Ω is to be matched to a 500-Ω source using a resonant circuit. The circuit values are $L = 159\ \mu H$ and 10 Ω, $C = 159$ pF. Show by diagram how direct coupling can be used, and calculate the inductance value between the common connection and the proper tap point.

17. A double-tuned circuit is to be used to match a 10,000-Ω source to its load. The tuned circuit values are $L_1 = 150\ \mu H$ and 12 Ω, $L_2 = 100\ \mu H$ and 10 Ω, $C_1 = 200$ pF, and $C_2 = 300$ pF. The load resistance is 15 Ω. Draw a circuit suitable for this impedance match and calculate at what value of coupling a proper match is obtained.

18. Repeat Problem 17 for an 8000-Ω load.

TYPICAL FCC QUESTIONS

3.140 Describe what is meant by "link coupling" and for what purpose(s) it is used.

3.449 What is meant by the term "unity coupling"?

3.450 Draw a diagram illustrating capacitive coupling between two tuned radio-frequency circuits.

3.451 Draw a diagram illustrating "inductive" coupling between two tuned radio-frequency circuits.

4.003 If the mutual inductance between two coils is 0.1 henry, and the coils have inductances of 0.2 and 0.8 henry, respectively, what is the coefficient of coupling?

4.096 What is a "low-pass" filter? A "high-pass" filter?

4.097 Draw a diagram of a simple low-pass filter.

6.461 How is the degree of coupling varied in a π network used to transfer energy from a vacuum-tube plate circuit to an antenna?

S-3.036 In general, why are filters used? Why are "band-stop," "high-pass," and "low-pass" filters used? Draw schematic diagrams of the most commonly used filters.

3

RF Voltage (or Small-Signal) Amplifiers

In many electronic circuits operating at radio frequencies, the RF signal level is often so weak that it must be amplified before it can be used. This is especially true in receiver operation, whether the receiver is an entertainment unit (radio or television), a radar receiver searching for an echo from a far-away target, or a telemetry receiver picking up some important data from an earth satellite. Generally, the signal voltage developed at the receiving antenna is only a few microvolts. Before the intelligence can be extracted from such a weak signal, it must be amplified. This function is accomplished in an RF amplifier. Since the signal level is low, such amplifiers are often referred to as *small-signal* amplifiers. Furthermore, since the power level is negligible, and the main purpose of these circuits is to increase the amplitude of the signal, they are also commonly called RF *voltage* amplifiers. In this chapter we will discuss the various circuits used for RF voltage amplifiers and the characteristics of these amplifiers.

When dealing with RF amplifiers, we are generally interested in a relatively narrow band of frequencies (the carrier, and the sidebands that contain the intelligence). Therefore the coupling network that transfers the signal energy from the antenna to the first amplifier, or from one amplifier stage to another, generally contains a resonant circuit. In this way, because of the Q effect, selectivity and gain can be obtained.

In the early days of radio, RF amplifiers used triode vacuum tubes. Because of their relatively high interelectrode capacitance C_{gp}, RF energy was fed back from the plate (output side) to the grid (input side), producing serious oscillation problems. These problems were solved with the inventions of the tetrode vacuum tube, and then the pentode. For many years the pentode was universally used for RF amplification. Then in the 1960s, with the improvements in transistor design, manufacturers began to use bipolar transistors in the RF stages—not because these transistors were better than the pentode, but rather because of the miniaturization possibility, the elimination of heater circuitry, the lower power consumption, and the (theoretically) infinite life span of the transistor. (Actually, as far as performance was concerned, the tube was still superior.) By the late 1960s no new designs used vacuum tubes as RF small-signal amplifiers. At about the same time, breakthroughs in the manufacturing processes made the FET commercially available. Their price was higher than for a bipolar, but their characteristics were sufficiently superior to warrant their use in the higher-quality receivers. The FET soon became the preferred type of RF small-signal amplifier. However, bipolar units are still being used in new products, and receivers with tubes are still in operation. Obviously, all three devices must be discussed.

Regardless of the active device used, it is necessary to couple RF energy into the stage and out to the next stage. The coupling network must:

1. pass the desired RF energy through with a minimum attenuation.
2. prevent (block) any dc component in the source voltage from affecting the bias, or operating point, of the next stage.
3. provide selectivity.

Three of the more commonly used coupling methods are impedance coupling, single-tuned inductive coupling, and double-tuned inductive coupling. Each of these will be discussed in this chapter.

IMPEDANCE-COUPLED RF AMPLIFIERS

Figure 3–1 shows an RF voltage amplifier employing *impedance coupling* between its drain and the gate of the next stage. This type of coupling is often used with multistage wideband amplifier units found in radar and television receivers. The coupling network itself (shown within the dashed lines) consists of a tank circuit (L_1C_2), a coupling capacitor (C_c), and a resistor R_g. Either the capacitor C_2 or the inductor L_1 is variable (or adjustable) so that the circuit can be tuned (or aligned) to some specific frequency, as desired. Sometimes, the positions of the resistor and the tank circuit are interchanged, as shown in Fig. 3–1(b). This diagram also introduces another point. The tank circuit is shown only as a coil. The black arrow alongside indicates that the inductance value is adjust-

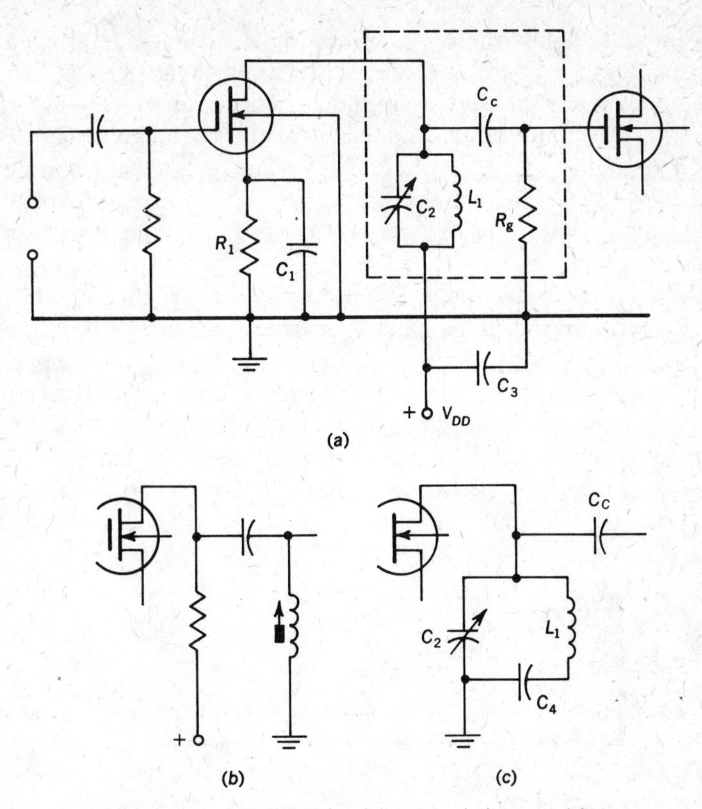

Figure 3-1 MOSFET impedance-coupled RF amplifier.

able (slug-tuned) and tunes this circuit to the desired frequency. However, no capacitor is shown across the inductor. This tank circuit uses for its capacitance the output capacitance of the previous FET, the input capacitance of the next FET, the distributed capacitance of the coil, and the stray shunt capacitance of the circuit. Such a technique is generally used only at relatively high frequencies (over 10 MHz). At these frequencies, the drain tank circuit of Fig. 3-1(a) might also employ the same technique.

Notice that the metal-oxide-semiconductor (MOS) or insulated gate (IG) FET is used in this circuit. This transistor has a very low feedback, or reverse transfer capacitance, C_{rss}. Consequently, there is no danger of oscillations well into the VHF band, and no special neutralizing circuitry is needed. Proper class-A bias is obtained by using the source bias resistor and bypass capacitor R_1C_1. Capacitor C_3 bypasses any RF component of the drain current to ground, keeping RF out of the power supply. Still better decoupling can be obtained by inserting an RF choke in the V_{DD} line just below the C_3 junction. With a capacitively tuned tank circuit, the variation shown in diagram (b) has an advantage. The rotor of the tuning capacitor C_2 can be grounded. This is of special importance when a two- or three-gang capacitor is used for single-control of a multistage RF amplifier. The tuning capacitor in diagram (a) can also be grounded—without shorting the power supply—if a fixed capacitor is inserted between the rotor of C_2 and coil L_1. This variation is shown in diagram (c).

Many other variations are possible. For example, a dual-gate MOSFET can be used. This has the added advantage of providing excellent AGC* action by feeding the control voltage to the second gate. Junction FETs can also be used. However, if the capacitive feedback effect is too great, neutralizing may become necessary.

Since the pentode RF amplifier is quite similar to the FET circuit, it is best to show it at this time. A typical circuit is shown in Fig. 3-2.

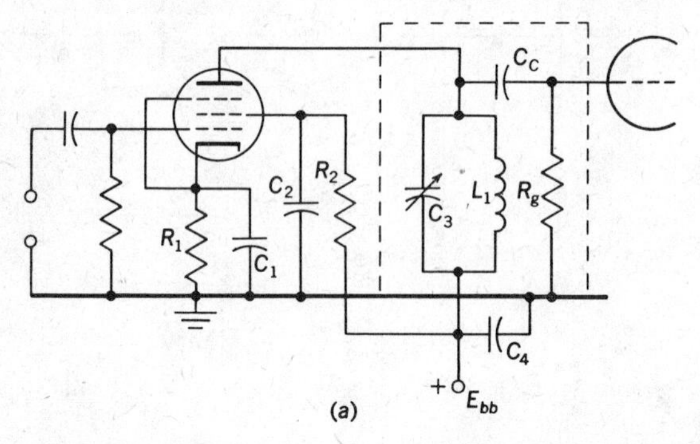

Figure 3-2 Pentode impedance-coupled RF amplifier.

The coupling network and the class-A bias circuitry are as before. The only "new" features are the supressor grid and screen wiring. The suppressor grid, as is common in pentode tube operation, is tied to the cathode. The screen grid must be operated at a suitable dc potential—but at RF ground. This grid is connected to the dc supply voltage through a dropping resistor R_2. The value of this resistor depends on the supply voltage. If the supply voltage is low enough (125 V or lower), the resistor may be omitted. The screen is kept at RF ground potential by the bypass capacitor C_2. This capacitor should have negligible reactance at the operating frequency.

Bipolar transistors are also used as small-signal RF amplifiers. Unfortunately, bipolar devices introduce two complications not encountered

*Automatic gain control (AGC) keeps the receiver volume constant as you tune from station to station. AGC circuits will be discussed in Chapter 8.

in FET or vacuum-tube circuits. One of these problems is caused by the relatively low input and output impedances of this transistor. Impedance matching techniques must be used to prevent excessive loading of the resonant tank circuits. The second problem is caused by the interaction between output and input circuits in the bipolar transistor. This high feedback factor would normally cause unwanted oscillations. The feedback action must be neutralized. However, in some transistors, the feedback may be small enough—particularly at low radio frequencies—to make neutralization unnecessary.

Figure 3–3 shows how impedance coupling can be used with bipolar transistors. These partial schematics show only the coupling network as an interstage device from the collector of one stage to the base of the next stage.

Notice that the base input circuit is never connected across the full tank circuit. This input impedance generally falls within a range of 300–2000 Ω. This is low enough to ruin the Q of the tank circuit, resulting in very low gain and poor selectivity. On the other hand, the output impedance is appreciably higher, ranging approximately 20,000–100,000 Ω. It is therefore permissible to connect the collector output across the entire tank circuit as shown in Fig. 3–3(a) or (b). However, for minimum loading and maximum selectivity, it is better to feed the collector output to a suitable tap as shown in Fig. 3–3(c). (The use of tap points for impedance matching was covered in Chapter 2.)

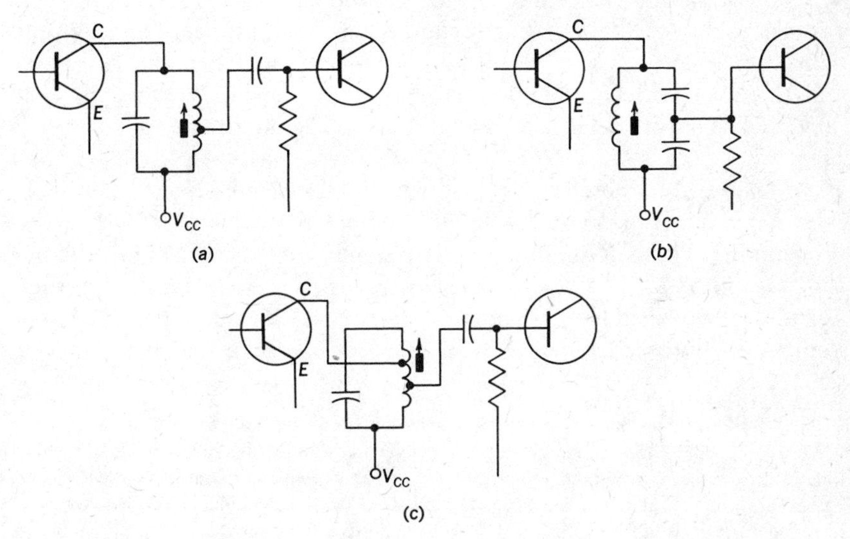

Figure 3-3 Impedance coupling methods for bipolar transistor RF amplifier.

Gain—Equivalent Circuit

To analyze the gain possibilities of an impedance-coupled RF amplifier, the circuit of Fig. 3–1 can be converted into its ac equivalent diagram. The constant-current form is shown in Fig. 3–4. Since the reverse transfer capacitance, C_{rss}, is very low, feedback can be neglected, and this simplified Norton equivalent circuit is valid. However, before proceeding with the gain calculations, some discussion of the parameters g_{fs} and r_d is in order. FET data sheets may present this information in a variety of ways. The parameter g_{fs}, known as the *forward transconductance,* may also be called *mutual conductance,* g_m, or just *transconductance* (as in vacuum tubes), or it may be equated with the *transadmittance,* y_{fs}. Sometimes the symbol RE(y_{fs}) is used, where the prefix RE stands for "the real part of." Actually, y_{fs}, the transmittance, combines the conductance component and the reactive (capacitive) component. Therefore, g_{fs} or RE(y_{fs}) is a more correct symbol for transconductance. However, since the capacitive component is quite small compared to the real or resistive component, this approximation ($y_{fs} \cong g_{fs}$) is permissible at RF values even into the VHF range.

The second parameter shown in Fig. 3–4 is r_d. This represents the *output resistance,* which in the common-source connection is also the drain resistance, hence r_d. In FET data sheets this parameter may be specified as r_{os} or—indirectly—as RE(y_{os}) or as g_{os}. (It should be realized that these last two symbols are for conductance rather than resistance values.) Sometimes only the output admittance, y_{os}, is given. Obviously, this combines the output conductance and susceptance. If the output capacitance, C_{oss}, is also given, the output conductance and resistance can be calculated. On the other hand, even into the VHF range, the susceptance value may be low enough that the approximation $y_{os} \cong g_{os}$ is valid.*

The third parameter shown in Fig. 3–4 is r_i, the input resistance of the next FET. For a common-source connection, this may be listed as r_{is} or, indirectly, as RE(y_{is}) or g_{is}. The same problems exist in evaluating this parameter as discussed earlier for r_d. In Fig. 3–4, the symbol for r_i is shown by dashed lines because in many cases its value is high enough, compared to R_g (particularly with MOSFET units), that it can be neglected.

*Exact evaluation of the output resistance (or conductance) from data sheets is further complicated in that the y_{os} or g_{os} value given in these sheets represents the maximum limit of this parameter. Minimum limits or typical values (which are not shown) can readily be one-tenth of the tabulated value (or even lower). For an exact value, direct measurement of this parameter (for the given transistor) at the desired operating frequency and bias conditions would be necessary.

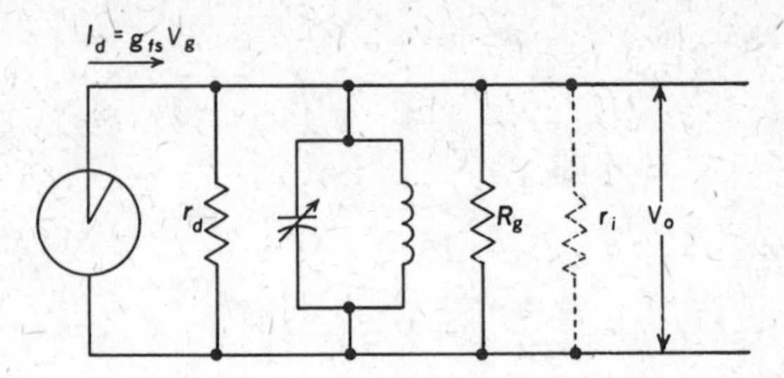

Figure 3-4 Equivalent circuit for a FET impedance-coupled RF amplifier.

Now, referring to Fig. 3–4, the input signal V_g will produce an ac (signal) component of drain current equal to $g_{fs}V_g$. The output voltage can then be calculated as the product of this current and the equivalent impedance Z_e, due to the parallel combination of r_d, the drain resistance, $Z_{\|}$, of the tank circuit,* R_g, the gate resistor of the coupling network, and (if applicable) the input resistance, r_i, of the next stage. Since the stage gain is the ratio of output voltage over input voltage,

$$A = \frac{V_o}{V_g} = g_{fs}Z_e \tag{3–1}$$

EXAMPLE 3–1

An impedance-coupled RF amplifier (see Fig. 3–1) has the following circuit values: a 2N4353 MOSFET with g_{fs} 4000 μmho, r_i of several megohms, and g_{os} 5.0 μmho; a tank circuit with $C = 300$ pF and with $L = 200$ μH and 8.33 Ω effective resistance; and a gate resistor of 220 kΩ. It is coupled to another similar MOSFET. Find (1) the effective load impedance Z_e, (2) the gain of the stage, (3) the output voltage for an input signal of 80 μV, and (4) the equivalent Q of the coupling network.

Solution

1. Find the equivalent load impedance.

(*a*) $r_d = \dfrac{1}{g_{os}} = \dfrac{1}{5.0 \times 10^{-6}} = 200 \text{ k}\Omega$

(*b*) The parallel combination of r_d and R_g is

$$\frac{r_d \times R_g}{r_d + R_g} = \frac{200 \times 220}{420}(10^3) = 105 \text{ k}\Omega$$

* Resistances R_g, r_d, and r_i are shunted across the tank circuit and load the tank circuit. The equivalent impedance Z_e is therefore the same as the impedance of the loaded tank circuit.

(*c*) $Z_{\|}$ (unloaded tank circuit) is

$$\frac{L}{CR} = \frac{200 \times 10^{-6}}{300 \times 10^{-12} \times 8.33} = 80.0 \text{ k}\Omega$$

(*d*) Since the tank acts as a pure resistance, and since r_i can be neglected, then

$$Z_e = \frac{Z_{\|} \times 105\text{k}}{Z_{\|} + 105\text{k}} = \frac{80 \times 105}{185} \times 10^3 = \mathbf{45.4 \text{ k}\Omega}$$

2. The gain of the stage is

$$A = g_{fs} Z_e = 4000 \times 10^{-6} \times 45.4 \times 10^3 = \mathbf{182}$$

3. The output voltage is

$$V_o = AV_g = 182 \times 80 \times 10^{-6} = \mathbf{14.6 \text{ mV}}$$

4. The equivalent Q of the coupling network is actually the loaded Q of the tank circuit. This is related to the load impedance by

$$Z_e = X_L Q_e \qquad \text{or} \qquad Q_e = \frac{Z_e}{X_L}$$

We must, therefore, find first the resonant frequency and then X_L.

(*a*) $f_0 = \dfrac{1}{2\pi\sqrt{LC}} = \dfrac{0.159}{\sqrt{200 \times 10^{-6} \times 300 \times 10^{-12}}} = 649 \text{ kHz}$

(*b*) $X_L = 2\pi fL = 6.28 \times 649 \times 10^3 \times 200 \times 10^{-6} = 816\ \Omega$

(*c*) $Q_e = \dfrac{Z_e}{X_L} = \dfrac{45{,}400}{816} = \mathbf{55.6}$

The indentical equivalent-circuit technique can be used with vacuum-tube RF amplifiers. Obviously, the three shunt resistors in Fig. 3–4 would now be: r_p the plate resistance, R_g the grid-leak resistor, and r_i the input resistance of the next tube. (For a class-A RF amplifier, no grid current flows, and even into the VHF range, the input resistance approaches infinity and can be omitted.

Data sheets for bipolar transistors intended for use as RF small-signal amplifiers often give *y*-parameter values—either tabulated or in curves. We can therefore again use the equivalent circuit of Fig. 3–4 to analyze such circuits. Obviously, the g_{fs} or y_{fs} of the FET diagram now becomes g_{fe}, RE(y_{fe}), or y_{fe}. Also, r_d or r_{os} becomes r_c, r_{oe}, or RE(y_{oe}); r_i more specifically is r_{ie}; and R_g is now called R_B. Unfortunately, the same problems in evaluating the FET parameters also exist with bipolar units.

There is one other complication. Notice in Fig. 3–3, that the feed to the next stage is connected to a tap on the tank circuit, rather than across the full tank. This is because of the very low input resistance of the next stage. Therefore, when analyzing its equivalent circuit, we must consider the impedance-transforming effect of these taps. Either the impedance on the collector (left side) must be stepped down, or the impedance on the next-stage ("secondary") side must be stepped up. Since the coils used are generally made with ferromagnetic (powdered iron) or ceramic cores, the coupling between turns may be considered close enough to unity, and the impedance transformation is proportional to the square of the turns ratio, n^2. The technique can be illustrated with a problem.

EXAMPLE 3–2

A 2N4874 bipolar transistor has the following parameters: RE(y_{fe}) 180 mmho, RE(y_{oe}) 0.008 mmho, and RE(y_{ie}) 16 mmho. It is impedance-coupled to a second 2N4874 (see Fig. 3–3(a)), through a tank circuit with $L = 200\ \mu\text{H}$ and an effective resistance of 8.33 Ω, and $C = 300$ pF. The coil is tapped at one-fifth up from the bottom. The base resistor R_B is 10 kΩ. Find: (1) the equivalent load impedance (primary side), (2) the signal voltage applied to Q_2 for an input signal of 80 mV at Q_1, and (3) the equivalent Q of the coupling network (primary side).

Solution

1. Find the equivalent load impedance

   (*a*) $r_{oe} = \dfrac{1}{g_{oe}} = \dfrac{10^3}{0.008} = 125\ \text{k}\Omega$

   (*b*) $Z_{\|}$ (unloaded tank) from Example 3–1 = 80.0 kΩ

   (*c*) r_{ie} of $Q_2 = \dfrac{1}{g_{ie}} = \dfrac{10^3}{16} = 62.5\ \Omega$

   (*d*) Compared to r_{ie}, R_B of 10 kΩ can be disregarded.

   (*e*) Transforming r_{ie} to the primary side, we have

   $$r'_{ie} = r_{ie}(n)^2 = 62.5 \times 25 = 1560\ \Omega$$

   (*f*) $Z_e = r_{oe}$, $Z_{\|}$ and r'_{ie} in parallel = **1510 Ω**

2. Find the signal voltage applied to Q_2.

   (*a*) Gain on the primary side:

   $$A = g_{fe}Z_e = 180 \times 10^{-3} \times 1510 = 272$$

   (*b*) Signal voltage across the tank:

   $$V_t = AV_{b1} = 272 \times 80 \times 10^{-3} = 21.8\ \text{mV}$$

   (*c*) Tank voltage stepped down by turns ratio = V_{b2}, or:

   $$V_{b2} = V_t \times \frac{1}{n} = \frac{21.8}{5} = 4.36\ \text{mV}$$

3. Find the equivalent Q of the coupling network:

(*a*) $f_0 = 649$ kHz (from Example 3–1)

(*b*) $X_L = 816\ \Omega$ (from Example 3–1)

(*c*) $Q_e = \frac{Z_e}{X_L} = \frac{1510}{816} = \mathbf{1.85}$

Obviously, this is a poorly designed circuit. The Q is too low. This shows that the same tank circuit values cannot be used with a bipolar as with a FET—or the tap point must be made further down to reduce loading. (See Problems 7 and 8 at the end of the chapter for possible remedies for this extremely low Q.)

Bandwidth

The tank circuit in the coupling network is the only frequency-selective component in an impedance-coupled amplifier. The selectivity and bandwidth of this stage are therefore dependent on the Q of this tank circuit. Obviously, the loaded Q must be used. The bandwidth can be found from Eq. (1–4) ($BW = f_0/Q_e$), and the full response curve can be readily obtained by the quick resonance-curve method described in Chapter 1. Again, the loaded Q must be used.

EXAMPLE 3–3

Find the bandwidth of the impedance-coupled circuit of Example 3–1 ($Q_e = 55.6$ and $f_0 = 649$ kHz).

Solution

$$BW = \frac{f_0}{Q_e} = \frac{649}{55.6} = \mathbf{11.6\ kHz}$$

Multistage Response

Quite often the gain and/or selectivity, as obtained from one stage, is not sufficient. In such cases, two or more stages are operated in cascade, that is, the output from the first stage is fed to the gate or base of the next stage, and so on.* The gain of such an amplifier is obviously the product of the gains of the individual stages. For example, if each stage of a three-stage amplifier has a gain of 40, the overall gain is $40 \times 40 \times 40$ or 64,000. As a further illustration, if a 10-μV signal is fed to the input of stage 1, the first stage output is 400 μV. This is fed as an input to stage 2, and the second stage output is 16,000 μV. When this, in turn, is applied

* When each stage is tuned to the same frequency, the amplifier is known as a *synchronous* tuned amplifier.

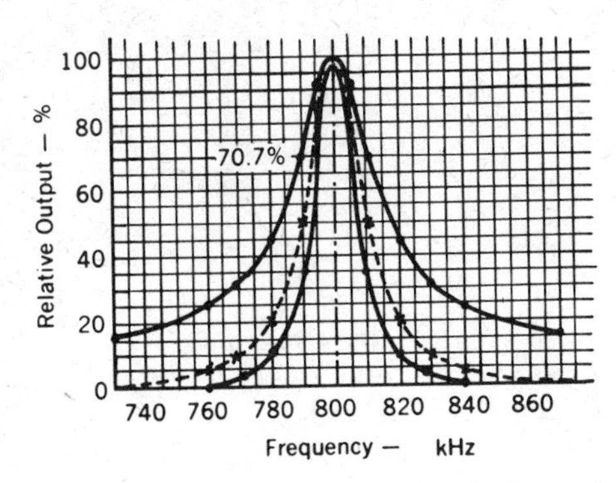

Frequency (kHz)	Relative Output (%)		
	One Stage	Two Stages	Three Stages
800	100	100	100
795 & 805	92	84	0.78
790 & 810	71	50	36
780 & 820	45	20	9.0
770 & 830	32	10	3.2
760 & 840	24	5.8	1.4
730 & 870	16	2.6	0.42

Figure 3-5 Effect of cascade stages on a response curve.

to stage 3, the final output is 640,000 μV. The gain, from 10 μV to 640,000 μV, is 64,000.

Because of the action of the tank circuit, the gain of any tuned amplifier circuit decreases for frequencies off resonance. For multistage amplifiers, the *product* of these decreasing gain values causes the overall response curve to drop off faster than the response curve of a single stage. In other words, by using cascade stages, we increase the selectivity or decrease the bandwidth. Mathematically this relationship* is given by

$$BW_n = BW_1\sqrt{2^{1/n} - 1} \tag{3-2}$$

where BW_n is the bandwidth of n stages and BW_1 is the bandwidth of a single stage. A graphical analysis will show this narrowing effect. Study Fig. 3-5 and the charted values carefully. Each stage, *taken individually,* has a bandwidth as shown for the single-stage response curve. For example, at a frequency of 780 kHz, the single-stage response has dropped to 45% of its maximum (resonant) value. In a two-stage circuit, the output will be 45% *of this 45% value* or only 20% of maximum. Similarly, in a three-stage circuit, the output will be further reduced to 45% of this 20% value or 9% of the maximum value. Notice also that the bandwidth of the single-stage circuit is 20 kHz (790–810 kHz); the bandwidth decreases to 13 kHz for two cascade stages and is down to 10 kHz for a three-stage circuit. This action can be interpreted in two ways. If a single-stage circuit cannot give sufficient selectivity, additional tuned circuits can be used in cascade until the desired selectivity is achieved. On the other hand, if several cascade stages are required for their gain effect only, *each stage's bandwidth must be made appreciably wider than the*

*S. Seeley, *Electronic Circuits* (New York, Holt, Rinehart and Winston, 1967), Chap. 7.

required bandwidth, so that the overall bandwidth will not be too narrow. In Fig. 3–5, if the desired three-stage bandwidth is 20 kHz, each stage must have an individual bandwidth of 39 kHz. Since each stage needs almost double the overall bandwidth, the individual circuit Q – and therefore the gain per stage – must be reduced.

EXAMPLE 3–4

A bandwidth of 35 kHz is to be obtained from a three-stage impedance-coupled amplifier at a frequency of 800 kHz. Find (*a*) the bandwidth for each individual stage and (*b*) the Q required for each stage. (*c*) Repeat (*a*) and (*b*) for a two-stage amplifier.

Solution

1. For a three-stage amplifier:

(*a*) $BW_n = BW_1\sqrt{2^{1/n} - 1} = 35$ kHz, from which

$$BW_1 = \frac{35}{\sqrt{2^{1/3} - 1}} = \frac{35}{\sqrt{1.26 - 1}} = \mathbf{68.7\ kHz}$$

(*b*) $$Q_e = \frac{f_0}{BW} = \frac{800}{68.7} = \mathbf{11.6}$$

2. For a two-stage amplifier:

(*a*) $$BW_1 = \frac{35}{\sqrt{2^{1/2} - 1}} = \frac{35}{\sqrt{0.414}} = \mathbf{54.3\ kHz}$$

(*b*) $$Q_e = \frac{f_0}{BW} = \frac{800}{54.3} = \mathbf{14.7}$$

Notice in Fig. 3–5 that the curves are plotted in terms of relative output – in percent. In each curve, the maximum value for that curve is used as the reference for that curve and is called the 100% point. This technique not only simplifies calculations, but – more important – it makes it possible to plot all three curves conveniently on a common ordinate scale of 0–100. If we wished to plot actual overall gain, first we must have the full circuit data needed to calculate gain, as in Example 3–1. Let us assume a gain of 80. The ordinate scale for one stage would then range from 0 to 80 (or rounded off to 100). At resonance, since the curves for all stages coincide, the overall gain for two stages would be 80 × 80 or 6400, and for three stages 80 × 80 × 80 or 512,000! On a common ordinate scale, it would be impossible to show the selectivity of one stage in comparison to the three-stage circuit. For the same reason it would also be impossible to plot actual output voltage vs frequency for a single stage and a three-stage amplifier on the same scale.

Stagger Tuning

Where wide bandwidth is necessary, the synchronous tuned amplifier has a serious drawback. The Q of the individual circuits must be dropped to a relatively low value, and the gain per stage is drastically reduced. This tends to defeat the reason for using multistage operation in the first place. Better overall gain and bandwidth can be obtained by slightly detuning each tank circuit to one side or the other of the desired center frequency. Figure 3–6 shows this effect graphically for a two-stage cascade amplifier. Each tank circuit (curves A and B) has a bandwidth of only 15 kHz. One is resonant at 790 kHz, the other at 810 kHz. The overall response curve has a center frequency of 800 kHz and a bandwidth of 35 kHz. Because of stagger tuning (10 kHz to each side), it is possible to use 15-kHz bandwidths for the individual circuits and still obtain a much wider overall response. The Q for the individual circuits is now 800/15 or 53.3. Compare this with the two-stage requirements of Example 3–4. Notice that the individual circuit Q can now be raised from 14.7 to 53.3, an increase of 3.64 times. However, the overall gain will not be 3.64 times greater. Because of stagger tuning, the maximum product (curve $A \times$ curve B) is only 0.37 (at 790 and 810 kHz) as compared to 1.0 (at 800) for the synchronous tuned circuit. The gain advantage for stagger tuning is therefore only 0.37×3.64, or 1.35 times greater than for synchronous single tuning.

The stagger-tuning technique is used to a great extent in the fixed-frequency wideband pulse amplifiers for radar and television. Some radar circuits may use as many as nine stagger-tuned circuits, at a mean frequency of 60 MHz. One television manufacturer, to obtain a 4-MHz bandwidth at a mean frequency of approximately 24 MHz, used four stagger-tuned circuits, resonant respectively at 25.3, 22.3, 25.2, and 23.4 MHz. Notice that the second tank is tuned to 22.3 instead of 25.2 MHz. With adjacent tank frequencies separated in this manner, the first stage

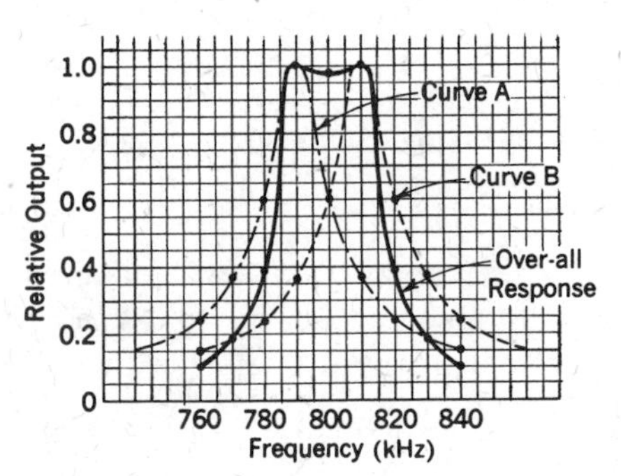

Frequency	Response Curves			
	Curve A	Curve B	Over-all	
			Product	Percent of Maximum
760	0.24	0.15	0.036	10
770	0.37	0.18	0.067	18
780	0.60	0.24	0.144	39
790	1.00	0.37	0.37	100
800	0.60	0.60	0.36	97.3
810	0.37	1.00	0.37	100
820	0.24	0.60	0.144	39
830	0.18	0.37	0.067	18
840	0.15	0.24	0.036	10

Figure 3-6 Effect of stagger tuning on a response curve.

input is tuned to 25.3 MHz, while its *output* is tuned to 22.3 MHz. Therefore, even if there is feedback from the output to the input, the phase of the feedback voltage will not sustain oscillations. Since the danger of oscillations is minimized, special circuitry such as neutralization* is not necessary.

Shielding

When dealing with multistage RF amplifiers, we find that the electrostatic and magnetic fields surrounding tubes, capacitors, and inductors of one stage may be strong enough to reach and interact with the components in another stage. Such feedback generally leads to oscillations. To prevent interaction, the offending components are usually enclosed in metal shields. Shielded coils are shown in Fig. 3–7. Tubes with glass envelopes are covered with close-fitting metal shields. On the other hand, some tube types and all bipolars and FETs are available with metal envelopes, eliminating the need for a separate shield. In high-gain or high-power circuits, each individual stage may be enclosed in a separate shielded compartment.

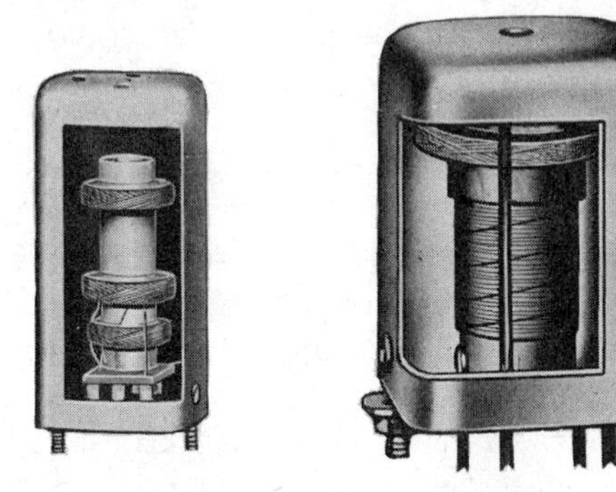

Figure 3-7 Shielded RF coils.

SINGLE-TUNED TRANSFORMER-COUPLED RF AMPLIFIERS

Another method for coupling RF amplifier circuits involves inductive coupling with an untuned primary and a tuned secondary. This type of coupling is generally found in receivers, to couple the antenna to the first RF amplifier, or as an interstage network between RF amplifiers—particularly where the resonant frequency has to be varied at will. Such a

*The problem of oscillation and the principles and techniques of neutralization are discussed later in this chapter.

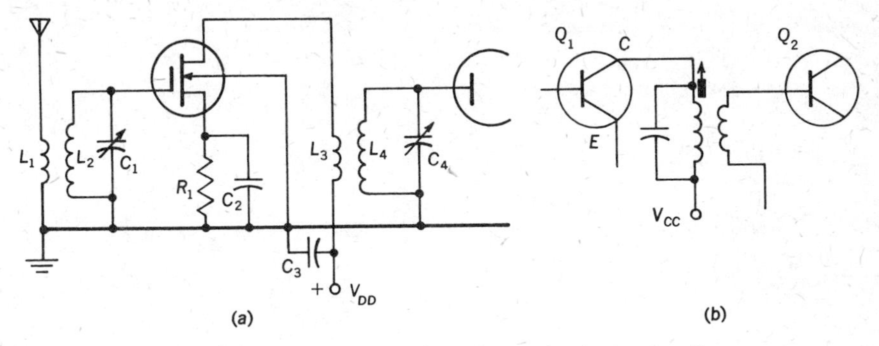

Figure 3-8 RF amplifier with single-tuned inductive coupling.

circuit is shown in Fig. 3–8(a). Single-tuned transformer coupling is shown both at the input and at the output of this stage. A MOSFET is used as the active element and, except for the coupling network, this circuit is identical to Fig. 3–1. No further discussion of the circuit components is therefore necessary. Also, since this is a single-tuned circuit, it has the same selectivity characteristics as the impedance-coupled amplifier. The bandwidth of a single stage is determined by its resonant circuit ($BW = f_0/Q$); multistage operation reduces the bandwidth, unless stagger tuning is employed.

Single-tuned inductive coupling is also extensively used with bipolar transistors. However, because of the low input impedance of the transistor, the tuned circuit is put on the primary side. The secondary winding has relatively few turns, providing an impedance match to the low base-input impedance. A typical transformer for use at 465 kHz (in the IF stages of a superheterodyne receiver) has a 155-turn primary and only an 18-turn secondary. A partial schematic is shown in Fig. 3–8(b).

A vacuum-tube diagram is not shown. The student should be able to develop this diagram from a comparison of Fig. 3–1 with 3–2, and Fig. 3–8(a).

Gain

To calculate the gain of this single-tuned inductive circuit, it is necessary to know the mutual inductance between the primary and secondary so that the secondary induced voltage can be determined. An example will illustrate this point.

EXAMPLE 3–5

The following values apply to the circuit of Fig. 3–8: $L_1 = 2$ mH; $L_2 = 250$ μH and 10 Ω; coupling coefficient between L_1 and $L_2 = 0.03$; $C_1 = 250$ pF; $L_3 = 10$ mH. The FET is a 40468A with $g_{fs} = 7500$ μmho. If the signal voltage developed at the antenna (to ground) is 48 μV, find the voltage at the gate of the FET, the output voltage across L_3, and the gain.

Solution

1. $\omega = 2\pi f = 2\pi \times \dfrac{1}{2\pi\sqrt{LC}} = \dfrac{1}{\sqrt{250 \times 10^{-6} \times 250 \times 10^{-12}}}$

   $= 4 \times 10^{6}$

2. $X_{L1} = \omega L_1 = 4 \times 10^{6} \times 2 \times 10^{-3} = 8000\ \Omega$

3. The resistance of coil L_1 is very much lower than its reactance, ωL_1, and can be neglected; so that

$$I_{L1} = \frac{E_{L1}}{X_{L1}} = \frac{48 \times 10^{-6}}{8 \times 10^{3}} = 6.00 \times 10^{-9}\ \text{A}$$

4. M (between L_1 and L_2) $= k\sqrt{L_1 L_2} = 0.03\sqrt{2 \times 10^{-3} \times 250 \times 10^{-6}}$

   $= 21.1\ \mu\text{H}$

5. The induced voltage in the tank circuit L_2C_1:

$$E_s = \omega M I_{L1} = 4 \times 10^{6} \times 21.2 \times 10^{-6} \times 6 \times 10^{-9} = 0.51\ \mu\text{V}$$

6. $X_{L2} = \omega L_2 = 4 \times 10^{6} \times 250 \times 10^{-6} = 1000\ \Omega$

7. Q (of tank circuit L_2C_1) $= \dfrac{X_{L2}}{R} = \dfrac{1000}{10} = 100$

8. The voltage at the gate of Q_1:

$$E_g = QE_s = 100 \times 0.51 \times 10^{-6} = \mathbf{51\ \mu V}$$

9. Drain current (signal component):

$$I_d = g_{fs}E_g = 7500 \times 10^{-6} \times 51 \times 10^{-6} = 38.2 \times 10^{-8}\ \text{A}$$

10. $X_{L3} = \omega L_3 = 4 \times 10^{6} \times 10 \times 10^{-3} = 40 \times 10^{3}\ \Omega$

11. E_o (across L_3) $= I_d X_{L3} = 38.2 \times 10^{-8} \times 40 \times 10^{3} = \mathbf{15.3\ mV}$

12. Gain (from antenna to across L_3):

$$A = \frac{E_o}{E_i} = \frac{15.3 \times 10^{-3}}{48 \times 10^{-6}} = \mathbf{318}$$

DOUBLE-TUNED TRANSFORMER-COUPLED RF AMPLIFIERS

From the discussion on coupled circuits in Chapter 2, you will recall that the double-tuned, inductively coupled circuit is capable of higher gain and (with slight overcoupling) better band-pass action. This makes it *theoretically* an ideal choice for interstage coupling. In practice, however, there are drawbacks that limit its application to circuits operating at a fixed frequency. These disadvantages can be made apparent from an examination of Fig. 3–9, which shows a junction FET RF amplifier using double-tuned inductive coupling for its input and output circuits. Assuming this to be a partial schematic for a receiver operating in the AM broadcast band, each of the *four* tank circuits must be tunable over the frequency range 540–1600 kHz. Since single-control tuning is desired, it is necessary to "gang" the four variable capacitors into a unit assembly with all the rotors on a common shaft. Two such stages, with double-tuned inputs and outputs, would require a *six-gang* capacitor! Such an assembly is bulky and expensive. Furthermore, for good band-pass action, the tuned circuits must be perfectly *aligned* (tuned to the proper frequency). This creates a serious practical problem. It is almost impossible to maintain perfect tracking alignment over a wide range of frequencies, since it requires exact duplication of all stages and precision construction of the tuning elements (variable C or L). On the other hand, for operation at one fixed frequency, these drawbacks disappear, because each tank circuit is individually adjusted for that one specific frequency.

The circuit of Fig. 3–9 introduces nothing new. Except for the addition of the extra tuning capacitors across the primary coils, this circuit is identical to that of Fig. 3–8. No further discussion is therefore necessary to explain the circuit component functions.

Double-tuned inductive coupling can also be used with bipolar transistors, but because of their low base-input impedance, the loading of the tuned secondary tank circuit can be excessive, and there is little advantage over single-tuned coupling. Consequently, double tuning is not popular. However when it is used, the connection to the base of the next stage is tapped down on the tank to reduce loading. A partial schematic is shown in Fig. 3–9(b). Notice, in this case, that the collector connecting point is also tapped down to further reduce loading and improve selectivity. (Tapping down of the collector or drain connection can also be done in any of the previous tuned circuits if the particular device output resistance is low and causes heavy loading.)

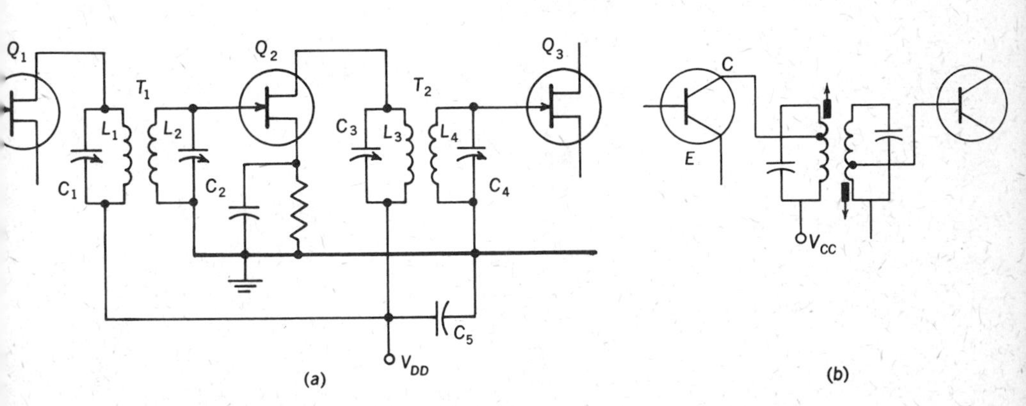

Figure 3-9 RF amplifier with double-tuned inductive coupling.

Bandwidth

The response curve obtained from a double-tuned inductively coupled RF amplifier depends on the characteristics of the coupling networks, as discussed in Chapter 2. The bandwidth is given by kf_0, and the shape of the response curve depends on whether the coils are under, critical, or overcoupled. (For best band-pass action, a coupling coefficient 50% greater than the critical value should be used.) Again, as with other types of coupling, the overall bandwidth will decrease when these circuits are used in a multistage amplifier. Compared to single-stage operation, the bandwidth for an n-stage amplifier is given by

$$\mathrm{BW}_{ndt} = \mathrm{BW}_{1dt}\sqrt[4]{2^{1/n} - 1} \tag{3–3}$$

where BW_{1dt} is the bandwidth of *one* double-tuned stage. To compare this relation with single-tuned circuitry, let us repeat Example 3–4, this time using double-tuned coupling networks.

EXAMPLE 3–6

A bandwidth of 35 kHz is to be obtained from a three-stage RF amplifier, using double-tuned inductive coupling, at a frequency of 800 kHz. Find (*a*) the bandwidth for each individual stage, (*b*) the k required for each stage, and (*c*) the Q required for optimum coupling (assume identical coils). (*d*) Repeat (*a*) for a two-stage amplifier.

Solution

1. For a three-stage amplifier:

(*a*) $\mathrm{BW}_{1dt} = \dfrac{\mathrm{BW}_{ndt}}{\sqrt[4]{2^{1/n} - 1}} = \dfrac{35}{\sqrt[4]{2^{1/3} - 1}} = \mathbf{49\ kHz}$

(*b*) $k = \dfrac{\mathrm{BW}_{1dt}}{f_0} = \dfrac{49}{800} = \mathbf{0.0613}$

(*c*) For optimum coupling:

(1) $k = 0.0613 = 1.5\ k_c$

(2) $k_c = \dfrac{0.0613}{1.5} = 0.0408$

(3) $Q = \dfrac{1}{k_c} = \dfrac{1}{0.0408} = \mathbf{24.5}$

2. For a two-stage amplifier:

$$\mathrm{BW}_{1dt} = \frac{\mathrm{BW}_{ndt}}{\sqrt[4]{2^{1/n} - 1}} = \frac{35}{\sqrt[4]{2^{1/2} - 1}} = \mathbf{43.8\ kHz}$$

Notice that, with double-tuned coupling networks, the decrease in overall bandwidth is not so drastic, so that each stage's bandwidth need not be as wide, and higher Q values can be used (24.5 for the above three-stage unit, as compared to 11.6 with the single-tuned circuits of Example 3–4). Yet, in spite of this superiority of the overcoupled double-tuned network, many manufacturers of multistage wideband amplifiers use single-tuned circuits (impedance coupled) and stagger-tune the stages. This is so because overcoupled circuits are more difficult to align because of the interaction of one winding on the other.*

Gain

Gain calculations for an RF amplifier with double-tuned inductive coupling follow the same general technique as shown above for single-tuned transformer coupling. There are, however, two additional points to be considered:

1. The current in the primary winding of the coupled circuit is appreciably higher than the drain or collector current, because of the Q rise of the resonant circuit.
2. In evaluating the Q for this Q rise, the reflected impedance from the secondary winding must be considered. This is so because the secondary is a closed circuit (tuned), and current will flow in the secondary winding when a voltage (E_s) is induced in this winding.

EXAMPLE 3–7

The following values apply to the circuit of Fig. 3–9: transformer T_2 has $L_3 = L_4 = 250\ \mu\text{H}$ and $10\ \Omega$; $C_3 = C_4 = 250$ pF; and $k = 0.015$. The transistor is a 2N4416 with $g_{fs} = 6200\ \mu$mho. If the signal voltage applied to the gate of this FET is $200\ \mu$V, find: (1) the Q of the primary, (2) the voltage at the gate of the next FET, and (3) the gain (gate-to-gate).

Solution

1. To find the primary Q, since $Q = \dfrac{X_L}{R_p + R_r}$:

(*a*) $\omega = 2\pi f = 4 \times 10^6\ \Omega$ (see Example 3–5)

(*b*) X_L (each winding) $= \omega L = 4 \times 10^6 \times 250 \times 10^{-6} = 1000\ \Omega$

(*c*) $M = k\sqrt{L_1 L_2} = 0.015\sqrt{250 \times 250 \times 10^{-12}} = 3.75\ \mu\text{H}$

(*d*) $R_r = \dfrac{(\omega M)^2}{R_s} = \dfrac{(4 \times 10^6 \times 3.75 \times 10^{-6})^2}{10} = 22.5\ \Omega$

*This is discussed more fully in Chapter 8, under Receiver Alignment.

(*e*) $Q_p = \dfrac{X_{L1}}{R_p + R_r} = \dfrac{1000}{10 + 22.5} = \mathbf{30.8}$

2. To find the voltage at the second gate, since $E_{g2} = QE_s$, and $E_s = \omega ML_1$:

(*a*) $I_d = g_{fs}E_g = 6200 \times 10^{-6} \times 200 \times 10^{-6} = 1.24\ \mu A$ (this is the drain current of the FET)

(*b*) $I_{L1} = Q_p I_d = 30.8 \times 1.24 \times 10^{-6} = 38.2\ \mu A$

(*c*) E_s (induced secondary voltage)

$$= \omega MI_{L1} = 4 \times 10^6 \times 3.75 \times 10^{-6} \times 38.2 \times 10^{-6} = 573\ \mu V$$

(*d*) $Q_s = \dfrac{X_{L2}}{R_s} = \dfrac{1000}{10} = 100$

(*e*) $E_{g2} = Q_s E_s = 100 \times 573 \times 10^{-6} = \mathbf{57.3\ mV}$

3. gain $= \dfrac{E_{g2}}{E_{g1}} = \dfrac{57.3 \times 10^{-3}}{200 \times 10^{-6}} = \mathbf{287}$

Neutralization

In any active device, semiconductor or tube, some capacitance exists between the output and the input circuits. In the bipolar transistor it is the capacitance between collector and base; in the FET it is the capacitance between drain and gate; in the tube it is the plate-to-grid capacitance. Up to this point we have assumed that this capacitance was small enough to have negligible effect. This is true, even into the VHF range, for pentodes, for dual-gate MOSFETs, and even for some other transistors specially designed for RF use. However, particularly with bipolar transistors, the capacitive reactance at RF may be low enough to cause complications. Oscillation can occur when voltage from the output of an amplifier is fed back to the input in such a phase as to reinforce the input.* The output voltage will increase; the feedback voltage will increase; the total input will increase. The action is cumulative and builds up so as to sustain itself. The output is no longer dependent on the input signal, and amplifier action is destroyed. To nullify the effect of such a positive feedback voltage, a second feedback path can deliberately be added. If the voltage fed back by this second path is *equal and opposite* to the original feedback, the two effects will cancel and normal amplifier operation is obtained. This technique is known as *neutralization*. Two neutralizing circuits are shown in Fig. 3–10. In the first circuit, Fig. 3–10(a), capacitor C_{BC} represents the interelement capacitance from base to collector. This capacitance allows part of the output voltage to be fed into the input circuit and will tend to cause oscillations. Capacitor C_N is the neutralizing

* Oscillators are discussed in detail in Chapter 5.

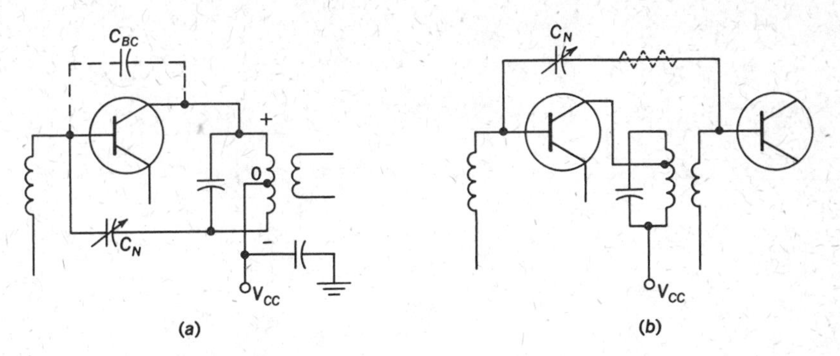

Figure 3-10 Neutralization of transistor RF amplifiers.

capacitor through which a voltage of opposite phase is fed into the base circuit. Notice how the opposite phasing is obtained. The supply voltage is fed to a tap on the tank circuit coil. Since the supply source is bypassed to ground, this tap point is at RF ground potential. At some instant when the top end of the tank circuit is at a positive RF potential compared to ground, the bottom end, being at a lower potential than the tap point, must be negative compared to ground. Feeding the supply voltage to a tap point therefore accomplishes two effects. It produces the opposite polarity for neutralization, and at the same time it reduces the loading on the tank circuit, since the transistor output circuit is shunted only across the upper portion of the tank circuit.

The circuit of Fig. 3–10(b) receives its neutralizing voltage from across the secondary winding of the coupling transformer. Since the secondary voltage is 180° out of phase with the primary voltage, the proper polarity is obtained. In both circuits the neutralizing capacitor is shown as variable. This allows for adjusting the magnitude of the neutralizing voltage until it equals the original feedback voltage. Since exact equalization is not necessary, this capacitor is often a fixed value. In other circuits a resistor is added in series with C_N, as shown by broken lines in Fig. 3–10(b). The resistance value is selected to effect a balance.

Similar neutralizing circuits can also be used with JFETs and single-gate MOSFETs if the feedback capacitance, C_{rss} (or the frequency of operation) is high enough to cause oscillation.

Bipolar Transistor RF Amplifier Circuits

So far, only partial schematics have been given to show methods of coupling and neutralization. When we also consider the various biasing and stabilizing techniques, it should be obvious that a variety of suitable amplifier circuits are possible. A typical bipolar transistor circuit is shown in Fig. 3–11. Most of this circuitry was explained in the preceding

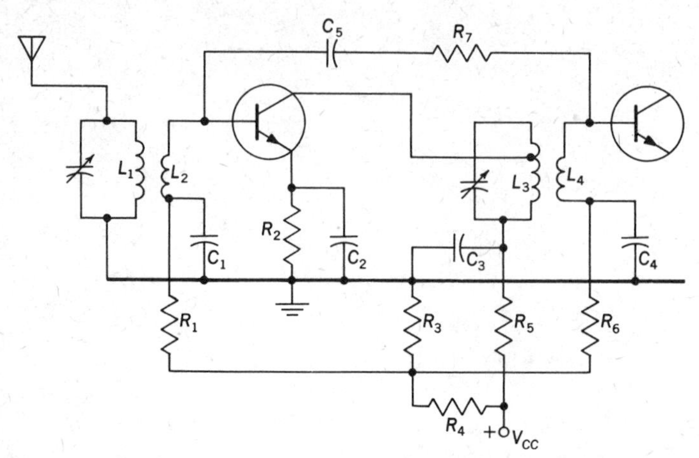

Figure 3-11 An *NPN*-transistor RF amplifier.

partial schematics. The remaining components are used to provide bias and stabilization. The base bias for this transistor is obtained from the voltage divider R_3 and R_4 and series resistor R_1. This series resistor, in combination with C_1, also serves as a decoupling network. Resistor R_2 in the emitter leg is used for dc bias stabilization. It is bypassed by C_2 to prevent ac degeneration. Resistor R_5 with capacitor C_3 forms a decoupling network for the collector circuit, while R_7 and C_5 are used for neutralization.

HIGH-FREQUENCY CIRCUITS

In a bipolar transistor, the small-signal current gain, h_{fe}, and therefore the forward transmittance, y_{fs}, fall rapidly at very high frequencies. To get more gain some manufacturers have used the common-base instead of the common-emitter connection. At the same time, since the reverse transfer admittance is appreciably lower with this change, neutralization is not needed. One manufacturer's version of a C-B RF amplifier is shown in Fig. 3-12. The C-B configuration has a very low input impedance (much lower than the C-E). To prevent loading of the tank circuit, three mutually-coupled coils are used. The low-impedance source (300 Ω) and the low-impedance transistor input (20 to 200 Ω) are transformed to suitably higher impedance values by the turns ratios of the respective coils. This three-coil separation also permits grounding of the rotor of the tuning capacitor C_1 without shorting the balanced input signal or any bias voltages. Notice also that the positive power supply terminal is connected to the emitter. This permits direct grounding of the rotor of the tuning capacitor C_2.

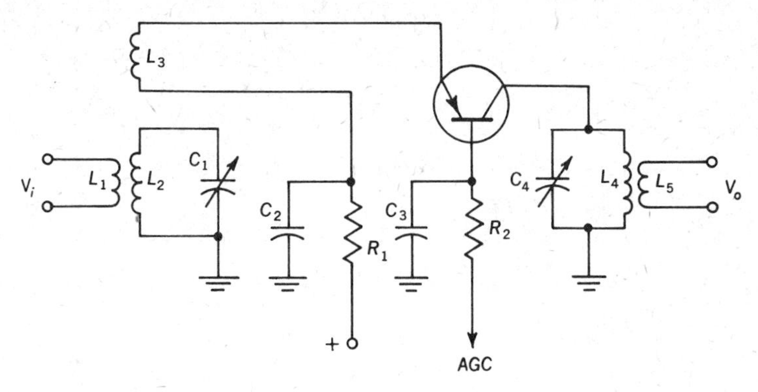

Figure 3-12 A grounded-base RF amplifier. (Courtesy Bogen Communications Div., Lear Siegler, Inc.)

In some applications, the low input impedance of the above circuit is undesirable. Furthermore, the gain from such a stage is still fairly low. Both of these disadvantages can be overcome by using a two-transistor circuit. One such circuit is the *emitter-coupled* amplifier. The input is fed to the base of the first unit, but the transistor is connected as an emitter-follower (or common-collector) circuit. As with any emitter-follower, this results in a much higher input impedance. The output from the emitter of the first transistor is directly coupled to the emitter of the second transistor, which is connected as a common-base circuit. Theoretically, an emitter-coupled pair has several advantages over a single-element circuit. Neither transistor needs neutralization; the input and output impedances are both high and suited for high-Q tank circuit applications; and the gain is higher than with a single-unit circuit. Unfortunately, because of dissimilar transistor characteristics, bias stability is poor. This is avoided by using integrated circuitry (IC). Transistors formed on the same chip can be matched very closely without difficulty. An IC schematic is given in the next section.

A variation of the circuit in Fig. 3–12 that provides even higher gain, with low noise and good stability, is the *cascode* circuit. Again two transistors are used. These could be bipolar units or FETs. A cascode circuit using junction FETs is shown in Fig. 3–13. The first transistor is a standard common-gate amplifier. Its output is impedance-coupled to the source of the second FET. Because of the low input impedance of Q_2, neutralization of Q_1 is not necessary. However, it is claimed that a properly adjusted neutralizing circuit will result in a much lower noise level. In this circuit neutralization is accomplished by L_2, R_1, and C_2. (Sometimes coil L_2 is variable and is adjusted for minimum noise output.) The second FET, being used in a common-gate circuit, has low feedback

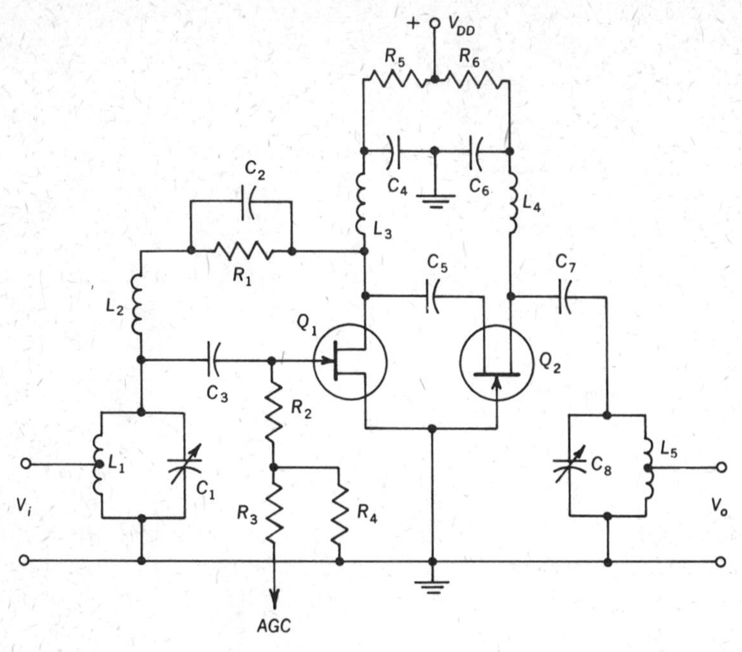

Figure 3-13 A cascode RF amplifier (Courtesy H. H. Scott, Inc.)

capacitance, C_{rss}, and again neutralization is not required. Notice that the drain circuit of Q_2 is shunt fed, thereby permitting direct grounding of the output tank tuning capacitor rotor.

Integrated Circuits (IC)

Integrated circuits offer tremendous advantages in size reduction, in improved performance, and—if manufactured in large quantities—in reduced costs. Unfortunately, the needs of any one receiver manufacturer are not of sufficiently large quantity to make specialized ICs economically practical. Then in the late 1960s the semiconductor industry came out with a basic IC building block that could be used in a variety of communication circuits, and that could be individually tailored by each user to suit his own circuit designs. Since then, ICs are fast appearing in new receiver designs. The essential features of this basic IC are shown in Fig. 3–14. Transistors Q_1 and Q_2 form a differential amplifier, with Q_3 acting as a constant-current source (or sink). With no ac input, and with equal dc bias on Q_1 and Q_2, the source current, I_{C3}, splits equally between I_{C1} and I_{C2}. On the positive swing of the ac input signal, Q_1 conducts heavier and I_{C2} decreases. Conversely, on the negative half of the input swing, Q_1 is driven toward cutoff, its current decreases, and the collector

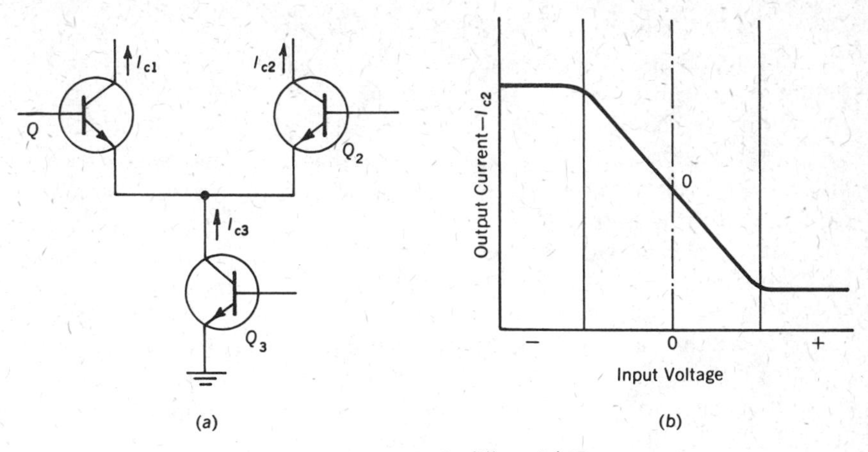

Figure 3-14 Basic differential IC.

current I_{C2} must decrease. Figure 3–14(b) shows the dynamic transfer characteristic (input voltage vs output current) for the IC. The curve is ideally linear for almost 100 mV of input signal swing.

In addition to the basic three transistors, commercial units also include resistors and diodes (or diode-connected transistors) for biasing purposes.* The total package is still no larger than for a single transistor. The schematic of one such unit is shown in Fig. 3–15(a). The diodes D_1,

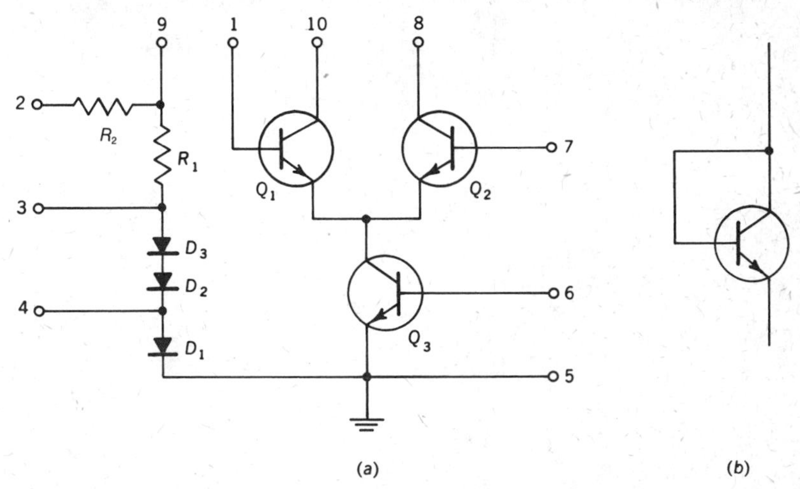

Figure 3-15 Internal schematic of a basic IC. (Courtesy Sylvania Electric Products, Inc.)

*The use of diodes and diode-connected transistors for biasing produces matched temperature characteristics and improved bias stability. (See *General Electronic Circuits Fig. 15–15.*)

D_2, and D_3 are actually diode-connected transistors, as shown in diagram (b). Circuit versatility is obtained by the external connections to the IC. Figure 3–16(a) shows this IC connected in a differential or emitter-coupled mode. Notice that the input is connected through the tuned circuit to the base of transistor Q_1. This transistor is connected as a common-collector circuit and emitter-coupled to Q_2. Q_2 is a common-emitter amplifier, with its output (from the collector) inductively coupled to the next stage. The drawing of a complete schematic is left as an exercise for the student. (See Problem 31 at the end of the chapter.)

Figure 3–16(b) shows the same IC reconnected into a cascode circuit. Notice this time that the input is fed to Q_3, connected as a common-emitter amplifier. The output from Q_3 is applied to Q_2 in a common-base circuit. Transistor Q_1 now serves as an AGC amplifier and control circuit. Again the drawing of a full schematic is left as a student exercise. (See Problem 32.)

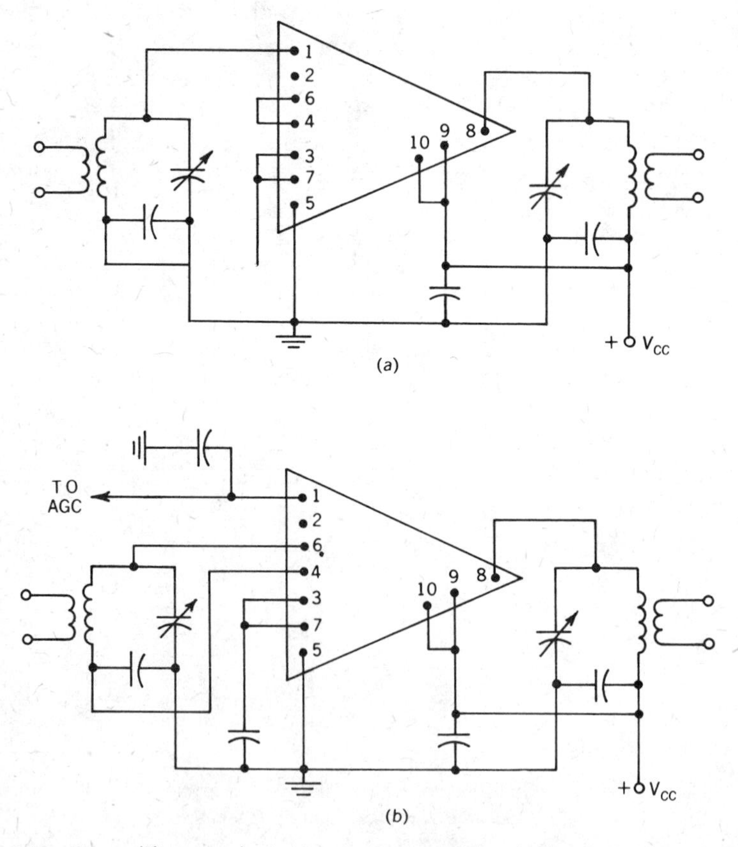

Figure 3-16 RF amplifier using basic IC unit. (Courtesy Sylvania Electric Products, Inc.)

REVIEW QUESTIONS

1. State two functions of an RF amplifier.

2. Refer to Fig. 3–1(a). (*a*) What type of interstage coupling is used? (*b*) Which components "belong" to the coupling network? (*c*) Are these components used in any other arrangement? Explain. (*d*) In what applications is this type of coupling commonly used? (*e*) What is the function of R_1? (*f*) What is the function of C_1?

3. Refer to Fig. 3–1(b). (*a*) What type of coupling is this? (*b*) What is the significance of the black arrow to the left of the coil? (*c*) Is there a tank circuit? (*d*) What constitutes the capacitance of this tank circuit? (*e*) At what frequencies is this type of tank circuit used?

4. When is the circuitry of Fig. 3–1(c) preferable to that in 3–1(a)? Explain.

5. Refer to Fig. 3–2. (*a*) What are each of the three grids called? (*b*) What is the function of R_1C_1? (*c*) of R_2C_2? (*d*) Is resistor R_2 always used? Explain.

6. What problem exists with bipolar transistor coupling that does not affect FET or vacuum-tube circuits? (*b*) Explain briefly how this problem is handled.

7. Refer to Fig. 3–3. (*a*) What type of coupling is used in Fig. 3–3(a)? (*b*) Why is the next stage connected to a tap on the coil? (*c*) Why is this technique necessary? (*d*) How does diagram (a) differ from (b)? (*e*) Which of the three circuits would produce the best selectivity? Why?

8. (*a*) What three factors determine the effective load impedance in an impedance-coupled amplifier? (*b*) How are these factors related?

9. Refer to Fig. 3–4. (*a*) What does g_{fs} stand for? (*b*) Give another symbol used for this parameter. (*c*) What does r_d represent? (*d*) Give another symbol for this quantity. (*e*) Give two other symbols by which this quantity could be found indirectly on data sheets. (*f*) Give the name for the quantity r_i and for its "opposite." (*g*) Give two symbols by which this quantity could be found indirectly on data sheets. (*h*) Why is r_i shown by dashed lines in this diagram?

10. What is the bipolar transistor symbol equivalent to each of the following FET parameters: g_{fs}, y_{fs}, RE(y_{os}), r_d, RE(y_{is}).

11. Give the relationship for the bandwidth of an impedance-coupled amplifier.

12. (*a*) If identical tank circuits are used in each stage of an impedance-coupled amplifier, how does increasing the number of stages affect the overall bandwidth? (*b*) Why is this so? (*c*) If each stage had the ideal rectangular response characteristic, would the above apply? Explain.

13. Refer to Fig. 3–5. (*a*) What is the relative output level for a single stage at a frequency of 775 kHz? (*b*) What is the mathematical relation between this value and the two-stage response? (*c*) Calculate the relative two-stage response. (*d*) What is the graphical value for the two-stage response at this frequency? (*e*) What is the mathematical relation between the single-stage and the three-stage response? (*f*) Check this from the graphical value.

14. For a given overall bandwidth, as more stages are used to increase gain, how does this affect (*a*) the bandwidth design for each stage? (*b*) the Q used for each stage? (*c*) the individual gain of each stage?

15. What is meant by a "synchronous single-tuned amplifier"?

16. (*a*) Instead of increasing the individual stage bandwidth, name another technique for obtaining wideband multistage operation using single-tuned circuits. (*b*) Give a numerical illustration of this technique. (*c*) For a given overall bandwidth, how does the gain obtained by this technique compare with a synchronous single-tuned amplifier of the same number of stages?

17. (*a*) Why is shielding needed in multistage RF amplifiers? (*b*) What components are most critical in this respect?

18. Refer to Fig. 3–8(a). (*a*) What type of coupling is used here? (*b*) Where is this type of coupling generally used? (*c*) How does the bandwidth of this type of circuit compare with impedance coupling?

19. (*a*) What type of coupling is used in Fig. 3–8(b)? (*b*) Why is the primary tuned instead of the secondary? (*c*) How is loading due to the input resistance of Q_2 reduced?

20. (*a*) Why are double-tuned RF transformers seldom used for amplifiers that are tunable over a wide frequency range? (*b*) For what type of service are they particularly suitable?

21. (*a*) In Fig. 3–9(a), if the Q of each tank circuit is increased, how does this affect the bandwidth of each stage? (*b*) What two factors determine this bandwidth? (*c*) How does Q affect the response? (*d*) What is the optimum Q value? Why?

22. How does multistage operation affect the overall bandwidth of an amplifier using double-tuned inductively coupled circuits?

23. For an equal number of stages and an equal *overall* bandwidth, compare the synchronous single-tuned circuit, the stagger-tuned circuit, and the overcoupled double-tuned circuit with respect to (*a*) bandwidth required per stage, (*b*) overall gain, and (*c*) ease of adjustment.

24. (*a*) Under what condition may oscillations occur in an RF amplifier? (*b*) If such a condition exists, how is it remedied? (*c*) Would this condition be prone to exist with pentodes? Explain. (*d*) Would this condition tend to exist with dual-gate MOSFETs? Explain. (*e*) Would this condition tend to exist with bipolar transistors? Explain.

25. What is the basic principle underlying any neutralization technique?

26. Refer to Fig. 3–10(a). (*a*) What does capacitor C_{BC} represent? (*b*) Why is it shown with dashed connections? (*c*) What is the effect of this capacitor? (*d*) Why is the supply voltage fed to a tap on the collector tank circuit? (Give two reasons.) (*e*) What is the function of capacitor C_N? (*f*) Why is this capacitor variable?

27. (*a*) How is an out-of-phase neutralizing voltage obtained in Fig. 3–10(b)? (*b*) What is the purpose of the dashed resistor in this diagram?

28. With reference to Fig. 3–11, what is the function of (*a*) R_1 and C_1, (*b*) R_3, (*c*) R_5 and C_3, (*d*) C_5, and (*e*) R_7.

29. Refer to Fig. 3–12. (*a*) Give two advantages of this type of circuit over a C-E configuration. (*b*) Why are three coils used in the input side? (*c*) Is this an NPN or PNP transistor? (*d*) What polarity should the collector be compared to base? (*e*) How is correct polarity obtained in this circuit? (*f*) What advantage, if any, does this technique have?

30. With respect to an emitter-coupled amplifier: (*a*) How many transistors does such a circuit have? (*b*) How is the first transistor connected? (*c*) In what configuration is the second transistor connected? (*d*) Give three advantages of this type of RF amplifier. (*e*) Why is this circuitry not used much with discrete transistors?

31. Refer to Fig. 3–13. (*a*) What type of transistors are Q_1 and Q_2? (*b*) Are these N- or P-channels? How can you tell? (*c*) In what configuration is Q_1 connected? (*d*) How is its output coupled to the next transistor? (*e*) In what configuration is Q_2 connected? (*f*) Is neutralization of Q_1 normally required? Explain. (*g*) Is neutralization of Q_2 normally required? Explain. (*h*) What are the functions of R_1, C_2, and L_2? (*i*) In view of the answer to (*f*), then why is Q_1 neutralized? (*j*) Why is it possible to ground the tuning capacitor rotor, C_8, without shorting the supply voltage?

32. Give two advantages of integrated circuits over discrete-component circuits.

33. Refer to Fig. 3–14(a). (*a*) Which transistors form the differential amplifier? (*b*) What is the function of Q_3? (*c*) What determines the value of the current I_{C3}? (*d*) What determines how much of I_{C3} goes to Q_1 and Q_2? (*e*) How should I_{C3} divide when the ac input signal is zero?

34. Refer to Fig. 3–14(b). With the base of Q_2 at ac ground and an input signal applied between base and ground of Q_1: (*a*) What determines the position of point 0 on the curve? (*b*) Why does I_{C2} increase when the input signal swings negatively?

35. Refer to Fig. 3–15. (*a*) Which of the components shown are not part of the IC? (*b*) What is the purpose of the resistors and diodes? (*c*) Is D_1 actually a diode? Explain. (*d*) Why are transistors used instead of actual diodes?

36. Refer to Fig. 3–16. (*a*) Are the coils and capacitors shown part of the IC? (*b*) Comparing the external connections in diagram (a) with the internal schematic of Fig. 3–15, what kind of circuit is this? (*c*) What kind of circuit does diagram (b) represent?

PROBLEMS AND DIAGRAMS

1. Draw the circuit diagram for an impedance-coupled RF amplifier using FETs.

2. Repeat Problem 1 using pentode tubes.

3. Repeat problem 1 using bipolar transistors.

4. An impedance-coupled amplifier has the following characteristics: a MOSFET with $g_{os} = 6.67$ μmho and $g_{fs} = 10{,}000$ μmho; a tank circuit with $L = 30$ μH and 12.5 Ω, and $C = 20$ pF; gate resistor $= 220$ kΩ. Find: (*a*) the effective load impedance, (*b*) the Q of the coupling network, (*c*) the resonant frequency and bandwidth of this circuit, (*d*) the gain of the stage, and (*e*) the output voltage for an input signal of 5 mV.

5. Repeat Problem 2(*a*)–(*e*), for a tank circuit of $L = 120$ μH and 7.07 Ω, and $C = 60$ pF.

6. To improve the selectivity of the coupling network of Example 3–2, the tank values are changed to $L = 50\ \mu H$ and $3.0\ \Omega$, and $C = 1200$ pF. Find (1), (2), and (3) as required in Example 3–2.

7. Repeat Example 3–2, with the original tank circuit values, but this time with the tap point one-tenth up from the bottom.

8. In a two-stage impedance-coupled amplifier, each stage has a bandwidth of 40 kHz. What is the overall bandwidth?

9. (*a*) Repeat Problem 8 for a three-stage amplifier. (*b*) Repeat Problem 8 for a four-stage amplifier.

10. A bandwidth of 200 kHz is required for FM operation at 10.7 MHz. If a four-stage amplifier is used, find (*a*) the bandwidth requirement for each stage and (*b*) the Q required for each stage. (*c*) Can the operating frequency be dropped to 2 MHz? Explain.

11. Each stage of a three-stage impedance-coupled amplifier has a Q of 50 at an operating frequency of 10 MHz. Using the general resonance-curve technique, (*a*) plot the response curve for one stage. (*b*) Plot, on the same scale, the response curve for two stages in percent of maximum value. (*c*) Repeat (*b*) for three stages.

12. Three tank circuits each have a Q of 100. Tanks 1 and 2 are tuned to 1000 kHz, tank 3 is resonant at 980 kHz. Consider the gain of each circuit as 100. (*a*) Plot the response curve for each circuit alone. (*b*) Plot the response curve for tanks 1 and 2 in cascade. (*c*) Plot the response curve for tanks 1 and 3 in cascade. (Use frequencies of 960, 970, 975, 977.5, 980, 985, 990, 995, 1000, 1002.5, 1005, 1010, and 1020 kHz.) (*d*) What is the bandwidth in each case? (*e*) What is the relative gain at resonance, in each case?

13. Each stage of a two-stage impedance-coupled amplifier has a bandwidth of 40 kHz. The first stage is tuned to 9.97 MHz and the second to 10.03 MHz. Using the general resonance-curve technique, (*a*) plot the response curve of each stage on the same axes. (*b*) Plot the overall response curve in percent of maximum value.

14. Draw the circuit diagram for a single-tuned inductively coupled RF amplifier, using FETs.

15. Repeat Problem 14 using bipolar transistors.

16. Repeat Problem 14 using pentode vacuum tubes.

17. In the circuit if Fig. 3–8, the FET has $g_{fs} = 5700\ \mu$mho; $L_3 = 10$ mH; $L_4 = 300\ \mu H$ and $16.9\ \Omega$; the coupling coefficient between coils is 0.02; $C_5 = 104$ pF. The input signal level at the gate of Q_1 is 5 mV. Find (*a*) the signal current through L_3, (*b*) the voltage induced in L_4, (*c*) the voltage applied to the gate of Q_2, and (*d*) the gain—gate-to-gate.

18. The circuit of Problem 17 has the following additional values: $L_1 = 1.5$ mH; $L_2 = 200\ \mu H$ and $11.3\ \Omega$; the coupling coefficient is $k = 0.035$; $C_1 = 156$ pF. If the voltage developed at the antenna (across L_1) is $100\ \mu V$, find (*a*) the signal voltage applied to the gate of Q_1, (*b*) the signal voltage applied to the gate of Q_2, (*c*) the bandwidth of the input circuit, (*d*) the bandwidth of the output circuit, and (*e*) the overall bandwidth.

19. Draw the circuit diagram for a double-tuned inductively coupled RF amplifier using FETs.

20. Repeat Problem 19 using bipolar transistors.

21. Repeat Problem 19 using pentodes.

22. In a two-stage RF amplifier using double-tuned circuits, each stage has a bandwidth of 40 kHz. Find the overall bandwidth.

23. A bandwidth of 200 kHz is required for FM operation at 10.7 MHz. If two stages of double-tuned amplifiers are used, find (*a*) the bandwidth required for each stage, (*b*) the coupling coefficient required for each transformer, and (*c*) the Q of each winding for critical coupling.

24. An RF amplifier using MOSFETs ($g_{fs} = 6100$) is coupled to the next stage by a double-tuned transformer whose constants are $L_1 = 159$ μH and 10 Ω, $C_1 = 159$ pF, L_2 and C_2 values the same as for the primary, and coupling coefficient $= 0.01$. The signal voltage applied to the first gate is 0.82 mV. Find (*a*) the drain current (ac component), (*b*) the primary coil current, (*c*) the induced secondary voltage, (*d*) the gate signal at the next FET, (*e*) the gain (gate-to-gate), and (*f*) the resonant frequency and bandwidth of this stage.

25. Repeat Problem 24(*a*)–(*f*) for a coupling transformer whose constants are $L_1 = L_2 = 0.5$ mH, $C_1 = C_2 = 240$ pF, and $k = 0.023$.

26. Draw a circuit diagram showing how neutralization can be applied to a transistor RF amplifier (*a*) with impedance-coupled output and (*b*) with inductive-coupled output.

27. Draw the circuit diagram for a two-stage transistor RF amplifier using PNP bipolar transistors.

28. Draw a diagram for an NPN grounded-base RF amplifier, without AGC, and with the "hot" side of the power supply applied to the collector. Keep the tuning capacitor's rotor grounded.

29. Draw the schematic for an emitter-coupled RF amplifier.

30. Draw the schematic for a cascode RF amplifier.

31. Draw the schematic for an IC emitter-coupled RF amplifier.

32. Draw the schematic for an IC cascode RF amplifier.

TYPICAL FCC QUESTIONS

3.101 Does a pentode vacuum tube usually require neutralization when used as a radio-frequency amplifier?

3.119 What is the principal advantage of a tetrode over a triode as a radio-frequency amplifier?

3.279 Why must some radio frequency amplifiers be neutralized?

3.307 Explain the purposes and methods of neutralization in radio-frequency amplifiers.

4.185 Why are grounded-grid amplifiers sometimes used at very high frequencies?

4.189 Draw a diagram of a grounded-grid amplifier.

6.177 Why must some radio-frequency amplifiers be neutralized?

6.570 Why are bypass capacitors used across the cathode bias resistors of a radio-frequency amplifier?

6.571 What is the purpose of shielding between radio-frequency amplifier stages?

6.572 Draw a simple schematic diagram showing a method of "impedance" coupling between two stages of a radio-frequency amplifier.

6.573 Draw a simple schematic diagram showing a method of inductive or transformer coupling between two stages of a radio-frequency amplifier.

4

RF Power Amplifiers

In contrast to the low power levels encountered in the receiver applications of Chapter 3, many commercial RF applications require tremendous power. For example, radio and television transmitters may have output levels as high as 50 kW. Loran navigational stations use as much as 1 MW, while search radar units may have a peak pulse power of 10 MW. Even industrial applications, such as dielectric or induction heating equipment, have power levels in the kilowatt region. When dealing with such high power levels, efficiency of operation is of prime importance. For this reason, class-C amplifiers are used whenever possible. Although this type of operation causes heavy distortion of *current* waveshapes, the use of resonant-circuit loads results in excellent output waveforms.

Theoretically, any active device – semiconductor or vacuum tube – could be operated in class C as an RF power amplifier. However, their usefulness is limited by their power-handling ability. Because of this, FETs are seldom used; bipolar transistors are used only at relatively low power levels (for small mobile transmitters or in the early stages of a larger unit); while the vacuum tube is the workhorse in the RF power field. Triodes are available with power output capabilities in the order of 400 kW. We will therefore start this discussion of power amplifiers with vacuum-tube circuits.

VACUUM-TUBE CLASS-C AMPLIFIERS

An amplifier is considered to operate in class C if it is biased well beyond cutoff, so that plate current flows for less than 180°. These amplifiers are also invariably driven into the positive region, and grid current flows.

(Strictly speaking, they should be considered as class C_2. However, class C_1 has no practical value and is not used.)

Voltage and Current Relations

Figure 4–1 is a simplified diagram of a class-C amplifier. It is used merely to assist in developing the current and voltage waveshapes. The bias supply voltage is more than twice the cutoff value, while the *excitation* voltage, e_g, must be sufficient to drive the tube out of cutoff and into the positive grid region. The grid potential at any instant is the phasor sum of the bias voltage and the instantaneous value of the input signal. Since the tube is biased beyond cutoff, plate current cannot flow except when the exciting voltage is positive enough to drive the grid out of cutoff. Obviously, the plate current flows in pulses of less than 180° duration. Furthermore, by making the amplitude of the exciting voltage greater than the bias voltage, the grid is driven positive, and grid current flows during the peak of the RF signal input. These conditions are shown in Fig. 4–2.*

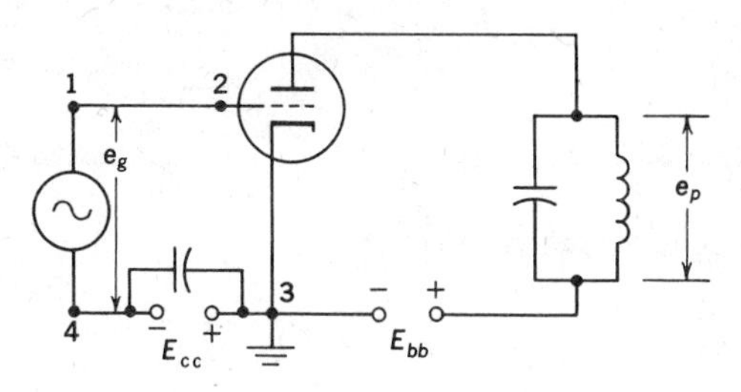

Figure 4-1 Basic Class-C amplifier.

Notice in Fig. 4–2 that the grid is above cutoff for 120°, and plate current flows for 120°. This current is a complex wave with a dc component I_b, a fundamental component i_p, and various harmonic components (not shown). Now notice the plate *voltage* waveform. There is a dc component E_b and a fundamental component e_p—but no distorted varying waveshape. Therefore there are no harmonic components in the output *voltage*. The distortion in the plate current waveshape is nullified by the resonant action of the plate tank circuit. This circuit is tuned to the frequency of the exciting voltage, so that it presents a high impedance to the fundamental frequency component of the plate current. The output voltage I_pZ at this frequency is high. On the other hand, at the harmonic frequencies, the tank is far off resonance, its impedance approaches zero, and the output voltage at any harmonic frequency is negligible. Because of the tuned-circuit load, distortion is eliminated without the use of push-pull circuitry.

*See Appendix (page 570) for explanation of letter-subscript symbols.

During time interval 2–3, the exciting voltage drives the grid positive. Grid current flows. In Fig. 4–2, grid current flows for approximately 70°. It should be obvious that the angle (or duration) of current flow (grid or plate current) is dependent on two factors—the dc bias value and the amplitude of the exciting voltage. For example, an increase in signal voltage will increase the angle of current flow, whereas an increase in the bias voltage (more negative) will reduce the angle of flow. Even more important is that with a more negative bias—even though the excitation is increased to maintain the original value of $E_{c(\max)}$—the angle of current flow will be less.

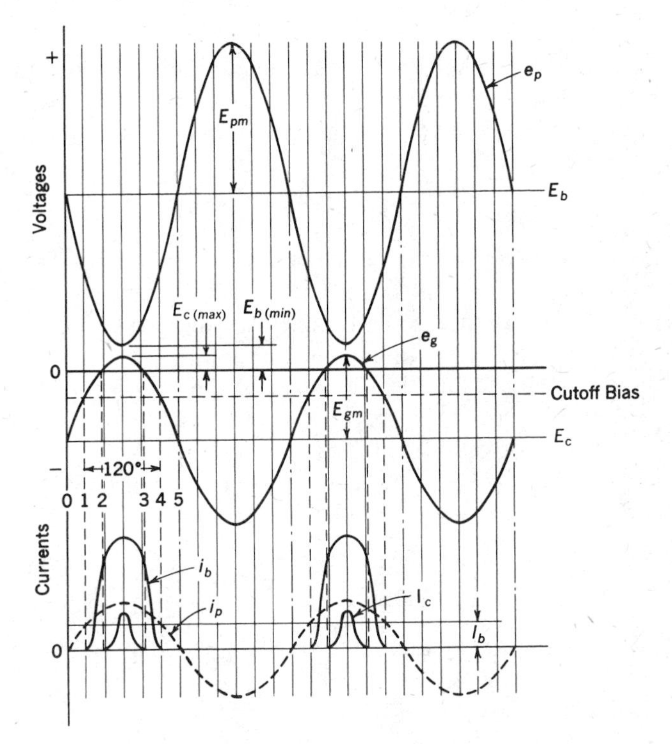

Figure 4-2 Current and voltage relations in a Class-C power amplifier.

Definition of Terms

Before going into a detailed analysis of class-C operation, let us clarify some of the terminology used with these circuits.

1. *Plate power input.* This refers to the power supplied to the plate circuit by the *dc* power supply, and it is equal to the product of the dc plate potential and plate current.

$$P_{\text{in}} = E_b I_b \tag{4-1}$$

2. *Plate power output.* This refers to the ac output from the tube and is equal to the product of the *ac* components of plate voltage and plate current.

$$P_o = E_p I_p = \frac{E_{pm} I_{pm}}{2} \tag{4-2}$$

3. *Plate dissipation.* This refers to the power lost inside the tube in the form of heat as a result of electron bombardment of the plate structure. Obviously, the difference between the dc power input to the tube and the ac power output is due to this power loss.

$$P_{\text{dis}} = P_{\text{in}} - P_o \tag{4-3}$$

4. *Plate efficiency.* As in any efficiency consideration, this is the ratio of output over input. In this case, it would be the ratio of ac power output to dc power input.

$$\eta_p = \frac{P_o}{P_{\text{in}}} \times 100 \tag{4-4}$$

5. *Grid driving power.* Since the grid of a class-C amplifier is driven positive, grid current flows and power is dissipated within the tube as well as externally in the bias source. This power must be supplied by the previous tube or driving source. The driving power at any instant is a function of the exciting voltage e_g and the grid current i_c. However, notice from Fig. 4–2 that grid current flows for only a small portion of each cycle of input. The average power can be obtained by integrating the product of the instantaneous values over a full cycle (0–2π) or

$$P_g = \frac{1}{2\pi} \int_0^{2\pi} e_g i_c \, d(\omega t)$$

This has been evaluated to a close approximation as*

$$P_g = 0.9 E_{gm} I_c \tag{4-5}$$

where E_{gm} is the peak signal voltage and I_c is the average or dc component of grid current.

Efficiency and Conduction Angle

The power dissipated within the tube at any instant is the product of the instantaneous values of the plate potential, e_b, and the plate current, i_b. Notice in Fig. 4–2 that, when the plate current i_b is a maximum, the

* H. P. Thomas, "Determination of Grid Driving Power in Radio Frequency Power Amplifiers," *Proceedings IRE*, **21** (August, 1933), p. 1134.

plate potential e_b is at its minimum value. Also, as the plate potential rises, the plate current rapidly drops to zero, so that plate current flows only while the plate potential is relatively low. This keeps the power loss to a low value and accounts for the high efficiency of class-C amplifiers. Well-designed amplifiers are capable of efficiencies as high as 85%. If the angle of plate current flow were reduced further, current flow would be limited to still lower values of the e_b curve. Theoretically, if plate current were restricted to a few degrees at the very peak of the excitation cycle, and if the load impedance and operating conditions were adjusted so that $e_{b(\min)}$ would be zero, then the power loss would be zero and a plate efficiency of 100% could be obtained.

In spite of the fact that plate efficiency improves with a decrease in the angle of plate current flow, there are several reasons why too low a conduction period is not desirable. As the angle of current flow decreases, the distortion content increases and the amplitude of the fundamental component decreases, thereby reducing the power output capability of the tube. This effect is shown in Table 4-1 and Example 4-1.

Table 4-1

EFFECT OF CONDUCTING ANGLE ON PLATE-CURRENT VALUES

CONDUCTION ANGLE	$\frac{I_{b(\max)}}{I_b}$	$\frac{I_p}{I_b}$	$\frac{I_{pm}}{I_b}$	CONDUCTION ANGLE	$\frac{I_{b(\max)}}{I_b}$	$\frac{I_p}{I_b}$	$\frac{I_{pm}}{I_b}$
180	3.14	1.11	1.57	120	4.6	1.27	1.80
170	3.35	1.14	1.61	110	5.0	1.29	1.82
160	3.55	1.17	1.66	100	5.5	1.31	1.85
150	3.75	1.20	1.70	90	6.1	1.33	1.88
140	4.00	1.22	1.73	80	6.9	1.35	1.91
130	4.25	1.25	1.77	40	13.4	1.41	2.00

EXAMPLE 4-1

A 6161 triode is operated as a class-C amplifier with a peak plate current $I_{b(\max)}$ of 750 mA and a conduction period of 150°. The plate load impedance is 6000 Ω at the input signal frequency. Find (1) the fundamental component of plate current and (2) the power output at this frequency. (3) Repeat (1) for a conduction period of 100°. (4) Repeat (2) for this reduced conduction period.

Solution

1. For 150° conduction:

(a) $I_b = \dfrac{I_{b(\max)}}{3.75} = \dfrac{750}{3.75} = 200 \text{ mA}$

(b) $I_p = 1.20 I_b = 1.20 \times 200 = \mathbf{240\ mA}$

2. $P_o = I_p^2 R = (0.24)^2 \times 6000 =$ **345 W**
3. For 100° conduction:

   (*a*) $I_b = \dfrac{I_{b(\max)}}{5.5} = \dfrac{750}{5.5} = 136$ mA

   (*b*) $I_p = 1.31 I_b = 1.31 \times 136 =$ **179 mA**
4. $P_o = I_p^2 R = (0.179)^2 \times 6000 =$ **192 W**

Notice in Example 4–1 that, because of the reduction in conduction angle, the power output is cut almost in half. An even more drastic loss of power output would result if the angle of plate current flow is decreased by reducing the exciting voltage or by using a higher (more negative) grid bias. In either case, the maximum positive grid signal swing $E_{c(\max)}$ is decreased, causing a sharp reduction in $I_{b(\max)}$. This in turn reduces the value of the fundamental current component and causes a drastic drop in power output. The preferred method for reducing the conduction angle is to use a more-negative bias, but at the same time, *raise* the excitation voltage so as to maintain $E_{c(\max)}$ (and therefore $I_{b(\max)}$) constant. (This technique was implied in Example 4–1.) Unfortunately, small conduction angles require very high bias and excitation voltages. This not only increases the grid driving power required from the previous stage, but may also exceed the maximum grid-voltage or grid-dissipation ratings of the tube. As a compromise, class-C amplifiers are generally operated with conduction angles of 150–120°.

Actually, if the power output level is to be maintained when the conduction angle is decreased, it is necessary to simultaneously increase the peak plate current value ($I_{b(\max)}$). This requires an increase in the maximum positive grid potential $E_{c(\max)}$. Here again, we run into complications. If this grid potential exceeds the minimum value of the plate potential $E_{b(\min)}$, electrons may be attracted to the grid in such great quantities that the plate current will decrease, reducing the power output. Meanwhile, the resulting increase in grid current will cause excessive grid dissipation. *For maximum efficiency,* $E_{c(max)}$ *should just about equal* $E_{b(min)}$.

Grid Bias Circuits

In the basic circuit of Fig. 4–1 the bias is obtained from a source E_{cc}. This type of bias would be a *fixed bias* and could be supplied by a battery, a generator, or some form of rectifier power supply. The bias source should have good regulation and low dc resistance; otherwise the flow of grid current could cause the bias voltage to vary, thereby upsetting the

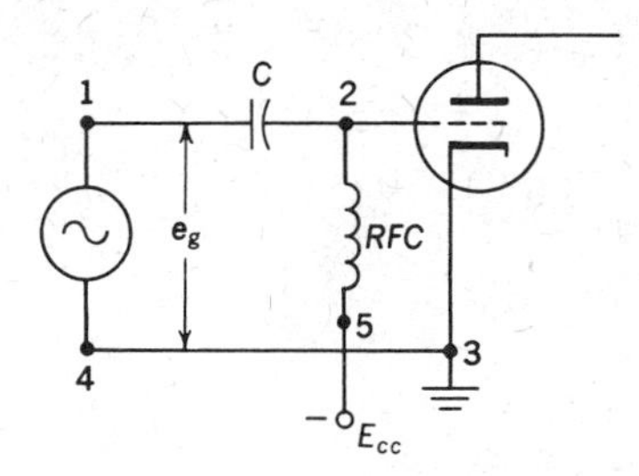

Figure 4-3 Shunt-fed fixed bias.

$E_{c(max)}$ to $E_{b(min)}$ relationship. In the basic schematic of Fig. 4–1, the bias source is in series with the signal voltage source. Such a connection is called a *series-fed* bias. An alternate connection is shown in Fig. 4–3. Notice that the bias voltage (grid to ground) is now in parallel or shunt with the signal voltage. This is a *shunt-fed* bias circuit. The positive side of the bias source (not shown) is understood to be connected to ground. Since the bias source has a low impedance, direct connection of E_{cc} to the grid of the tube would practically short-circuit the RF signal to ground. This is prevented by the addition of an RF choke in series with the bias source. The high reactance of the choke eliminates the above loading effect. To maintain good regulation, the choke should have a low dc resistance. Notice capacitor C in series with the signal source. Its use is necessary to prevent short-circuiting of the dc supply through the low dc resistance of the signal source.

Although fixed bias is featured in the larger or better-quality commercial equipment, class-C amplifiers often use *grid-leak bias,* wherein the bias voltage is obtained by rectification of the driving signal voltage. Grid-leak bias action can be examined from the basic schematic and waveshapes of Fig. 4–4. Capacitor C must have a low reactance at the signal frequency. During the first positive half-cycle of the input signal (e_i), the grid is driven positive (compared to ground), and grid current flows from ground, through the tube, through the capacitor, and back to the source e_i.* The capacitor begins to charge with the polarity as shown. However, the charging time constant is relatively long compared to the period of the signal, so that during time interval 0–1 of Fig. 4–4(b), the capacitor has only partially charged. At this instant, the signal voltage level just equals the capacitor voltage, and the grid potential e_g is zero. The capacitor stops charging, and grid current stops flowing. Throughout time interval 1–2, the grid potential ($\dot{e}_g = \dot{e}_C + \dot{e}_i$) is negative and grid current cannot flow. Meanwhile, the capacitor discharges through resistor R. The discharge time constant (by choice of R value) is deliberately

* Some current also flows up through resistor R.

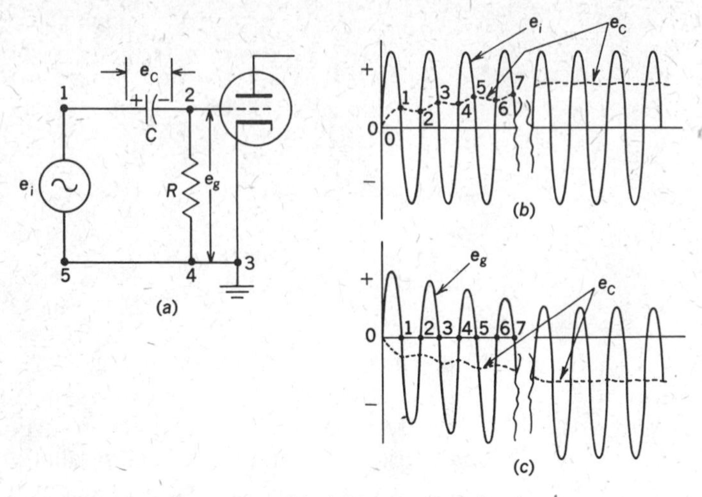

Figure 4-4 Grid-leak bias circuit and action.

made longer than the charging time constant, so that the capacitor discharges only a slight amount. At time instant 2, the input voltage rises above the capacitor voltage value, driving the grid positive. Again, grid current flows, and the capacitor begins to recharge. It continues to charge during interval 2–3, reaching a higher value than in the first cycle. But again during time interval 3–4, the capacitor discharges slightly as the signal voltage falls below the capacitor voltage. This cumulative integrating action of the *R-C* circuit continues for several cycles. Notice, however, that, as the capacitor voltage rises, the charging time interval is getting shorter and the rate of charge is decreasing. On the other hand, the discharge time interval is getting longer. A condition is soon reached when the amount of charge is balanced by the amount of discharge. From now on, the capacitor voltage remains essentially constant. This is the condition shown in Fig. 4–4(b) after the first four cycles.*

Figure 4–4(c) shows the instantaneous grid potential, e_g, with respect to ground. It should be realized that the grid potential at any instant is the phasor sum of the input signal, e_i, and the capacitor voltage, e_C. Notice that e_C in this diagram is shown with a negative polarity. This time, since we are interested in the grid potential, we are plotting the potential of point 2 with respect to point 1. From this diagram it can be readily seen that, after the first few transient cycles, the action in a grid-leak bias circuit is quite similar to that of the class-C amplifier with fixed bias, as described earlier.

*Actually the rate of charge and discharge in these first four cycles is exaggerated to show the integrating action.

In a grid-leak bias circuit, the bias voltage is mainly dependent on the excitation voltage. With a higher excitation voltage, the capacitor will charge to a higher final value and the bias voltage will increase. The component values C and R are not too critical. However, if the grid-leak resistor is made too small, the heavier discharge and the resulting higher grid current will cause the grid-circuit losses to increase. Conversely, if R is made too large, the integrating action may allow the capacitor to charge to the peak value of the RF input, causing the grid to be clamped at ground potential with consequent loss in power output.

Grid-leak bias has one serious disadvantage: Should the excitation fail, the bias would drop to zero and the tube would be ruined. With large power tubes, this can be a very expensive drawback. Consequently, when using this type of bias, safety measures are incorporated into the circuit. One technique is to insert overload devices such as fuses or relays into plate and screen supply lines. Another technique is to provide sufficient auxiliary cathode bias so as to limit the zero-signal plate and screen currents to safe values, in the event that the excitation fails or is accidentally removed.

Grid Excitation Factors

In a class-C amplifier, the source of the grid exciting voltage is obviously the previous stage. This could be a previous RF power amplifier or the oscillator stage. In either case, the amplitude of the exciting voltage must be sufficient to drive the grid to the required positive value ($E_{c(\max)} = E_{b(\min)}$). In addition, the driving source must be capable of supplying all of the power dissipated in the grid–cathode circuit, in the bias circuit, and in the components and wiring, and—at higher frequencies—the additional losses due to transit time effects.* Depending on frequency, this *total* grid circuit power loss may range 2–10 times the driving power given by Eq. (4–5), or as listed in tube manuals. Finally, the driving source must have good regulation, so as to maintain its waveform in spite of the drastic variation in loading as the grid circuit of the driven stage goes in and out of conduction. (See Problem 16.)

Because of their higher power sensitivity, pentodes and beam power tubes have less demanding grid-circuit requirements than triodes, for a given power output. This can be seen from the manufacturer's listed data as shown in Table 4–2.

Many operating circuits have provisions for measuring the dc component of grid current. This is readily done by inserting a dc milliammeter

*These effects cannot be ignored if the time it takes an electron to travel from cathode to plate approaches the time for one-quarter cycle of the input signal.

Table 4–2
COMPARISON OF GRID CIRCUIT REQUIREMENTS

	TRIODE	BEAM POWER
dc plate voltage	1500 V	1500 V
dc grid voltage	−120 V	−90 V
from grid resistor	4000 Ω	7500 Ω
Peak RF grid voltage	200 V	175 V
dc grid current (approximate)	30 mA	12 mA
Grid driving power (approximate)	6.5 W	1.9 W
Power output (approximate)	190 W	210 W

between the RF choke and the bias supply, or between the grid-leak resistor and ground. A grid-current reading is useful in several ways. It indicates the presence of the excitation voltage and the relative value of this voltage. The proper excitation is obtained when the grid current is at its "rated" value. A grid-circuit milliammeter also serves as an indicator when the plate tank circuit of the previous stage (or its own grid tank circuit if one is used) is tuned to resonance. The Q rise in the tank circuit at resonance produces a maximum exciting voltage, and this in turn produces a maximum grid current. The grid current indication can also be used when "neutralizing" a triode RF amplifier circuit to counteract the effect of interelectrode feedback. Tuning and neutralizing adjustments are discussed in more detail later in this chapter.

Plate Power Supply

The basic circuit of Fig. 4–1 shows a dc supply E_{bb} fed to the plate of the tube. For small mobile equipment, such as a "walkie talkie," batteries are most appropriate. Larger mobile units may use dynamotors or even generators. On the other hand, when ac mains are available as the primary source of power, rectifier power supplies are preferred. Three-phase, full-wave circuits are generally used if the dc power requirement is high (above 1 kW). Since the current drain of the class-C amplifier varies drastically (from zero when the tubes are nonconducting to many amperes at the peak of the plate current pulses), the power supply must be capable of supplying high peak currents and must have good regulation. The use of silicon rectifiers and choke-input filters generally satisfies these requirements.

In Fig. 4–5(a), the dc supply is fed to the plate of the tube through a tank circuit. It is therefore called a *series-fed* plate circuit. Although it is the simpler of the two circuits shown, it creates serious insulation and safety problems, particularly if used with high-power tubes where the plate supply voltage may run as high as 20 kV. Notice that neither side

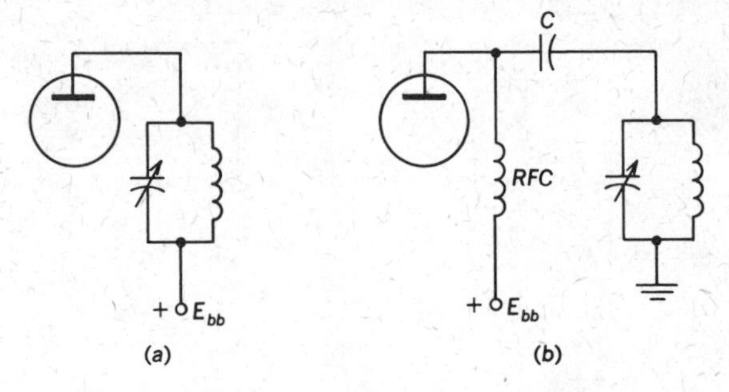

Figure 4-5 Alternate methods of supplying plate voltage.

of the tuning capacitor is at ground potential. The rotor frame must therefore be insulated against the full dc supply voltage, and the rotor shaft must be well insulated for tuning. Also, as can be seen from the waveshapes of Fig. 4–2, the instantaneous potential at the plate side of the tank circuit will approach twice the supply voltage value. This adds to the insulation problems of the tank circuit.

These problems are reduced by using the *shunt-fed* circuit of Fig. 4–5(b). The plate supply voltage is fed to the tube through an RF choke, in parallel with the tank circuit. (The choke must have a high impedance at the operating frequency.) The direct voltage is kept off the tank circuit by the blocking capacitor C. Since both sides of the tank circuit are at *dc* ground potential, the capacitor frame and the tuning shaft can be physically grounded. When RF is applied, the top of the tank circuit will again be "hot," but this time the maximum instantaneous potential cannot exceed the E_{bb} value. Meanwhile, the tuning shaft remains at ground potential.

Tank Circuit Considerations

The tank circuit shown in Fig. 4–5 must satisfy several specific but conflicting requirements. The most obvious of these is that the circuit should resonate at the operating frequency. This criterion is readily met by using the correct LC product. Obviously, there are an infinite variety of possibilities in selecting values for L and C. Secondly, since the plate current in a class-C amplifier has a high harmonic content, the tank circuit must discriminate against these harmonics. The impedance of the tank must be low (and reactive) at the harmonic frequencies. This would result in low output voltage (IZ) at such frequencies, while the harmonic power output would be even lower because of low voltage and a low power factor. The required discrimination can be obtained by using a high-Q coil, a low-loss capacitor, and low-resistance wiring. The resulting high-Q tank circuit is known as the *unloaded* tank circuit. In Fig.

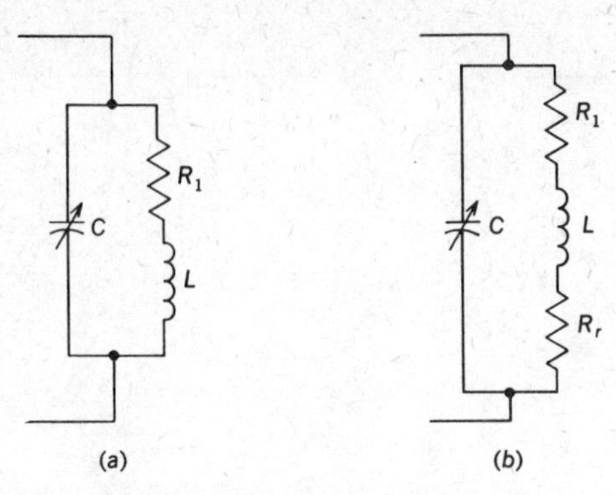

Figure 4-6 The equivalent circuit of a loaded and unloaded tank circuit.

4–6(a), R_1 represents all the resistive effects in the unloaded tank circuit. Any power dissipated by this resistance (I^2R_1), because of the high circulating tank current, is wasted as heat. Obviously, this quantity should be kept to a minimum.*

However, the tank circuit must perform two other functions. It must transfer power from the tube to the load circuit, and to do this efficiently, it must match the tube with the proper load impedance. The impedance (L/CR_1) of the unloaded tank circuit of Fig. 4–6(a) approaches infinity; thus there would be a severe mismatch. Very little power would be transferred for this condition of operation. Now consider a load coupled to the tank circuit, so that a resistive effect R_r is reflected into the circuit. This effect is shown in Fig. 4–6(b). Because of the increased circuit resistance, the tank circuit impedance drops and this new impedance is called the *loaded* impedance. By adjusting the coupling between the load and the tank circuit, it is possible to reflect a resistance value R_r such that the loaded impedance of the tank circuit will match the tube. The power "dissipated" in this resistive effect represents the power transferred to the actual load. Obviously, for maximum efficiency, the reflected resistance value should be as high as possible compared to the resistance R_1. This, in turn, means that the unloaded Q should be as high as possible, whereas the Q of the loaded tank circuit should be as low as possible.

EXAMPLE 4–2

A tank circuit has $X_L = 1000\ \Omega$ and $R_1 = 10\ \Omega$ at the operating frequency. When load is coupled, a reflected resistance of $10\ \Omega$ is added. Find (1) the unloaded Q, (2) the loaded Q, and (3) the tank circuit efficiency for this load. (4)

*Coils in high-power stages are often wound with tubing to keep the effective resistance as low as possible.

If the loading is changed so that the reflected resistance is increased to 190 Ω, what is the loaded Q and the tank circuit efficiency?

Solution

1. $Q_{\text{unloaded}} = \dfrac{X_L}{R_1} = \dfrac{1000}{10} = 100$

2. $Q_{\text{loaded}} = \dfrac{X_L}{R_1 + R_r} = \dfrac{1000}{10 + 10} = 50$

3. The total power is "dissipated" across $R_1 + R_r$, but only the portion across R_r is useful. Therefore

$$\text{Tank eff} \propto \frac{R_r}{R_1 + R_r} = \frac{10}{10 + 10} \times 100 = 50\%$$

4. $Q_{\text{loaded}} = \dfrac{X_L}{R_1 + R_r} = \dfrac{1000}{10 + 190} = 5.0$

$$\text{Tank eff} \propto \frac{R_r}{R_1 + R_r} = \frac{190}{10 + 190} \times 100 = 95\%$$

This example shows the improvement in efficiency obtained by reducing the loaded Q. Yet this requirement is in direct contradiction to the need for a high Q to reject harmonics. A compromise is necessary. Q values between 10 and 12 are generally used in commercial applications.

Selection of L and C Values

Example 4–2 brings out another problem. When the Q is reduced, as in part 4, the efficiency is improved. But does the tank circuit now present the proper impedance to the tube? Actually, there are three conditions that the tank circuit must meet—Q, impedance, and resonant frequency—and they are all interrelated with the component values L, C, and R. An example will show how proper choice of these values will satisfy all three requirements.

EXAMPLE 4–3

The data sheet for the 833A power tube gives the following "typical operating" values as a class-C amplifier:

dc plate voltage (E_b)	3000 V
dc grid voltage (E_c)	–200 V
Peak RF grid voltage (E_{gm})	360 V
Power output (approximate) (P_o)	1000 W

Calculate the tank circuit values for a Q (loaded) of 10 and an operating frequency of 16 MHz.

Solution

1. The maximum positive grid swing is

$$E_{c(\max)} = E_{gm} + E_c = 360 + (-200) = +160 \text{ V}$$

2. For good plate-circuit efficiency:

$$E_{b(\min)} = E_{c(\max)} = 160 \text{ V}$$

3. The peak value of the output voltage is

$$E_{pm} = E_b - E_{b(\min)} = 3000 - 160 = 2840 \text{ V}$$

4. Using this *peak* voltage value, $P_o = E_{pm}^2/2R_L$, and substituting the tank circuit impedance $Z_{\|}$ for R_L, we have

$$Z_{\|} = \frac{E_{pm}^2}{2P_o} = \frac{(2840)^2}{2 \times 1000} = 4040 \ \Omega$$

5. From Eq. (1–6), $Z_{\|} = QX_L$, or

$$X_L = \frac{Z_{\|}}{Q} = \frac{4040}{10} = 404 \ \Omega$$

$$L = \frac{X_L}{2\pi f} = \frac{404}{2\pi \times 16 \times 10^6} = 4.04 \ \mu\text{H}$$

6. The capacitive reactance must also be 404 Ω, and

$$C = \frac{1}{2\pi f X_C} = \frac{0.159 \times 10^{12}}{16 \times 10^6 \times 404} = 24.6 \text{ pF}$$

Coupling Methods

Any of the coupling techniques discussed and shown in Chapters 2 and 3 can be used for interstage coupling between class-C power amplifiers. In the impedance-coupled circuit of Fig. 3–1(a), resistor R_g would be the grid-leak bias resistor. If fixed bias is desired, this resistor must be replaced with an RF choke. With single-tuned inductive coupling, the tuned circuit is generally in the plate circuit of the first tube rather than in the grid circuit of the next tube (as shown in Fig. 3–6). Double-tuned inductively coupled circuits are also used. In addition, if the two stages are not in close proximity (or are in separate chassis), the link coupling technique of Fig. 2–3(b) can be used.

Coupling between the final class-C amplifier and its load follows the principles and techniques discussed in Chapter 2. However, although direct coupling (Figs. 2–12 and 2–13) can be used, inductively coupled circuits are preferred. For low-impedance loads, the series-tuned output circuit (Fig. 2–14) is used. High-impedance loads necessitate shunt-tuned outputs (Fig. 2–15), while the π networks (Fig. 2–16) are generally used with intermediate load values. Regardless of the type of coupling,

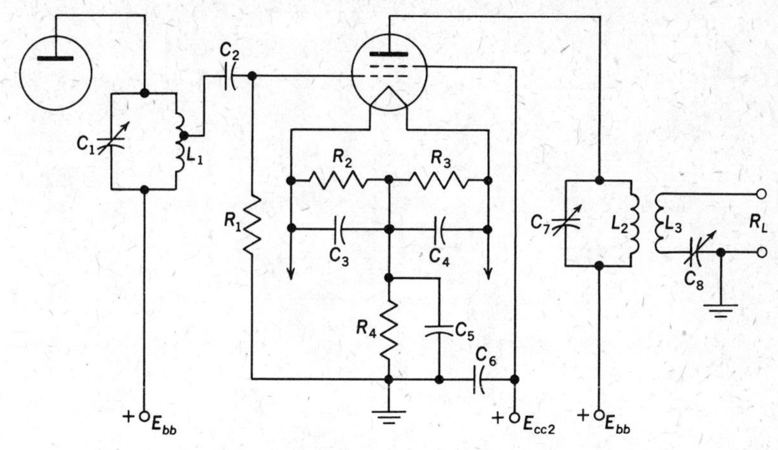

Figure 4-7 A beam-power class-C amplifier.

the circuits—grid and/or plate—can be shunt-fed or series-fed. A typical circuit diagram is shown in Fig. 4–7. Analysis of this diagram and additional circuit diagrams are left as student exercises. (See Question 48 and Problems 22–30.)

Parallel Operation of Tubes

If a higher power output is desired than a given type of tube can safely deliver, two or more such tubes can be operated in parallel. Two tubes can deliver twice the power output of a single tube. When using tubes in parallel, the output impedance of the stage decreases in proportion to the number of tubes used, and the load impedance must be reduced accordingly to maintain proper match. For example, using two tubes, the proper load impedance is one-half the recommended single-tube value. On the other hand, the dc plate current requirement increases in proportion. This in turn increases the ac component value. Again, for two tubes, I_p is doubled and the power output $I_p^2R_L$ is doubled.

Since the tubes are in parallel, no change in excitation *voltage* is necessary to drive the tubes to their full power output capability. However, grid current increases, grid losses increase, and higher driving power is required. Parallel operation has a drawback in that the danger of parasitic oscillation* increases, and the precautions to prevent this unwanted oscillation must be increased.

Push-Pull Operation of Tubes

Increased power output can also be obtained by operating a given tube type in push-pull. The circuitry follows the same basic principles

*This effect is discussed in more detail later in this chapter.

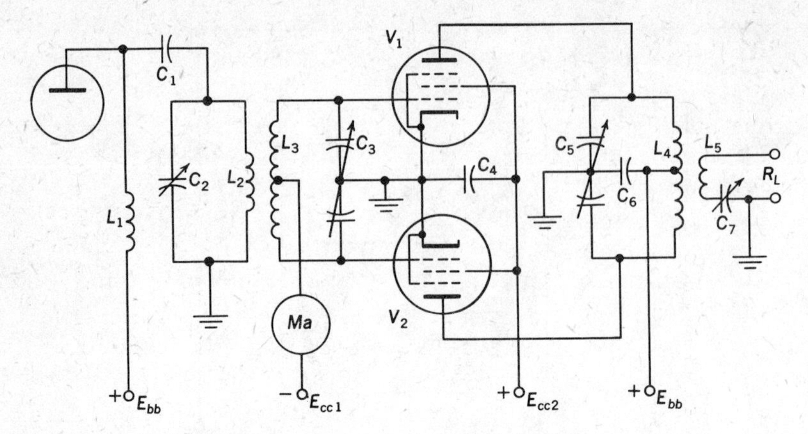

Figure 4-8 A push-pull class-C amplifier.

that apply to untuned push-pull amplifiers,* except that RF coupling methods are now used. A typical circuit is shown in Fig. 4–8. This circuit uses series-fed, fixed-bias, and series-fed plate-tank circuits, and a series-tuned load. Notice the *split-stator* capacitors used in the grid and plate tank circuits of the push-pull stage. The effect is similar to connecting two capacitors in series. This permits the rotors to be grounded.

Since the two tubes are effectively in series, the output impedance (plate-to-plate) is double the single-tube value. Also, each tube's plate current flows through its normal load value, so that each tube will develop its full power output, making the total output twice the single-tube value. To deliver its full output, each tube requires full excitation voltage and driving power. Obviously, the total exciting voltage and the driving power requirement must be twice the single-tube value.

Both push-pull and parallel operation double the power output, but push-pull requires twice the exciting voltage. On the other hand, since push-pull action cancels even harmonics, it is possible to use a lower Q tank circuit for higher efficiency.

NEUTRALIZATION IN CLASS-C AMPLIFIERS

A difficulty arises when triodes are used as class-C amplifiers. Because of the relatively high interelectrode capacitance between their grids and plates (C_{gp}), sufficient energy can feed back from the output to input to cause the triode to act as an oscillator. Such action is undesirable, and it is prevented by using *neutralizing* circuits, wherein a second voltage is

*J. J. DeFrance, *General Electronics Circuits* (New York, Holt, Rinehart and Winston, Inc.), Chap. 11.

deliberately fed back from the output to the input. If the neutralizing voltage is adjusted so as to make it equal in magnitude but opposite in phase to the original feedback voltage through C_{gp}, the two effects cancel. In tetrodes, pentodes, and beam power tubes, the interelectrode capacitance C_{gp} is much lower, and generally neutralization is not necessary. For this reason triode circuits are not popular, and their use is generally restricted to the higher power output levels where suitable tetrodes are not available. (Triodes are also more desirable than multigrid tubes as class-C RF-modulated amplifiers because of the relative ease in getting high modulation levels with them.) On the other hand, at higher frequencies (VHF and UHF), the amount of feedback even with multigrid tubes can be sufficient to cause oscillations; so neutralization is necessary.

Neutralizing Circuits

The three most commonly used neutralizing circuits involve *plate, grid,* and *cross* neutralization. The plate neutralization circuit is shown in Fig. 4–9. It is also known as the *Hazeltine* circuit, after its inventor. Al-

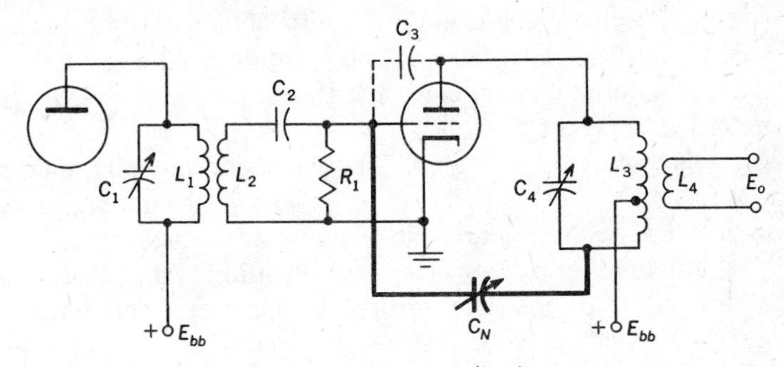

Figure 4-9 Plate neutralization.

though many class-C circuit variations are possible, the essential connections for plate neutralization are shown by heavy lines. Notice that the plate tank circuit L_3C_4 is tapped and the plate voltage is fed to this tap. This places the coil tap at RF ground potential. With respect to this point, the ends of the tank circuit have RF potentials 180° out of phase. Therefore, at some instant when the top of the tank has a positive RF potential, a positive voltage is fed back through C_{gp} to the grid of the tube. Simultaneously, a negative voltage is fed to the grid through the neutralizing capacitor C_N. By adjusting the value of this capacitor, the two feedback voltages can be made equal in magnitude, thus canceling their effects.

In the grid (Rice) neutralization circuit of Fig. 4–10, the tapped inductor is in the grid circuit. This time, the feedback voltages via C_{gp} and C_N are of the same polarity. However, notice that they are fed to opposite

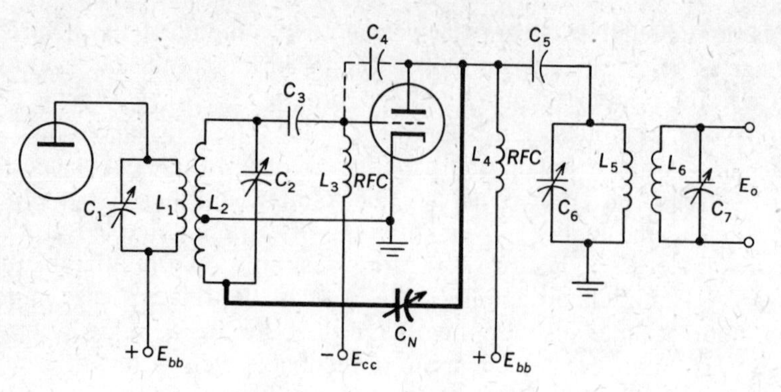

Figure 4-10 Grid neutralization.

ends of the grid tank circuit. Since the tap point is at ground potential, the two feedback voltages have opposite effects, and they cancel. Notice also in Figs. 4–9 and 4–10 that the tap points are not shown in the center. Symmetry or center-tapping is not necessary. The feedback voltages will be 180° out of phase regardless of the location of the tap point. Only their relative magnitude is affected, and this is balanced by changing the value of C_N with respect to C_{gp}.

When push-pull class-C amplifiers are used, the plate and grid circuits are already center-tapped. Also the two grid inputs and the two plate outputs (respectively) are out of phase with each other. We have all the "ingredients" for grid or plate neutralization. All that is necessary is to feed the output of one tube (through a capacitor) back to the input of the other tube. Such a circuit, shown in Fig. 4–11, is called a *cross-neutralization* circuit.

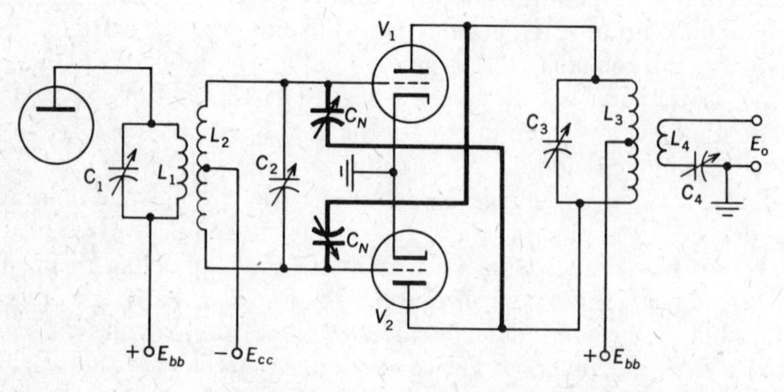

Figure 4-11 Cross neutralization.

Neutralizing Techniques

If the plate supply voltage is removed from a class-C amplifier, plate current cannot flow in the tank circuit, even though full excitation is applied. There should be no RF voltage across the plate tank. Therefore, in a well-designed amplifier, the presence of an RF output voltage is an indication of improper neutralization. This condition is corrected by adjusting the neutralizing capacitor until the RF output voltage is zero.* As an RF indicator, a vacuum-tube voltmeter, an oscilloscope, a dc milliammeter with a rectifier, or even a flashlight bulb can be used. The indicator is connected in series with a few turns of wire, and this "probing coil" is loosely coupled to the plate tank coil.

When a circuit has provision for metering grid current, the indication of this instrument provides another means for checking and adjusting for proper neutralization. The procedure is as follows: Remove the plate supply voltage, rock the tuning capacitor back and forth through its resonant setting, and watch the grid current reading. If the amplifier is not properly neutralized, the grid current reading will vary. The neutralizing capacitor should then be adjusted – while still rocking the tuning capacitor through resonance – until the grid current reading remains steady.

Parasitic Oscillations

It is possible for an amplifier to break into oscillation, generating a frequency that has no relation to the frequency of operation. Such oscillations are known as *parasitic* oscillations because they absorb power from the circuits in which they occur. This either overloads the tubes or reduces the power output at the operating frequency. In either case, the performance and efficiency are impaired. Parasitics commonly occur when some components in the input and output circuits are resonant to the same frequency. This produces a *tuned-plate, tuned-grid* oscillator.† In a class-C amplifier, such action must be prevented.

Low-frequency parasitics (below the operating frequency) are generally the result of an improper choice of RF chokes and bypass capacitors in the plate and grid circuits. (This is especially true when shunt-feed is used in both the plate and grid circuits.) This effect can be seen from Fig. 4–12(a) and (b). Tank circuits C_1L_1 and C_3L_2 are resonant at the normal operating frequency. The RF chokes have a much higher inductance than the tank coils L_1 and L_2. At some much lower frequency, the reactance of L_1 and L_2 becomes negligible. Also, the lower end of

* In a poorly or cheaply designed circuit, stray coupling may exist between the output tank circuit and previous stages. A condition of zero RF output cannot be obtained. In such cases, the neutralizing capacitor is adjusted for a minimum RF amplitude.

†Oscillator circuits are discussed in detail in Chapter 5.

these coils is at RF ground potential because of the power-supply filter capacitors. The circuit reduces to the tuned-plate, tuned-grid schematic of Fig. 4–12(b). If the resonant frequencies of the tank circuits match, parasitic oscillations will result. This effect can be prevented either by eliminating one of the RF chokes or by altering any of these critical component values so as to avoid the matching resonant frequencies.

Parasitic oscillations are even more likely to occur above the operating frequency as a result of lead inductances, stray capacitances, and interelectrode capacitances. An example of such a case is shown in Fig. 4–12(c). The inductances L_g and L_p represent the inductances of the grid and plate leads, respectively, including the leads inside the tube envelope. (At these higher frequencies, the normal tank circuits have negligible impedances because of the low reactances of C_1 and C_3. They therefore play no part in the parasitic equivalent circuit.) These inductances are tuned to resonance by interelectrode and stray capacitances. If the input and output resonant frequencies are matched, oscillations will result.

It is often possible to prevent such oscillations by altering lead

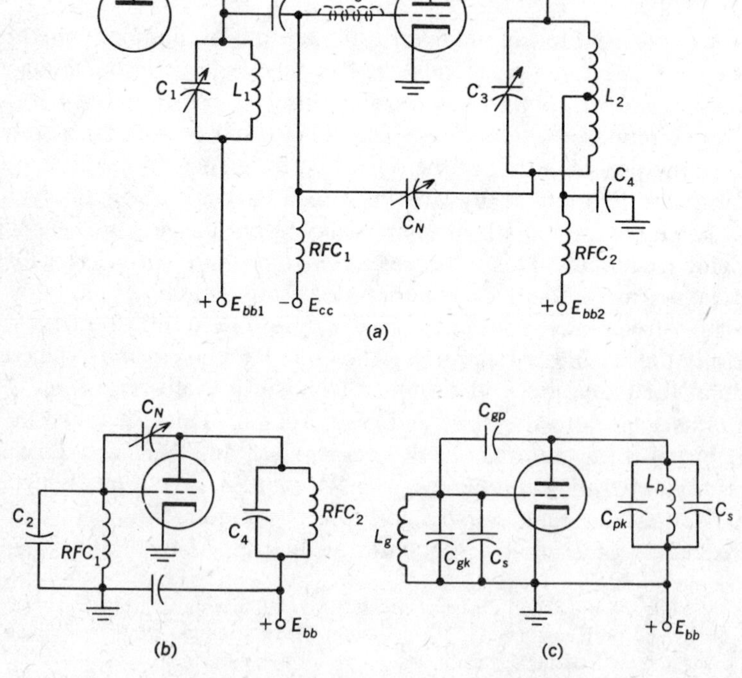

Figure 4-12 Development of parasitics.

lengths (preferably by shortening leads) and by changing lead dress, so as to change the stray capacitance effects. Sometimes, however, these techniques merely shift the frequency of the parasitic. Another method used to suppress oscillations is to insert a noninductive resistor in series with the grid lead, plate lead, or both—*as close to the socket terminal as possible.* These resistors rapidly dissipate parasitic oscillations but do not appreciably affect performance at the operating frequency. Resistance values used may range 10–100 Ω. In high-current circuits, a small RF choke is shunted across the resistor to minimize power loss.

To check for the presence of parasitic oscillations, the stage is first neutralized, and then the excitation is removed. If grid-leak bias is used, sufficient fixed bias must be applied to prevent the power input ($E_b I_b$) from exceeding the plate dissipation rating of the tube. Now an RF indicator is coupled to the plate tank circuit. If RF is present, it must be parasitic. A grid-dip (frequency) meter can then be used to check the frequency of the oscillation. Grid current can also be used as a check for parasitics. Since there is no excitation, the grid current should be zero. If there is any grid current, it indicates that the circuit is oscillating.

Tuning Adjustments

To obtain maximum output from a class-C amplifier, the tank circuits must be tuned to the operating frequency and the output load must be matched to the tube. Tuning and adjustment of a multistage RF power amplifier starts with the input stage and proceeds stage by stage to the output. Regardless of the type of tube or the specific circuit details, the procedure used is generally the same. Before we discuss this procedure, let us review some pertinent aspects of theory. The impedance of a tank circuit, off resonance, is low, but it rises to a maximum at resonance. For a high-Q, *unloaded* tank circuit the impedance at resonance approaches infinity. If we observe the plate current (the dc average value), it will drop to a minimum, approaching zero, when the unloaded tank circuit is tuned to resonance, and it will rise to a much higher value if the tank circuit is detuned to either side of resonance. Such a curve of plate current versus frequency for a class-C power amplifier is shown in Fig. 4–13. Notice the sharp dip in plate current for the unloaded circuit, at resonance. However, this circuit will not take power from the power amplifier.

As load is coupled into the tank circuit, its impedance at resonance decreases, and the plate current at resonance increases. If we tune the tank circuit through resonance, again the current will dip at the resonant frequency, but this time, the dip will not be as pronounced. This condition is shown in Fig. 4–13 by the dashed curve. When the correct load coupling has been achieved, the plate current, at its resonant dip, will

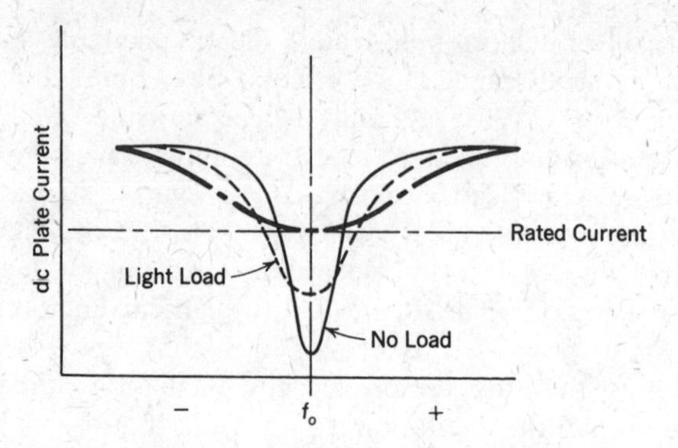

Figure 4-13 Plate-current resonance curves.

have reached its rated value. Beyond this point, any additional load coupled into the tank circuit constitutes an overloaded and overcoupled condition. The tank impedance is too low, the plate current is too high, the power output decreases, and the plate dissipation increases; furthermore, if either the plate current or plate dissipation rating is exceeded, the tube will be damaged.

Now we are ready to consider the actual tuning adjustments. Figure 4–14 shows a three-stage CW transmitter consisting of an oscillator stage V_1 (details not shown), an intermediate power amplifier V_2, and the final power amplifier V_3. Its output would normally be fed to the antenna, but to prevent transmission during these adjustments, the antenna has been replaced by a *dummy load* having the same impedance as the antenna itself. In a properly designed antenna this is a pure resistive value. As mentioned earlier, tuning and adjustment start with the first stage—the oscillator. A typical procedure can be itemized using the circuit of Fig. 4–14 as a guide.

1. Remove plate and screen voltages from all stages following the oscillator. This precaution is necessary to prevent excessive plate current as a result of low excitation or a detuned plate tank circuit.
2. Apply filament or heater power to all stages.
3. Apply reduced plate voltage to the oscillator, tune the plate tank C_1L_1 to resonance, and tune the following grid tank C_2L_2 to resonance. Because of reflected impedance, it is often necessary to retune C_1L_1. Now apply full plate voltage and recheck both tank circuits for resonance.
4. If the plate current I_{b1} is not at its rated value, adjust the coupling between L_1 and L_2, while rechecking each tank for resonance, until the desired I_b value is obtained. (I_{c2} should now also be at its rated value.)

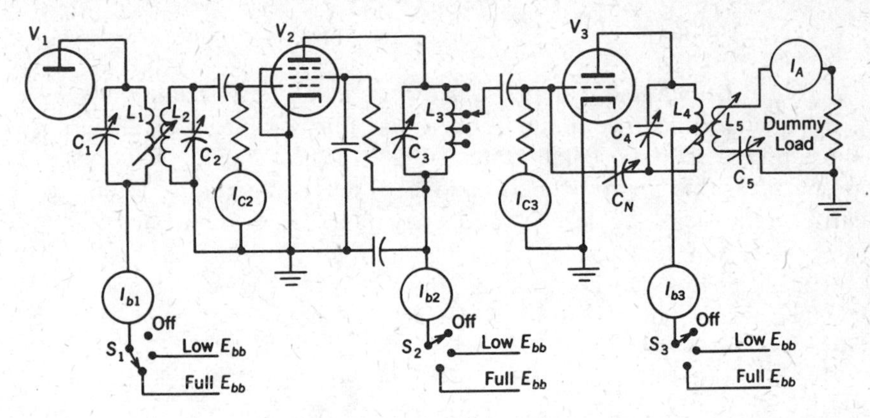

Figure 4-14 A CW transmitter.

5. Now apply reduced plate voltage to V_2, and tune its plate tank C_3L_3 to resonance.
6. Switch to full plate and screen voltage for V_2, and recheck C_3L_3 for resonance. If the plate current is not at its rated value, adjust the coupling to V_3 (change tap connection) and retune for resonance, until the desired I_b value is obtained. This also applies correct excitation to V_3. I_{c3} will be at its rated value.
7. While the plate voltage to V_3 is still off, check this stage for neutralization, and neutralize it, if necessary, as described earlier.
8. With the coupling between L_4 and L_5 at some low value, apply reduced plate voltage to V_3 and tune its tank C_4L_4 to resonance.
9. Now it is safe to apply full E_{bb}. Again plate circuit resonance should be checked. Because of the loose coupling between L_4 and L_5, the plate current I_{b3} at resonance is low, and the antenna current I_A is very low. If this current is readable, tune the secondary C_5L_5 to resonance. Since this is a series-resonant circuit, tune for *maximum* antenna current.
10. To transfer power to the load, the coupling must be tightened. As this is done—gradually, and in small steps—I_{b3} and I_A will increase. At each step, both the primary and secondary circuits must be retuned to resonance. This process is continued until the plate current and the antenna current are at their rated values.

If the coupling is tightened further, the plate tank impedance decreases and the plate current increases above its rated value. On the other hand, the mismatch caused by low plate-load resistance will decrease the power output, thus decreasing the antenna current I_A. This can be used as a check on the tuning and adjustment of the amplifier.

Frequency Multipliers

It has been found that an oscillator has much better stability when it is designed for the lower radio frequencies. Furthermore, as is discussed in Chapter 5, crystal oscillator circuits have the best stability, but crystals are not available for frequencies above 30 MHz. Therefore, when high carrier frequencies are desired, frequency multiplying circuits are used to raise the basic frequency of the oscillator circuit. Any amplifier that generates harmonics can be used for this purpose. A class-C amplifier is ideally suited to this because of the high harmonic content in its plate current.

Earlier in this chapter, it was shown that for "straight-through" service class-C amplifiers are operated with conduction angles ranging between 150° and 120°. This is a compromise among efficiency, power output, driving requirements, and harmonic distortion. But in a frequency multiplier, higher harmonic content is an advantage, and therefore lower conduction angles are desirable. However, for a given value of peak plate current, there is a conduction angle at which the harmonic component is a maximum. The optimum conduction angle is in the region around 120° for a frequency doubler, 80° for tripler action, and 60° for quadruplers. Unfortunately, the harmonic content—in spite of optimized angles—still decreases with the order of the harmonic. In practical applications the multiplication in any one stage is generally restricted to doubler or tripler action. When higher multiplication is required, several stages are used.

To obtain the low conduction angle for multiplier service, the class-C amplifier must be operated at appreciably higher negative biases, compared to a "straight-through" amplifier. For a given peak plate current value, this requires a much higher excitation voltage and driving power. Even so, because of the reduced conduction time, the power output will decrease. Consequently, the plate circuit efficiency for a doubler is at best approximately 60% of the straight-through efficiency and decreases rapidly with the order of multiplication.

From a circuit diagram, it would not be possible to distinguish a straight-through amplifier from a doubler or a tripler. Any of the circuitry discussed earlier in the chapter can be used. The only changes are the bias, or bias component values, and the constants of the plate tank circuit. The tank is now resonant at the desired harmonic frequency. One circuit simplification is that neutralization is not necessary since the input and output circuits are at different frequencies. Another difference is that push-pull circuits cannot be used for doublers, since the push-pull action would cancel the even harmonics.

If you reexamine the current and voltage waveshapes for a straight-through class-C amplifier (Fig. 4–2), you will notice that there is one pulse

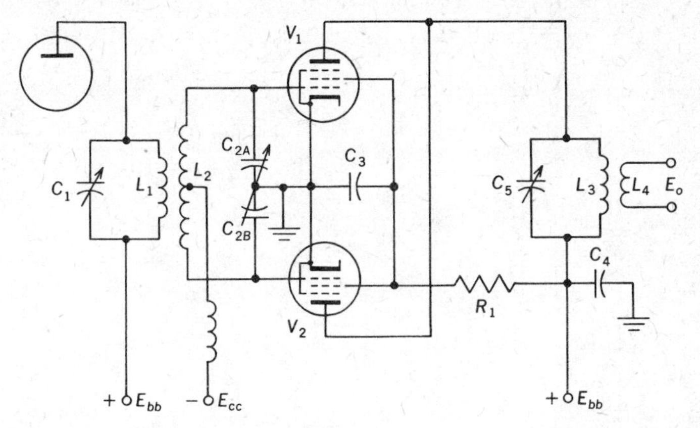

Figure 4-15 A push-push frequency coupler.

of plate current for each cycle of RF voltage output. These pulses occur each time the excitation drives the grid above the cutoff value. In frequency multiplier service the excitation is not changed, and the frequency of the plate current pulses is not changed, but the frequency of the RF output voltage is doubled or tripled. Therefore, we now have only one current pulse for every two or three cycles of RF output, depending on whether the circuit is a frequency doubler or tripler. If the Q of the plate tank circuit is too low, it may tend to produce a damped output voltage wave. For frequency doubling, this condition can be remedied by using the *push-push* circuit of Fig. 4–15. Notice that the plates are connected in *parallel,* whereas the input circuit is push-pull. In this circuit, the tubes conduct alternately on successive half-cycles of the input signal, producing two pulses of plate current for each excitation cycle, or one current pulse for each cycle of the RF output voltage. Because we have doubled the number of plate current pulses, this circuit also produces twice the power output of a single tube.

High-Efficiency Harmonic Resonator Amplifiers

From Fig. 4–2 and the discussion that followed, it was shown that the high efficiency of a class-C amplifier is a result of the reduced conduction angle and the relatively low plate potential during the conduction time interval. These plate voltage-plate current waveshapes are shown again in Fig. 4–16(a). If these waveshapes could be "squared out," as shown by the solid-line waveshapes, it would be obvious that the power output would increase. Also, since the plate potential remains at its $E_{b(\min)}$ value for the full duration of the plate current flow, plate dissipation is reduced, and efficiency is increased. Such squared waveshapes can be obtained by inserting third-harmonic resonator tank circuits in series with

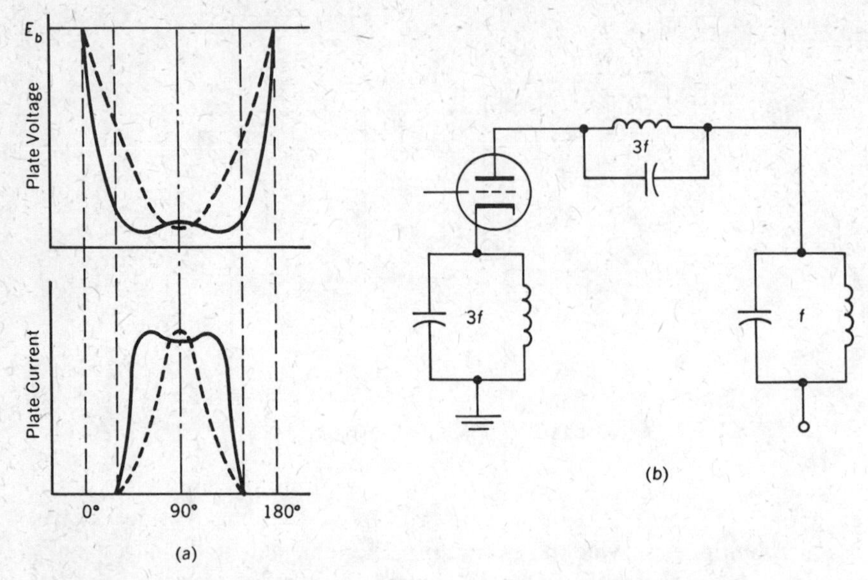

Figure 4-16 Effect of third-harmonic resonator on plate voltage and plate current waveshapes.

the plate and cathode leads. A partial schematic illustrating the use of these resonators is shown in Fig. 4–16(b). One transmitter manufacturer claims a plate efficiency of 94% using this technique.

CLASS-B LINEAR AMPLIFIERS

Class-C operation is generally used for RF power amplifiers because of its high efficiency (up to 85%). Unfortunately, its use is restricted to applications wherein the exciting voltage is essentially constant in amplitude. This is fine for unmodulated waves, frequency-modulated waves, or constant-amplitude pulse-modulated waves. However, there are many cases wherein the amplitude of the wave varies in accordance with the intelligence it contains. Such an *amplitude modulated* (*AM*) wave is shown in Fig. 4–17(b), compared to the unmodulated carrier in Fig. 4–17(a). Notice that this AM wave varies in amplitude from zero to twice the unmodulated value. This corresponds to 100% modulation.*

For optimum class-C operation, the dc bias is set to more than twice the cutoff value, and the excitation drives the grid into the positive region, such that $E_{c(\max)} = E_{b(\min)}$. Using the unmodulated carrier of Fig. 4-17(a) for excitation, the circuit is adjusted so that this condition is obtained at

*The details of amplitude modulation are covered in Chapter 7.

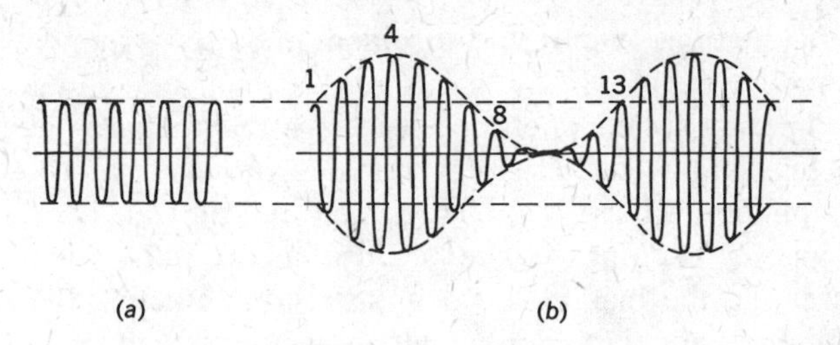

Figure 4-17 Comparison between an unmodulated and an amplitude-modulated wave.

the peak of each RF cycle. A pulse of plate current flows *for each RF cycle,* and the fundamental component of this current produces the output voltage across the tank circuit.

Figure 4–18 shows the effect of applying the 100% modulated signal to a class-C amplifier. The optimum operating condition of $E_{c(\max)} = E_{b(\min)}$ must be established for RF cycle 4, because this has the maximum (crest) amplitude. For RF cycles 4 through 7 there is no problem. Each cycle drives the grid out of cutoff, plate current flows, and an output voltage is produced. Since the plate-current pulses vary in amplitude, the output varies accordingly, resulting in linear amplification. But notice that, because of amplitude modulation, RF cycles 8 through 12 do not drive the tube out of cutoff. Plate current does not flow; there is no output voltage. Obviously, the output waveshape will be distorted.

This condition can be remedied by reducing the dc bias to the cutoff value. Now, the slightest amplitude of excitation voltage will drive the

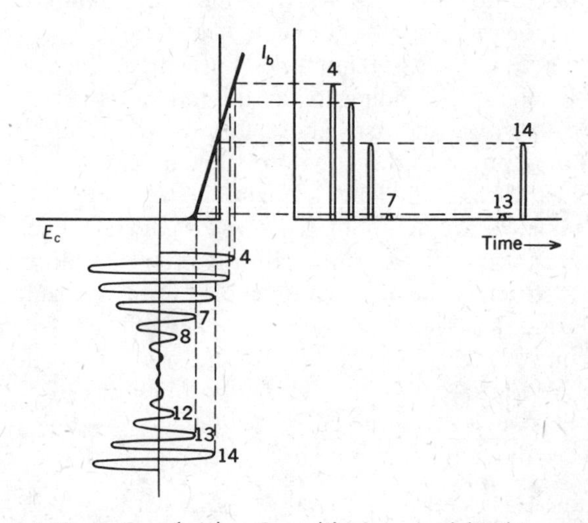

Figure 4-18 Action of a class-C amplifier on a modulated input signal.

tube into conduction. A current pulse and an output voltage will be produced for each cycle of RF. With this new bias value, it should be recognized that the stage is now operating as a class-B amplifier. Except for the change in bias value, the circuitry remains the same as discussed for class-C amplifiers. There is one restriction. Grid-leak bias cannot be used with an AM exciting voltage because, as the amplitude of the input signal decreases, the bias would decrease, counteracting the decreased signal level and tending to maintain all current pulses equal in amplitude. This would remove the modulation from the output voltage.

The efficiency of a class-B amplifier—with modulated input—is appreciably lower than for class-C operation, for two reasons. Optimum operation is obtained only at the crest of the modulation cycle (RF cycle 4 in Figs. 4–17 and 4–18). Even so, since plate current flows for 180°, as compared to 120–150° for class C, the plate dissipation is higher and the efficiency is lower than with class C. The maximum efficiency obtainable—at this instant—in a class-B amplifier is about 65%. Secondly, as the amplitude of the exciting voltage drops to zero (RF cycle 10), the plate current pulse amplitude drops to zero, the output drops to zero, and the efficiency drops to zero. Therefore the *average* efficiency for a class-B amplifier, with a 100% modulated signal, is at best between 30 and 35%. If the input signal is less than 100% modulated, the amplifier can be operated with a more negative bias (between B and C) for increased efficiency.

The Doherty High-Efficiency Amplifier

An RF power amplifier specifically designed for amplifying modulated waves—at high efficiency—is the *Doherty amplifier.** This circuit uses two tubes essentially connected in parallel. One tube, V_1, is designed to operate as a class-B amplifier, but with maximum efficiency at a grid signal level corresponding to the unmodulated carrier amplitude. When the excitation voltage rises above this value (cycles 1 through 7 of Fig. 4–17), the output from this tube is distorted because the tube begins to saturate, but the plate efficiency remains high. On the other hand, for RF amplitudes below the unmodulated carrier level (RF cycles 8 through 13), the amplification is linear, but again the efficiency drops off to zero at the trough (bottom) of the modulation cycle. However, since the power level is low during these RF cycles, the lessened efficiency is not serious.

The second tube, V_2, is used to correct for the distortion produced when V_1 is driven into saturation. This tube functions as a class-C amplifier with a bias such that plate current begins to flow only as the exciting voltage rises above the unmodulated carrier level. The operating con-

*A. V. Eastman, *Fundamentals of Vacuum Tubes,* 3rd ed. (New York, McGraw-Hill, Inc., 1949).

ditions are further adjusted so that maximum efficiency occurs at the crest of the modulation cycle. Since this tube conducts only while the exciting voltage is relatively high, its average efficiency remains high. With the Doherty circuit, efficiencies of 60 to 65% can be obtained even with fully modulated signals.

TRANSISTOR RF POWER AMPLIFIERS

Although the discussion thus far has dealt with vacuum-tube circuits, much of it applies equally well to transistor circuitry. The coupling methods are generally the same; tank circuit considerations are quite similar. Transistors can be operated in parallel, in push-pull, or both, and danger of parasitic oscillation also exists. However, as mentioned earlier in this chapter, transistors are limited as to power level and frequency of operation. On the other hand, the transistor has unique advantages in the smaller portable and mobile transmitters such as the walkie-talkie. Let us now examine some of the special features of transistor circuits.

Transistor Bias Voltages

As in any transistor application, the collector-base junction must be reverse-biased. Ideally, for maximum collector circuit efficiency, class-C operation should be used. This would imply that the emitter-base junction should also be reverse-biased. However, several other factors must be considered when using transistors. First, the classical vacuum-tube treatment of classes A, B, and C does not exactly apply to transistors. Because of turn-off transition and turn-off delay,* the collector current does not follow the input signal, and conduction may exceed 180° even with reverse bias. Furthermore, transistors are relatively low-impedance devices. (The input impedance in the common-emitter configuration may be less than 5 Ω.) Consequently, they cause heavy loading of the tank circuits, and Q values are generally below 10. Such tank circuits will not discriminate against the high harmonic content produced with deep class-C operation. Class-B operation might even be preferable. Finally, it has been found that a dc biasing effect is produced in the input circuit of a transistor when an ac signal is applied. This voltage (V_{BE}) is a reverse bias, and quite often it is sufficient to produce a class-C type of operation.

With transistor RF power amplifiers, any one of three input biasing techniques (or combinations thereof) may be seen. These are shown in the partial schematics of Fig. 4–19. Many amplifiers, particularly with large driving signal levels, will use the circuit of Fig. 4–19(a). Bias for

* DeFrance, *General Electronic Circuits*, p. 452.

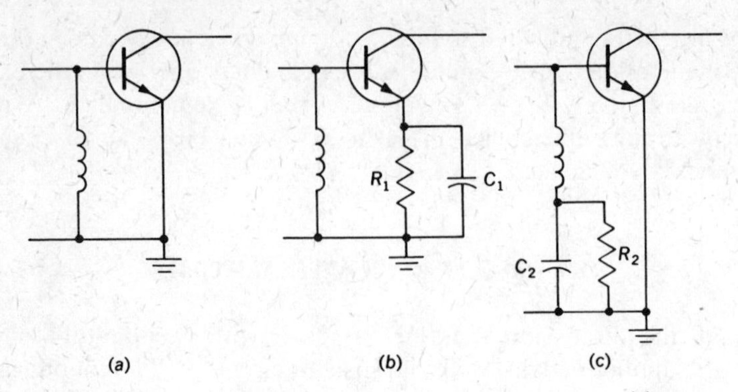

Figure 4-19 Biasing techniques in transistor RF power amplifiers.

class-C operation (V_{BE}) is obtained through rectification of the input signal. With silicon transistors this bias may sometimes be excessive, and a slight amount of *forward bias* may be needed to improve gain and power output.

In other applications, the above bias technique may not be sufficient. Additional bias can be obtained by inserting an *R-C* network in the emitter leg, as shown in Fig. 4–19(b). When the transistor conducts, current (electrons) flowing up through the resistor makes the emitter positive compared to ground, or the base negative compared to the emitter. For the NPN transistor shown, this is reverse bias and drives the transistor deeper into cutoff. An alternate method is shown in Fig. 4–19(c). Here the bias network is in the base loop, and it produces a negative bias in the same manner as the grid-leak bias circuit of a vacuum tube.

Regardless of the bias circuitry, the transistor is already turned off even before the driving signal is applied, and no external bias source is needed to produce this condition. This brings out one of the distinct advantages of a transistor transmitter—should the excitation be lost, the power amplifier merely cuts off and sustains no damage.

Design Considerations

For high-efficiency class-C operation, the voltage relations in transistor power amplifiers will closely resemble the vacuum-tube curves in Fig. 4–2. Typical voltage-time curves for a transistor are shown in Fig. 4–20. Notice that the collector voltage approaches the breakdown voltage, BV_{CEO}, on its positive swing, and zero on its negative swing. The closer it approaches these limits, the greater the output voltage, and the greater the power output that can be obtained from a given transistor. Since the positive swing should not exceed the breakdown voltage, it follows that the power supply voltage, V_{CC}, must not be greater than $\frac{1}{2}BV_{CEO}$.

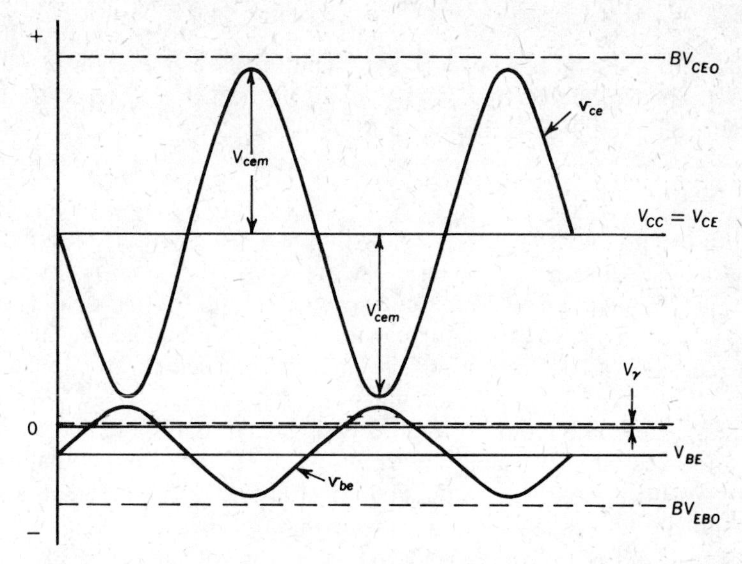

Figure 4-20 Voltage relations in an *NPN* transistor class-C power amplifier.

Notice also that the peak value of the output voltage, V_{cem}, is approximately equal to the power supply voltage.* Therefore, the power output is given by:

$$P_o = \frac{(V_{cem})^2}{2R_L} \cong \frac{V_{CC}^2}{2R_L} \tag{4–6}$$

Obviously, for a given supply voltage, as the load resistance is decreased, the obtainable power output increases. But there is a limit! As the load resistance is decreased, the collector current is increased and may exceed the maximum collector current rating – or the product of current and voltage may exceed the transistor power rating. (It should be realized that with a PNP transistor the voltage relations would be the same, but all polarities would be reversed.)

Equation (4–6) can also be used to find the value of load needed for a desired power output, once the supply voltage is fixed by the breakdown voltage BV_{CEO}.

On the input side, the excitation voltage swings positive for a portion of the cycle. However, base current and collector current do not begin to flow until the total emitter-base voltage, v_{BE}, rises above the value shown as V_γ. This is referred to as the *cut-in or threshold voltage*. (Typical

*Actually, the peak output voltage is always less than V_{CC} by about two volts. This is so because $V_{CE(\min)}$ must not fall below $V_{BE(\max)}$, and $V_{BE(\max)}$ is (approximately) three times more positive than the cut-in or threshold voltage V_γ.

values are 0.6 V for silicon transistors and 0.2 V for germanium units.) Notice the limitation imposed by the emitter-base breakdown voltage. The total instantaneous base voltage swing, v_{BE}, must not exceed BV_{EBO}—or the peak excitation is limited to $BV_{EBO} - v_{BE}$.

Neutralization

In the previous chapter we saw the need for neutralization in transistor RF small-signal amplifiers. Theoretically, the same problem also arises in RF power amplifiers. Given proper conditions of gain and phase, feedback from output to input can produce oscillations. As in small-signal amplifiers, neutralization circuits can be used to counteract this effect. However, RF power amplifiers do not generally need to be neutralized. The combination of high driving-signal amplitude and low collector-load resistance usually overrides any tendency toward instability. To obtain any appreciable amount of power output, a transistor must operate with a load resistance much lower than its own output impedance. This is a result of the transistor's low collector voltage rating. The peak value of the output voltage, at best, can only approach the collector supply voltage, and the power output would then be $V_{CC}^2/2R_L$. With a low supply voltage, even moderate power output requires the use of a low value of load resistance. For example, with a 12-V supply, if 24 W of power output is desired, a load resistance of approximately 3 Ω must be used. This mismatch reduces the stage gain. Obviously, gain must be sacrificed to obtain power output. Meanwhile, the mismatch and the low net gain increase stability, making feedback neutralization circuits unnecessary.

Typical Circuits

Many circuit variations are possible for transistor RF power amplifiers. These include not only the bias circuitry as shown in Fig. 4–19, but also the tuning circuits. For example, single-tuned, double-tuned, or π-type circuits can be used; the tuned circuits could be series-tuned or parallel-tuned, and series-fed or shunt-fed. Two typical circuits are shown in Fig. 4–21. In the first circuit, Fig. 4–21(a), single-tuned impedance coupling is used between stages. Each collector connects to a tap on its tank circuit to reduce loading effects. The transistor seems to be operating at zero bias. Yet class-C operation results from the junction voltage V_{BE} developed by the excitation voltage. Capacitors C_3 and C_6 serve not only to keep the collector voltage off the base, but also to transform the transistor input resistance to a higher resistance in parallel with the tank circuits. This improves the loaded Q of the tanks.

In the circuit of Fig. 4–21(b), a double-tuned circuit is used at the input. Again the collector of transistor Q_1 is connected to a tap to reduce

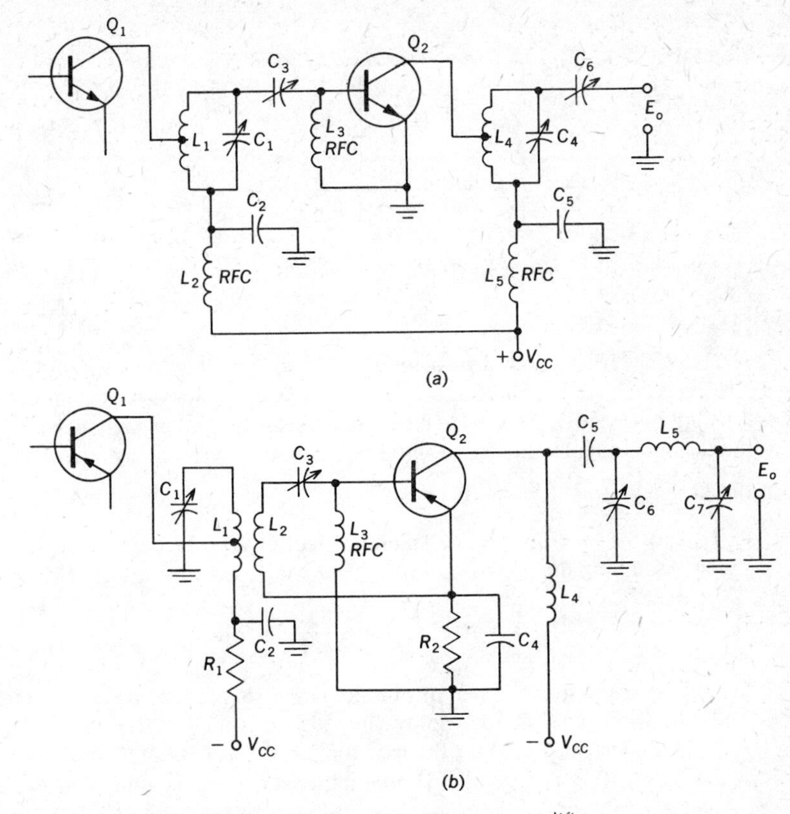

Figure 4-21 Transistor RF power amplifiers.

loading. On the secondary side, a series-resonant circuit is used for better match to the low-impedance input circuit. In the output, a shunt-fed collector circuit is used together with a π coupling network. This transistor uses emitter bias for class-C operation. The dc resistance between base and emitter should be low to prevent buildup of additional reverse dc bias. This is accomplished by returning the base to ground through the RF choke L_3.

Because of the extremely low values of collector and load impedances encountered with bipolar transistors, the above coupling networks are at times unsuitable. The design of such circuits may yield values of L and C that are impractical to obtain. Consequently (particularly in the VHF range), when matching from a 50-Ω source or load to a transistor output impedance of less than 50 Ω, a T coupling network may be preferable. Such a network is shown in Fig. 4–22, and a typical design situation is given in Example 4–4.

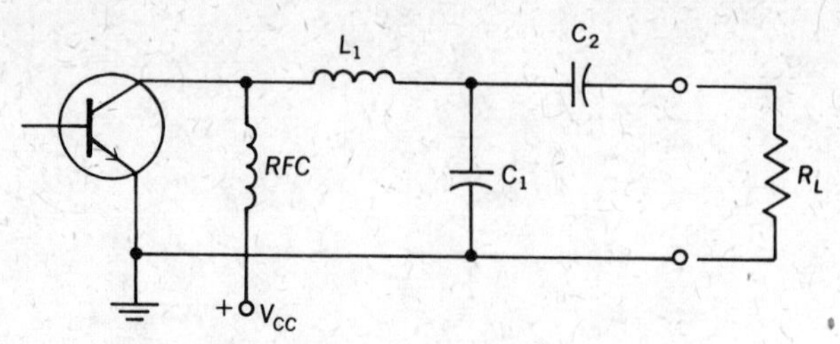

Figure 4-22 A T coupling network.

EXAMPLE 4–4

A 2N3950 NPN silicon transistor RF power amplifier is to supply 50 W to a 50-Ω (resistive) load at 50 MHz. The manufacturer's data sheet recommends a power supply voltage of 28 V and gives the transistor output capacitance as 180 pF. Find the component values for a suitable T coupling network, using a Q of 10.

Solution

1. To find the load impedance value needed for a 50-W output:
   (*a*) Assuming the peak output voltage $\cong V_{CC}$, then from Eq. (4–6)

$$R'_L = \frac{V^2_{CC}}{2P_o} = \frac{(28)^2}{2 \times 50} = 7.84\ \Omega$$

   (*b*) The transistor can be considered as the source feeding the coupling network, and the total source impedance is this 7.84-Ω resistance in parallel with the 180-pF capacitor (or at 50 MHz = 17.7 Ω).
   (*c*) Converting this parallel source impedance into its equivalent series value:

$$R_s = \frac{R_p}{1 + (R_p/X_p)^2} = \frac{7.84}{1 + (7.84/17.7)^2} = 65.5\ \Omega$$

$$X_s = R_s \frac{R_p}{X_p} = 6.55 \times \frac{7.84}{17.7} = 2.9\ \Omega$$

$$Z_s = 6.55 - j2.9$$

2. For an impedance match, the T network must have a conjugate value of $6.55 + j2.9$.
3. To calculate the value of L_1: For a Q of 10, since $Q = X_L/R$ and the resistive component is 6.55 Ω

$$X_L = QR = 10 \times 6.55 = 65.5\ \Omega$$

$$L = \frac{X_L}{2\pi f} = \frac{65.5}{2\pi \times 50 \times 10^6} = 0.208\ \mu\text{H}$$

4. To calculate the values of C_1 and C_2:
   (*a*) The coupling network can be redrawn as in Fig. 4–23(a). The impedance of the parallel section is obviously $Z_s - X_L$, or

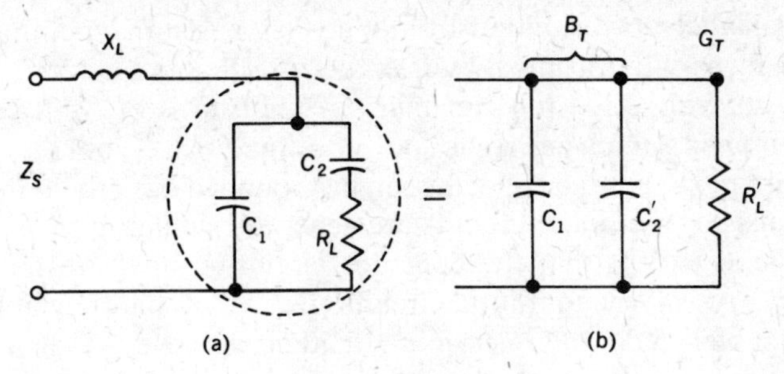

Figure 4-23 Example 4-4.

$$(6.55 + j2.9) - (+j65.5) = 6.55 - j62.4$$

(*b*) The parallel section can be further redrawn as a "full" parallel, and admittances values calculated. (See Fig. 4–23(b).)

$$G_T = \frac{R}{Z_T^2} = \frac{6.55}{3940} = 0.00166 \text{ mho}$$

$$B_T = \frac{X}{Z_T^2} = \frac{62.4}{3940} = 0.0158 \text{ mho}$$

(*c*) But from branch 2 of diagram (a)

$$G_2 = \frac{R}{Z_2^2} = \frac{R}{R^2 + X_{C2}^2} = \frac{50}{2500 + X_{C2}^2}$$

and the G values in diagrams (a) and (b) are equal; so that

$$G = \frac{50}{2500 + X_{C2}^2} = 0.00166 \text{ mho}$$

$$X_{C2}^2 = \frac{50}{0.00166} - 2500 = 27{,}600\ \Omega$$

$$X_{C2} = 164\ \Omega$$

$$C_2 = \frac{1}{2\pi f X_{C2}} = \frac{10^6}{2\pi \times 50 \times 164} = 19.4 \text{ pF}$$

(*d*) Still in branch 2 of diagram (a)

$$B_2 = \frac{X_{C2}}{Z_2^2} = \frac{164}{(50)^2 + (164)^2} = \frac{164}{29{,}500} = 0.00556 \text{ mho}$$

(*e*) $B_1 = B_T - B_2 = 0.0158 - 0.00556 = 0.0102$ mho

$$X_{C1} = \frac{1}{B_1} = \frac{1}{0.0102} = 98\ \Omega$$

$$C_1 = \frac{1}{2\pi f X_{C1}} = \frac{10^6}{2\pi \times 50 \times 98} = 32.5 \text{ pF}$$

For higher power output from a given transistor type, units can be operated in parallel, in push-pull, or both. Suitable heat sinks must be provided to realize the full capability of the transistor. Still better heat dissipation, and somewhat higher power output, can be obtained by altering the circuit so that the collector can be grounded directly to the chassis.

Transistor multiplier circuits are generally similar to the basic class-C amplifiers, except that the collector circuit is tuned to the harmonic frequency. As in vacuum-tube circuits, frequency multiplication is limited to doubler and tripler service. In either case the bias is adjusted to produce the highest output at the desired harmonic frequency.

Varactor Multipliers

As the frequency of operation is increased, the gain and power output obtainable from transistor circuits decreases rapidly, so that in the UHF region a transistor "power amplifier" can actually become an attenuator circuit. Higher power outputs can be obtained at these high frequencies by using *varactor diodes** as frequency multipliers. Although as diodes these devices have no gain, they act as nonlinear *reactances*, and therefore the losses or attenuation with such circuits is a minimum. Efficiencies of 50–90% are possible with single-stage units used as doublers, triplers, and quadruplers. For example, two cascade varactor quadruplers can deliver an output of 5 W at 640 MHz from a 20-W, 40-MHz input.

REVIEW QUESTIONS

1. Name four applications requiring the use of RF power amplifiers.

2. (*a*) Why is class-C operation preferred for RF power amplifiers? (*b*) Doesn't this cause excessive distortion of the output waveshape? Explain.

3. What are the characteristics of class-C operation, using vacuum tubes, with regard to (*a*) grid bias, (*b*) angle of plate current flow, and (*c*) grid current?

4. Why isn't a numerical subscript used to further classify class-C operation?

5. Refer to Fig. 4–1. (*a*) Is this a practical operating circuit? Explain. (*b*) What does the source e_g represent? (*c*) What name is generally given to this voltage? (*d*) What does E_{cc} represent? (*e*) What value is used for E_{cc} in this circuit? (*f*) Which is greater, this E_{cc} value or the peak value of e_g? Explain. (*g*) Will grid current flow in this circuit? Explain. (*h*) Will plate current flow in this circuit? When? (*i*) Why does plate current flow for less than 180°?

6. In Fig. 4–2, what do each of the following symbols represent: (*a*) E_c, (*b*) E_{gm}, (*c*) $E_{c(max)}$, (*d*) E_b, (*e*) E_{pm}, (*f*) $E_{b(min)}$, (*g*) i_c, (*h*) i_b, (*i*) i_p, (*j*) I_b.

*A varactor is a variable-reactance PN junction diode. When reverse-biased, its capacitance is a nonlinear function of the bias voltage.

7. Refer to Fig. 4–2. (*a*) At what time instant does the tube come out of cutoff? Why? (*b*) At what instant does it go back beyond cutoff? (*c*) What is the angle of plate-current flow? (*d*) At what instant does the grid go positive? (*e*) At what instant does the grid go negative again? (*f*) What is the duration of grid-current flow? (*g*) Since the plate current flows in pulses, why is the output voltage shown as a pure sine wave? (*h*) If the plate tank circuit had a very low Q (2 or 3), would this have any effect on the *shape* of e_p? Explain.

8. For a given dc bias, how does an increase in excitation voltage affect the duration of grid current flow? Why?

9. For a given excitation voltage, how does a less-negative bias affect the angle of grid current flow? Why?

10. In converting a "straight-through" class-C amplifier into a frequency doubler, the bias is increased by 40 V, and the amplitude of the exciting voltage is also increased by 40 V. (*a*) What happens to the $E_{c(\text{max})}$ value? Explain. (*b*) What happens to the angle of grid-current flow? Why? (*c*) What happens to the angle of plate-current flow? (*d*) What happens to the peak value of plate current? Explain.

11. (*a*) Is *plate power input* an ac or a dc quantity? (*b*) What supplies this power? (*c*) Give the equation for this quantity.

12. (*a*) Is *plate power output* an ac or dc quantity? (*b*) Give an equation for this quantity, based on peak values.

13. (*a*) What causes the effect known as *plate dissipation?* (*b*) How can we tell that such an effect exists? (*c*) How can this quantity be evaluated? (*d*) In making adjustments to a class-C amplifier, the load impedance is decreased to below the matching value. What happens to the plate dissipation? Why? (*e*) What effect could this have on the tube?

14. (*a*) Why is grid driving power needed for a class-C amplifier? (*b*) Where does this power come from? (*c*) Where is it dissipated? (*d*) What two quantities determine the power dissipated in the grid circuit?

15. (*a*) For a given power output, what relation exists between grid driving power and plate efficiency? Explain. (*b*) Give the equation for plate efficiency.

16. Refer to Fig. 4–2. (*a*) What two quantities account for plate dissipation at any instant? (*b*) What accounts for the high plate efficiency of this type of amplifier? (*c*) If the angle of plate current flow were decreased further, how would this affect efficiency? Why?

17. Refer to Table 4–1. (*a*) For an average plate current (dc) of 100 mA, what would the peak plate current be for a conduction angle of 160°? for 90°? (*b*) What is the fundamental component value for each of the above cases?

18. With reference to Table 4–1, for a peak plate current of 400 mA, what is the dc component of plate current for a conduction angle of (*a*) 140°? (*b*) 110°? (*c*) Explain why I_b changes in this fashion. (*d*) For a given load impedance, which of these two cases would produce the higher power output? Explain.

19. Refer to Example 4–1. (*a*) In the solution step 1(a), what does 3.75 represent and where does it come from? (*b*) In step 1(b), what does 1.20 represent, and where does it come from? (*c*) What happens to the power output capability of a given tube as the conduction angle is decreased?

20. With reference to Fig. 4–2, we wish to reduce the conduction angle to some value such as 90° while maintaining $E_{c(max)}$ at the value shown. (*a*) How can this be done? (*b*) How will such a change affect the grid driving power requirement? Why?

21. (*a*) State an advantage of class-C operation at a reduced angle of plate current flow. (*b*) State two disadvantages.

22. Refer to Fig. 4–1. (*a*) What type of bias circuit is this? (*b*) State three possible sources for the bias voltage. (*c*) State two requirements for the bias sources. (*d*) How is this bias circuit further classified?

23. Refer to Fig. 4–3. (*a*) Why is this called a shunt-fed bias? (*b*) Why is capacitor *C* necessary? (*c*) Why is the RF choke used?

24. (*a*) Can class-C bias be obtained by using a resistor in the cathode circuit? Explain. (*b*) For a given plate supply voltage, how would this affect the power output? (*c*) Name another technique for obtaining class-C bias without using a separate bias supply. (*d*) Where does the bias supply power come from?

25. Refer to Fig. 4–4(a). (*a*) Is this a series-fed or shunt-fed bias circuit? Why? (*b*) How can it be converted to the other type? (*c*) When does capacitor *C* charge? (*d*) Trace the path of the charging current. (*e*) When does the capacitor discharge? (*f*) Trace the path of the discharge current. (*g*) What requirement must the capacitance value satisfy? (*h*) What determines the charging time constant? (*i*) Is this time constant long or short compared to the period of the incoming signal? (*j*) Is the discharge time constant longer, equal, or shorter than the charging time constant? (*k*) How is this relation obtained?

26. Refer to Fig. 4–4(b) and (c). (*a*) What does e_i represent? (*b*) What does e_C represent? (*c*) Why is e_C shown with opposite polarity in the two diagrams? (*d*) Why doesn't the e_C curve rise to the peak e_i value during the first quarter cycle? (*e*) Why does it drop during time interval 1 to 2? (*f*) Why does it continue to climb during the first few cycles? (*g*) Why does it level off and remain essentially constant after these first few cycles?

27. (*a*) What single factor is mainly responsible for the bias developed in a grid-leak bias circuit? (*b*) What would be the effect of too large a grid-leak resistor value? (*c*) What is the disadvantage of too low a resistor value?

28. A three-stage class-C amplifier unit with grid-leak bias in each stage is used to raise the power level of an RF oscillator. The second stage develops a malfunction and delivers no output. What happens to the bias value of (*a*) the first stage? (*b*) the second stage? (*c*) the third stage? (*d*) What is the danger of such a situation? (*e*) State two ways by which this disadvantage can be corrected.

29. (*a*) Why must the source feeding a class-C amplifier have good regulation? (*b*) What would happen to $E_{c(max)}$ if the excitation were obtained from a high-resistance source?

30. With reference to Table 4–2, what is the $E_{c(max)}$ value for (*a*) the triode? (*b*) the beam power tube? (*d*) Which type of tube has the lower excitation requirements? Give substantiating data.

31. We wish to measure the dc component of grid current in a class-C amplifier.

Where should the milliammeter be placed, and to which side is the positive terminal of the instrument connected in (*a*) Fig. 4–1? (*b*) Fig. 4–3? (*c*) Fig. 4–4?

32. (*a*) What condition could cause a grid current reading to fall below its rated value? Why? (*b*) Give a second reason for low grid current.

33. (*a*) Why must the plate power-supply sources for class-C amplifiers have good regulation? (*b*) What type of rectifier is preferable for such units? (*c*) What type of filter is used? Why?

34. What type of plate circuit is shown in (*a*) Fig. 4–5(a)? (*b*) Fig. 4–5(b)? (*c*) Why is it so called? (*d*) Why is an RF choke used in the latter diagram? (*e*) Why is capacitor C also necessary? (*f*) State two advantages of the circuit in Fig. 4–5(b) over the circuit in Fig. 4–5(a).

35. State three functions of the tank circuits in Fig. 4–5.

36. Refer to Fig. 4–6(a). (*a*) What does R_1 represent? (*b*) If R_1 is a very low value, will the tank circuit impedance match the tube? (*c*) Will much power be transferred to the tank circuit? (*d*) If the resistance value R_1 is deliberately increased, can the tank circuit impedance be made to match the tube? (*e*) Will this produce maximum power transfer *to the tank circuit*? (*f*) Is this the best value for R_1? Explain.

37. Refer to Fig. 4–6(b). (*a*) What does R_r represent? (*b*) What is the effect of this added quantity? (*c*) How can this tank circuit impedance be made to match the tube? (*d*) When power is transferred to this tank circuit, what happens to the power dissipated in R_1? (*e*) What happens to the power "dissipated" in R_r? (*f*) For high efficiency, what relation should exist between the values of R_1 and R_r?

38. (*a*) With respect to efficiency, should the unloaded tank circuit Q be high or low? Why? (*b*) Should the loaded tank Q be high or low? (*c*) With respect to harmonic suppression, should the loaded tank Q be high or low? (*d*) What Q values are commonly used? (*e*) How is this range of values arrived at?

39. Compare the two circuits of Example 4–2 ($R_r = 10\ \Omega$ and $R_r = 190\ \Omega$), with respect to (*a*) efficiency and (*b*) harmonic suppression.

40. Two tubes are to be used as class-C amplifiers. The tube types are different, the operating potentials and power outputs are different, but the operating frequencies are the same. Can identical values be used for the plate tank L and C values? Explain.

41. (*a*) Can the coupling shown in Fig. 3–1(a) be used for a class-C power amplifier? (*b*) What type of plate circuit would this be? (*c*) What would the function of R_g be?

42. Would the coupling shown in Fig. 3–1(b) be used for a class-C amplifier? Explain.

43. (*a*) Can the *type* of coupling shown in Fig. 3–6 be used with class-C amplifiers? (*b*) How is this coupling modified for class-C service?

44. Is the coupling shown in Fig. 3–7 used between class-C stages?

45. What type of coupling is used when the class-C stages are not on the same chassis?

46. The coupling network shown in Fig. 2–14 is to be used between the final class-C amplifier and its load. To what are the leads marked *From Source* connected (*a*) for a series-fed plate circuit? (*b*) for a shunt-fed plate circuit?

47. What type of coupling circuit is best suited for matching the final stage into (*a*) a high-impedance load? (*b*) a moderate-impedance load?

48. Refer to Fig. 4–7. (*a*) What type of interstage coupling is being used? (*b*) What added function does R_1 serve? (*c*) Is any other type of bias used? If so, what type? (*d*) What components produce this? (*e*) Why is this feature used? (*f*) What is the function of C_6? (*g*) Why are components R_2, R_3, C_3, and C_4 needed? (*h*) For what type of load is this circuit suited? Why? (*i*) Could C_8 be shifted into the top lead of L_3? (*j*) Which connection would be preferable? Why?

49. For what reason would a class-C stage use more than one tube, connected in parallel?

50. Compare the operation of two tubes in parallel to a single tube, with regard to (*a*) the plate current requirement (dc), (*b*) the plate supply voltage requirement, (*c*) the load impedance requirement, (*d*) power output, (*e*) the excitation voltage requirement, and (*f*) grid driving power.

51. What is a disadvantage of using tubes in parallel?

52. Refer to Fig. 4–8. (*a*) What type of plate circuit is used in the driver stage? (*b*) What is L_1? (*c*) Should its impedance match the tube? Explain. (*d*) Should the instrument marked mA be an ac or dc instrument? What does it measure? (*e*) What type of bias circuit is used with V_1 and V_2? (*f*) Describe capacitor C_4. (*g*) What is the advantage of using this type over a standard variable capacitor? (*h*) What type of plate circuit is used in the push-pull stage? (*i*) What type of load will this circuit match?

53. The circuit of Fig. 4–8 is to be changed to grid-leak bias. (*a*) To what should the center tap of L_3 be connected? (*b*) How many grid-leak resistors are needed? (*c*) Between what points should they be connected? (*d*) Are any other components needed? (*e*) Where are these connected?

54. The circuit of Fig. 4–8 is to be converted to shunt-fed fixed bias. (*a*) To what should the center tap of L_3 be connected? (*b*) Are any capacitors needed? How many? (*c*) Between what points should they be connected? (*d*) Are any other components needed? (*e*) Between what points are these connected?

55. Compare the operation of a push-pull stage with the same tubes used in parallel, with respect to (*a*) excitation voltage needs, (*b*) driving power needs, (*c*) plate current (dc) needs, (*d*) plate supply voltage needs, (*e*) power output, and (*f*) harmonic suppression.

56. Which circuit (parallel or push-pull) can be designed for higher plate-circuit efficiency? Why?

57. (*a*) What tube types require neutralization when used at radio frequencies? (*b*) What parameter of the tube makes neutralization necessary? (*c*) What is the effect of this parameter? (*d*) How does neutralization "cure" this effect?

58. Refer to Fig. 4–9. (*a*) Give another name for this circuit. (*b*) What is C_3? (*c*) Is it a "real" capacitor? (*d*) Where is it located? (*e*) What is C_N? (*f*) Why is the supply voltage fed to a tap on coil L_3 instead of at the bottom? (*g*) For optimum results, must the tap be at the center of the winding? Explain. (*h*) If the tap is at the center, what value of capacitance is required for C_N?

59. Refer to Fig. 4–10. (*a*) Give another name for this circuit. (*b*) What type of bias circuit is this? (*c*) What type of plate circuit is this? (*d*) For what type of load is this circuit best suited? (*e*) What is C_4? (*f*) Compare the phase of the voltages fed back through C_4 and C_N. (*g*) How is cancellation effected?

60. Refer to Fig. 4–11. (*a*) Why are both capacitors in the neutralizing circuit made variable and marked C_N? (*b*) Must either or both taps (L_2 and L_3) be at the center of the windings? Explain. (*c*) What type of bias circuit is this? (*d*) What type of plate circuit is this? (*e*) Could split-stator capacitors be used for C_2 or C_3?

61. (*a*) When checking a class-C amplifier for proper neutralization, what value of plate supply voltage should be used? (*b*) Should excitation be applied? If so, how much? (*c*) Under this condition, what should the RF output be? Why? (*d*) Name four devices that can be used to indicate the presence of RF in the output tank. (*e*) How are these devices energized? (*f*) Using such a device, how is proper neutralization obtained? (*g*) What plate potential should be applied to any stages following the stage being checked? Why?

62. A class-C amplifier is being neutralized and an output indicator is attached. Zero output cannot be obtained. (*a*) What could cause such a condition? (*b*) How is proper neutralization obtained?

63. (*a*) What indication—other than output—can be used to check for neutralization? (*b*) What value of plate supply voltage is used with this technique? (*c*) How is neutralization effected?

64. (*a*) What relation exists between a parasitic oscillation and the operating frequency? (*b*) Are parasitics desirable in a class-C amplifier? Why? (*c*) What, basically, is responsible for this action?

65. (*a*) Can parasitic oscillations occur at frequencies below the operating frequency? (*b*) What type of circuit components are mainly responsible for this effect?

66. Refer to Fig. 4–12(b). (*a*) Is this the high-frequency or low-frequency *equivalent* diagram of Fig. 4–12(a)? (*b*) Why is the lower end of RFC_1 shown as at ground potential? (*c*) Why can the bottom (or left) side of C_2 be shown at ground potential? (*d*) Why can C_4 be shown as across RFC_2? (*e*) Why is the tank circuit C_3L_2 ignored in this diagram? (*f*) Under what conditions would this circuit produce oscillations? (*g*) How can such oscillations be prevented?

67. State three circuit effects that can contribute to parasitic oscillations at high frequencies.

68. Refer to Fig. 4–12(c). (*a*) What does this circuit represent? (*b*) At what relative frequency does this apply? (*c*) What do each of these quantities represent: L_g? C_{gk}? C_{gp}? C_s? C_{pk}? L_p? (*d*) Under what conditions will this circuit produce oscillations? (*e*) State two ways by which oscillations may be prevented (without resorting to additional components).

69. When resistors are used as parasitic suppressors, (*a*) where would they be connected in the circuit of Fig. 4–12(a)? (*b*) What *physical* location is best?

(*c*) What range of resistance value is used for this purpose? (*d*) How is excessive power loss prevented?

70. When checking for parasitics, (*a*) what other *prior* adjustment is necessary? Why? (*b*) What value of excitation is used for this check? Why? (*c*) Is the plate potential removed? Why? (*d*) If the stage uses grid-leak bias, what precaution is necessary? Why? (*e*) Can the grid current reading be used as an indication of parasitics? Explain. (*f*) Is any other indication possible? Explain.

71. When a class-C stage has been properly designed, (*a*) what two adjustments are necessary for maximum power output? (*b*) In a multistage amplifier, what is the proper sequence of adjustments?

72. Refer to Fig. 4–13. (*a*) Why is the plate current high for the no-load condition when off resonance? (*b*) Why does the no-load current drop to such a low value at resonance? (*c*) Why does the resonant value of current increase at light load? (*d*) What condition does the heavy (dot-dash) curve represent? (*e*) Why is the plate current value, off resonance, not affected by change in load?

73. If excessive loading is applied to a properly tuned class-C stage, what happens to (*a*) the plate current? (*b*) the power output? (*c*) the plate dissipation?

74. When making tuning adjustments, (*a*) why is transmission of signals undesirable? (*b*) What device is used to prevent such transmissions? (*c*) Where is it connected? (*d*) What does it consist of?

75. Refer to Fig. 4–14. (*a*) What is the function of V_1? V_2? V_3? (*b*) When starting the adjustment, what should be the position of switch S_1? S_2? S_3? (*c*) Explain why these settings are necessary. (*d*) How can we tell when C_1 has been properly set? Explain. (*e*) Can S_1 now be set for full E_{bb}? Explain. (*f*) What value of I_{c2} indicates proper setting of C_2? Explain. (*g*) Can any other indication be used in setting C_2? Explain. (*h*) Is there any reason to recheck the C_1 setting? Explain. (*i*) Can S_1 now be set to the full E_{bb} position? Explain. (*j*) Is there any need to recheck the C_1 and C_2 settings? Why? (*k*) Can S_2 now be set for full E_{bb}? Explain. (*l*) After C_3 has been adjusted, what determines if the tap position on L_3 should be changed? (*m*) Is this check made on low or full E_{bb}? Explain. (*n*) If a change in the tap position is necessary, should the C_3 adjustment be rechecked? Explain.

76. Assume that the grid circuit and neutralizing adjustments in Fig. 4–14 have been properly made and we are ready to adjust the output circuit of V_3. (*a*) What is the proper setting for S_2? Why? (*b*) What is the proper setting for S_3? Why? (*c*) When the adjustment is begun, how is the couplıng between L_4 and L_5 set? (*d*) What is adjusted now? (*e*) How can we tell when it is properly set? (*f*) Is it safe to apply full plate voltage now? Give two reasons. (*g*) As the coupling between L_4 and L_5 is tightened, what other circuit components must be adjusted? (*h*) What is each of these components adjusted for? (*i*) When proper coupling has been achieved, what should I_{b3} and I_A register? (*j*) If the coupling is increased further, what happens to I_{b3}? Why? (*k*) What happens to I_A? Why?

77. (*a*) What is meant by a "straight-through" amplifier? (*b*) If an amplifier is not of this type, to what other classification does it belong? (*c*) What class of operation is best suited for this latter service?

78. (*a*) When a class-C stage is used for frequency multiplication, what relation exists between the order of multiplication and the operating bias? (*b*) Why is this so? (*c*) What is the optimum bias for operation as a doubler? tripler? quadrupler?

79. An oscillator's frequency is to be raised by a factor of 12. (*a*) Can it be done in one frequency multiplier stage? Explain. (*b*) In a practical case, how many stages would be used, and how much multiplication would be required in each stage?

80. Compare "straight-through" and frequency-multiplier operation with regard to (*a*) excitation requirements. (Why is this so?) (*b*) power output, for the same peak plate current. (Why is this so?) (*c*) plate efficiency.

81. Refer to Fig. 4–7. (*a*) What *circuit* changes would be necessary to convert this to a frequency multiplier? (*b*) What *component* changes would be necessary?

82. Refer to Fig. 4–10. (*a*) What circuit changes would be *necessary* to convert to multiplier service? (*b*) What circuitry change can be made? (*c*) What component and/or supply values should be changed?

83. (*a*) Is the circuit of Fig. 4–8 suitable for frequency doubling? Explain. (*b*) If the output level desired is greater than a single tube can deliver, what circuitry can be used? (*c*) What other advantage does this latter circuit have?

84. Refer to Fig. 4–15. (*a*) What type of circuit is used on the input side? (*b*) What type is used on the output side? (*c*) How many plate current pulses are produced for each cycle of output?

85. Refer to Fig. 4–16(a). (*a*) What do the dashed-line waveshapes represent? (*b*) What do the solid-line waveshapes represent? (*c*) Which waveshapes are preferable, dashed or solid? Why?

86. Refer to Fig. 4–17. (*a*) Which is the AM wave? (*b*) Compared to the unmodulated wave, what is the maximum amplitude of this wave? (*c*) What is the minimum amplitude? (*d*) What percent modulation does this represent?

87. Refer to Fig. 4–18. (*a*) What value of bias (qualitative) is used? (*b*) What class of operation is this? (*c*) What type of input signal is used? (*d*) Which cycles of the input signal, if any, will produce an undistorted output? Why? (*e*) Which cycles, if any, will result in distortion of the output? Why? (*f*) What change in the operating conditions would eliminate this distortion? (*g*) Why is this so? (*h*) What class of operation would this be now?

88. (*a*) Is the circuit of Fig. 4–10 suitable for use as a class-B RF power amplifier? (*b*) What component and/or supply values must be changed? (*c*) Is the circuit of Fig. 4–9 suitable for class-B use? Explain.

89. In a class-B amplifier with an AM input signal, (*a*) Does $E_{c(\max)} = E_{b(\min)}$ for each RF cycle? Explain. (*b*) For which cycle of Fig. 4–17(b) does this condition apply? (*c*) How does this affect the plate efficiency? (*d*) What is the angle of conduction of the plate current pulses? (*e*) How does this affect the plate efficiency? (*f*) Compare the average efficiency of this amplifier to class C with unmodulated input.

90. (*a*) Name an amplifier specifically designed for high efficiency with an amplitude-modulated input. (*b*) How many tubes are used in one stage of this type of amplifier? (*c*) What class of operation is used with each tube? (*d*) Which tube conducts for signal levels below the unmodulated carrier level? (*e*) Which tube conducts for signal levels above this value? (*f*) What is the average efficiency obtainable with this amplifier?

91. (*a*) In what applications are transistor RF power amplifiers most likely to be found? (*b*) Why is their use limited to such service?

92. In transistor RF power amplifiers, (*a*) what class of operation would theoretically give best efficiency? (*b*) Is such operation completely obtainable in practice? Explain. (*c*) Why may some lower bias level be better?

93. Refer to Fig. 4–19. (*a*) What class of operation is intended in each of these three cases? (*b*) How is the proper bias obtained in diagram (a)? (*c*) What type of transistor is shown in Fig. 4–19(b)? (*d*) What is the function of R_1C_1 in this diagram? (*e*) Explain how it accomplishes this result. (*f*) Would reverse bias be obtained if this were a PNP transistor? Explain. (*g*) What is the function of R_2C_2 in Fig. 4–19(c)? (*h*) Explain how it accomplishes this result.

94. In the circuit of Fig. 4–19(b), what would happen to the bias and to the transistor itself, if the excitation voltage were accidentally removed? Explain.

95. Refer to Fig. 4–20. (*a*) What does BV_{CEO} stand for? (*b*) To get more power output, is it permissible to increase the V_{CC} and/or the V_{cem} values? Explain. (*c*) Would the conduction angle be increased, decreased, or remain unaffected if the dc bias, V_{BE}, were made more negative and the excitation voltage, v_{be}, increased to maintain the same $V_{BE(\mathrm{max})}$? Explain. (*d*) Is this technique for increasing efficiency permissible in this case? (*e*) What is V_γ called? (*f*) What does it represent? (*g*) Give two typical values for V_γ.

96. (*a*) Give an equation for the (approximate) power output obtainable from a transistor for a given collector supply voltage and load resistance. (*b*) Can the power output exceed this value? Explain. (*c*) In a typical transistor circuit, what limitation does this relationship impose on load resistance values? (*d*) Does the above equation apply to vacuum-tube amplifiers? (*e*) Does the load resistance limitation also apply to vacuum tubes? Explain.

97. (*a*) What type of neutralizing circuit is generally used with transistor RF power amplifiers? (*b*) Why is this so?

98. Refer to Fig. 4–21(a). (*a*) Why are the collectors connected to taps on their respective tank circuits? (*b*) What type of coupling is used between units? (*c*) What class of operation is used with transistor Q_2? (*d*) How is this bias obtained? (*e*) State two functions for capacitor C_3.

99. Refer to Fig. 4–21(b). (*a*) What type of coupling is used between units? (*b*) What type of circuit is used on the secondary side? Why? (*c*) What class of operation is used with transistor Q_2? (*d*) What base polarity—with respect to emitter—would produce this effect? (*e*) What circuit components contribute to this bias? (*f*) Explain how. (*g*) What other effect contributes to this bias? (*h*) What is the purpose of L_3? (*i*) Is its dc resistance of any importance? Explain. (*j*) What type of output circuit is used?

100. Refer to Fig. 4–22. (*a*) What type of circuit is used to connect the transistor to its load, R_L? (*b*) When is this type of coupling preferred?

101. (*a*) How does the *schematic* for a transistor frequency doubler differ from a straight-through class-C amplifier? (*b*) State two respects in which operating values would differ.

102. (*a*) What is a varactor diode? (*b*) What property makes it useful as a frequency multiplier? (*c*) How much gain will this circuit provide? (*d*) When and why is this circuit preferable to a transistor frequency multiplier?

PROBLEMS AND DIAGRAMS

1. A class-C amplifier is operated with a grid bias of −125 V and a peak excitation voltage of 255 V. Through what values will the instantaneous grid voltage swing?

2. A class-C amplifier is operated with a bias of −120 V, a plate potential of 600 V, and an excitation voltage (rms) of 140 V. Find (*a*) the maximum value of the grid voltage swing and (*b*) the optimum value of the output voltage E_p (rms). (Save this answer for Problem 7.)

3. In Fig. 4–2, the bias value is −200 V, the plate potential is 3000 V, and the grid is to be driven to +160 V. (*a*) What value of excitation voltage will be needed? (*b*) What is the optimum value of output voltage obtainable? (Save this answer for Problem 8.)

4. A class-C RF power amplifier is operated under the following conditions: $E_{bb} = 3300$ V, $E_{cc} = -600$ V, $I_b = 300$ mA, $I_c = 40$ mA. It produces an RF output of 780 W with a grid driving power of 34 W. Find (*a*) the plate power input, (*b*) the plate dissipation, and (*c*) the plate efficiency.

5. The following values are read from a class-C amplifier: plate voltage = 1000 V, grid bias = −175 V, excitation voltage = 315 V, plate current = 150 mA, grid current = 20 mA, power output = 105 W. Calculate (*a*), (*b*), and (*c*) as in Problem 4.

6. The following data are listed in a tube manual for a class-C power amplifier:

dc plate voltage	400 V	dc plate current	95 mA
dc screen voltage	250 V	dc screen current	9 mA
dc grid voltage	−50 V	dc grid current	2.5 mA
Peak RF grid voltage	80 V	Power output	30 W

Calculate (*a*), (*b*), and (*c*) as in Problem 4.

7. The tube in Problem 2 draws a plate current (dc) of 200 mA and produces a fundamental current component of 350 mA peak. Find (*a*) the plate power input, (*b*) the plate power output, (*c*) the plate dissipation, and (*d*) the plate efficiency.

8. The tube in Problem 3 draws a plate current of 410 mA (dc) and produces a power output of 1000 W. Find (*a*) the plate power input, (*b*) the plate efficiency, (*c*) the plate dissipation, and (*d*) the fundamental component of plate current (rms value).

9. A class-C amplifier draws 320 mA from the dc supply. The conduction period is 140°. Find (*a*) the peak value of plate current and (*b*) the rms value of the fundamental component.

10. The amplifier of Problem 9 is to be converted to service as a frequency tripler. Its conduction angle is accordingly reduced to 80°. (*a*) Find the new value of peak plate current if the dc value is maintained at 320 mA. (*b*) What is the value of the fundamental component now?

11. A tube has a peak plate-current rating of 1.60 A. Find the ac component of the plate current when the tube is operated at its peak rating and at conduction angles of (*a*) 150°, (*b*) 120°, and (*c*) 80°.

12. In Problem 4, calculate the value of resistor needed for (*a*) the grid resistor, if the full bias is obtained by the grid-leak method and (*b*) the grid resistor, if one-half the bias is obtained from a fixed-bias source.

13. Calculate the value of resistor needed in Problem 5 for (*a*) the grid resistor for full grid-leak bias, (*b*) the grid resistor if one-third of the bias is obtained by cathode bias, and (*c*) the cathode resistor needed in step (*b*).

14. Using the data of Problem 6, repeat steps (*a*), (*b*), and (*c*) of Problem 13.

15. A 5770 power triode is operated as a class-C amplifier, with $E_{bb} = 17{,}000$ V, $E_{cc} = -1450$ V, $E_{gm} = 2375$ V, and a differential of 75 V between $E_{c(\max)}$ and $E_{b(\min)}$. (*a*) Tabulate the instantaneous values of plate and grid potentials for 10° intervals between 0–90°. (*b*) From the plate characteristics of Fig. 4-24, find the instantaneous values of plate current for each of the above angles. Take an additional step halfway between the last point of zero current and the first current reading. (*c*) Plot current versus time for 180°. (*d*) What is the angle of plate current flow? (*e*) Find the values of I_b and I_p. (*f*) Calculate the power output. (*g*) Calculate the dc power input. (*h*) Calculate the plate efficiency. (*i*) Find the load impedance of the tank circuit for full power output. (*j*) Using a Q of 10 and an

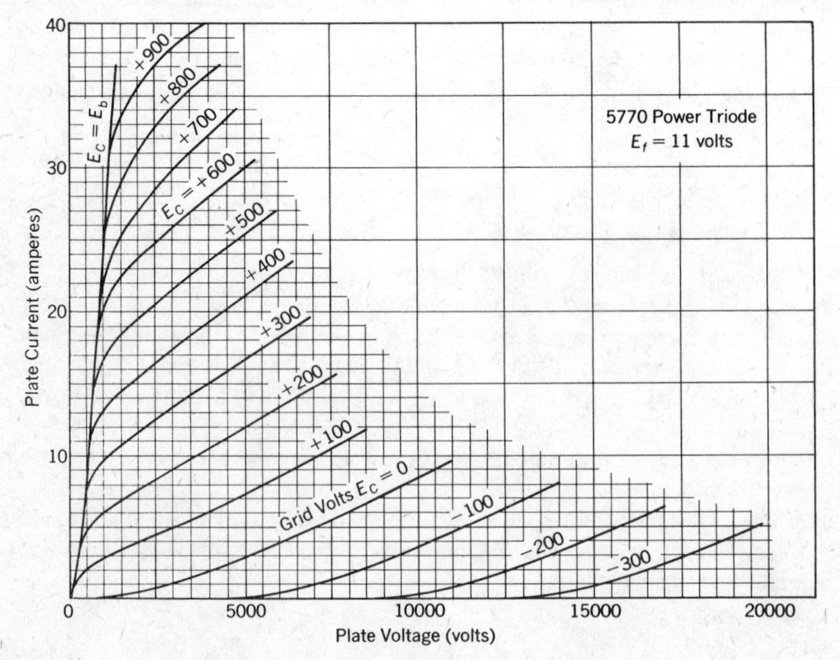

Figure 4-24 Plate characteristics for problem 15.

operating frequency of 800 kHz, solve for the proper values of L and C for the tank.

16. A driver stage with an output resistance of 5000 Ω feeds 180 V of peak excitation to a class-C amplifier. The input resistance of the class-C stage drops from (approximate) infinity to 800 Ω when the grid is driven positive. Treating this as a simple series circuit (neglecting the dc bias action), find the grid input signal (peak) when the grid is (*a*) negative and (*b*) positive. (*c*) What value of driver stage output resistance would keep the regulation to within 10%?

17. A class-C amplifier plate tank coil has a reactance of 1200 Ω and an effective resistance of 8 Ω at 2000 kHz. When load is applied, a reflected resistance of 12 Ω is introduced into the tank circuit. Find (*a*) the unloaded Q, (*b*) the loaded Q, and (*c*) the tank efficiency. (*d*) What value of reflected resistance would produce a Q of 12? (*e*) What is the tank efficiency now?

18. A tank circuit designed to operate at 10 MHz has an inductance of 5.0 μH and an effective resistance of 1.96 Ω. The tank circuit impedance when loaded is 6000 Ω. Find (*a*) the reflected resistance due to loading and (*b*) the tank efficiency.

19. The following typical operating data are listed for a GL-810 RF power amplifier, class-C telegraphy:

dc plate voltage	2000 V	dc grid current	40 mA
dc grid voltage	−160 V	Driving power	19 W
Peak RF grid voltage	330 V	Power output	375 W
dc plate current	250 mA		

Find (*a*) the output voltage, allowing a 20-V safety differential between maximum grid and minimum plate potentials, (*b*) the required tank circuit impedance, (*c*) the tank circuit inductance for a Q of 12 and an operating frequency of 1200 kHz, and (*d*) the tank circuit capacitance.

20. Repeat (*a*), (*b*), (*c*), and (*d*) in Problem 19, using the tube data of Problem 6 and an operating frequency of 600 kHz.

21. Repeat (*a*) through (*d*) of Problem 19 for a 5770 power triode, using a safety differential of 75 V and an operating frequency of 2.5 MHz.

dc plate voltage	17,000 V	dc grid current	1.1 A
dc grid voltage	−1450 V	Driving power	2300 W
Peak RF grid voltage	2375 V	Power output	105 kW
dc plate current	8.5 A		

22. Draw the circuit diagram for a beam power tube used as a class-C RF power amplifier with single-tuned inductively coupled input, series-fed grid-leak bias, and a shunt-fed plate circuit coupled to a low-impedance load.

23. Repeat Problem 22 using impedance-coupled input, shunt-fed fixed bias, and a series-fed plate circuit coupled to a high-impedance load.

24. Repeat Problem 22 using a double-tuned inductively coupled input circuit, with shunt feed to the previous stage, series-fed fixed bias, and a π network coupling to the load.

25. Draw the circuit diagram for a push-pull beam power class-C stage using double-tuned inductive-coupled input, grid-leak bias, shunt-fed plate tank, and coupled to a low-impedance load.

26. Draw the circuit diagram for a class-C triode RF power amplifier using impedance-coupled input, fixed bias, plate neutralization, and a series-fed plate tank.

27. Repeat Problem 26 using a double-tuned inductively coupled input, grid-leak bias, plate neutralization, and a shunt-fed plate tank.

28. Repeat Problem 27 using grid neutralization.

29. Draw the circuit diagram for a push-pull triode class-C RF power amplifier with neutralization, using grid-leak bias and a series-fed tank circuit.

30. Repeat Problem 29 using shunt-fed fixed bias and a shunt-fed plate tank.

31. Draw the circuit diagram of a frequency-tripler circuit using a beam power tube, with single-tuned inductively coupled input, shunt-fed fixed bias, and a series-fed plate tank.

32. Draw the circuit diagram of a push-push triode frequency doubler using grid-leak bias and a shunt-fed plate tank.

33. Draw the circuit diagram for a class-B RF power amplifier to be used with AM input signals and featuring single-tuned inductive coupling in the input and a shunt-fed plate tank with high-impedance load in the output.

34. An MSA 8506 NPN silicon power transistor has a breakdown voltage of 36 V. It is desired to produce a power output of 25 W. (*a*) What is the maximum supply voltage that can be used? (*b*) What value of load resistance will produce the desired power output?

35. Repeat (*a*) and (*b*) of Problem 34 for a 3TE611 transistor, with a breakdown voltage of 60 V, and for a power output of 75 W.

36. Repeat (*a*) and (*b*) of Problem 34 for an MM1558 transistor, with a breakdown voltage of 65 V, and for a power output of 20 W.

37. An RF power amplifier is to be constructed for an output of 100 W. What approximate value of load resistance would be needed, using (*a*) a transistor with a 15-V supply and (*b*) a vacuum tube with a 200-V supply.

38. Draw the circuit diagram for a PNP transistor RF power amplifier using impedance coupling and base-return bias.

39. Draw the circuit diagram for an NPN transistor RF power amplifier using double-tuned inductive coupling and internal (V_{BE}) biasing.

40. Draw the circuit diagram for a transistor RF power amplifier with inductive-coupled input, emitter bias, and with collector at chassis ground for maximum dissipation.

41. An MSA 8506 transistor is used as an RF power amplifier at 115 MHz. The power supply voltage is 18 V. It is to deliver a power output of 22 W to a 50-Ω resistive load through a matching T network (see Fig. 4–22). The transistor output capacitance is 40 pF. Calculate the component values for the T network using a Q of 12.

42. Find the values needed for a matching T network with a Q of 10.5, using a 3TE611 transistor (output capacitance 125 pF), a 28-V power supply, and to deliver 50 W at 88 MHz to a 72-Ω antenna (resistive) load.

43. Find the values for a matching T network for an MM1558 ($C_o = 55$ pF), a 28-V supply, and a power output of 20 W at 60 MHz to a 50-Ω resistive load. Use a Q of 10.0.

TYPICAL FCC QUESTIONS

3.083 Draw a simple schematic diagram showing a method of coupling the radio-frequency output of the final power amplifier stage of a transmitter to an antenna.

3.090 Draw a simple schematic circuit of a radio-frequency doubler stage indicating any pertinent points which will distinguish this circuit as that of a frequency doubler.

3.104 Describe the characteristics of a vacuum tube operating as a class-C amplifier.

3.105 During what approximate portion of the excitation voltage cycle does plate current flow when a tube is used as a class-C amplifier?

3.107 Describe the characteristics of a tube operating as a class-B amplifier.

3.108 During what portion of the excitation voltage cycle does plate current flow when a tube is used as a class-B amplifier?

3.110 What is the meaning of the term "maximum plate dissipation"?

3.122 Draw a grid voltage-plate current characteristic curve of a vacuum tube and indicate the operating points for class-A, class-B, and class-C amplifier operation.

3.126 Why is it important to maintain transmitting tube filaments at recommended voltages?

3.127 When an alternating current filament supply is used, why is a filament center-tap usually provided for the tube plate and grid return circuits?

3.138 Draw a circuit of a "frequency doubler" and explain its operation.

3.139 For what purpose is a "doubler" amplifier stage used?

3.142 Given the following vacuum tube constants: $E_b = 1000$ volts, $I_b = 150$ mA, $I_c = 10$ mA, and grid leak = 5000 ohms, what would be the value of dc grid bias voltage?

3.143 Explain how you would determine the value of cathode bias resistance necessary to provide correct grid bias for any particular amplifier.

3.277 What is the approximate efficiency of a class-A vacuum tube amplifier? class-B? class-C?

3.279 Why must some radio-frequency amplifiers be neutralized?

3.307 Explain the purposes and methods of neutralization in radio-frequency amplifiers.

3.362 Draw a simple schematic diagram of a system of neutralizing the grid-plate capacitance of a single electron tube employed as a radio-frequency amplifier.

3.365 What are the characteristics of a "frequency-doubler" stage?

3.368 What is the primary purpose of a grid-leak in a vacuum-tube transmitter?

3.370 What may be the result of parasitic oscillations?

3.371 How may the production of harmonic energy by a vacuum-tube radio-frequency amplifier be minimized?

3.372 What is a definition of "parasitic oscillations"?

3.373 What is the purpose of a "Faraday" screen between the final tank inductance of a transmitter and the antenna inductance?

3.374 How may the distortion effects caused by class-B operation of a radio-frequency amplifier be minimized?

3.380 What is the purpose of a "dummy antenna"?

3.381 In a class-C radio-frequency amplifier stage of a transmitter, if plate current continues to flow and radio-frequency energy is still present in the antenna circuit after grid excitation is removed, what defect would be indicated?

3.383 What are some possible causes of overheating vacuum-tube plates?

3.439 What are the advantages of using a resistor in series with the cathode of a class-C radio-frequency amplifier tube to provide bias?

3.440 How may the generation of even harmonic energy in a radio-frequency amplifier stage be minimized?

3.441 What tests will determine if a radio-frequency power amplifier stage is properly neutralized?

3.442 Why is the plate circuit efficiency of a radio-frequency amplifier tube operating as class C higher than that of the same tube operated as class B? If the statement above is false, explain your reasons for such a conclusion.

3.447 Why does a screen grid tube normally require no neutralization when used as a radio-frequency amplifier?

3.448 What instruments or devices may be used to adjust and determine that an amplifier stage is properly neutralized?

3.463 In the adjustment of a radiotelephone transmitter, what precautions should be observed?

3.469 How would loss of radio-frequency excitation affect a class-C modulated amplifier when using a grid-leak bias only?

3.470 What is the purpose of a center-tap connection on a filament transformer?

3.471 What would be the result of a short circuit of the plate radio-frequency choke coil in a radio-frequency amplifier?

3.473 What class of amplifiers is appropriate to use in a radio-frequency doubler stage?

3.481 Compare the design and operating characteristics of class-A, class-B, and class-C amplifiers.

3.484 Explain how grid bias voltage is developed by the grid leak in an oscillator.

3.485 Explain why radio-frequency chokes are sometimes placed in the power leads between a motor-generator power supply and a high power radio transmitter.

3.487 In what circuits of a radio station are three-phase circuits sometimes employed?

3.507 Explain the process of neutralizing a triode radio-frequency amplifier.

4.017 In a class-C radio-frequency amplifier, what ratio of load impedance to dynamic plate impedance will give the greatest plate efficiency?

4.060 Draw a schematic diagram of a final amplifier with capacity coupling to the antenna which will discriminate against the transfer of harmonics.

4.081 Draw a diagram of a class-B push-pull linear amplifier using triode tubes. Include a complete antenna coupling circuit and antenna circuit. Indicate points at which the various voltages will be connected.

4.084 If an oscillatory circuit consists of two identical tubes, the grids connected in push-pull and the plates in parallel, what relationship will hold between the input and output frequencies?

4.087 If, upon tuning the plate circuit of a triode RF amplifier, the grid current undergoes variations, what defect is indicated?

4.088 A 50-kilowatt transmitter employs six tubes in push-pull parallel in the final class-B linear stage, operating with a 50-kilowatt output and an efficiency of 33 percent. Assuming that all of the heat radiation is transferred to the water cooling system, what amount of power must be dissipated from each tube?

4.120 Indicate, by a simple diagram, the shunt-fed plate circuit of a radio-frequency amplifier.

4.121 Indicate, by a simple diagram, the series-fed plate circuit of a radio-frequency amplifier.

4.124 In adjusting the plate tank circuit of a radio-frequency amplifier, should minimum or maximum plate current indicate resonance?

4.128 Under what circumstances is neutralization of a triode radio-frequency amplifier not required?

4.129 Why is it necessary or advisable to remove the plate voltage from the tube being neutralized?

4.140 What is the last audio-frequency amplifier stage which modulates the radio-frequency stage termed?

4.148 Define the plate input power of a broadcast station transmitter.

4.160 What is the purpose of neutralizing a radio-frequency amplifier stage?

4.188 What is the principal advantage of a class-C amplifier?

4.205 What is a Doherty amplifier?

6.098 What is meant by the "fly-wheel" effect of a tank circuit?

6.100 Define "parasitic oscillations."

6.101 What is the effect of parasitic oscillations?

6.150 What is the approximate efficiency of a class-A vacuum-tube amplifier? class-B? class-C?

6.171 Describe the characteristics of a vacuum tube operating as a class-C amplifier.

6.172 During what approximate portion of the excitation voltage cycle does plate current flow when a tube is used as a class-C amplifier?

6.174 Why are tubes, operated as class-C amplifiers, not suited for audio-frequency amplification?

6.177 Why must some radio-frequency amplifiers be neutralized?

6.188 When an alternating-current filament supply is used, why is a filament center-tap usually provided for the vacuum-tube plate and grid return circuits?

6.189 Explain how you would determine the value of cathode bias resistance necessary to provide correct grid bias for any particular amplifier.

6.190 Given the following vacuum-tube constants, $E_b = 1000$ volts, $I_b = 150$ mA, $I_c = 10$ mA, and grid leak = 5000 ohms, what would be the value of dc grid bias voltage?

6.191 A triode transmitting tube, operating with plate voltage of 1250 volts, has filament voltage of 10, filament current of 3.25 amperes and plate current of 150 mA. The amplification factor is 25. What value of control-grid bias must be used for operation as a class-C stage?

6.197 For what purpose is a "doubler" amplifier stage used? Draw the circuit and explain its operation.

6.198 Does a pentode vacuum tube usually require neutralization when used as a radio-frequency amplifier?

6.239 What is the purpose of a "choke" coil?

6.247 During what portion of the excitation voltage cycle does plate current flow when a tube is used as a class-B amplifier?

6.258 Why is it not feasible to employ a vacuum tube operated class-C as an audio amplifier, either singly or in push-pull?

6.261 What is the purpose of decoupling networks in the plate circuits of a multi-stage RF amplifier?

6.416 What is the primary function of the power amplifier stage of a marine radio-telegraph transmitter?

6.417 What is the purpose of a buffer amplifier stage in a transmitter?

6.419 What class of amplifier should be employed in the final amplifier stage of a radiotelegraph transmitter for maximum plate efficiency?

6.420 Under what "class" of amplification are the vacuum tubes in a linear radio-frequency amplifier stage, following a modulated stage, operated?

6.421 If a final radio-frequency amplifier, operated as class-B linear, were excited to saturation with no modulation, what would be the effects when undergoing modulation?

6.422 Define a class-C amplifier.

6.426 What is the function of a grid leak in a class-C amplifier?

6.427 Describe how a radio-frequency amplifier stage may be neutralized. What precautions must be observed.

6.443 In a series-fed plate circuit of a vacuum-tube amplifier, what would be the result of a short circuit of the plate bypass capacitor?

6.444 In a shunt-fed plate circuit of a vacuum-tube amplifier, what would be the result of a short circuit of the plate RF choke coil?

6.448 Name three instruments which may be used as indicating devices in neutralizing a radio-frequency amplifier stage of a transmitter.

6.449 Draw a simple schematic circuit of a radio-frequency doubler stage, indicating any pertinent points which will distinguish this circuit as that of a frequency doubler. Describe its operation.

6.451 What factors are most important in the operation of a vacuum tube as a frequency doubler?

6.453 What precautions should be observed in tuning a transmitter to avoid damage to components?

6.454 Draw a complete schematic diagram of a system of inductive coupling between the output of a radio-frequency amplifier and an antenna system.

6.455 What is the result of excessive coupling between the antenna and output circuits of a self-excited type of vacuum-tube transmitter?

6.457 In a transmitter involving a master-oscillator, intermediate amplifier, and final amplifier, describe the order in which circuits should be adjusted in placing this transmitter in operation.

6.458 Should the antenna circuit of a master-oscillator, power-amplifier type of transmitter be adjusted to the resonant frequency before the plate tank circuit of the final stage? Give the reason(s) for your answer.

6.459 What may cause a radio-frequency amplifier tube to have excessive plate current?

6.464 In a radio-frequency amplifier employing fixed bias, as the plate circuit is varied in adjustment from a point below resonance to a point above resonance, what effect will be observed on the grid current?

6.465 In a self-biased radio-frequency amplifier stage having a plate voltage of 1250, a plate current of 150 mA, a grid current of 15 mA, and a grid leak resistance of 4000 ohms, what is the value of operating grid bias?

6.466 In a series-fed plate circuit of a vacuum-tube amplifier, what would be the effect of a short circuit of the plate supply bypass capacitor?

6.467 In a shunt-fed plate circuit of a vacuum-tube amplifier, what would be the effect of an open circuit in the plate radio-frequency choke?

6.468 Explain how you would determine the value of cathode-bias resistor for a specific amplifier stage.

6.469 Draw a simple schematic diagram showing a method of "link" coupling between two radio-frequency amplifier stages.

6.470 What is the advantage of link coupling between radio-frequency amplifier stages?

6.473 Why should a transmitter be tuned initially at reduced power?

6.474 What is meant by "split tuning"?

6.496 What effect upon the plate current of the final amplifier stage will be observed as the antenna circuit is brought into resonance?

6.497 How may instruments used to indicate various direct currents and voltages in a transmitter be protected against damage due to stray RF energy?

6.521 What is meant by harmonic radiation?

6.523 What is the primary reason for the suppression of radio-frequency harmonics in the output of a transmitter?

6.527 Why is an artificial antenna sometimes used in testing a transmitter? By what other names is this instrument known?

6.577 What is the purpose of a center-tap connection in a filament-supply transformer?

6.579 Draw a simple schematic diagram showing a method of coupling between two triode vacuum tubes in a tuned radio-frequency amplifier and a method of neutralizing to prevent oscillation.

6.584 Draw a diagram showing a method of obtaining grid bias for an indirectly heated cathode-type vacuum tube by use of resistance in the cathode circuit of the tube.

S-3.059 Why is the efficiency of an amplifier operated class C higher than one operated class A or class B?

S-3.120 What is an RFC? Why are they used?

S-3.121 What are the advantages of using a resistor in series with the cathode of a class-C radio-frequency amplifier tube to provide bias?

S-3.122 What is the difference between RF voltage amplifiers and RF power amplifiers with respect to applied bias? What type of tube is generally employed in RF voltage amplifiers?

S-3.123 Draw schematic diagrams of the following circuits and give some possible reasons for their use?

(*a*) Link coupling between a final RF stage and an antenna. (Include a low-pass filter.)

(*b*) Capacitive coupling between an oscillator stage and a buffer amplifier.

(*c*) A method of coupling a final stage to a quarter-wave Marconi antenna other than link or transmission line.

S-3.124 Draw a schematic diagram of a grounded-grid RF amplifier, and explain its operation.

S-3.125 Explain the principle involved in neutralizing an RF stage.

S-3.126 State some indications of, and methods of testing for, the presence of parasitic oscillations in a transmitter.

S-3.127 Draw a circuit diagram of a push-pull (triode) final power amplifier with transmission-line feed to a shunt-fed quarter-wave antenna, and indicate a method of plate neutralization.

S-3.128 Explain, step-by-step, at least one procedure for neutralizing an RF amplifier stage.

S-3.129 Draw a circuit diagram of a "push-push" frequency multiplier, and explain its principle of operation.

S-3.130 Push-pull frequency multipliers normally produce what order of harmonics, even or odd?

S-3.131 Draw a schematic diagram and explain the operation of a harmonic-generator stage.

S-3.132 What class of amplifier is appropriate to use in a radio-frequency doubler stage?

5

RF Oscillators

In discussing resonant and coupled circuits, we spoke of selectivity—the ability of the circuit to select one specific frequency and reject others. We also briefly mentioned how intelligence is raised to a high frequency—via the carrier. The generation of these carrier frequencies is the function of an *oscillator*. Although we speak of oscillators as "generating" a frequency, it should be emphasized that they do not create energy, but merely convert energy from a dc supply source into ac energy at some specific frequency.

Oscillators have a variety of applications in communications equipment. As implied above, they are used in the *exciter* section of a transmitter to generate the carrier frequency. In receivers we will see how a *local oscillator* is used in the *superheterodyne* circuit, or as a *beat frequency oscillator* for detection of code signals. A similar *frequency translation* function is performed by oscillators in one type of frequency-modulated transmitter. However, their use is not restricted to the communications field. Many other electronic devices also require a source of energy at some specific frequency. Some examples of such needs are diathermy machines in medicine, and dielectric or induction heating equipment and timing controls in industry.

GENERAL CONSIDERATIONS

Tank Circuit Action

An RF oscillator generally uses a parallel-resonant *L-C* circuit (tank circuit) to fix the frequency of oscillation. The action of this tank circuit can be appreciated better if we analyze the similar but more

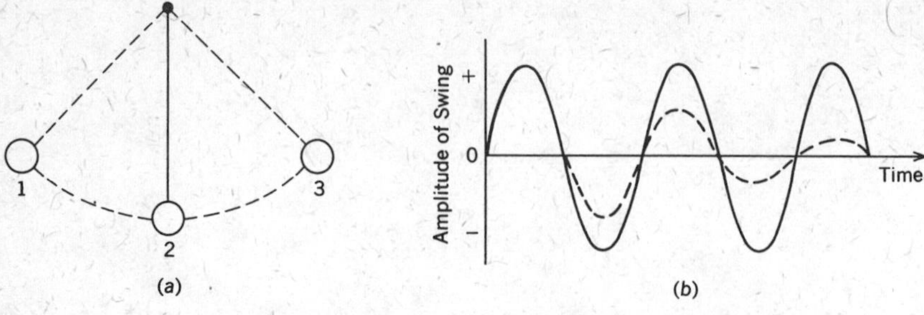

Figure 5-1 Pendulum action.

familiar action that takes place with the pendulum shown in Fig. 5–1(a).

If the pendulum is raised from position 2 to position 1, the pendulum acquires energy because of the work done. This is *energy of position* or *potential energy.* Now, if the pendulum is released, it gains speed as it swings down toward position 2. In losing height, the pendulum loses potential energy, but because of its speed it acquires *kinetic energy* or *energy of motion.* When it reaches position 2, it is back to its original position and the potential energy of the pendulum is zero. However, it is now traveling at its maximum speed. The kinetic energy is a maximum. This energy carries the pendulum beyond position 2. If there were no friction or windage losses, the pendulum would reach the peak of its swing at a position equal in height to position 1, but at the opposite side (that is, position 3). At this point all of the energy is again potential energy, and in the ideal case (no losses) this energy is equal to the original energy at position 1. The pendulum would then swing back to position 1, and the process would be repeated *ad infinitum.* This oscillatory motion of the pendulum is represented graphically by the solid-line sinusoidal plot in Fig. 5–1(b).

If the system had losses (friction at the pivot and windage resistance to the motion), some energy would be lost during each swing and the amplitude of each half-cycle would decrease. This effect is shown by the dashed-line curve in Fig. 5–1(b). This latter condition is known as a *damped* oscillation. The greater the losses, the heavier the damping, and the sooner the pendulum will come to rest.

Now let us consider the effect of introducing energy into an *L-C* circuit. In Fig. 5–2(a) energy is introduced by charging the capacitor as shown. Because of the accumulation of electrons, there is a voltage across the capacitor and the capacitor has *potential energy.* When the switch is thrown to position 1, current starts to flow as the capacitor begins to discharge. Due to the inductive effect, the current builds up slowly toward a maximum value. Maximum current occurs when the capacitor is fully discharged. At this instant, the potential energy of the system is

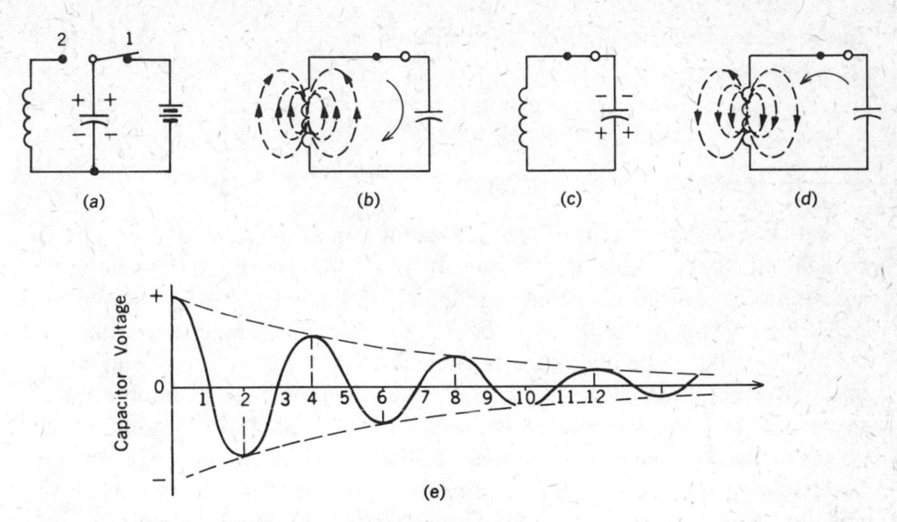

Figure 5-2 Damped oscillations in an *L*/*C* circuit.

zero, but because the electron *motion* is greatest (maximum current), the magnetic field energy around the coil is a maximum. This condition is shown by Fig. 5–2(b). Obviously, this magnetic energy corresponds to the kinetic energy of the pendulum system.

Once the capacitor is fully discharged, the magnetic field will begin to collapse, but the counter EMF will keep current flowing in the same direction, charging the capacitor with opposite polarity as shown in Fig. 5–2(c). As the charge builds up, the current value decreases and the magnetic field energy decreases. In an ideal case, the magnetic field energy drops to zero when the capacitor charges to the value it had in condition (a). Once again all the energy is potential energy. The capacitor now begins to discharge, with the current flowing in the *opposite* direction. Figure 5–2(d) shows the capacitor fully discharged and maximum current flowing. Again all the energy is in the magnetic field. This interchange or oscillation of energy between *L* and *C* is repeated over and over. In a practical circuit, because of losses (resistive and radiation losses in the coil and dielectric losses in the capacitor), the amplitude of these oscillations will die down (see Fig. 5–2(e)). A tank circuit by itself will produce damped oscillations.

In either case—pendulum or tank circuit—the frequency of these oscillations is determined by the "*constants* of the circuit." In the pendulum system, it is the mass of the pendulum and the length of the pendulum arm. The height to which the pendulum is raised has no bearing on the time of swing. In the electrical system, the constants are the inductance and capacitance values. The actual frequency of oscillation is the resonant frequency of the tank circuit, or

$$f_0 = \frac{1}{2\pi\sqrt{LC}} \tag{5-1}$$

The Continuous Wave (CW)

If we wish to maintain continuous motion of the pendulum in a practical situation, it is necessary to supply additional energy to the pendulum from time to time to overcome the effect of the losses. A light tap at the peak of its swing *on either side* should be sufficient. The energy supplied by this tap should be just equal to the energy lost from friction and windage. The pendulum will then maintain a constant height (amplitude) of swing. If the energy imparted is more than is needed, the pendulum will swing to greater heights. In doing so, the losses increase. A balance is reached at some new height when once more the energy of the tap is equalled by the losses. Obviously, the energy supplied will determine the height at which the system stabilizes. Furthermore, it should be realized that the timing (or phasing) of the tap is very important. A tap downward while the pendulum is still on its upswing can actually stop the swing.

The same effect can be produced in a tank circuit. By supplying a spurt or pulse of energy *at the right time* in each cycle, the amplitude of the oscillation can be maintained at a constant level. The resulting "undamped" wave is known as a *continuous wave* (CW). Before we study practical circuits to see how this energy is supplied, let us analyze this action of the tank circuit from ac circuit principles.

L-C Circuit Analysis

To maintain a constant amplitude of oscillation, a pulse of current can be fed to the capacitor *once in each cycle* to recharge it to its original value. Also, since this pulse must be properly timed and must remain properly timed, the pulse must have a specific repetition rate. Such a current pulse is a complex wave having a dc component, a fundamental frequency component, and harmonics of this fundamental frequency. Each of these ac components is a full sine wave of current. It should now be obvious that proper timing of one pulse per cycle is obtained when the fundamental frequency of this current pulse corresponds to the resonant frequency of the tank circuit.

At its resonant frequency, the impedance of the parallel *L-C* circuit is a maximum. The voltage across the tank circuit, IZ, is a maximum. Since this component of current is a sine wave, this voltage is also a sine wave. To any of the harmonic components of current, the tank circuit is far off resonance, and its impedance is practically zero. Therefore, even if the harmonic *current* content is high, the voltage developed at these harmonic frequencies is negligible. The only voltage across the tank circuit is at the resonant frequency of the circuit and is a pure sine wave.

General Requirements

Many varieties of *L-C* oscillator circuits have been devised. Yet regardless of the circuit differences, three features are common to all circuits:

1. They must contain a device (tube or semiconductor) that will amplify, that is, take energy from a dc supply and convert it to increased signal energy in its output circuit, as compared to its input circuit.
2. There must be *positive* feedback from the output circuit to the input circuit. The energy fed back must be in phase with the "signal" energy in the input circuit.
3. The amount of feedback must be large enough to overcome the losses.

BASIC OSCILLATOR CIRCUITS

Most of the circuits that follow were originally designed as vacuum-tube circuits. Since the FET has characteristics very similar to the vacuum tube's, these circuits are also well suited to FET devices. However, with bipolar transistors, because of their relatively low impedance and the resulting loading effect, some modifications were necessary. A few of the circuits shown are specifically designed for the bipolar units. Since the late 1960s, most new low-power designs use semiconductors (FET or bipolar). Yet, tube circuits are still in use and cannot be neglected. In fact, for high power levels, the vacuum tube is the only suitable device. Therefore all three types of circuits will be shown.

The Armstrong Oscillator

Figure 5–3 shows a commonly used oscillator circuit. It is named the Armstrong circuit (after its inventor, Major Armstrong). It is also known as the "tickler feedback circuit" because of the action of coil L_2. The active device here is an N-channel FET. (The original Armstrong circuit used a vacuum tube, with the elements grid, plate, and cathode, respectively, replacing the gate, drain, and source of the FET.) A P-channel unit could be used merely by reversing the power supply polarity.

When the circuit is energized, since the gate bias is zero, the drain current starts to rise to a high value. This rising drain current, flowing through the "tickler" coil L_2, creates an expanding magnetic field that reaches and cuts coil L_1. The coils are so phased that the voltage induced in L_1 at this instant makes the top of L_1 positive. The gate of the FET is

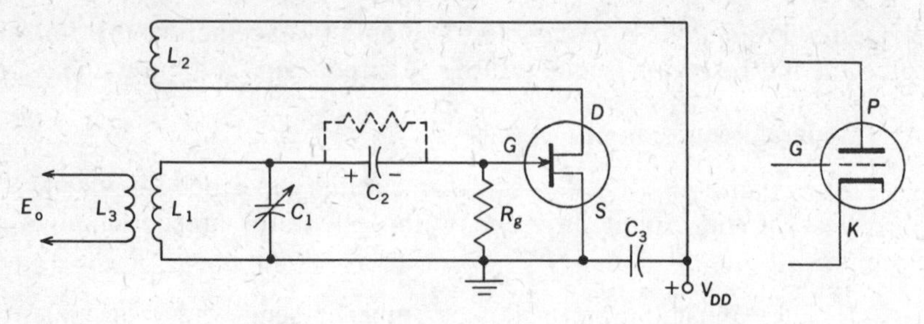

Figure 5-3 Armstrong or tickler feedback oscillator.

driven positive, the drain current increases at a faster rate, and the induced voltage increases. The process is cumulative. Almost immediately a strong positive voltage is built up across the tank circuit and capacitor C_1 charges with a positive polarity on its top plate. At the same time, since the gate of the FET is driven positive, gate current flows, and since the forward-biased gate-source resistance is low, capacitor C_2 tends to charge to the peak value of this induced voltage with the polarity as shown.

As the FET begins to approach saturation, the *rate of rise* of drain current starts to decrease, and even though the drain current is still increasing, the induced voltage will decrease. Capacitor C_2 must discharge. Electrons flowing out of the right plate and down through R_g drive the gate of the FET negative. This action is abetted by a series of almost instantaneous reactions. The drain current starts to decrease, the collapsing magnetic field induces a *negative* voltage in coil L_1, the gate of the FET is driven further negative, and the drain current is cut off. The discharge of capacitor C_2 through the gate resistor R_g keeps the FET well beyond cutoff. Effectively, then, a pulse of drain current is produced, and because of the coupling between L_1 and L_2, energy is fed back from the drain circuit into the tank circuit. Meanwhile, the sharp collapse of the magnetic field when the FET is cut off produces such a strong negative voltage spike in coil L_1 that it discharges the tank capacitor and recharges it with opposite polarity. Therefore, when the FET is cut off, capacitor C_1 is charged with a negative polarity on its top plate, and the induced voltage is zero.

Now, the action of the tank circuit proceeds as described earlier in the chapter, producing a sine wave at its resonant frequency. However, as the positive voltage developed by the oscillations in the tank circuit rises toward its peak value, the FET will come out of cutoff. Once again, conduction starts. The rise in drain current induces a voltage in the tank circuit, recharging C_1 and C_2 to the original values and supplying the energy to overcome the losses. After a few transient cycles, the circuit action is stabilized. The waveshapes can be represented as shown in Fig. 5–4.

Gate-Leak Bias

Notice that gate-leak bias is used in this oscillator circuit (Fig. 5–3). Therefore, when the circuit is first energized, and since there is no input signal, the instantaneous bias value e_g is zero. Then, during the first few transient cycles, and as the oscillations build up to their stabilized level, capacitor C_2 charges to a higher and higher value and finally stabilizes as shown. After this, the action is similar to that of a class-C RF power amplifier, the input signal in this case being supplied from the feedback winding L_2. This self-bias circuit has several advantages. In the first place, it makes the oscillator circuit self-starting. It should be realized that if fixed bias were used, with the FET biased beyond cutoff, drain current would not flow and oscillations could never begin, unless an external signal were applied. Also, if for any reason (such as an overload) the oscillations tend to die down in this self-bias circuit, the bias will decrease. With the lower bias, this weaker oscillation can still drive the gate into the positive region, and the drain current pulse tends to remain at a constant amplitude. In other words, gate-leak bias tends to stabilize the amplitude of the oscillations because the operating point varies with the signal level.

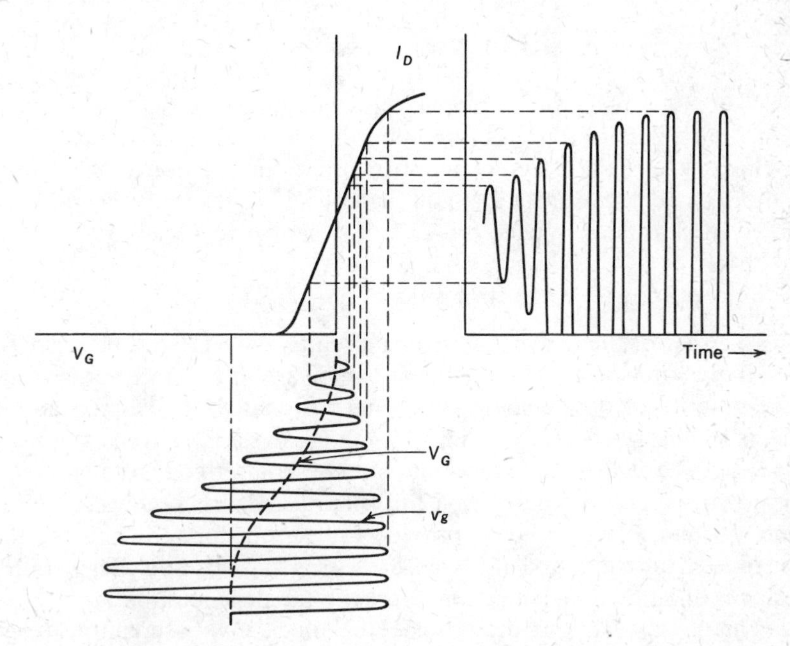

Figure 5-4 Oscillator gate voltage and drain current waveshape.

Shunt-Fed and Series-Fed Circuits

In the Armstrong circuit of Fig. 5–3, the gate-leak resistor is shunted across the tank circuit, from grid to ground. The dc bias is therefore in parallel with the ac signal voltage developed across the tank. This type of bias connection is obviously a *shunt-derived* or *shunt-fed* bias circuit. Resistor R_g could also be connected across capacitor C_2, as shown by the dashed lines. The action of the circuit is quite similar, but now the bias voltage is in series with the signal voltage. This alternate connection is a *series-bias* or *series-fed* bias circuit.

If we examine the drain circuit of Fig. 5–3, we notice that the dc supply is connected in series with the drain coil L_2. The dc component and the ac component of drain current both flow through this coil. This circuit is therefore a *series-fed* drain circuit. A shunt-fed version of this circuit is shown in Fig. 5–5. The dc path is through the RF choke to the

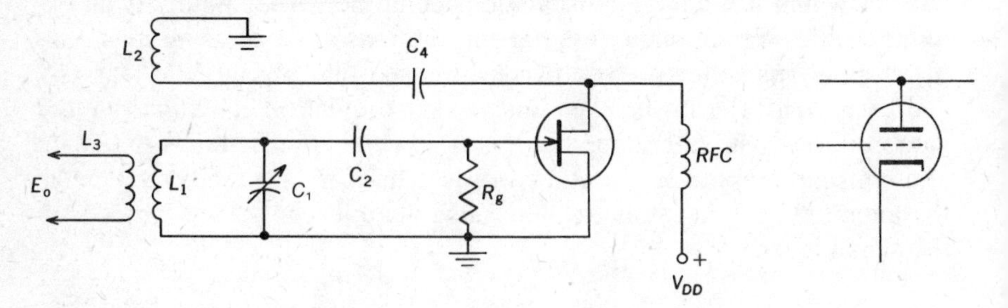

Figure 5-5 A shunt-fed Armstrong oscillator.

power supply. Capacitor C_4 blocks this component of the drain current out of the tickler winding. On the other hand, the high reactance of the RF choke keeps the ac component of drain current out of the power supply. The ac path is from drain through C_4 and L_2 to ground and back to source. In all other respects, this circuit is similar to Fig. 5–3.

Bipolar Transistor Armstrong Oscillator

In either of the above Armstrong circuits, the FET can be replaced by a bipolar transistor. The transistor can be an NPN or a PNP, so long as the polarity of the dc source is changed to match. The circuit action remains as described before. Figure 5–6(a) shows a PNP transistor used with series-fed collector circuit and common-emitter configuration. Notice the strong similarity between this circuit and that of Fig. 5–3. Resistors R_1 and R_2 are used to provide the proper forward bias for the base-emitter junction, and R_E is used for bias stabilization. The emitter resistor is shunted by a capacitor to prevent ac degeneration.

This circuit has one drawback. Due to the low base-emitter resistance, the transistor loads the tank circuit, reducing its Q. This in turn can cause drift in the oscillator frequency and poor output waveform because of the decreased discrimination against harmonics. To reduce this loading effect, the transistor is often connected to a tap on the tank circuit, as shown by the dashed connection. An even better technique used to reduce loading of the tuned circuit is to shift the tank to the collector circuit, and use an untuned circuit between base and emitter. Such a circuit is shown in Fig. 5-6(b). The circuit components are given the same legends for direct comparison with Fig. 5-6(a). Notice the use of the additional component, C_3. This capacitor is necessary if it is desired to ground the rotor of the variable capacitor, C_1, while still maintaining a complete tank circuit, and without shorting the power supply to ground. Meanwhile, because the tank is on the collector side, the higher impedance of the collector-base junction reduces the loading. Other circuit variations

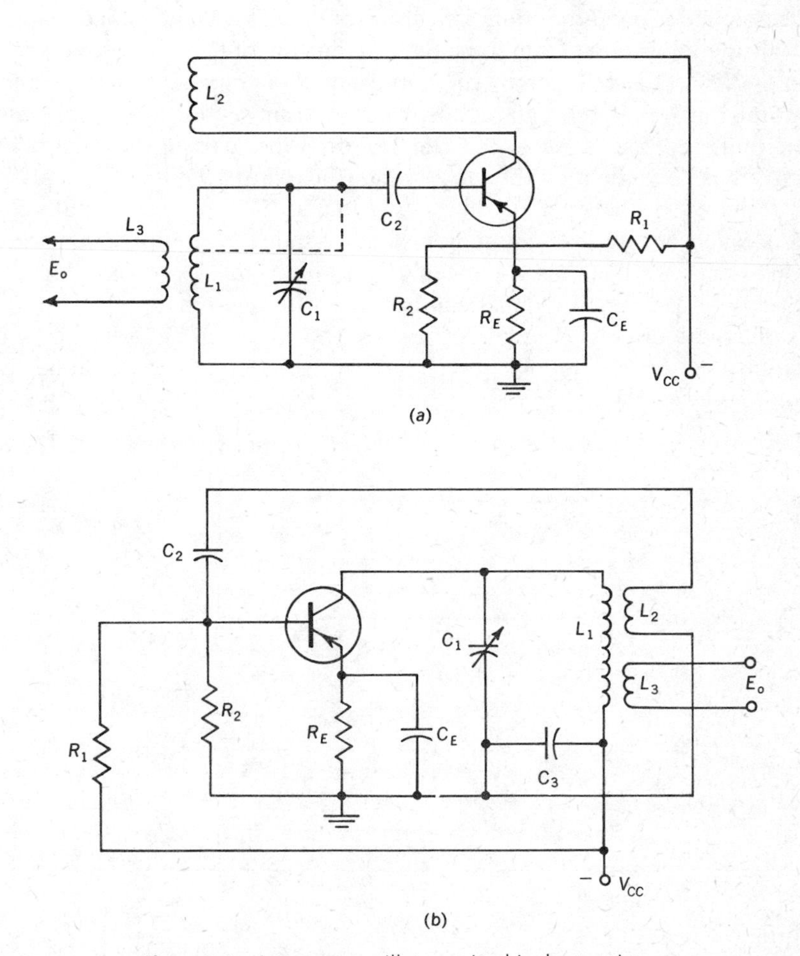

Figure 5-6 Armstrong oscillators using bipolar transistors.

are possible, such as the use of a shunt-fed collector circuit, other biasing arrangements, and the use of NPN transistors.

The Hartley Oscillator

In the shunt-fed Armstrong circuit of Fig. 5–5, notice that one end of each coil, L_1 and L_2, is grounded. Instead of using two separate coils, it is therefore possible to use one coil with a tap corresponding to the common junction of these two grounded points. Such a "revised" circuit is called a *shunt-fed Hartley circuit* and is shown in Fig. 5–7(a). A comparison of this circuit with Fig. 5–5 reveals one other difference. The tuning capacitor C_1 is now shunted across the entire coil, $(L_1 + L_2)$. Also notice that, since the coil *tap* remains grounded, the rotor of this variable capacitor can no longer be grounded. Sometimes it is desirable, or even necessary, to ground the frame of capacitor C_1. This can be done readily by shifting the ground from the tap to the bottom of the coil, as shown in Fig. 5–7(b). This changes the dc path. The dc component of drain current must now flow through the feedback or drain section of the coil, and the circuit becomes a *series-fed* Hartley. In either circuit, the amount of feedback is dependent on the tap point, or the relative values of L_1 and L_2.

One other change should be noted in Fig. 5–7(b). The output voltage is taken out by *R-C* coupling instead of inductive coupling as in the previous cases. This has no specific bearing on the series-fed Hartley. Either type of output coupling can be used with any circuit.

In these circuits (Fig. 5–7(a) or (b)) the FET can be replaced by a vacuum tube or by a bipolar transistor. In the latter case it would be de-

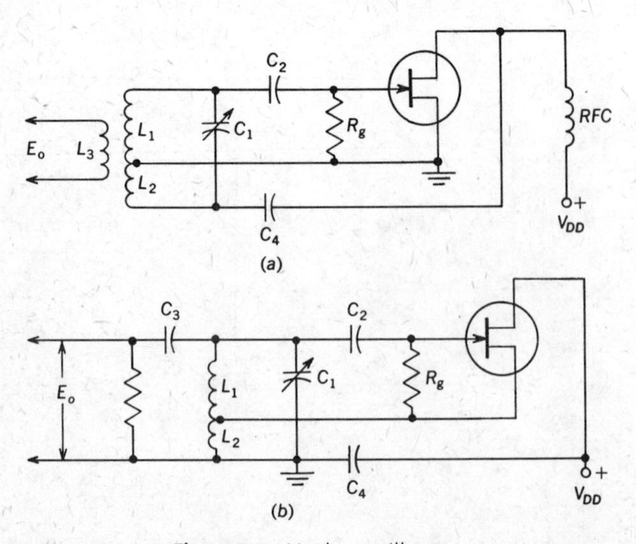

Figure 5-7 Hartley oscillators.

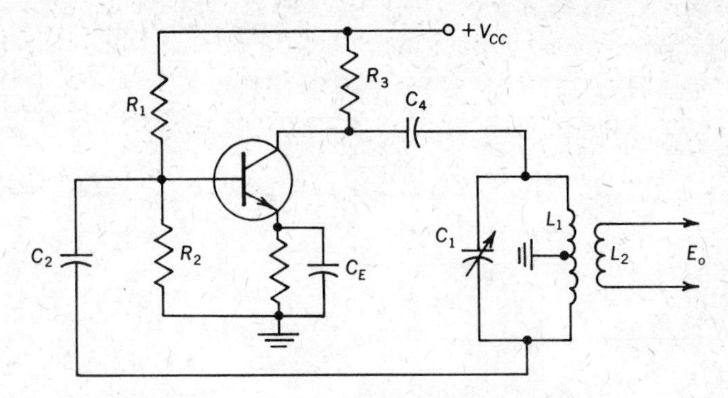

Figure 5-8 Bipolar transistors Hartley oscillator.

sirable, for bias stabilization, to add either an emitter resistor and capacitor (as shown in Fig. 5–6(a)) or a voltage-feedback stabilization resistor between collector and base.

As in the Armstrong circuit, when using bipolar transistors it is common to place the tank circuit on the collector side. Such a Hartley circuit is shown in Fig. 5–8. Notice that this is a shunt-fed collector circuit with resistor R used in place of an RF choke, RFC. When the collector (drain or plate) current is low, a suitable resistance value can provide the same "choking" effect as the reactance of an RF choke, with negligible power loss and at much lower cost. Again, as in Fig. 5–6, resistors R_1 and R_2 provide proper base bias, and resistor R_E provides bias stabilization. Notice also that this time an NPN transistor is used, requiring grounding of the negative side of the dc supply and feeding a positive voltage (V_{CC}) to the collector. Again, a number of circuit variations are possible.

The Colpitts Oscillator

In the above Hartley circuits, feedback is obtained by returning the drain, plate, or collector circuit to one end of the tank circuit and connecting the source, cathode, or emitter to a tap *on the inductive branch.* A similar effect can be obtained by connecting the source, cathode, or emitter to a "tap" on the capacitive branch. This "tap" can be obtained by using two capacitors in series, their junction forming the tank circuit tap. Two such circuits are shown in Fig. 5–9. They are called *Colpitts oscillators.* The FET version, Fig. 5–9(a), is drawn in identical manner to the Hartley circuit of Fig. 5–7(a). Notice that, except for the method of tapping the tank circuit, these two schematics are identical. The amount of feedback in the Colpitts circuit is dependent on the relative capacitance values of C_{1A} and C_{1B}. The smaller the capacitance C_{1B}, the greater the feedback. As the tuning is varied, both capacitance values increase or decrease simultaneously, but the capacitance ratio remains fixed.

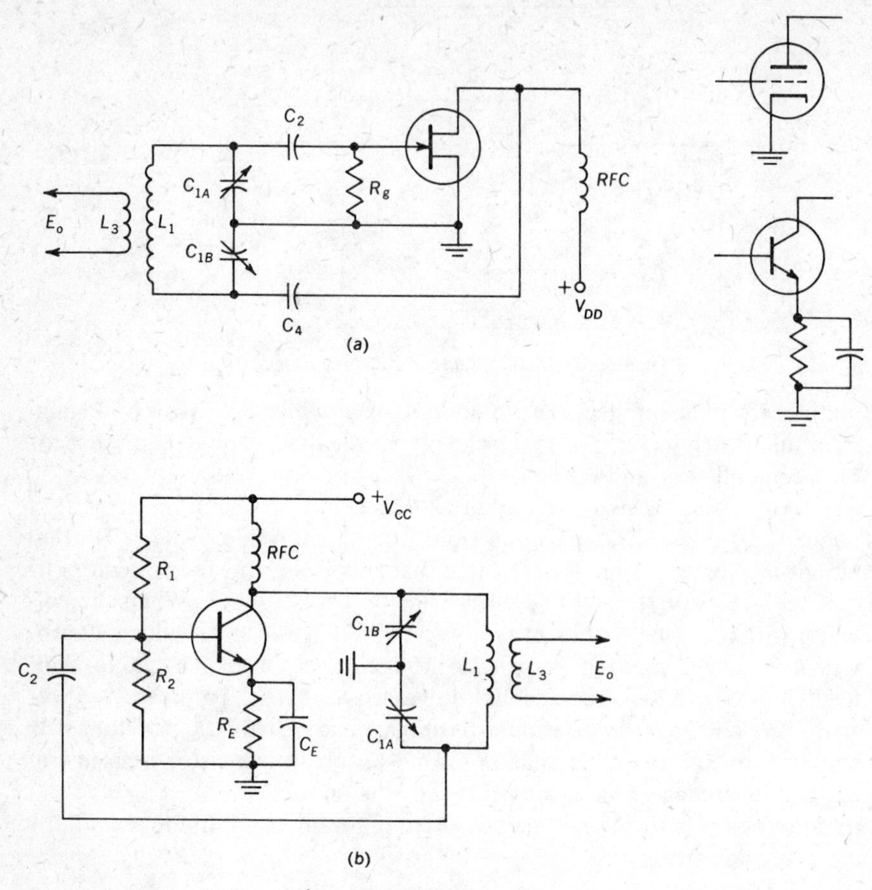

Figure 5-9 Colpitts oscillators.

Here again, in Fig. 5–9(a), the FET can be replaced by a tube or a bipolar transistor (as shown by the inserts). In the bipolar circuit, emitter bias stabilization is recommended.

Because of the heavier loading when a bipolar transistor is used, the circuit of Fig. 5–9(b) is preferable. The tank is now on the collector side. Compare this circuit with the one shown in Fig. 5–8. Except for the split-capacitor tap, it is practically identical to the Hartley shunt-fed circuit. Notice, however, the absence of capacitor C_4 between the collector and the tank circuit. Since the midpoint of the capacitive branch is grounded, there is no dc path between ground and collector, and the omission of C_4 will not short-circuit the power supply. The same is true of the FET and tube version, but since we may be dealing with higher voltage sources, capacitor C_4 serves as a safety measure to keep any high direct voltages off the tuning unit.

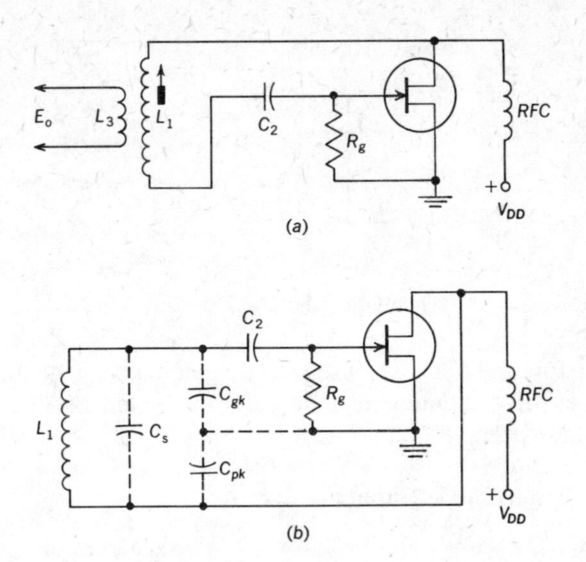

Figure 5-10 The ultra-audion oscillator.

The Ultra-Audion or Dow Oscillator

At very high frequencies (VHF) and into the ultrahigh region (UHF), a modification of the Colpitts circuit is used. At these high frequencies, the inductance and capacitance values needed for the tank circuit are much lower. It is therefore possible to eliminate the tuning capacitances altogether and resonate the circuit by varying the inductance, using stray capacitance and transistor (or tube) capacitances for the second branch.* Such a circuit is shown in Fig. 5–10(a). It is called the ultra-audion (or Dow) oscillator. To show its similarity to the Colpitts, the circuit is redrawn in Fig. 5–10(b) including all the effective capacitances.

The Clapp Oscillator

Another version of the Colpitts oscillator, used with bipolar transistor circuits, is known as the *Clapp* oscillator. In this circuit, fixed capacitors are used for C_{1A} and C_{1B}, and a third capacitor C_5 is added *in series* with the inductive branch, as shown in Fig. 5–11. The feedback is still dependent on the relative values of C_{1A} and C_{1B}. However, now these capacitors are made much larger than C_5, so that the frequency of the tank circuit is essentially dependent on C_5 and L_1. Because the low impedance of this *series* resonant circuit matches the low input impedance

*At still higher frequencies, butterfly tuners, transmission line tuners, and cavity resonators are used in place of conventional *L-C* circuits.

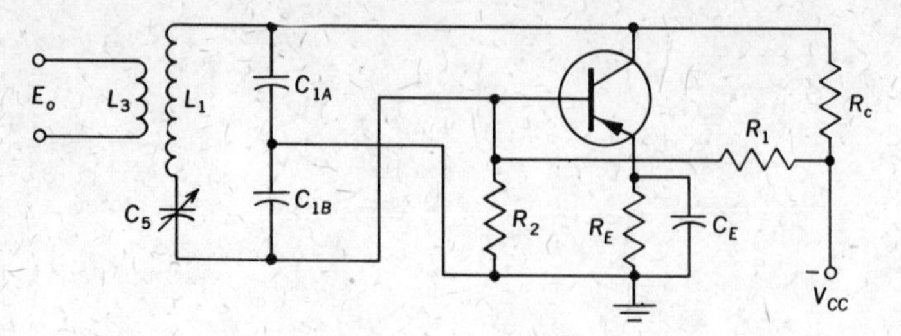

Figure 5-11 The Clapp oscillator.

of the transistor, and because of the large capacitance values C_{1A} and C_{1B}, the frequency of oscillation is relatively unaffected by changes in the transistor parameters.

Tuned-Input Tuned-Output Oscillators

Figure 5–12 shows an oscillator with two tank circuits. Depending on the active device used, this circuit is known as a *tuned-drain tuned-gate* (FET), or *tuned-plate tuned-grid* (tube), or *tuned-collector tuned-base* (bipolar) oscillator. The feedback from output to input is via the interelectrode capacitance C_{gd} or C_{gp} or C_{bc} as appropriate. If this interelectrode capacitance does not provide sufficient feedback, a capacitor can be added externally between these two elements. To remove any high direct voltage from the output tank, the output circuit can be shunt fed instead of as shown. This is more important with vacuum-tube circuits, especially for high-power outputs using beam power tubes instead of triodes.

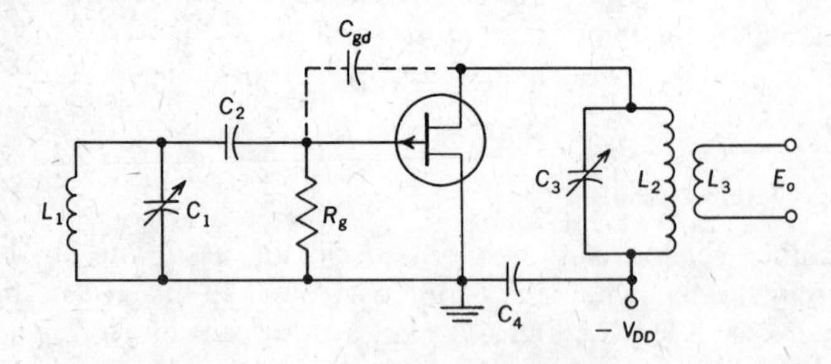

Figure 5-12 The tuned-drain tuned-gate oscillator.

Frequency of Oscillation

From the development of the oscillator action, it was inferred that the frequency of oscillation is the resonant frequency of the tank circuit. Although this is *approximately* correct, the exact value is influenced by

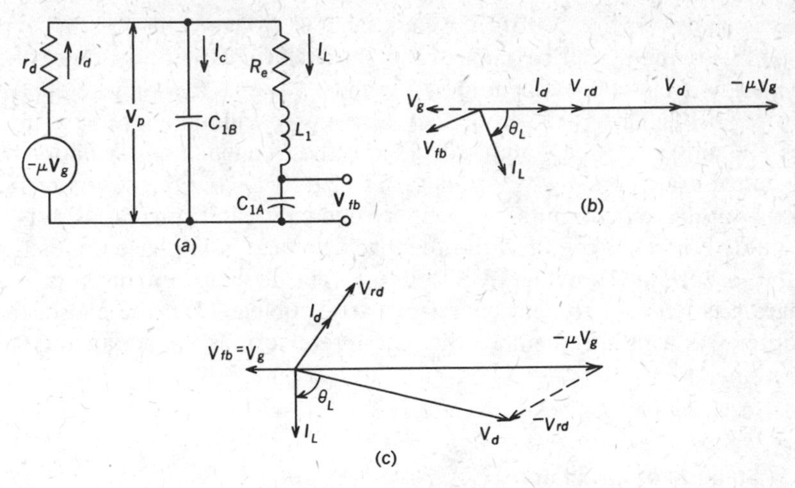

Figure 5-13 Colpitts oscillator analysis.

other factors, such as the resistance in the tuned circuit, the load on the circuit, and the transistor or tube parameters. The ultimate criterion is that the voltage fed back must be at the proper phase. This can be shown most readily by analyzing the phasor relationships in one of the previous circuits. Figure 5–13(a) shows the *equivalent* circuit of the Colpitts oscillator of Fig. 5–9(a). The phasor $-\mu V_g$ is used as the reference. If the tank circuit is resonant at the oscillator frequency, the total load is resistive, and the total current I_d is in phase with the source voltage μV_g. The internal resistance drop V_{rd} is in phase with this current. Subtracting this value from μV_g produces the output voltage V_d, and this voltage is also in phase with μV_g. These phasor relations are shown in Fig. 5–13(b). The total current I_d now breaks into two components: I_c through C_{1B}, and I_L through the outer branch. Since coil L_1 is resonant with the series combination of C_{1A} and C_{1B}, the inductive reactance is greater than the reactance of C_{1A} alone. This branch is inductive, but because of the resistance R_e, the current lags by less than 90°.* Since the feedback voltage V_{fb} is developed across the capacitor, it must lag the current by 90°. Notice that this feedback voltage is *not* in phase with the grid voltage V_g required to produce the source voltage $-\mu V_g$.

On the other hand, the proper phase relations are established if the frequency of oscillation is taken as *slightly higher* than the resonant frequency of the tank circuit. Now the tank circuit acts as a capacitive load, and I_d leads $-\mu V_g$ by some angle less than 90° (because of r_d). As before, V_{rd} is in phase with I_d, and V_d must equal $\mu V_g - V_{rd}$, *taken vectorially*.

* R_e represents the resistance of the tank circuit and any coupled resistance resulting from the load.

This is shown in Fig. 5–13(c). Again, current I_L lags behind V_d by some angle less than 90°, but because of the phase shift in V_d, the current *at this frequency* is exactly 90° behind $-\mu V_g$, and V_{fb} is exactly in phase opposition to $-\mu V_g$. This is the frequency at which oscillations are sustained.

In contrast to the Colpitts oscillator, the frequency of oscillation for the tuned-drain tuned-gate oscillator is slightly *below* the resonant frequency of the tank circuits. At such a frequency, each tank circuit acts as an inductive reactance, and the feedback voltage is in phase opposition to the $-\mu V_g$ of the equivalent FET circuit. The development of the phasor diagram is left as a student exercise. (See Problem 11 at the end of this chapter. As a hint, remember that the interelectrode capacitance C_{gd} is very small.)

Stability of Oscillators

Many applications of oscillator circuits in industrial electronics and in communications require that the generated frequency remains constant to very close tolerances. For example, an FM broadcasting station in the 88–108 MHz band must maintain its frequency to within ±2 kHz. This is in the order of two parts in one hundred thousand or 0.002%. (In an AM transmitter, the maximum allowed carrier shift is only ±20 Hz.) There are many factors that can affect this stability. The most obvious are those that concern the tank circuit directly, such as variation of L and C values because of temperature; variation of input and output capacitances of the tube or transistor due to temperature, change in operating point, aging, or replacement. Stability is also affected by changes in the power supply voltage. This can affect the operating point, causing changes not only in capacitances, but also in the effective internal resistance (r_d in the phasor diagram of Fig. 5–13). Finally, the resistance coupled into the tank circuit by the load can also affect the frequency of oscillation.

The stability of any oscillator can be greatly improved by preventing or minimizing the above effects. For example, voltage regulation can be used in the dc supply to the oscillator. This prevents any change in frequency that might be caused by a change in the operating point. The LC product of the tank circuit can be maintained constant to very close tolerances by using special temperature-compensating capacitors* as part of the total tank circuit capacitance. By selecting a suitable capacitor with an opposite temperature coefficient, it is possible to compensate for drift of the tank circuit resonant frequency. To reduce the effects of varying shunt capacitance, oscillators are often designed with higher C/L

* *Silver mica* and ceramic (TC) capacitors are available in many sizes and with various (positive or negative) temperature coefficients.

ratios than would be used for amplifiers. Because of the higher tank circuit capacitance, any stray capacitances would have less effect on the oscillator frequency. Also, since resistance affects the frequency of oscillation, high-Q tank circuits are very desirable. In a high-Q circuit, only a very slight change in oscillator frequency, as compared to the resonant frequency of the tank circuit, will produce the proper phase relations to sustain oscillations. Toward this end, where frequency stability is important, it is preferable to design the oscillator for low power output and to use loose coupling between the oscillator and its load. This will reduce the effective resistance coupled into the tank circuit by the load. When high power levels are needed, power amplifiers are used between the oscillator and the load. For maximum stability, the first stage after the oscillator should be a class-A amplifier, or a class-B with little or no excursion into the positive grid region.* This minimizes the loading effect on the oscillator circuit. The remaining amplifier stages are generally class-C amplifiers for maximum circuit efficiency. Such a system is often referred to as a *master oscillator power amplifier* (MOPA).

CRYSTAL OSCILLATORS

When the ultimate in frequency stability is desired, a quartz crystal is used in place of an L-C tank circuit. The property that makes this possible is the same piezoelectric effect that makes crystal microphones possible. Many crystalline substances exhibit this effect. Rochelle salts, as used for microphones, have the strongest activity, but they are too unstable for frequency control. Quartz crystals have been found to be electrically superior.

Crystal Cuts

To take advantage of the piezoelectric effect it is necessary to cut small "plates" from the natural raw crystal. The plane in which these plates are cut, with respect to the axes of the raw crystal, affects the electrical properties of the finished plate. A crystal and some of the cuts are shown in Fig. 5–14. The raw crystal is hexagonal in cross section and comes to a point at top and bottom. The longitudinal axis is called the Z axis; the lines through *diagonally opposite* corners are the X axes; and the lines through the opposite faces and perpendicular to the faces are the Y axes. One X axis and one Y axis are shown in Fig. 5–14(a).

*Such a stage is called a *buffer amplifier*.

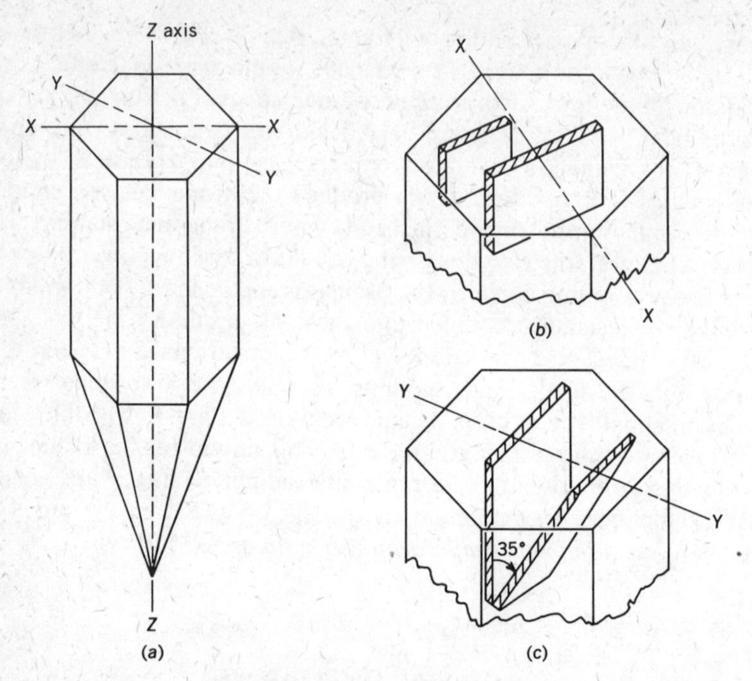

Figure 5-14 Crystal and crystal cuts.

When a plate is cut parallel to the Z axis, with its faces perpendicular to the X axis, it is called an *X cut*. This type of cut is shown in Fig. 5–14(b). The *Y cut* is obtained by making the faces of the plate perpendicular to the Y axis, as shown in Fig. 5–14(c). By rotating the plane of the cut around one or more axes, a variety of cuts such as the *AT*, *BT*, *CT*, *DT*, *GT*, and *NT* are obtained. For example, if the *Y*-cut plate of Fig. 5–14(c) is rotated clockwise away from the Z axis, an *AT* cut results. The type of cut and mode of vibration (such as length or thickness) have a great effect on the natural resonant frequency of the plate. By proper selection of the type of cut, dimensions of the plate, and mode of vibration, it is possible to obtain crystal frequencies from as low as 6 kHz (*NT*-cut flexural mode) to as high as 30 MHz (*AT* thickness shear). For operation at higher frequencies, the plates would become quite thin and fragile. However, the upper frequency range can be extended by finishing the plates so that they can be excited at the third or fifth harmonic of their fundamental frequency. By this technique, *overtone crystals* are available for operation up to 100 MHz.

The finished quartz plate is mounted in a suitable holder and the entire assembly is popularly termed a "crystal." The term "crystal" will therefore now be used to signify the finished unit rather than the raw material. Three such units are shown in Fig. 5–15.

The Effect of Temperature

The natural frequency of a crystal is influenced to some extent by operating temperature. This effect is expressed as the number of hertz change per megahertz of crystal frequency per degree Celsius (formerly called centigrade) variation in temperature, and it is known as the *temperature coefficient of frequency.* A positive coefficient indicates that the crystal frequency increases with an increase in temperature, and vice versa. Temperature coefficients may be expressed in hertz per megahertz per degree Celsius, or more simply in parts per million (ppm), with the per degree Celsius understood. The change in frequency for any crystal operation is then given by:

$$\Delta f = k(f_{\mathrm{MHz}} \times \Delta °\mathrm{C}) \tag{5-2}$$

This equation implies that the temperature coefficient of frequency is constant. Actually it varies, depending not only on the type of cut but also on the temperature. The "fixed" k values given are only average

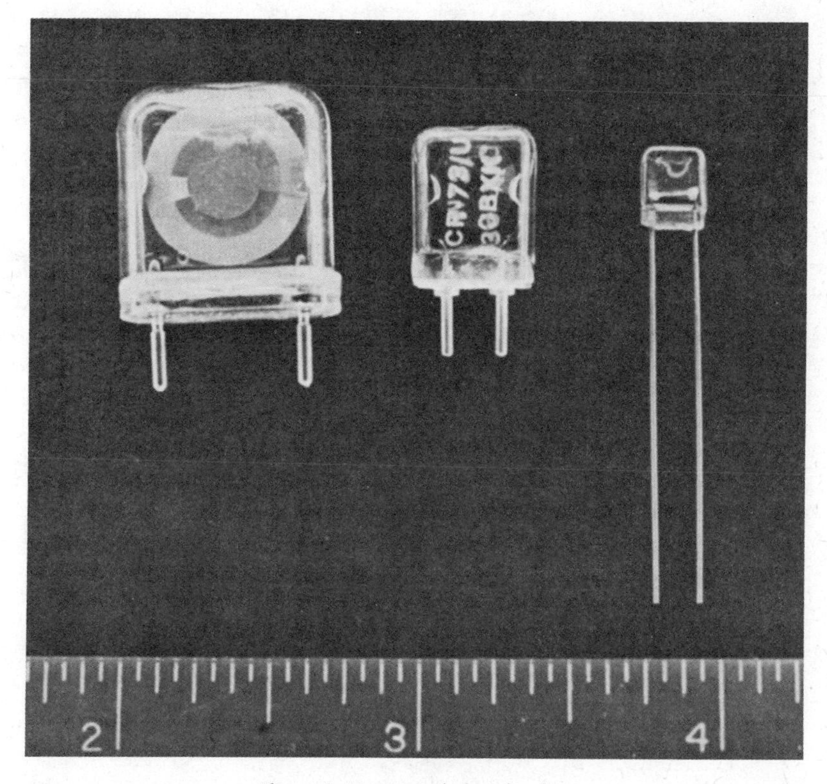

Figure 5-15 Typical crystal units.

values, generally over a temperature range of 20°–55°C.

For accurate design information, the manufacturers provide curves of frequency deviation (ppm) versus temperature. Because of their relatively high temperature effects, *X*- and *Y*-cut crystals have been almost entirely superseded by the so-called zero temperature coefficient crystals. These units have low temperature coefficients that vary (with temperature) from some small negative value to a small positive value. For a small range of in-between temperature values, the coefficient is zero. When stability is of paramount importance, the crystal plate is enclosed in a temperature-controlled oven holder.

Electrical Characteristics

The electrical action of an oscillating quartz crystal can be analyzed by use of an equivalent electrical network. Such an equivalent circuit is shown in Fig. 5–16, together with the crystal symbol and its impedance characteristics. The inductance L represents the mass of the crystal; the capacitor C_1, the resilience or elasticity; the resistance R, the frictional losses; and the capacitor C_2 is the actual capacitance formed by the crystal electrodes, with the crystal itself as the dielectric. The inductance of quartz crystals is very high, ranging from 0.1 H to well over 100 H, depending on the cut and dimensions of the unit. On the other hand, the resistance is relatively low. This results in extremely high Q factors. Values of Q in the 10,000–50,000 range are quite common, and Q values up to two million are claimed for some units.

The capacitance values C_1 and C_2 are relatively low (less than 1 pF for C_1 and between 4–40 pF for C_2). Therefore, at low frequencies, the impedance of this network is high and capacitive (negative). At some definite frequency, for a given crystal, the reactances X_L and X_{C1} will be equal. This is the series-resonant frequency and corresponds to the natural frequency of the crystal. At a slightly higher frequency, the net

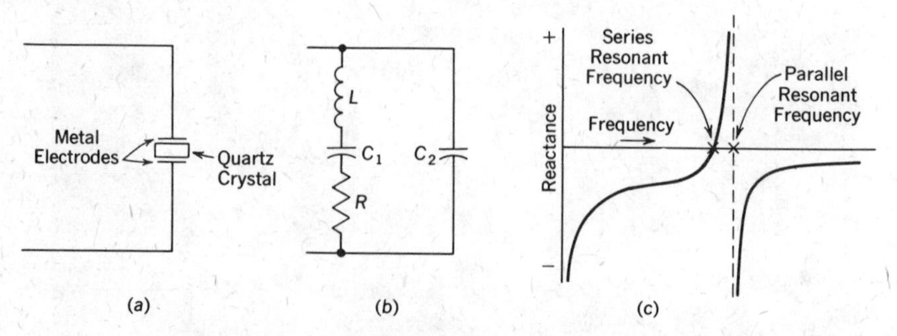

Figure 5-16 Electrical characteristics of a crystal.

reactance of this branch becomes inductive and equal to X_{C2}. The crystal now acts as a parallel-resonant circuit. Notice the steepness of the resonance curve shown in Fig. 5–16(c). This results from the high Q of the circuit and the close proximity of the two resonant frequencies. This feature is responsible for the excellent stability of the crystal oscillator. Only a very small shift in circuit frequency is needed to maintain the proper phase relations to compensate for varying voltages, aging of components, or other causes.

Crystal Excitation

When an alternating voltage is applied across a crystal, and this voltage is at or near the natural resonant frequency, the crystal is "excited" into mechanical vibration. The intensity of the vibration varies directly with the magnitude of the exciting voltage. If too high a voltage is applied, and the amplitude of vibration becomes too great, the internal stresses created can shatter the crystal. Since the amplitude of vibration is also a function of the current through the crystal, it is accepted practice to rate a crystal's power limitation by specifying the maximum safe crystal current.*

Tuned-Output Crystal-Input Oscillators

Tuned-output tuned-input oscillators can be converted into crystal-controlled oscillators by replacing the input tank circuit with a crystal. (Bipolar circuits are not suited for such conversion due to the low input resistance of these transistors.) Figure 5–17 shows a FET used in a tuned-drain crystal-gate circuit. Compare this circuit with that of Fig. 5–12. Because of the high Q of the crystal, an RF choke has been added in series with the gate resistor, R_g, to reduce the shunting action. For

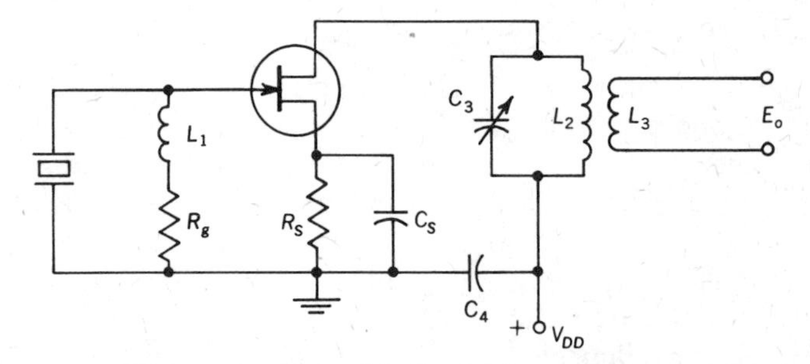

Figure 5-17 A tuned-drain crystal-gate oscillator.

* In some crystal circuit applications, a suitable pilot light (2–6.3 V) is connected in series with the crystal, the lamp brilliance serving as an indication of the crystal current. Also, in case of over-excitation, the lamp will burn out, thus protecting the crystal.

safety, some source bias is also generally included. Notice also that the gate-leak capacitor, C_2, is omitted, since the crystal itself blocks the dc path.

In vacuum-tube crystal oscillator circuits, pentodes and beam power tubes are preferred. Because of their higher power sensitivity, a lower grid voltage is needed for a given power output (as compared to using a triode). This reduces the excitation applied to the crystal and the danger of crystal failure.

If the drain current of these crystal oscillator circuits is observed while the output tank capacitor is tuned from *maximum* to minimum, it will be noticed that the current will take a sharp drop as the crystal goes into oscillation and builds up a bias. This condition is shown as point *A* in Fig. 5–18, and it represents the condition for maximum output. At this setting, the output tank is tuned to just enough above the frequency of oscillation to produce the proper phase relations. However, this condition is unstable, and it is recommended that the tank capacitance be further decreased so as to operate at point *B*. The loss of power output is slight, but now circuit variations are less likely to pull the circuit out of oscillation.

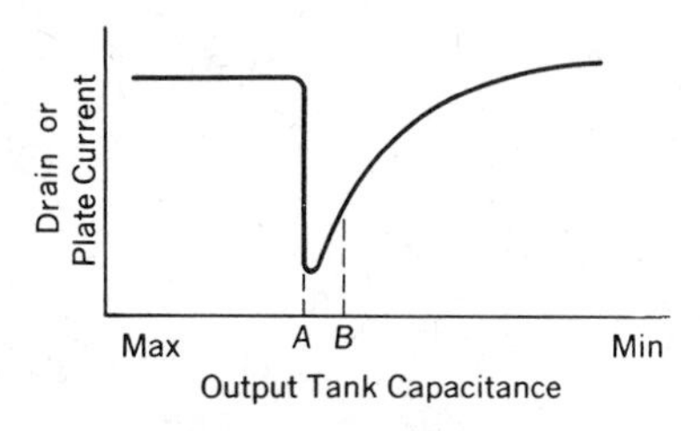

Figure 5-18 Tuning characteristics of the tuned-output crystal-input oscillator.

The Pierce Oscillator

In the Pierce oscillator, Fig. 5–19, the crystal is connected between the drain and the gate of the FET. Careful analysis will reveal that this is the ultra-audion circuit, with a few simple changes. The tank circuit is replaced by the crystal; a small amount of additional source bias has been added; capacitor C_2 may be used on the drain side of the "tank" circuit so as to keep any high supply voltage off the crystal; and capacitor C_1 has been added. As in the basic ultra-audion, feedback is obtained via the interelectrode capacitances C_{gs} and C_{ds}. Since the crystal requires less excitation, C_1 has been added across C_{gs} to decrease the feedback level.

The Pierce circuit has a distinct advantage if the operating frequency is to be changed quickly (or often). Since there are no tuned circuits, such changes are made simply by replacing (or switching) crystals. On

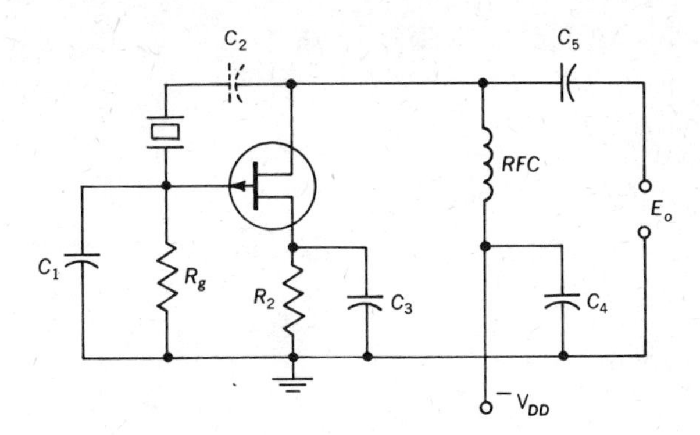

Figure 5-19 FET Pierce oscillator.

the other hand, because of the lack of tuned circuits, harmonic-type (overtone) crystals cannot be used, since the crystal automatically would work at its most active frequency, the fundamental.

BIPOLAR TRANSISTOR CRYSTAL OSCILLATORS

Crystals can also be used with bipolar transistors to improve the stability of oscillator circuits. However, special care must be taken so that the low impedance of the transistor does not load down the crystal.

The Armstrong Crystal Oscillator

In discussing the transistor version of the Armstrong oscillator (Fig. 5–6), it was mentioned that reduced loading would be obtained if the tank circuit were shifted to the collector side. This change is even more important if crystal control is desired. Figure 5–20 shows the Armstrong circuit adapted for crystal control. For convenience of comparison with Fig. 5–6(b), the same symbols have been used to identify common circuit

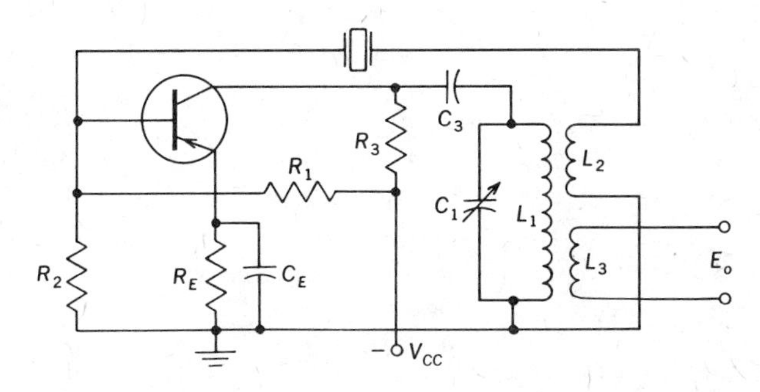

Figure 5-20 A crystal-controlled transistor oscillator.

components. The crystal in this circuit is in series with the tickler coil L_2. Maximum energy transfer between L_1 and L_2 occurs at the series-resonant frequency of the crystal.

The Clapp Crystal Oscillator

Another transistor oscillator that is readily converted to crystal control is the Clapp oscillator of Fig. 5–11. In this circuit, the frequency of oscillation is primarily a function of the series-resonant frequency of L_1C_5. For crystal control, these components are replaced with a crystal. One other change is obviously necessary. The output is now obtained by *R-C* coupling.

REVIEW QUESTIONS

1. (*a*) What is the function of an oscillator? (*b*) Basically, how is this function accomplished? (*c*) Give four applications of these circuits.

2. Refer to Fig. 5–1(a). (*a*) What kind of energy does the pendulum have at position 1? (*b*) How does it get this energy? (*c*) What kind of energy does it have at position 2? (*d*) How does it get this energy? (*e*) What kind of energy does it have at a location between 2 and 3? Explain. (*f*) What kind and how much energy does it have at position 3, in an ideal case?

3. Refer to Fig. 5–1(b). (*a*) Explain the significance of the solid curve. (*b*) Explain the significance of the dashed curve. (*c*) What is this latter case called? (*d*) What determines the "frequency" of these waves? (*e*) If the pendulum had been raised to a higher position, how would this affect the timing?

4. Refer to Fig. 5–2. (*a*) In condition (a): (1) What kind of energy is present? (2) Where is this energy located? (3) How was this energy introduced into the circuit? (4) What happens when the switch is closed? (*b*) When is the current in this circuit a maximum? (*c*) What kind(s) of energy is (are) present at this time? (*d*) Where is this energy located? (*e*) What happens to the current flow after the capacitor is fully discharged? (*f*) Explain the current and energy relations in condition (c). (*g*) Explain the current and energy relations in condition (d). (*h*) What causes the damping of the wave shown in (e)? (*i*) Which time instants in curve (e) correspond to condition (a)? (b)? (c)? (d)? (*j*) What factors determine the frequency of oscillation? (*k*) Give the equation for this frequency.

5. (*a*) What is meant by the term *continuous wave*? (*b*) Basically, how can a damped-wave operation be made to produce continuous waves?

6. (*a*) What components are present in a recurring pulse wave? (*b*) If current pulses are fed to an *L-C* circuit, describe the waveshape of the resulting tank circuit voltage. (*c*) Explain why this waveshape is obtained. (*d*) What is the frequency of this voltage? (*e*) If the tank circuit were a low-*Q* circuit, what would be the effect on the voltage waveshape?

7. State three general requirements for any oscillator circuit.

8. Refer to Fig. 5–3. (*a*) How is feedback obtained? (*b*) What determines the phase of the feedback voltage? (*c*) What determines the amount of the feedback voltage? (*d*) If the circuit is *not* oscillating, what reading (polarity and relative value) would be indicated by a high-resistance voltmeter connected across R_g? Explain. (*e*) If a dc milliammeter were inserted into the source lead (or the drain supply lead), what change, *if any,* would be noted as the circuit is made to oscillate or stop oscillating? Explain. (*f*) If the circuit is oscillating, would reversal of the L_2 connections have any effect? Explain. (*g*) If the circuit is oscillating, explain what would happen to the dc gate voltage as coil L_2 is rotated with respect to coil L_1.

9. Refer to Fig. 5–3 again. (*a*) Does drain current flow for the full 360° of the cycle? (*b*) Explain why this is so. (*c*) What class of operation is this? (*d*) What is this bias system called? (*e*) How is the bias produced? (*f*) What is the significance of the dashed resistor across capacitor C_2? (*g*) How are these variations distinguishable in terminology?

10. (*a*) What is the basic difference between a series-fed and shunt-fed drain circuit? (*b*) Which type is used in Fig. 5–3?

11. Refer to Fig. 5–5. (*a*) Trace the path of the dc component of the drain current. (*b*) Trace the path of the ac component.

12. Refer to Fig. 5–6(a). (*a*) What type of transistor is used? (*b*) What is the function of R_1? R_E? C_E? (*c*) What is the significance of the dashed connection between C_2 and L_1? (*d*) What other technique can be used to alleviate this condition? (*e*) Is this a series- or shunt-fed collector curcuit? (*f*) What is the main difference between this circuit and the one in Fig. 5–6(b)? (*g*) What is the purpose of capacitor C_2 in diagram (b)? (*h*) If capacitor C_3 in diagram (b) were omitted, how would this affect the tank circuit? (*i*) If C_3 were replaced by a direct wire connection, what other problem would arise?

13. Refer to Fig. 5–7(a). (*a*) How is feedback obtained between the output and input? (*b*) How can the amount of feedback be increased? (*c*) State two functions for capacitor C_4.

14. (*a*) Compare the output coupling technique of Fig. 5–7(a) with that of Fig. 5–7(b). (*b*) What other difference is there between these circuits? (*c*) Trace the path for the dc component of drain current in Fig. 5–7(b).

15. Refer to Fig. 5–8. (*a*) Does this circuit use a series- or shunt-fed collector circuit? Explain by tracing the current path. (*b*) Why does this circuit use a positive power supply voltage, whereas negative supply voltages were used in Fig. 5–7? (*c*) How can the loading effect of the transistor be further reduced?

16. What is the major difference between a Colpitts and a Hartley circuit?

17. Refer to Fig. 5–9. (*a*) What determines the amount of feedback? (*b*) What change in capacitor C_{1B} would increase this amount? (*c*) Which of these circuits (if any) has a series-fed dc supply? Explain. (*d*) Why is capacitor C_4 used in one circuit and not in the other?

18. Name a high-frequency counterpart of the Colpitts oscillator circuit.

19. Refer to Fig. 5–10(a). (*a*) How is feedback obtained? (*b*) Where is the tank circuit? (*c*) What is the significance of the arrow alongside the coil?

20. (*a*) Name a variation of the Colpitts circuit that is particularly suited to transistor operation. (*b*) What is the major difference between this circuit and the previous oscillator circuits? (*c*) Why is this change advantageous in transistor circuits?

21. In Fig. 5–11, what components are primarily responsible for the frequency of the output voltage? Explain.

22. Refer to Fig. 5–12. (*a*) How is feedback obtained? (*b*) What is the general frequency relation between the two tank circuits? (*c*) Is this a series- or shunt-fed drain circuit? (*d*) Give a disadvantage of this type of feed.

23. Refer to Fig. 5–13(b). (*a*) What is the tank circuit resonant frequency compared to the frequency of voltage $-\mu V_g$? (*b*) Why is I_d drawn in phase with $-\mu V_g$? (*c*) What does voltage V_{rd} represent? (*d*) Why is it in phase with I_d? (*e*) How is voltage V_d obtained? (*f*) Why is current I_L shown lagging V_d? (*g*) What is the phase of V_{fb} with respect to I_L? Why? (*h*) Is this the proper phase to sustain oscillations? Explain.

24. Refer to Fig. 5–13(c). (*a*) To what frequency is the tank circuit tuned? (*b*) Why does I_d lead $-\mu V_g$? (*c*) Why is this angle less than 90°? (*d*) How is voltage V_d obtained? (*e*) How should angle θ_L compare with the same quantity in diagram (b)? Explain. (*f*) Is the feedback voltage of proper phase to sustain oscillations? Explain. (*g*) What conclusion can be drawn from these phasor diagrams?

25. In a tuned-drain tuned-gate oscillator, what is the relation between the tank-circuit resonant frequencies and the frequency of oscillation?

26. Give three *general* factors that can affect the stability of an oscillator.

27. (*a*) What is a silver mica capacitor? (*b*) What is a ceramic (TC) capacitor? (*c*) Of what value are these in oscillator circuits?

28. Why is the stability of an oscillator improved by (*a*) using high C/L ratio tank circuits? (*b*) using voltage regulated power supply? (*c*) using high-Q tank circuits? (*d*) low power and loose coupling to the load?

29. (*a*) What is a buffer amplifier? (*b*) How does it apply to oscillator circuits?

30. (*a*) What property of a quartz crystal makes it useful in oscillator circuits? (*b*) Over what frequency range are these crystals usable? (*c*) Why are different cuts used? (*d*) When a crystal is specially designed to operate on a harmonic mode, what is it called? (*e*) How high a frequency can be achieved by such crystals?

31. (*a*) What type of cut is indicated in Fig. 5–14(b)? (*b*) What two cuts are shown in Fig. 5–14(c)? (*c*) Which is which?

32. (*a*) What is the effect of temperature on the natural frequency of a crystal? (*b*) Is it the same for all crystals? Explain. (*c*) How is this temperature effect expressed? (*d*) Is there actually a zero-temperature coefficient crystal? Explain. (*e*) How can the effect of temperature be almost completely nullified?

33. Refer to Fig. 5–16. (*a*) What does diagram (b) represent? (*b*) What does each component in this diagram represent? (*c*) Why is the impedance of the crystal high and reactive (negative) at low frequencies? (*d*) What determines the series-resonant frequency? (*e*) What is the crystal impedance (nature and magnitude) at this frequency? (*f*) Why does the impedance rise and become positive, slightly beyond this frequency? (*g*) How is parallel resonance produced? (*h*) What is the impedance (nature and magnitude) now? (*i*) Why is the impedance decreasing and negative beyond this frequency?

34. (*a*) How does the Q of a crystal compare with conventional L-C circuits? (*b*) Give a typical range of crystal Q values.

35. (*a*) What limits the power level of a crystal oscillator circuit? (*b*) In general, how is this limitation of the crystal specified? (*c*) With respect to power limitation, what type of tube is preferable with crystal circuits? Why?

36. Refer to Fig. 5–17. (*a*) What is the major difference between this circuit and the TDTG oscillator? (*b*) Why is L_1 added in crystal oscillators? (*c*) Why isn't a gate capacitor used? (*d*) How is feedback obtained?

37. (*a*) In Fig. 5–18, at what tank capacitance setting would maximum power output be obtained? (*b*) At what setting is the circuit generally operated? (*c*) Why is this so?

38. Refer to Fig. 5–19. (*a*) What is the basic "noncrystal" equivalent of this circuit? (*b*) How is feedback obtained? (*c*) What is the function of capacitor C_2? (*d*) What is the function of capacitor C_1?

39. (*a*) Give an advantage of the Pierce oscillator over other oscillator circuits. (*b*) Can this circuit be used at frequencies of around 80 MHz? Explain.

40. Refer to Fig. 5–20. (*a*) What basic type of oscillator circuit is this? (*b*) Explain how the crystal controls the oscillator frequency. (*c*) Why is this mode of crystal operation particularly suited to bipolar transistor circuits?

41. Refer to Fig. 5–11. To convert this to crystal control (*a*) what components must be removed? (*b*) Where is the crystal now connected? (*c*) How is output obtained?

PROBLEMS AND DIAGRAMS

1. (*a*) Draw the circuit diagram for an Armstrong FET oscillator using series bias and a series-fed drain circuit. (*b*) Repeat, using a vacuum tube.

2. (*a*) Repeat for shunt bias and a shunt-fed drain circuit. (*b*) Repeat, using a vacuum tube.

3. Draw the circuit diagram of an Armstrong oscillator using an NPN transistor, a common-emitter connection, a shunt-fed collector circuit, and some provision for reduced loading on the tank circuit.

4. (*a*) Draw the circuit diagram of a FET Hartley oscillator with a shunt-fed drain circuit and an R-C coupled load. (*b*) Repeat using a tube.

5. Draw the circuit diagram for a Hartley oscillator using an NPN transistor, a common-emitter connection, a shunt-fed collector circuit, provision for reduced loading on the tank, and an R-C coupled load.

6. (*a*) Draw the circuit diagram of a Colpitts FET oscillator. (*b*) Repeat using a tube.

7. Draw the circuit diagram of a Colpitts oscillator using an NPN transistor, a common-emitter connection, reduced loading provision, and an *R-C* coupled load.

8. Draw the circuit diagram of an ultra-audion oscillator using an NPN transistor.

9. Draw the circuit diagram of the Clapp oscillator using an NPN transistor.

10. (*a*) Draw the circuit diagram of a TDTG oscillator, with a shunt-fed drain circuit. (*b*) Repeat using a tube.

11. Draw the equivalent circuit for a tuned-drain tuned-gate oscillator, with both tank circuits resonant to a somewhat higher frequency than the frequency of oscillation. Develop the phasor diagram for this condition to show the phase relation between the feedback voltage and the $-\mu V_g$ voltage.

12. A crystal has an average temperature coefficient of frequency +30 ppm. If its frequency is rated as 2500 kHz at 20°C, how much drift would occur at 50°C, and what would the final frequency be?

13. A crystal operating at a temperature of 45°C produces a frequency of 842 kHz. If its rated frequency is 843.25 kHz at 20°C, find the temperature coefficient for this unit.

14. Draw the circuit diagram for a tuned-drain crystal-gate oscillator, with a shunt-fed drain circuit.

15. (*a*) Repeat Problem 14 using a triode. (*b*) Repeat using a pentode.

16. Draw the circuit diagram of a Pierce oscillator.

17. Draw the circuit diagram of an Armstrong oscillator with crystal control, using an NPN transistor, a common-emitter connection, and an *R-C* coupled load.

18. Draw the circuit diagram of a Clapp crystal oscillator using an NPN transistor.

TYPICAL FCC QUESTIONS

3.069 Draw a simple schematic diagram showing a tuned-plate tuned-grid oscillator with series-fed plate. Indicate polarity of supply voltages.

3.070 Draw a simple schematic diagram showing a Hartley triode oscillator with shunt-fed plate. Indicate power supply polarity.

3.071 Draw a simple schematic diagram showing a tuned-grid Armstrong-type triode oscillator with shunt-fed plate. Indicate power supply polarity.

3.072 Draw a simple schematic diagram showing a tuned-plate tuned-grid triode oscillator with shunt-fed plate. Indicate polarity of supply source.

3.073 Draw a simple schematic diagram of a crystal-controlled vacuum-tube oscillator. Indicate power supply polarity.

3.074 Draw a simple schematic diagram showing a Colpitts-type triode oscillator with shunt-fed plate. Indicate power supply polarity.

3.075 Draw a simple schematic diagram showing a tuned-grid Armstrong-type triode oscillator with series-fed plate. Indicate power supply polarity.

3.076 Draw a simple schematic diagram of an electron-coupled oscillator, indicating power supply polarities where necessary.

3.077 Draw a simple schematic diagram of a pentode-type tube used as a crystal-controlled oscillator, indicating power supply polarities.

3.248 What crystalline substance is widely used in crystal oscillators?

3.249 Why is the crystal in some oscillators operated at constant temperature?

3.250 What is meant by "negative temperature coefficient" of a quartz crystal when used in an oscillator?

3.280 Describe how a vacuum tube oscillates in a circuit?

3.297 Why is a high ratio of capacity to inductance employed in the grid circuit of some oscillators?

3.298 What is the purpose of a buffer amplifier stage in a transmitter?

3.360 Draw a simple schematic diagram indicating a link coupling system between a tuned-grid tuned-plate oscillator stage and a single electron-tube neutralized amplifier.

3.366 What are the advantages of a master oscillator-power amplifier type of transmitter as compared to a simple oscillator transmitter?

3.367 What are the differences between Colpitts and Hartley oscillators?

3.369 By what means is feedback coupling obtained in a tuned-grid tuned-plate type of oscillator?

3.411 Draw a simple schematic diagram of a quartz-crystal-controlled oscillator, indicating the circuit elements necessary to identify this form of oscillatory circuit.

3.413 Draw a simple schematic diagram of an electron-coupled oscillator, indicating the circuit elements necessary to identify this form of oscillatory circuit.

3.414 What does the expression "positive temperature coefficient" mean, as applied to a quartz crystal?

3.415 Draw a simple schematic diagram of a crystal-controlled vacuum-tube oscillator using a pentode-type tube. Indicate power supply polarity where necessary.

3.416 What will result if a dc potential is applied between the two parallel surfaces of a quartz crystal?

3.417 What does the expression "negative temperature coefficient" mean, as applied to a quartz crystal?

3.418 What does the expression "low temperature coefficient" mean, as applied to a quartz crystal?

3.419 What is the function of a quartz crystal in a radio transmitter?

3.420 What may result if a high degree of coupling exists between the plate and grid circuits of a crystal-controlled oscillator?

3.421 What is the purpose in maintaining the temperature of a quartz crystal as constant as possible?

3.422 Why is a separate source of plate power desirable for a crystal oscillator stage in a radio transmitter?

3.423 What are the principal advantages of crystal control over tuned-circuit oscillators?

3.424 What is the approximate range of temperature coefficients to be encountered with *X*-cut quartz crystals?

3.425 Is it necessary or desirable that the surfaces of a quartz crystal be clean? If so, what cleaning agents may be used which will not adversely affect the operation of the crystal?

3.427 List the characteristics of an electron-coupled oscillator.

3.430 If a frequency meter having an overall error proportional to the frequency is accurate to 10 hertz when set at 600 kilohertz, what is its error in hertz when set at 1110 kilohertz?

3.475 Draw a diagram of a Hartley oscillator and a Colpitts oscillator.

3.484 Explain how grid bias voltage is developed by the grid leak in an oscillator.

4.080 Draw a diagram of a crystal oscillator.

4.105 What precautions should be taken to ensure that a crystal oscillator will function at one frequency only?

4.107 A 600-kilohertz *X*-cut crystal, calibrated at 50 degrees Celsius, and having a temperature coefficient of −20 parts per million per degree, will oscillate at what frequency when its temperature is 60 degrees Celsius?

4.108 Why are quartz crystals in some cases operated in temperature-controlled ovens?

4.166 Draw a diagram and describe the electrical characteristics of an electron-coupled oscillator circuit.

4.187 For maximum stability, should the tuned circuit of a crystal oscillator be tuned to exact crystal frequency?

4.214 What is the maximum carrier shift permissible at a standard broadcast station?

4.249 What is the frequency tolerance of an FM broadcast station?

6.095 What is meant by "shock" excitation of a circuit?

6.098 What is meant by the "fly-wheel" effect of a tank circuit?

6.176 Describe how a vacuum tube oscillates in a circuit?

6.194 What is an electron-coupled oscillator? Explain its principle of operation.

6.199 What is the function of a quartz crystal in a radio transmitter?

6.200 Name four advantages of crystal control over tuned-circuit oscillators.

6.201 Why is the temperature of a quartz crystal usually maintained constant? What does the expression a "low temperature coefficient crystal" mean?

6.202 Why is a separate source of power sometimes desirable for the crystal oscillator unit of a transmitter?

6.203 What does the statement "the temperature coefficient of an X-cut crystal is negative" mean?

6.204 What will be the effect of applying a dc potential to the opposite plane surfaces of a quartz crystal?

6.205 What does the statement "the temperature coefficient of a Y-cut crystal is positive" mean?

6.228 What crystalline substance is widely used in crystal oscillators?

6.229 Why is the crystal in some oscillators operated at constant temperature?

6.399 Draw a simple schematic diagram showing a Colpitts-type triode oscillator with shunt-fed plate. Indicate power supply polarity.

6.400 Draw a simple schematic diagram of an electron-coupled oscillator, indicating power supply polarities where necessary.

6.401 Draw a simple schematic diagram showing a Hartley triode oscillator with shunt-fed plate. Indicate power supply polarity.

6.402 Draw a simple schematic diagram showing a tuned-grid Armstrong-type triode oscillator with shunt-fed plate. Indicate power supply polarity.

6.403 Draw a simple schematic diagram showing a tuned-plate tuned-grid triode oscillator with shunt-fed plate. Indicate polarity of supply voltages.

6.404 Draw a simple schematic diagram of a crystal-controlled triode vacuum-tube oscillator. Indicate power supply polarity.

6.405 Draw a simple schematic diagram of a pentode-type tube used as a crystal-controlled oscillator, indicating power supply polarities.

6.406 Draw a simple schematic diagram showing a tuned-plate tuned-grid oscillator with series-fed plate. Indicate polarity of supply voltages.

6.407 Draw a simple schematic diagram of a crystal-controlled vacuum-tube oscillator using a tetrode-type tube. Indicate power supply polarity where necessary.

6.408 What will be the effect of a high degree of coupling between the plate and grid circuits of a quartz crystal oscillator?

6.409 Draw a simple schematic diagram of a crystal-controlled oscillator and means of coupling to the following radio-frequency amplifier stage, showing power supply polarities.

6.410 What type of oscillator depends upon secondary emission from the anode for its operation?

6.412 Why is an additional plate-grid feedback capacitor sometimes necessary in a crystal oscillator?

6.413 Draw a simple schematic diagram of a Pierce oscillator.

6.414 What is the principal advantage to be gained by the use of a crystal-controlled oscillator in a marine radiotelegraph transmitter?

6.415 Discuss the advantages and disadvantages of self-excited oscillator and master oscillator-power amplifier transmitters.

6.417 What is the purpose of a buffer amplifier stage in a transmitter?

6.460 What are the disadvantages of using a self-excited oscillator-type of transmitter for shipboard service?

6.471 What is the effect of excessive coupling between the output circuit of a simple oscillator and an antenna?

6.499 A station has an assigned frequency of 8000 kilohertz and a frequency tolerance of plus or minus 0.04%. The oscillator operates at one-eighth of the output frequency. What is the maximum permitted deviation of the oscillator frequency, in hertz, which will not exceed the tolerance?

6.500 A transmitter is operating on 5000 kilohertz, using a 1000-kilohertz crystal with a temperature coefficient of −4 hertz/megahertz/degree Celsius. If the crystal temperature increases 6°C, what is the change in the output frequency of the transmitter?

6.504 What is the crystal frequency of a transmitter having three "doubler" stages and an output frequency of 16,870 kilohertz?

6.634 What precautions should be used when an absorption-type frequency meter is used to measure the output of a self-excited oscillator?

6.635 What is the meaning of "zero beat" as used in connection with frequency-measuring equipment?

6.636 If a wavemeter having an error proportional to the frequency is accurate to 20 hertz when set at 1000 kilohertz, what is its error when set at 1250 kilohertz?

6.637 What precautions should be taken before using a heterodyne-type of frequency meter?

6.638 What are the advantages and disadvantages of using an absorption-type wavemeter in comparison to other types of frequency meters?

6.639 Draw a simple schematic diagram of an absorption-type wavemeter.

6.641 A certain frequency meter contains a crystal oscillator, a variable oscillator, and a detector. What is the purpose of each of these stages in the frequency meter?

S–3.086 Draw circuit diagrams of each of the following types of oscillators (include any commonly associated components). Explain the principles of operation of each. (*a*) Armstrong. (*b*) Tuned plate-tuned grid (series fed and shunt fed, crystal and *LC* controlled). (*c*) Hartley. (*d*) Colpitts. (*e*) Electron coupled. (*g*) Pierce (crystal controlled).

S–3.087 What are the principal advantages of crystal control over tuned-circuit oscillators?

S–3.088 Why should excessive feedback be avoided in a crystal oscillator?

S–3.089 Why is a separate source of plate power desirable for a crystal oscillator stage in a radio transmitter?

S–3.090 What may result if a high degree of coupling exists between the plate and grid circuits of a crystal-controlled oscillator?

S–3.091 Explain some methods of determining if oscillation is occurring in an oscillator circuit.

S–3.092 What is meant by parasitic oscillations; how may they be detected and prevented?

S–3.093 What determines the fundamental frequency of a quartz crystal?

S–3.094 What is meant by the temperature coefficient of a crystal?

S–3.095 What are the characteristics and possible uses of an "overtone" crystal? A "third mode" crystal?

S–3.096 Explain some of the factors involved in the stability of an oscillator (both crystal- and *L-C*-controlled).

S–3.097 Is it necessary or desirable that the surfaces of a quartz crystal be clean? If so, what cleaning agents may be used which will not adversely affect the operation of the crystal?

S–3.098 What is the purpose of a buffer amplifier stage in a transmitter?

S–3.190 Draw a simple schematic diagram of a Colpitts-type transistor oscillator, and explain its principle of operation. Use only one voltage source; state typical component values for low-power, MHz operation.

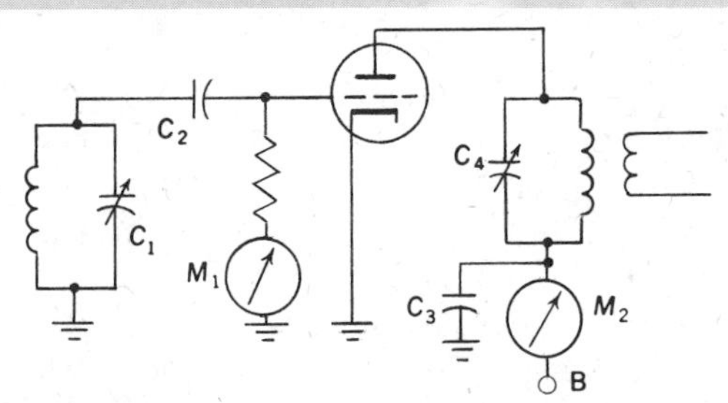

Figure 5-21 For Question S-3.190.

S-A3.1 If C_1 should short: (*a*) meter M_1 would read higher; (*b*) meter M_1 would read lower; (*c*) meter M_2 would read lower; (*d*) meter M_2 would read 0. (*e*) None of the above are true. Both meters would read normal.

S-A3.2 If R_1 should burn out: (*a*) meter M_1 would read 0; (*b*) meter M_2 would read lower; (*c*) meter M_1 would read higher; (*d*) meter M_2 would read higher. (*e*) Both (*a*) and (*b*) are true.

S–4.013 If a standard broadcast station is licensed to operate at a frequency of 1260 kHz, what are the minimum and maximum frequencies at which it may operate and still be within the proper limits established by the Commission's Rules?

S–4.042 Draw the approximate equivalent circuit of a quartz crystal.

S–4.043 What factors affect the resonant frequency of a crystal? Why are crystal heaters often left on all night even though the broadcast station is not on the air?

6

Amplitude Modulation—Transmitters

If we attempt to transmit "intelligence"—music, speech, picture information, or coded signals—via radio waves directly at its own frequency span, we run into two problems. First, only one communication can be transmitted at any one time. Otherwise, since the frequencies of each transmission would overlap, there would be no means for separating one communication from another. Bedlam would result. Secondly, it is extremely difficult to transmit low-frequency radio waves. Intelligence signals often contain frequency components ranging from below 100 Hz to several megahertz. Not only would the efficiency of transmission vary drastically over such a wide range, but also it is physically impossible to radiate energy at frequencies as low as 100 Hz. Both of these difficulties are solved by raising the frequency of the intelligence to some much higher radio-frequency band before transmission. The process by which this is done is called *modulation*, and the specific frequency used as a vehicle for "carrying" the intelligence is called the *carrier*. Modulation is effected by varying some characteristic of the carrier. Since the carrier is a sine wave, its equation can be given as

$$e_c = E_{cm} \sin (\omega_c t + \phi) \tag{6–1}$$

Modulation can be made to affect one of three factors: the amplitude E_{cm}, the frequency f_c contained in the term ω_c, or the phase ϕ. In this chapter we are concerned only with first of these, known as *amplitude modulation* (AM).

ANALYSIS OF AM WAVES

The Effect of the Modulating Signal

In an amplitude-modulated wave, the amplitude of each cycle of the resultant (the *modulated* wave) varies in accordance with the intelligence (or modulating signal). If the modulating signal is a complex or nonrecurring wave, as would be produced from voice or music, the variations of the resultant are quite erratic and difficult to analyze. Therefore, when studying modulation effects, a simple sine wave is used as the modulating or intelligence signal.

The effects of a modulating signal on a carrier are shown in Fig. 6–1. The unmodulated carrier is shown in Fig. 6–1(a). Notice that all cycles of this wave have identical amplitudes. Two modulating signals are shown in Fig. 6–1(b) as E_{s1} and E_{s2}. Waveshape (c) is the resultant obtained by modulating the carrier (a) with the modulating signal E_{s1}.* The first three cycles of this resultant are of constant amplitude—the modulating signal has not yet been applied. For the remainder of this waveshape (time interval T), notice that, as E_{s1} increases positively, the amplitude of each RF cycle increases above the value of the unmodulated carrier amplitude. Conversely, as E_{s1} decreases in amplitude, reverses, and rises to its negative peak value, the RF cycles decrease in amplitude, falling below the unmodulated carrier level to a *minimum* value. The variation in the amplitude of the modulated wave follows the rise and fall of the intelligence signal.

Now examine the waveshape of Fig. 6–1(d). This is obtained when the carrier is modulated by signal E_{s2}. Again the variation in the amplitude of the resultant wave follows the change in the intelligence signal, but this time the *amount* of change in amplitude (ΔE_{cm}) is twice as great. This should be expected from a comparison of the amplitudes of E_{s1} and E_{s2}. Notice that in each case—waveshapes (c) and (d)—the envelopes are replicas of the modulating signal.

Figure 6–1(e) shows two additional intelligence signals, E_{s3} and E_{s4}, whose amplitudes are the same as signal E_{s2}. Therefore, the amount of change in amplitude (ΔE_{cm}) of their corresponding resultants, waveshapes (f) and (g), is the same as in waveshape (d). Notice, however, that the frequencies of these three modulating signals differ; E_{s3} is the lowest and E_{s4} the highest frequency signal. A comparison of the corresponding modulated waves (d), (f), and (g), shows that the *rate* of change in ampli-

*The dashed line joining the positive peaks (and negative peaks) of this modulated wave is known as the *envelope*. It should be emphasized that this envelope is not another waveshape, nor is it a component of the resultant. It is simply a means of analyzing the variation of the resultant wave.

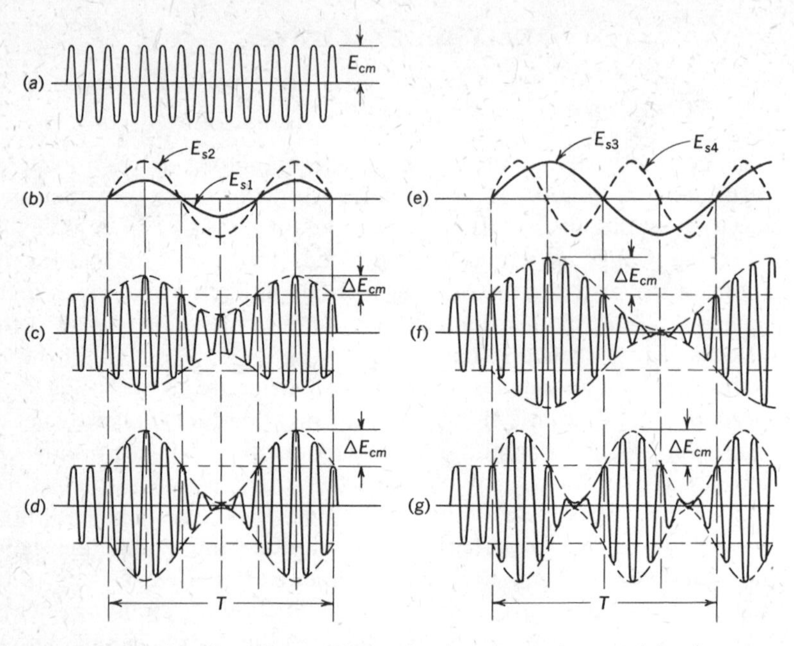

Figure 6-1 The effect of a modulating signal on the resultant modulated wave.

tude differs in a like manner. Waveshape (f), modulated by E_{s3}, has the lowest rate of change (one cycle during time interval T); waveshape (g) has the highest rate of change (two cycles during the same time period).

Percent Modulation

In Figs. 6–1(a), (c), and (d), we see three different levels or degrees of modulation. Waveshape (a) has no variation in amplitude, and the modulation level is zero. Waveshape (d) has a high degree of modulation, and waveshape (c) has some intermediate modulation level. A more exact specification of the degree of modulation is given by the term *modulation factor* (m), which measures the amount of change in amplitude as compared to the original unmodulated carrier amplitude. Quite often the term *percent modulation* is also used. This is merely the modulation factor expressed as a percentage (× 100). Many equations are used to evaluate this quantity. Sometimes one form is more convenient than another. However, they are all based on the relation

$$m = \frac{\text{rise (or fall) above the unmodulated carrier level}}{\text{carrier amplitude}}$$

To see how these variations can arise, examine Fig. 6–2. E_{cm} represents the amplitude of the unmodulated carrier. $E_{\max}$ represents the peak amplitude of the modulated wave. $E_{\min}$ represents the minimum amplitude

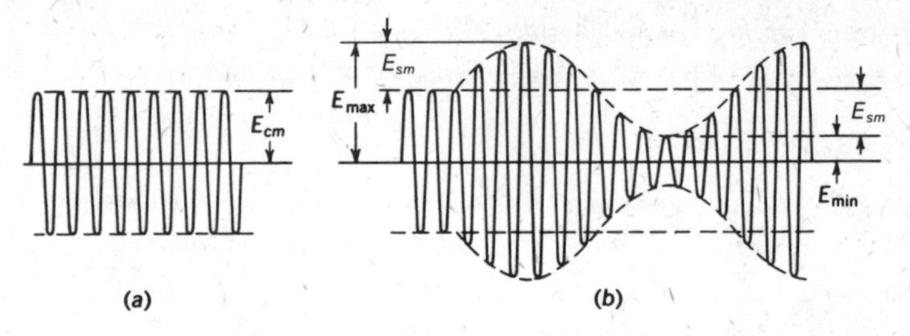

Figure 6-2 Evaluation of modulation level.

of the modulated wave. E_{sm} represents the amplitude of the modulating signal. Notice that the term ΔE_{cm} has been replaced by E_{sm}. This change should be obvious, since the amount of change in amplitude (ΔE_{cm}) is directly dependent on the amplitude of the modulating signal. Now, applying our basic definition for the modulation factor, we get

$$m = \frac{E_{sm}}{E_{cm}} = \frac{E_s}{E_c} \tag{6–2}$$

For sine-wave modulation, and if the modulation process is linear (no distortion), the amount of rise in amplitude is equal to the amount of fall in amplitude, so that

$$E_{sm} = \tfrac{1}{2}(E_{\text{max}} - E_{\text{min}})$$

and

$$E_{cm} = \tfrac{1}{2}(E_{\text{max}} + E_{\text{min}})$$

therefore

$$m = \frac{E_{\text{max}} - E_{\text{min}}}{E_{\text{max}} + E_{\text{min}}} \tag{6–3}$$

If the modulating signal is nonsinusoidal (or if the modulation process produces distortion), it is necessary to analyze the rise and fall effects separately:

$$\text{Positive peak modulation,} \quad m+ = \frac{E_{\text{max}} - E_{cm}}{E_{cm}} \tag{6–4}$$

$$\text{Negative peak modulation,} \quad m- = \frac{E_{cm} - E_{\text{min}}}{E_{cm}} \tag{6–5}$$

EXAMPLE 6–1

A carrier of 100 V and 1500 kHz is modulated by a 60-V, 1200-Hz sine-wave signal. (1) Calculate the modulation factor. (2) Express this as a percentage.

Solution

1. $m = \frac{E_s}{E_c} = \frac{60}{100} = 0.6$

2. $m = 0.6 \times 100 = 60\%$

(In this problem, it does not matter whether the voltages given are peak values (amplitudes) or rms values, so long as both are of the same type.)

EXAMPLE 6–2

The waveshape shown in Fig. 6–2(b) is seen on an oscilloscope. E_{max} is evaluated as a height of 15 units and E_{min} as 3 units. Find the percent modulation.

Solution

$$m = \frac{E_{max} - E_{min}}{E_{max} + E_{min}} \times 100 = \frac{15 - 3}{15 + 3} \times 100 = 66.7\%$$

In Figs. 6–1(d), (f), and (g), the amplitude of the intelligence signal has been made equal to the carrier amplitude, producing a modulation level of 100%. Notice that the resultant modulated wave rises to a crest value of double the unmodulated carrier amplitude and falls to zero at the trough of the modulation cycle. If the intelligence signal is stronger than the carrier, a condition known as *overmodulation* will result. This is shown in Fig. 6–3. During the positive portion of the modulation cycle, the process is linear and the resultant rises to its crest amplitude ($E_{cm} + E_{sm}$). However, for the full interval t_1, while E_{sm} *exceeds* E_{cm}, the resultant amplitude ($E_{cm} - E_{sm}$) is zero and the intelligence is lost. This type of distortion should be avoided.

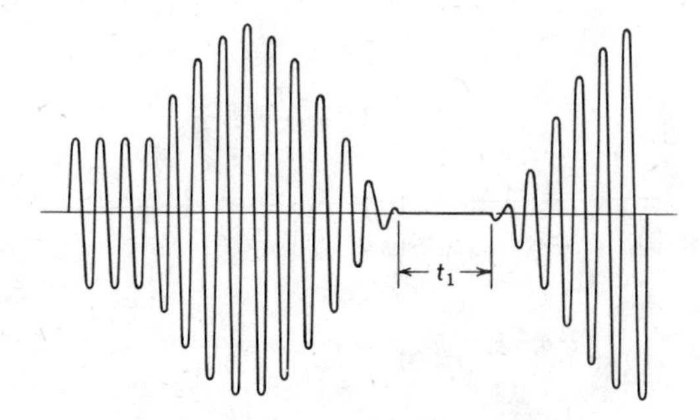

Figure 6-3 An overmodulated wave.

Frequency Components in an AM Wave

Many circuit variations have been used to modulate a carrier, but regardless of the circuit details, the process generally consists of feeding the carrier and the intelligence signal to some nonlinear device. This could be a vacuum tube or semiconductor device biased to operate on the nonlinear portion of its dynamic characteristic. Because of this nonlinearity, the plate (or collector) current is distorted and will contain a dc component, a component at the carrier frequency and at the intelligence frequency, harmonics of each of these original frequencies, and components at the sum and difference of the original frequencies. This complex current is made to flow through a tank circuit. The resonant frequency and bandwidth of this tank are deliberately chosen so as to develop an output voltage only at the carrier frequency and at the sum and difference frequencies. The output *voltage*—the modulated wave—therefore contains only three components: (1) the carrier frequency (f_c), (2) the *upper side frequency* ($f_u = f_c + f_s$), and (3) the *lower side frequency* ($f_l = f_c - f_s$).

EXAMPLE 6–3

A carrier frequency of 1210 kHz is modulated by a pure 4000-Hz audio signal. (1) What components are present in the plate current? (2) What components are present in the output voltage from the modulated RF amplifier?

Solution

1. Direct current, 1210 kHz, 4 kHz, harmonics of 1210 kHz, harmonics of 4 kHz, 1214 kHz, and 1206 kHz.
2. Only 1206, 1210, and 1214 kHz.

The above analysis can be supported mathematically as follows. The carrier can be represented by Eq. (6–1), which, if $e = 0$ when $t = 0$, reduces to

$$e_c = E_{cm} \sin \omega_c t$$

Similarly, the modulating signal equation is

$$e_s = E_{sm} \sin \omega_s t$$

But, depending on the degree of modulation m, the intelligence signal amplitude is mE_{cm}, or

$$e_s = mE_{cm} \sin \omega_s t \tag{6–6}$$

The *amplitude* of the modulated wave is the sum $E_{cm} + mE_{cm} \sin \omega_s t$, so that the equation of this wave becomes

$$\begin{aligned} e_{\text{mod}} &= (E_{cm} + mE_{cm} \sin \omega_s t) \sin \omega_c t \\ &= E_{cm}(1 + m \sin \omega_s t) \sin \omega_c t \end{aligned} \tag{6–7}$$

Multiplying out and expanding for the product of two sines* gives

$$e_{\text{mod}} = \underbrace{E_{cm} \sin \omega_c t}_{1} - \underbrace{\frac{mE_{cm}}{2} \cos (\omega_c + \omega_s)t}_{2} + \underbrace{\frac{mE_{cm}}{2} \cos (\omega_c - \omega_s)t}_{3} \tag{6–8}$$

* $\sin x \sin y = \frac{1}{2} \cos (x + y) + \frac{1}{2} \cos (x - y)$

The first component of this equation is obviously the original carrier. The second component is the upper side frequency ($f_c + f_s$), and the third component is the lower side frequency. Notice that the amplitude of each side frequency is related to the carrier amplitude by the factor $m/2$. For example, at 100% modulation, each side frequency has one-half the amplitude of the carrier.

EXAMPLE 6–4

A carrier with an amplitude of 140 V is modulated by an intelligence signal with an amplitude of 80 V. (1) What is the percent modulation? (2) What is the amplitude of the lower side frequency?

Solution

1. $m = \frac{E_{sm}}{E_{cm}} \times 100 = \frac{80}{140} \times 100 = 57\%$

2. $E_{lm} = \frac{mE_{cm}}{2} = \frac{0.57 \times 140}{2} = 40 \text{ V}$

Notice that the amplitude of the side frequency is one-half the amplitude of the modulating signal. This is true for any degree of modulation below 100%.

Phasor Representation

If we examine the three components of Eq. (6–8), we see that not only are the relative amplitudes and frequencies given, but also the phase relations. At the instant $t = 0$, the upper side-frequency component *lags* 90° behind the carrier, and the lower side-frequency component *leads* by 90°. These instantaneous phase relations are shown in Fig. 6–4(a) for a condition of 100% modulation (that is $E_l = E_u = 0.5E_c$). The resultant of these phasors, *at this instant,* is equal to E_c. However, these are waves of different frequencies, and their phase relations will not stay as shown. On a time basis, all phasors are considered to rotate counterclockwise. Yet, if we hold E_c steady and use it as our reference frequency, E_u (the higher-frequency component) is rotating faster, or still counterclockwise. On the other hand E_l, the lower-frequency component, rotates slower. Compared to E_c, it rotates backward or clockwise.

Now consider a time instant when the phasor E_u has gained 45° with respect to E_c. Simultaneously E_l must have fallen back by 45°, as shown in Fig. 6–4(b). Notice how the resultant has increased in amplitude. At some instant later, E_u has gained 90° and is in-phase with E_c. At the same time, E_l has lost 90° and is also in-phase with E_c. This is the condition for maximum amplitude of the modulated wave, reaching (for $m = 1$) $2E_c$. Now consider a time when E_u has swept through an angle of 150°. It is now to the left of E_c by 60°. Phasor E_l is now to the right, also by 60°. The resultant has decreased in amplitude. When each side frequency has rotated through 180°, the phasor representation will appear as in Fig.

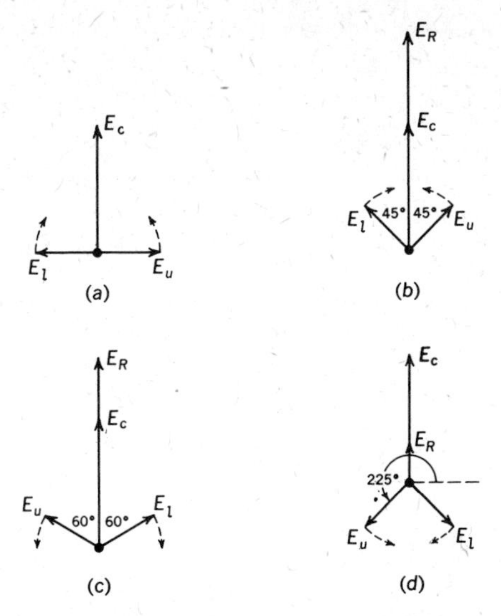

Figure 6-4 Phasor relations in a modulated wave.

6–4(a), but with the side frequencies reversed, and once again the resultant's amplitude will equal the carrier amplitude.

At some instant later, the upper side frequency has advanced 225° and will appear as shown in Fig. 6–4(d). Simultaneously, E_l will have rotated clockwise to a position 135° behind E_c. Now, the phasor sum of $E_l + E_u$ subtracts from E_c. The resultant falls to a low value. It reaches its minimum value (zero, for 100% modulation) when each side frequency has rotated through 270° and is now in direct opposition to the carrier. Finally, as each side frequency completes its rotation through 360°, the phase relations are back as shown in Fig. 6–4(a). This completes one cycle of modulation.

Sidebands

So far, for the sake of simplicity in analysis, we have limited our discussion to modulating signals that are pure sine waves. Yet most intelligence signals—voice, music, picture details, and control signals—are complex waves containing many frequencies. Each of these modulating signal components, in mixing with the carrier, produces its own side frequencies. Consequently, instead of only two side frequencies, one above and one below the carrier (as explained earlier), we now have many side frequencies on each side of the carrier. These are called *sidebands,* and the intelligence that has been raised or translated to the carrier-frequency level is contained in these sidebands.

EXAMPLE 6–5

A 2500-kHz carrier is modulated by audio (music) signals with a frequency span of 40–15,000 Hz. (1) Specify the frequencies of the upper and lower sidebands. (2) What bandwidth is necessary to handle the output of the modulated RF amplifier?

Solution

1. The modulating signal covers 0.04–15 kHz. The side frequencies produced range from $f_c \pm 0.04$ kHz to $f_c \pm 15$ kHz, or the upper sideband = 2500.04 to 2515 kHz, and the lower sideband = 2499.96 to 2485 kHz. (These can be specified more simply as 2500 ± 15 kHz.)
2. The bandwidth requirement is 2485–2515, or 30 kHz. (Notice that the bandwidth requirement is twice the frequency of the highest modulating-signal frequency.)

Power Distribution in an AM Wave

The power dissipated in any circuit is a function of the square of the voltage across the circuit and the effective resistance of the circuit. If we reexamine Eq. (6–8), we see that the voltage developed across the output tank circuit of the modulated RF amplifier stage contains three components. Obviously, the power output must be distributed among these components. Notice that the voltage output at the carrier frequency has a peak value of E_{cm}. But if the modulating signal were removed, the output voltage at this frequency would still be E_{cm}. Since the carrier itself is unaffected by the modulating signal, *there is no change* in the carrier power level, and no intelligence is conveyed by the carrier.* From the basic power relations, the carrier power is

$$P_c = \frac{E_c^2}{R} \quad \text{(using rms values)} \tag{6–9}$$

or

$$P_c = \frac{E_{cm}^2}{2R} \quad \text{(using peak values)} \tag{6–10}$$

As in any RF power amplifier, this power is supplied by the plate dc power source.

Notice also from Eq. (6–8) that each side frequency has an amplitude of $mE_{cm}/2$. Therefore, *each* side frequency has a power level of

$$P_{sf} = \frac{(mE_c)^2}{4R} \quad \text{or} \quad \frac{(mE_{cm})^2}{8R} \tag{6–11}$$

*This leads to the *suppressed carrier* transmission system discussed in Chapter 12.

The power level of these side frequencies increases with the degree of modulation, until at 100% modulation the power in each side frequency is equal to one-fourth of the carrier power. Obviously, then, the intelligence is contained entirely in these side frequencies—or, if we use a complex modulating signal, the intelligence will be translated entirely into the sidebands. For 100% modulation, the *total* sideband power (both sides) is equal to 50% of the carrier power. For any other degree of modulation, the total sideband power is

$$P_{sb} = \tfrac{1}{2}m^2P_c \tag{6-12}$$

EXAMPLE 6–6

A 50-kW carrier is to be modulated to a level of 85%. (1) What is the carrier power after modulation? (2) What is the sideband power?

Solution

1. $P_c = 50$ kW and is not affected by modulation.
2. $P_{sb} = \tfrac{1}{2}m^2P_c = \tfrac{1}{2}(0.85)^2 \times 50 = 18.06$ kW

In the above problem, 18 kW of power must be supplied to the sidebands. For 100% modulation, 25 kW would be needed. This power is not supplied by the dc plate supply, but instead comes from the modulating signal. Since this power must come through the modulation stage, the plate efficiency of this RF amplifier must also be considered. The modulating signal amplifier must therefore have a power output in excess of 25 kW.

EXAMPLE 6–7

How much audio power is required to fully modulate a 40-kW carrier if the efficiency of the modulated RF amplifier is 72%?

Solution

A block diagram will clarify the power relations. (See Fig. 6–5.)

1. $P_{sb} = \dfrac{1}{2}n^2P_c = \dfrac{40}{2} = 20$ kW

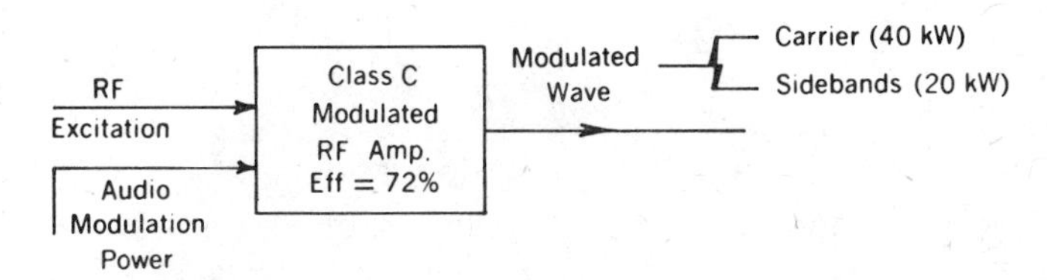

Figure 6-5 Power relations for Example 6-7.

2. To provide 20 kW of sideband power through a circuit that is 72% efficient,

$$P_{\text{audio}} = \frac{20}{72} \times 100 = 27.8 \text{ kW}$$

As modulation is applied to a carrier, it is obvious that the total power in the modulated wave must increase, since it is the sum of the carrier power and the sideband power. Expressed mathematically, this is

$$P_T = P_c + \frac{1}{2}m^2P_c = P_c\left(1 + \frac{m^2}{2}\right) \tag{6–13}$$

At 100% modulation, the total power would be 150% of the unmodulated (carrier) level. This increase in power level would cause the antenna current to rise. The change in antenna current can therefore be used as a measure of the modulation level.

At the crest of the modulation cycle, the amplitude of the modulated wave can rise to double the unmodulated value (100% modulation). The instantaneous peak power, in such a case, will be four times the unmodulated value. It is therefore necessary to reduce the ratings of tubes and semiconductors when they are used as modulated RF power amplifiers, compared to their service as unmodulated class-C RF power amplifiers. For example, the 6161 power triode has maximum ratings of 1600 V for dc plate voltage and 250 W for plate dissipation as an unmodulated class-C amplifier, but only 1300 V and 167 W as a modulated amplifier.

When intelligence is transmitted by simple amplitude modulation, it is desirable to keep the degree of modulation as close to unity as possible. The reason for this is seen from the next example.

EXAMPLE 6–8

The 40-kW carrier of Example 6–7 is modulated to a level of 10%. What is the sideband power?

Solution

$$P_{sb} = \tfrac{1}{2}m^2P_c = \tfrac{1}{2}(0.1)^2 \times 40 = 200 \text{ W}$$

In Example 6–7, we generated and transmitted a 40-kW carrier in order to send 20 kW of intelligence. Now the same carrier level—40,000 W—is used to send a mere 200 W of intelligence. Obviously, the efficiency of operation decreases rapidly for low degrees of modulation.

We are now ready to analyze the circuitry used to produce a modulated wave. Basically, as mentioned earlier, we need a nonlinear device—vacuum tube or semiconductor. In this respect, even a simple diode can be used. However, if a triode or pentode tube or a transistor is used, we will also get additional RF amplification.* For amplitude modulation, the modulation process is generally performed at high power levels. Obviously, FETs are not suitable devices, and bipolar transistors would be used only for relatively low-power transmitters (below the 50-W level). For higher-power transmitters, tubes must be used. The discussion on modulator circuits will therefore start with vacuum-tube circuitry.

PLATE MODULATION

In the most commonly used modulation circuit, the intelligence signal is fed to the plate of the RF power amplifier, in series with the dc plate supply voltage. This method of producing a modulated wave is therefore known as *plate modulation.*

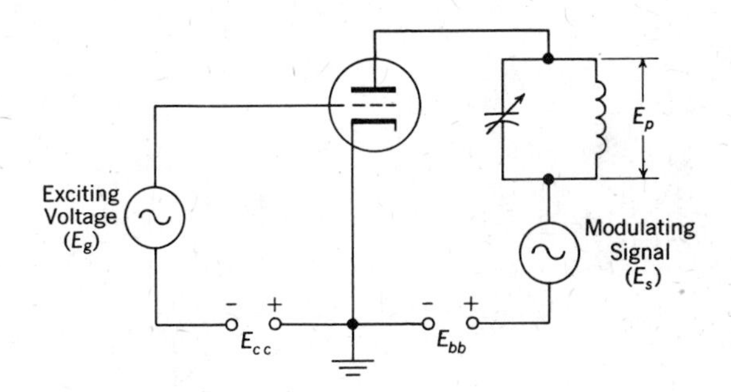

Figure 6-6 The basic circuit for plate modulation.

Basic Principles

The simplified circuit of Fig. 6–6 can be used to explain the operating conditions and the waveshapes in a plate-modulated circuit. The tube is biased well beyond cutoff, as in any class-C amplifier. The exciting voltage (the unmodulated carrier) should have sufficient amplitude to drive the grid slightly positive. In the absence of a modulating signal, the plate current will be a series of pulses of constant amplitude, and the RF output voltage will also be constant in amplitude. These waveshapes are shown in Fig. 6–7(a)–(d) during the time interval 0–1. Notice also that, during this no-modulation interval, the average plate potential (per RF cycle) is constant at the supply value.

*Diodes are used at VHF and UHF where their lower noise level is more important than the lack of gain.

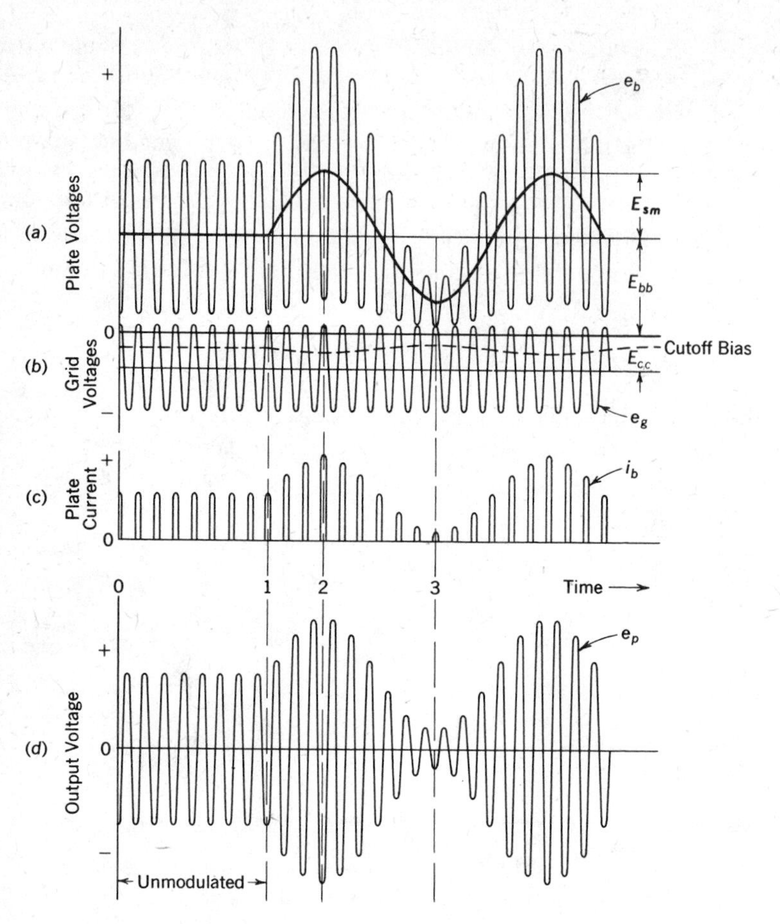

Figure 6-7 Waveshapes in a plate-modulated class-C amplifier.

When the modulating signal is applied, the plate potential must rise and fall in accordance with the intelligence voltage. At any instant, the *effective* plate potential now equals $\dot{E}_{bb} + \dot{e}_s$. Meanwhile, since the modulating signal has no effect on the grid circuit, the grid circuit waveshapes remain the same. However, because of the change in the effective plate potential, the plate current pulses will vary in amplitude. Notice in Fig. 6–7(c) that the maximum pulse amplitude is obtained at the positive peak of the modulating voltage, and that the current pulses drop to a minimum amplitude as the modulating signal reaches its maximum negative value.* Since the fundamental component of plate current is dependent on this pulse amplitude, it too will vary in accordance with the

*The waveshapes shown are for a condition of approximately 75% modulation.

intelligence signal. Consequently, the output voltage e_p developed across the tank circuit will vary in a like manner. These waveshapes, resulting from the application of a modulating signal, are shown in Fig. 6–7 beyond the time interval 0–1.

Notice the difference between the RF output voltage e_p in Fig. 6–7(d) and the instantaneous plate potential e_b as shown in diagram (a). This latter voltage is the sum of the RF voltage across the tank, the dc supply voltage E_{bb}, and the slowly varying modulating signal voltage E_s. In this resultant waveshape, $E_{b(\min)}$ remains essentially constant compared to the plate supply voltage.* Since the ideal class-C operating conditions are maintained throughout the modulation cycle, the efficiency of a plate-modulated stage remains at the same high level as for a straight class-C amplifier.

In the above diagrams, the plate current pulses (and the output voltage) do not rise to twice the unmodulated value, nor drop to zero. The modulation level shown is somewhat below 100%. Also notice in Fig. 6–7(a) that the effective plate potential does not rise to twice E_{bb}, nor fall to zero. However, if the modulating signal amplitude were increased, this effect could be obtained. The plate current pulses would then rise to a slightly higher amplitude at the modulation crest and drop to zero at the modulation trough, and the output voltage would be 100% modulated. Obviously, the ratio of E_{sm} to E_{bb} determines the degree of modulation, or

$$m = \frac{E_{sm}}{E_{bb}} \tag{6–14}$$

Heising Modulation

One of the earliest circuits used for plate modulation is the Heising, or choke-coupled, modulator of Fig. 6–8. The RF carrier is fed to the grid of the tube† and the modulating signal is applied to the plate, in series with E_{bb}. Tube V_2 is the modulating signal (audio) power amplifier,‡ and its output voltage appears across the low-frequency choke coil L_4. The RF choke L_3 and the capacitor C_3 form an RF bypass to keep RF voltages out of the power-supply and V_2 circuits. Capacitor C_3 must have a low reactance at the carrier frequency, but a high reactance at the highest modulating signal frequency. Notice that a common dc source is used for V_2 and V_1. This is actually a disadvantage. Since the peak

*For 100% modulation, $E_{b(\min)}$ would vary from approximately $E_{c(\max)}$ at no modulation, to $2E_{c(\max)}$ at the modulation crest, and to zero at the modulation trough.

†The grid circuitry is not shown. Any of the previously discussed coupling and bias methods can be used.

‡In industry, this final audio amplifier stage (whose output is fed to the RF power amplifier) is called the *modulator*.

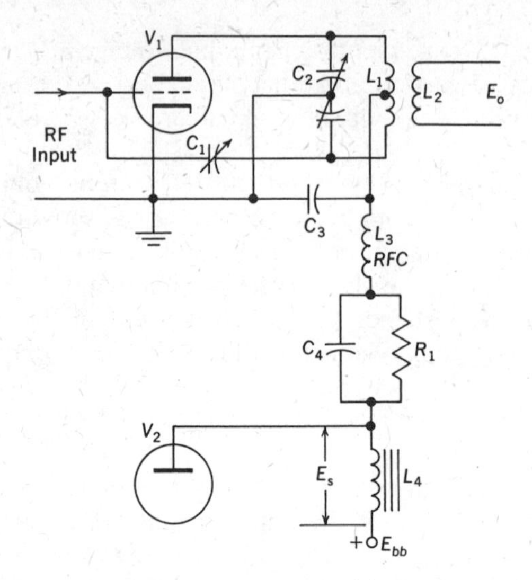

Figure 6-8 Heising modulation.

modulating signal value can never equal E_{bb}, it would be impossible to obtain 100% modulation if the full E_{bb} voltage is also applied to V_1. The voltage dropping resistor R_1 must therefore be used to reduce the plate potential of V_1. To prevent simultaneous reduction of the intelligence-signal amplitude, this resistor is bypassed by capacitor C_4. Its reactance must be low at the lowest modulating frequency. For high-power modulators, the power loss in R_1 can be serious. A second disadvantage is the problem of impedance matching for the intelligence-signal power amplifier, V_2. Its load consists of the impedance of the choke coil L_4 in parallel with the output impedance of tube V_1. There is very little flexibility in effecting an impedance match.

A Typical Plate-Modulation Circuit

Both of the above disadvantages are overcome by the more commonly used transformer-coupled circuit of Fig. 6–9. The output of the modulating-signal power amplifier (V_2 and V_3 in push-pull) is coupled to the class-C RF amplifier stage by a *modulation-matching transformer* T_1, and again the modulating signal E_s is fed to the plate of V_1 in series with the dc plate-supply voltage. Impedance match is obtained by using the proper turns ratio in transformer T_1. The output impedance of V_1 acts as the load across the secondary of the transformer. When the circuit is properly adjusted, the output impedance of this tube—as for any class-C amplifier—is resistive and equal to E_b/I_b. The turns ratio of transformer T_1 is chosen so as to match this value of load resistance to the requirement

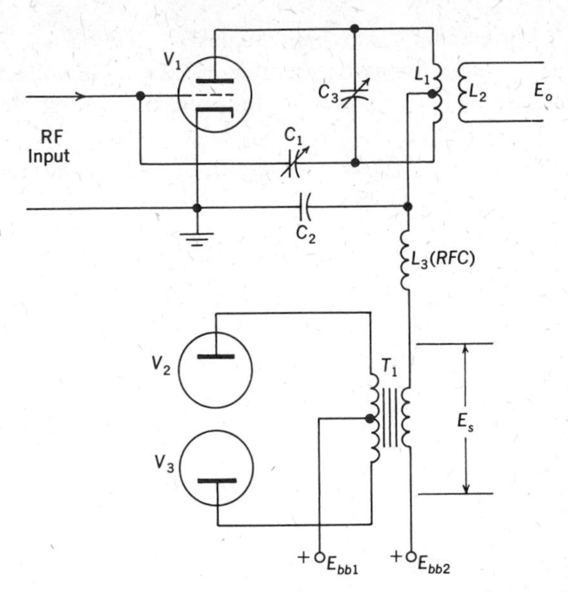

Figure 6-9 A triode plate-modulated circuit.

of the modulating-signal power amplifier. Simultaneously, the *gain* of the modulating signal amplifier is adjusted to produce the desired degree of modulation. For 100%, the output voltage E_s, across the secondary of T_1, must equal the dc supply voltage E_{bb2}. Notice also that transformer coupling allows the use of a separate power supply for each stage. This eliminates the need for voltage-dropping resistors and their concomitant power loss.

Many variations of the circuit in Fig. 6–9 are possible. The RF amplifier could be push-pull; its plate circuit could be shunt-fed. The grid circuit (details not shown) could be impedance-coupled to the previous class-C amplifier, or it could be inductively coupled, single-tuned or double-tuned. The bias could be fixed, grid-leak, or a combination, and it could be series-fed or shunt-fed. The intelligence power amplifier could be single-ended. The drawing of such circuits is left as an exercise for the student.

Multigrid Plate Modulators

Tetrodes, pentodes, and beam power tubes can also be used in plate-modulation circuits. The circuits are generally similar to the triode circuitry just discussed. Of course, depending on the frequency of operation, neutralization may not be needed. However, for 100% modulation, it is necessary to modulate the screen as well as the plate. Several techniques for doing this are shown in the partial schematics of Fig. 6–10. In Fig. 6–10(a), a common power supply is used for the plate and screen, and the modulating signal is fed to the screen through the voltage-dropping resistor R_1. Capacitor C_2 is the bypass capacitor used to keep the screen at RF ground. On the other hand, its reactance should be high for all modulating signal frequencies.

In Fig. 6–10(b), separate power sources are used for plate and screen. A proper choice for the screen supply voltage eliminates the necessity for a voltage-dropping resistor. The modulating signal is applied to the screen grid in series with the dc supply source by means of a tertiary

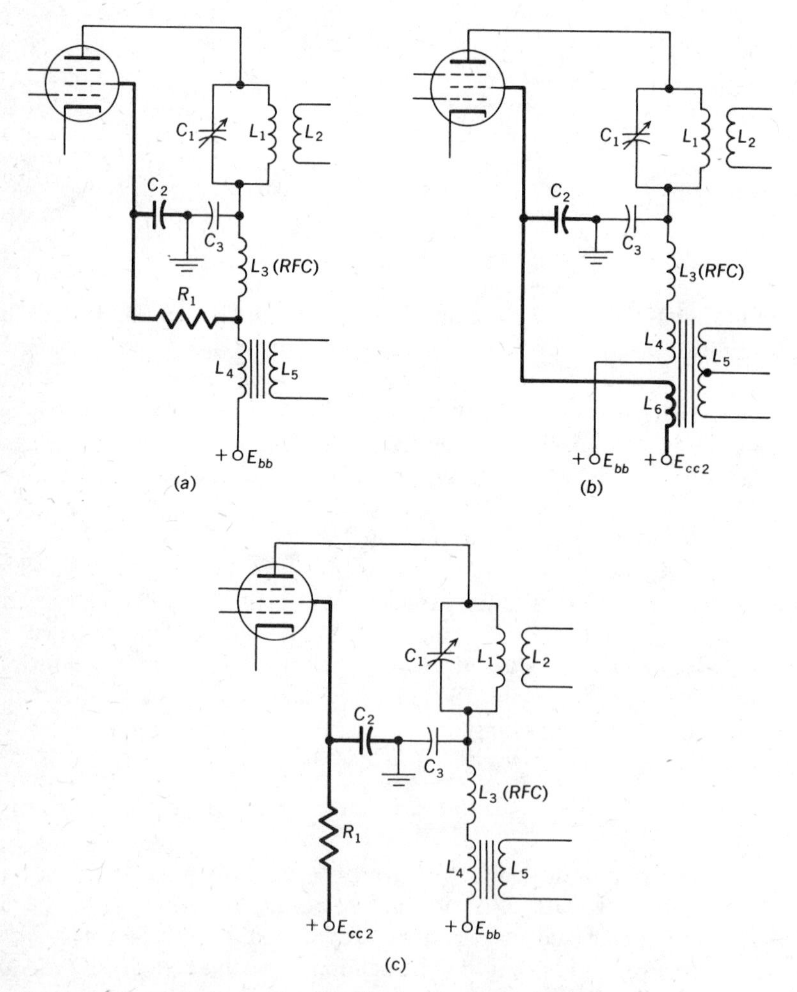

Figure 6-10 Pentode plate-modulation circuits.

winding on the modulation-matching transformer. In either case, Fig. 6–10(a) or (b), the modulating signal power for the screen circuit is supplied by the modulating-signal power amplifier.

Now examine Fig. 6–10(c). At first glance, it might appear that the screen grid is not modulated. However, in this circuit, capacitor C_2 is deliberately selected to have a high reactance at the modulating frequency, so that the screen grid is not "grounded" at this frequency. Consequently, as the modulating signal swings positively and the "effective dc" plate potential increases, the screen current decreases, the drop across resistor R_1 also decreases, and the screen potential rises. Conversely, when the modulating signal opposes the dc plate supply voltage, reducing the effective plate potential, the screen current rises and the screen potential drops. Notice that the screen potential varies in accordance with the modulating signal. In this case, the modulating power for the screen circuit is supplied from the supply source E_{cc2}. The power requirement can be reduced by replacing the resistor with a choke having a high impedance at the intelligence frequency.

OTHER MODULATION CIRCUITS

In the above plate-modulation circuits, high modulating power levels are needed to achieve 100% modulation. Sometimes it is either inconvenient or uneconomical to provide such modulating sources. For example, the modulation-matching transformer for a 100-kW AM transmitter is shown in Fig. 6–11. In addition to its obvious bulk, this "component" weighs 2100 pounds. Where low bulk and low weight for the modulating amplifier is a prime requirement, the modulating signal is fed to some tube element *other than the plate*. Yet, regardless of the injection point, modulation still takes place in the plate circuit. Therefore the circuits that follow are, strictly speaking, plate-modulation circuits.

Grid-Bias Modulation

A circuit for grid-bias modulation is shown in Fig. 6–12. The important feature is that the modulating signal and the carrier signal are both applied to the grid of the tube. This circuit uses double-tuned inductively coupled input, series-fed bias, plate neutralization, and a shunt-fed plate circuit. Any of the variations shown for class-C amplifiers can be used here. Also a tetrode, pentode, or beam power tube can be used in place of the triode, and additional modulation would not be necessary for the screen circuit.

The waveshapes obtained with grid-bias modulation can be seen in Fig. 6–13. Notice that the grid voltage waveshape e_c is not a modulated

Figure 6-11 Riggers using crane to set modulation transformer on its pad. (Courtesy Gates Radio Co.)

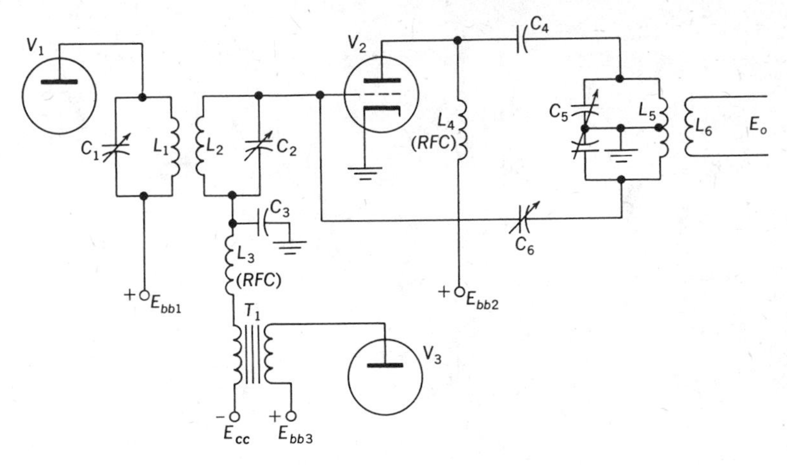

Figure 6-12 Grid-bias modulation.

wave. The excitation voltage merely rides up and down *following* the modulating signal, its amplitude remaining constant. On the other hand, the plate current waveshape does vary in amplitude so that both i_b and the resulting output voltage e_p are modulated waves.

A drawback of this modulation system is that it is rather critical in adjustment. The tube must be biased beyond cutoff. Usually, the bias voltage E_{cc} is approximately twice the cutoff value. The excitation voltage, in the absence of modulation (time interval 0–1), is generally set so as to drive the grid just slightly positive.* The excitation voltage for 100% modulation should be adjusted so that, at the crest of the modulation cycle, the plate-current pulse will double in amplitude (compared to its unmodulated value). At this instant (t_2) the ideal class-C condition $[E_{b(\min)} = E_{c(\max)}]$ should be realized. One further condition to be met is that at the modulation trough (t_3) the instantaneous value of the grid potential should just equal the cutoff bias value. This sets the peak swing of the unmodulated (carrier) voltage at halfway between the $E_{c(\max)}$ and the cutoff value.

To maintain these conditions for distortionless (linear) modulation, all three grid voltages (E_{cc}, e_c, and e_s) must maintain constant amplitudes. The bias source must have good regulation; grid-leak bias should not be used. The excitation stage must have good regulation; it is recommended that its power output capability should be two to three times the required grid driving power. The modulating amplifier should have good regulation. Positive grid operation should be held to a minimum, so as to reduce the variable loading effect.

*Somewhat better linearity (less distortion) is obtained if the grid is not driven positive. However, this sacrifices power output and efficiency.

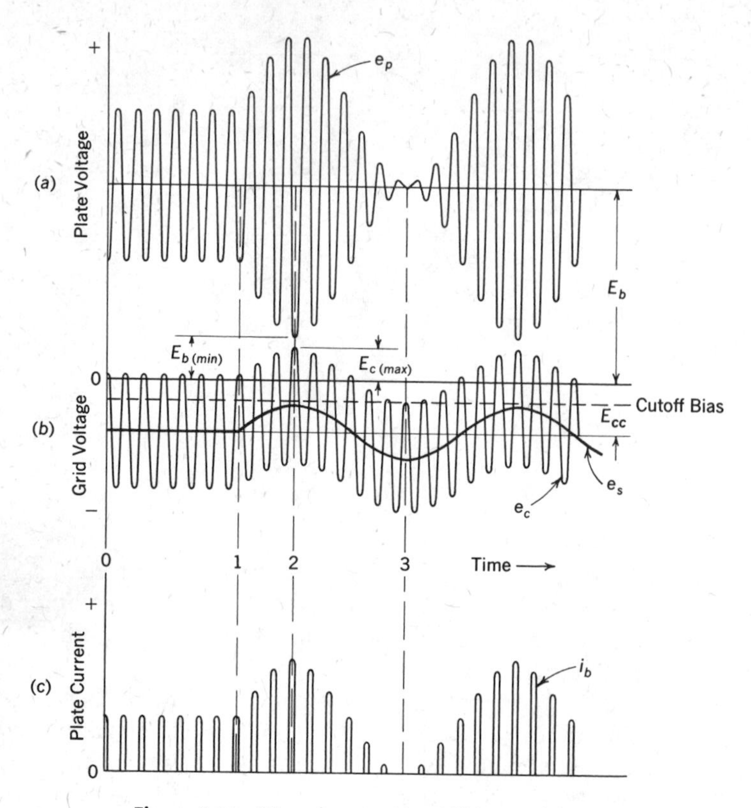

Figure 6-13 Waveshapes with grid-bias modulation.

If we examine the power output capability and efficiency of a grid-bias modulator, two other weaknesses of this circuit can be noted. For example, at the crest of the modulation cycle, the operating conditions are similar to typical class-C operation. The plate current pulse amplitudes for these two cases are equal. But notice in Fig. 6–13(c) that the plate current pulse for a no-modulation condition is only half this ideal value. Therefore the carrier power output is limited to only one-quarter of the rated power that the *same* tube could deliver as a class-C amplifier.

Now study the $E_{b(\min)}$ values in Fig. 6–13(a). At the modulation crest—*and only for 100% modulation*—the $E_{b(\min)}$ value is low, equal approximately to $E_{c(\max)}$, and the high efficiency of a class-C amplifier is realized. For any other condition, $E_{b(\min)}$ rises, the plate dissipation increases, and the plate efficiency falls off. The average efficiency over a modulation cycle is (approximately) only half the efficiency obtainable as a straight class-C amplifier.

If we compare grid-bias modulation with plate modulation, the following salient features should stand out. The grid-bias circuit has a lower plate efficiency; it produces a lower power output for the same

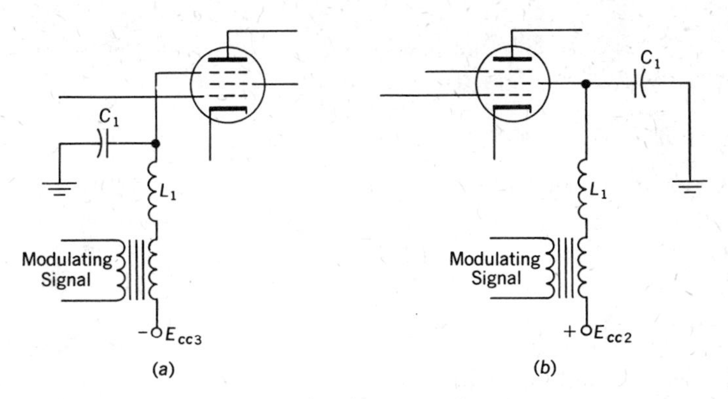

Figure 6-14 Suppressor-grid and screen-grid modulating techniques.

tube; it is more difficult to adjust; it has poorer linearity and thereby a higher distortion level; but it requires much lower modulating power. In this respect it should be recalled that when the modulating signal is injected into the plate circuit, an amplitude of $E_s = E_b$ is required for 100% modulation. By injecting the modulating voltage into the grid circuit, the amplification factor of the tube reduces this requirement to $E_s = E_b/2\mu$.*

Suppressor-Grid Modulation

A pentode class-C amplifier can also be modulated by injecting the modulating signal into the suppressor grid circuit, as shown by the partial schematic of Fig. 6–14(a). The RF excitation is applied to the control grid, using any of the previously discussed coupling methods. Fixed bias or grid-leak bias can be used, with shunt-fed or series-fed circuitry. However, instead of returning the suppressor grid to ground, it is connected to a negative bias source, in series with the modulating signal. The suppressor dc bias is adjusted so as to reduce the RF output voltage to about one-half its normal class-C amplitude, and the modulating signal amplitude is adjusted so as to drive the grid to zero potential. Therefore, the full class-C output and efficiency is obtained only at the crest of the modulation cycle. An RF filter, C_1L_1, may be used to keep RF out of the modulating source. As in previous circuits, the reactance of C_1 must be low at RF, but high at the modulating frequencies.

The power output, plate efficiency, and linearity obtained by injecting the modulating signal into the suppressor grid are about the same as with grid-bias modulation. Yet, because the RF and modulating signals are fed to different grids, this circuit is less critical in adjustment. On the other hand, a higher modulating signal amplitude is needed because of the lower amplification factor of the suppressor as compared to the

*F. E. Terman, *Electronic and Radio Engineering*, Section 15–3.

control grid. The modulating power requirement is very low, so long as the suppressor grid is not driven positive.

Screen-Grid Modulation

A very similar type of circuit is used to inject the modulating signal into the screen-grid circuit of a multigrid tube. This is shown in the partial schematic of Fig. 6–14(b). For the rest of the circuit, any of the variations discussed for class-C amplifiers can be used. However, to prevent exceeding any tube ratings, the screen voltage (dc) is reduced to approximately one-half its normal class-C operating value. Then, at the peak of the modulating voltage, the screen potential will reach its full class-C value. Obviously, the full power output and plate efficiency of a class-C circuit will be obtained only at the crest of the modulation cycle. This is similar to grid-bias and suppressor-grid modulation circuits. However, because of the higher screen voltage and screen current, this circuit requires more modulating power than a grid-bias modulator.

Cathode Modulation

The modulating signal could also be injected between cathode and ground. Such a technique, called *cathode modulation*, is sometimes used. It should be obvious that in this case the modulating signal will affect both the input (grid-cathode) and the output (plate-cathode) voltages. Therefore the characteristics of this circuitry will be in between those for grid and plate modulation.

TRANSISTOR MODULATORS

Modulation can be applied to transistor RF power amplifiers in much the same way as was shown for vacuum-tube circuits. The intelligence signal can be injected into any of the three transistor elements, giving rise to emitter-modulation, base-modulation, and collector-modulation circuits.

Collector Modulation

The most commonly used modulation technique is to feed the modulating signal to the collector, in series with the dc collector supply voltage, in much the same fashion as is done for vacuum-tube plate-modulation circuits. One such circuit is shown in Fig. 6–15. Notice that it is quite similar to the transistor class-C amplifier of Fig. 4–21(b), except that now a modulating signal is also applied to the collector. Notice that the collector circuit is shunt-fed. Transformer T_1 is the modulation matching transformer and matches the output of the modulating signal power amplifier to its load—the class-C RF power amplifier, Q_2. Capa-

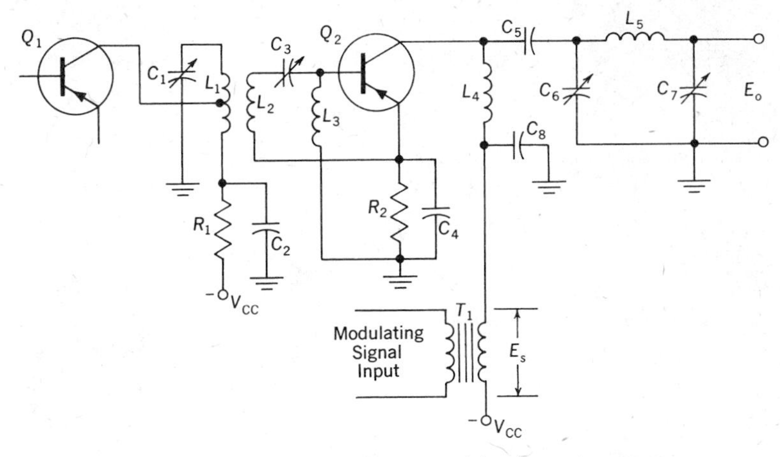

Figure 6-15 A collector-modulated circuit.

citor C_8 serves to keep RF out of the modulating circuit and out of the power supply. As in plate-modulation circuits, C_8 must have a low impedance to RF, but a high impedance to all intelligence frequencies.

Using vacuum-tube equivalence, 100% modulation should be obtained when E_{sm} equals V_{cc}. Unfortunately this condition is not realized in the above circuit. Because of saturation effects, the transistor output voltage does not rise in proportion to the increase in collector voltage. If high levels of modulation are desired, it is necessary to modulate the driving transistor as well as the final stage. Such a circuit is shown in Fig. 6-16.

Collector modulation circuits require high modulating signal power. For 100% modulation, this power should be equal to one-half the carrier power. However, collector efficiency is maintained at a high value, the distortion is low, and circuit adjustments are relatively easy.

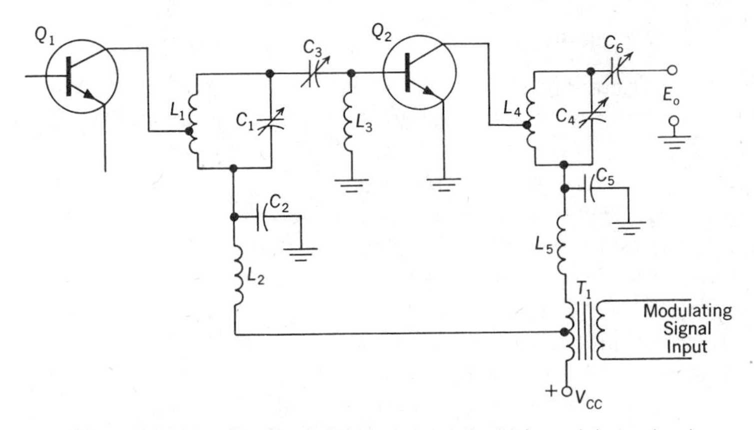

Figure 6-16 A collector modulation circuit for high modulation levels.

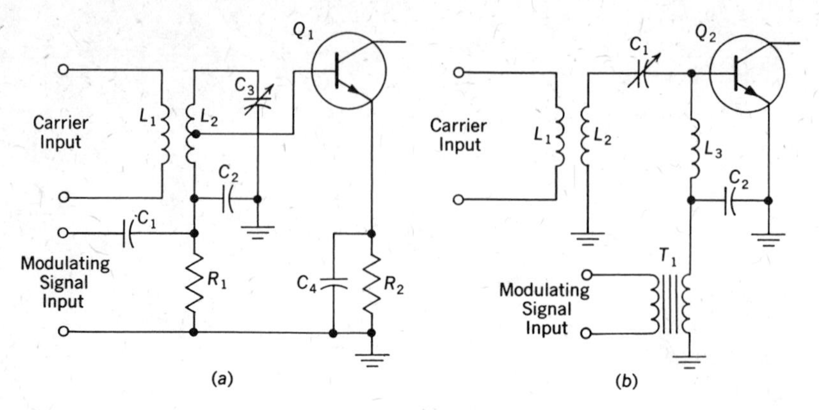

Figure 6-17 Base modulation circuits.

Base Modulation

An appreciable reduction in the modulating power requirement can be obtained if the modulating signal is injected into the base circuit of the transistor. The intelligence signal can be series-fed as shown in the partial schematic, Fig. 6–17(a), or shunt-fed as in Fig. 6–17(b). In either case *R-C* or transformer coupling could be used, with transformer coupling providing better impedance match. With *R-C* coupling, the resistance value R_1 (and the time constant R_1C_2) should not be too large. Otherwise excessive reverse bias could be produced by "grid-leak" action. With regard to operation and waveshapes, base modulation is very similar to its vacuum-tube counterpart—grid-bias modulation—and has the same drawbacks. Compared to collector modulation, this circuit has lower collector efficiency, lower power output for the same transistor, poorer linearity, and is more difficult to adjust.

Emitter Modulation

Characteristics in between those of collector and base modulation are obtained with emitter-modulation circuits, wherein the intelligence signal is injected into the emitter leg. Two such circuits are shown in Fig. 6–18. In diagram (a), resistor R_1 serves the dual function of injecting the modulation signal into the emitter-base circuit (in combination with capacitor C_1) and developing the proper bias (in combination with C_2). Obviously, capacitor C_2 should have a high impedance at the modulating signal frequencies. The transformer-coupled circuit of Fig. 6–18(b) will provide better impedance match, while the separation of coupling and bias circuitry makes adjustment easier.

In the above circuits, tube and transistor, the tank circuits have been shown with variable capacitance for tuning adjustments. It should be realized that, depending on operating frequency, it may be preferable to use fixed capacitance values and variable inductances.

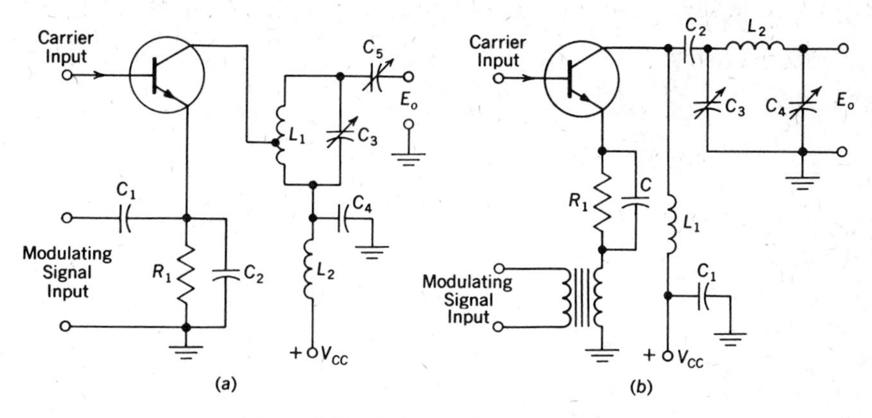

Figure 6-18 Emitter modulation circuits.

OPERATIONAL CHECKS

The action of a modulation circuit can be checked most readily for degree of modulation and for linearity (or distortion) by using an oscilloscope. Plate and collector current indications are also valuable in linearity checking.

Oscilloscope Testing

The simplest oscilloscope technique is to feed the modulated RF wave to the vertical deflection plates of the oscilloscope, and to apply the internal sawtooth sweep voltage (linear time base) to the horizontal plates. Pickup of the RF signal is generally made by coupling a probing coil to the output tank circuit of the modulated stage.* If the modulating signal is a recurrent waveshape, and if the oscilloscope sweep frequency is adjusted to the modulating signal frequency (or some submultiple of this frequency), the modulated wave will be seen on the screen as a stationary pattern. On the other hand, more accurate analyses can be made from *trapezoidal patterns* obtained by setting the oscilloscope for *external sweep* and using the modulating signal as the sweep voltage. Such a connection is shown in Fig. 6–19. The potentiometer P_1 (in combination with capacitor C) is used to adjust the phase of the sweep voltage. The proper phasing is obtained when the screen pattern does *not* have any three-dimensional effect.

*This technique was previously described for neutralization.

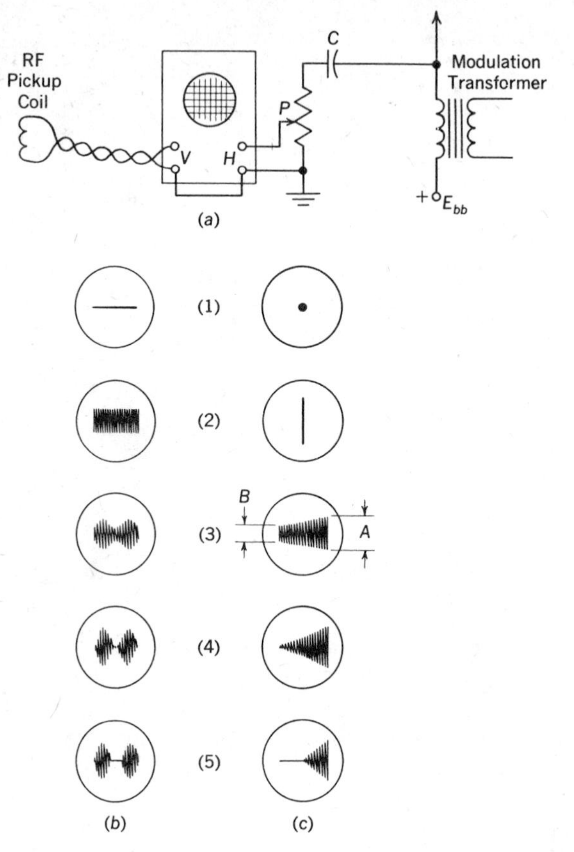

Figure 6-19 Modulation patterns.

The patterns obtained with internal and external sweep are shown in Fig. 6–19(b) and (c) respectively. Five separate conditions are represented. The first case (1) represents a condition of no RF carrier and no modulating signal. A horizontal line is seen in (b1), since the horizontal deflection voltage is supplied by the oscilloscope internal sweep generator. Condition (2) represents the carrier alone (no modulation). Since there is no modulating voltage, the trapezoidal method has no sweep voltage. Diagrams (b3) and (c3) show a modulated wave with less than 100% modulation. The degree can be found by using Eq. (6–3). The trapezoidal diagram is more convenient to use, the heights A and B corresponding to E_{max} and E_{min}. Condition (4) represents 100% modulation. Notice that the small base of the trapezoid has become a point. Overmodulation is represented in condition (5).

The trapezoidal diagrams have a distinct advantage when checking for linearity of modulation. If linearity is good, the trapezoid sides will

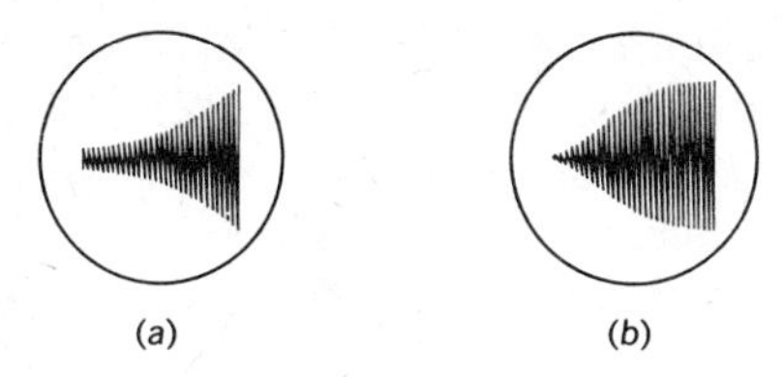

Figure 6-20 Effects of poor linearity.

be straight, as shown in Fig. 6–19(3) and (4). Poor linearity will produce curvature of these sides. Two such conditions are shown in Fig. 6–20. The pattern in Fig. 6–20(a) has an "*extended tip*." The point has been extended farther than would have been the case if the sides were straight lines. Obviously, the power output does not decrease rapidly enough as the modulation trough is approached. Such a condition can arise from improper neutralization or because of stray coupling from a previous stage. Figure 6–20(b) shows curvature at the large end of the trapezoid. This is caused by improper operating conditions and is usually due to excessive bias or insufficient excitation. In either case, $E_{c(\max)}$ (or $V_{BE(\max)}$) is not sufficiently positive, the plate (or collector) current is low, and the power output at the modulation crest is low.

Plate, Collector, and Antenna Current Indications

With linear, symmetrical modulation, the increase in the plate- (or collector-) current pulse amplitudes during the positive half-cycle of modulation is balanced by their decrease during the negative half-cycles, so that the average or direct-current reading will remain constant. Therefore, if the modulation is undistorted, the plate or collector current reading will not change. On the other hand, because of the added sideband power with modulation, the antenna current should increase.

With a condition of overmodulation, the amplitude of the plate or collector current pulses will continue increasing as the modulation crest is approached, but it cannot decrease below zero at the modulation trough. Obviously, the plate or collector current (dc) reading will begin to rise as the modulation level exceeds 100%. Such a condition can be caused by excessive modulating-signal amplitude or by too low a dc supply voltage. With overmodulation, the sideband power continues to increase, and the antenna current also continues to increase. Therefore, antenna current cannot be used to indicate this condition.

One of the most common malfunctions, known as *downward modulation*, is indicated by a decrease in the plate or collector current as modulation is applied. Such a condition is generally caused by insufficient excitation. Let us consider an extreme condition of downward modulation. Using a vacuum-tube circuit, assume that the plate potential is sufficient to cause saturation—before modulation is applied. Then, on the positive half of the modulation cycle, as the effective plate voltage increases with modulation, the plate current pulse amplitude cannot increase. Yet, on the negative half of the modulation cycle, the current pulses will decrease in amplitude. Obviously, the power output will decrease as modulation is applied. Consequently, downward modulation will also be indicated by a decrease in antenna current as modulation is applied. On the other hand, if the RF excitation is increased and the grid is driven more positive, the saturation effect is delayed, and the RF output will increase normally with modulation.

For similar reasons, low heater voltage or weak emission (old tube) may produce the same effect. Less frequent causes of downward modulation are overcoupling to the load and excessive bias. In the latter case this effect occurs at the higher levels of modulation. The antenna current will increase normally up to a point, and then decrease as the critical modulation level is passed.

BASIC TRANSMITTERS

At this point, we have covered the individual circuits used in the various stages of a basic AM transmitter. All that remains now is to put the necessary stages together. This is most readily done in block diagram form, rather than repeating the circuit details. The typical setup is to use a master oscillator followed by a number of power amplifiers and/or frequency multipliers, until the desired frequency and power output are obtained. The output from this final stage is fed to the antenna. Modulation may be applied to any of the power amplifiers. Depending on the stage of application, AM transmitters are classified as employing high-level or low-level modulation.

High-Level Modulation

Commercial transmitters are generally of the high-level modulation type, wherein the modulation is applied to the final power amplifier. A block diagram of such a transmitter is shown in Fig. 6–21. The master oscillator is used to generate the carrier frequency. However, if the operating frequency is high, better stability is obtained by designing the oscillator for some lower (submultiple) frequency and then using frequency multipliers to obtain the desired frequency value. If stability is a prime consideration, a crystal oscillator circuit will be used—particularly if the transmitter is to operate at a fixed frequency. Where ability to change the operating frequency is desired, a variable-frequency oscillator (VFO) circuit is selected.

The next block, a buffer amplifier, may or may not be used. The determining factor again is stability. For utmost stability, the load on the

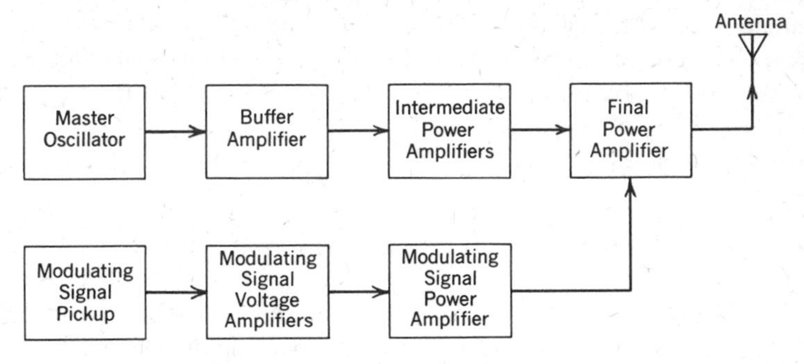

Figure 6-21 Block diagram of a high-level AM transmitter.

oscillator circuit should be light—*and constant.* A class-C power amplifier would not be a suitable load. Instead, a buffer amplifier is used. This stage is biased either class B or—for even lighter loading—class A. Using semiconductors, a field-effect transistor is preferable to a bipolar because of its much higher input impedance. With bipolar transistors, high feedback levels or an emitter-follower type circuit can be used to reduce the loading.

The intermediate power amplifiers are operated in class C for maximum efficiency. Since the signal is unmodulated, this class of operation can be used. This is the advantage of high-level modulation. The number of stages used depends on the power required to drive the final power amplifier. If multiplication of the oscillator frequency is desired, any of these stages may be operated as frequency doublers or triplers.

Modulation takes place in the final power amplifier. Notice the two input signals (carrier and intelligence). Maximum efficiency—class-C operation—is obtained by using plate or collector modulation, with the modulating signal in series with the dc supply voltage. The output of this stage is then fed through a transmission line to the antenna.

If this diagram were to represent a voice or music transmitter, the pickup device would be a microphone, and the following amplifiers would be audio amplifiers. For a television transmitter, the pickup would be a camera, and the amplifiers video amplifiers. If this were a telemetry transmitter, the pickup would be a transducer to convert the pressure, temperature, or other quantity being measured into an electrical signal, and the following stages would be some form of wideband amplifiers.

Low-Level Modulation

In a low-level modulation transmitter, the modulating signal is applied to one of the intermediate power amplifiers. The earlier the stage, the lower the carrier power level, and the modulating-signal power requirement decreases. However, since the RF signal is now modulated, all subsequent intermediate power amplifiers and the final power amplifier must be operated in linear class B, with a consequent loss in efficiency and power-handling ability.

CW Transmitters

An RF carrier can be made to convey intelligence by turning the transmitter on and off in accordance with some prearranged plan or code. A transmitter of this type is called a *continuous wave* (*CW*) transmitter. The term continuous, in this case, refers to the *characteristics* of the wave, which are not changed by the intelligence signal. However, because transmission of the wave is interrupted, it is also referred to as *interrupted continuous wave* (*ICW*). The earliest application of this type was in the transmission of Morse code signals (radio telegraphy), and the RF energy was turned on and off by means of a telegraph key. The term *keying* is still used to describe this process, regardless of the device and circuitry used to control the RF transmission. Since those early days, the applications of CW transmitters have come to include teletype, facsimile,* and telemetry. Telemetering systems have proved invaluable for both missile guidance and the returning of coded information from a space satellite to earth stations.

Keying Requirements

Whether a transmitter's output is controlled by a hand-operated (telegraph) key, by a remote-controlled relay, by punched tape or some other automatic system, several requirements must be met by the keying circuit. The transmitter should develop full power output when the key is closed (key-down position) and the output should drop to zero in the key-up (open) position. The key-down position produces a *mark* (dot or dash in telegraphy), while the key up gives the *space* between marks. If the keying system is not effective, some energy may still reach the antenna when the key is open. This *back wave* will be heard during the space period, and depending on its strength (relative to a mark) it may destroy the intelligibility of the transmission. The danger of a strong back wave is increased if the final power amplifier is keyed.†

At first thought it might seem that, in an ideal situation, the transmitter power output should rise to maximum and fall to zero instantaneously as it is keyed on and off. Such a waveshape is shown in Fig. 6–22(a) to represent the letter *R* (· – ·) in Morse code. This same waveshape would also be obtained if a carrier is fully modulated by the rectangular wave of Fig. 6–22(b). This modulating signal is a complex wave which, because of its steep sides, has a high harmonic content. Consequently,

*Transmission of pages of printed matter or still pictures.

†Incomplete neutralization can aggravate this condition.

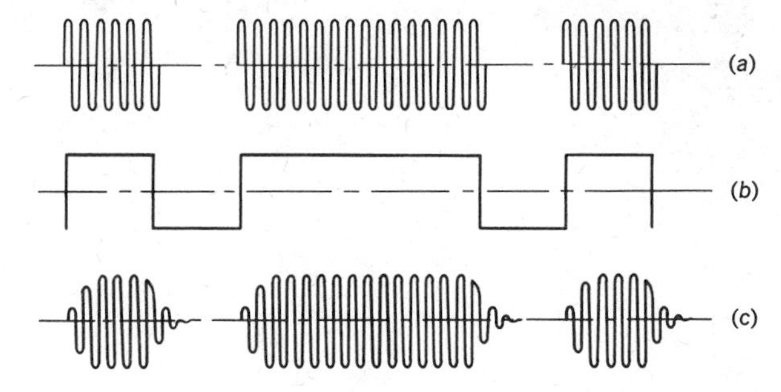

Figure 6-22 Keyed CW waveforms.

the modulated wave, Fig. 6–22(a), will contain many side frequencies, extending appreciably beyond the frequency of the carrier itself. This results in two undesirable effects. The high sideband frequencies produce disturbing clicking sounds in the receiver, making the signal difficult to read, while the wide bandwidth of the transmitted signal can cause interference to other receivers tuned to adjacent carrier frequencies. It is preferable for the power in the transmitted wave to build up gradually and decay gradually (see Fig. 6–22(c)).

Interference to nearby receivers—even when tuned to appreciably different frequencies—can also be caused by *arcing* at the key contacts. These sparks are damped oscillations and may be radiated directly by the wiring or may modulate the output of the transmitter.

These defects can be greatly reduced by using suitable filters and time-delay circuits. Such a circuit is shown in Fig. 6–23. The RF chokes and capacitor C_1 form an RF low-pass filter, preventing the high-frequency spark signals from getting into the keying line. The second part, L_3C_2, is a time-delay or *lag circuit*. The inductance will slow down the rise time of the output wave, while the capacitor increases the decay time. The effect of such a filter on the keyed output is shown in Fig. 6–22(c). On the other hand, if too much lag is used, the end of one character can run into the start of the next character, and the "space" may be lost. This results in indistinct signals that may be impossible to read. Unfortunately, no one lag filter is suitable for all conditions. The optimum

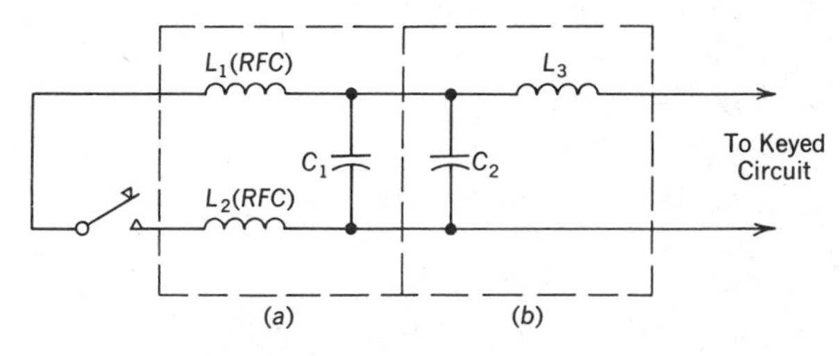

Figure 6-23 Keying filters.

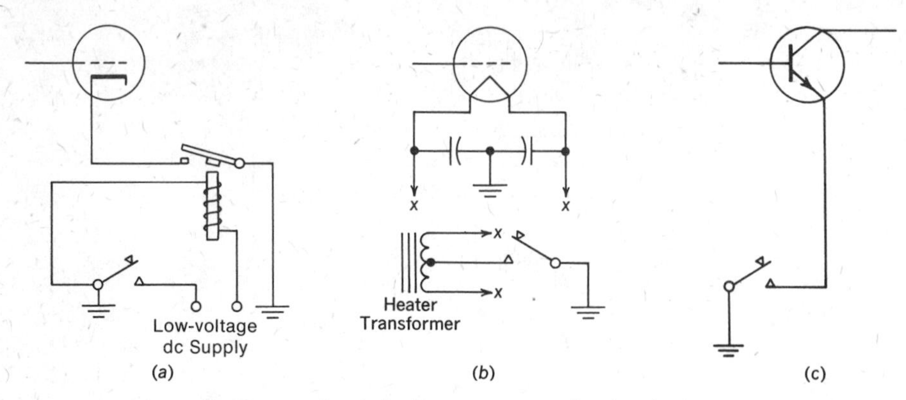

Figure 6-24 Cathode, center-tap, and emitter keying.

component values depend on the current and voltage being interrupted. In general, the higher the current, the lower the inductance value, and the higher the voltage, the lower the capacitance.

A keying circuit must meet three other criteria. It should not cause even a slight shift in the carrier frequency. Such a shift would be heard as a "chirp" in the receiver and would impair the intelligibility of the signals. For this reason keying circuits are seldom applied to an oscillator or buffer amplifier, since changes in temperature or changes in loading due to keying could result in a frequency shift. For operator safety, the voltage across the key itself, in hand-keyed circuits, should be low and, preferably, the keying bar should be grounded. Where high voltages exist, a relay-operated system should be used wherein the grounded hand key energizes the relay from a low-voltage source, and the relay opens and closes the transmitter circuit (see Fig. 6–24(a)). Finally, for safety of the equipment, the key-up condition should not cause any tube or transistor ratings to be exceeded. In this respect it is preferable in vacuum-tube circuits that all stages following the keyed stage use fixed bias. Otherwise, if grid-leak biased, they should be provided with sufficient safety bias to limit the plate current to safe values during key-up periods. Actually, full cutoff is preferable, since this reduces the danger of a back wave. In this respect, the transistor circuits have a decided advantage. Full cutoff is obtained when the base-emitter bias is zero, and no added precautionary circuits are needed.

Keying Circuits

Theoretically, a key (or relay) can be inserted at any point where interrupting the circuit will cut the RF power output to zero. However, only a relatively few locations are used commercially. Generally, best results are obtained when keying the final stage. In high-power transmitters, this means that the keyed stage is a vacuum tube.

CATHODE, CENTER-TAP, OR EMITTER KEYING. One of the more

commonly used basic keying circuits is shown in Fig. 6–24.* The cathode circuit, or (in a directly heated tube) the heater center-tap, or the emitter circuit is broken to stop transmission. In Fig. 6–24(a), a relay is used to break the cathode circuit. When the key is closed, the relay is energized, pulling down the armature and completing the cathode circuit. Notice that the relay armature is always at ground potential. It might therefore seem that direct keying could be used with complete safety. Yet when the relay is open, the cathode potential rises to the full E_{bb} value. This would make a key extremely dangerous to handle. With transistors, the power supply voltage may be low enough so that a relay would not be necessary.

PRIMARY KEYING. A very simple method of keying a small transmitter is to interrupt the primary circuit of the *plate* (*or main*) power transformer. Because of the high inductance and capacitance in the power supply filter, lag circuits are not needed. In fact, keying is generally too "soft," so that it can be used only with relatively slow-speed hand keying.

PLATE- OR COLLECTOR-CIRCUIT KEYING. In Fig. 6–25, the power output is cut off by breaking the dc supply line to the final power amplifier. Since the key location is beyond the power supply filters, adequate key-click filtering is necessary. Although the key could be inserted into either the positive or negative supply line, it is usually connected as shown. In either case, when tubes are used, a high-voltage problem exists, and remote relay control should be used.

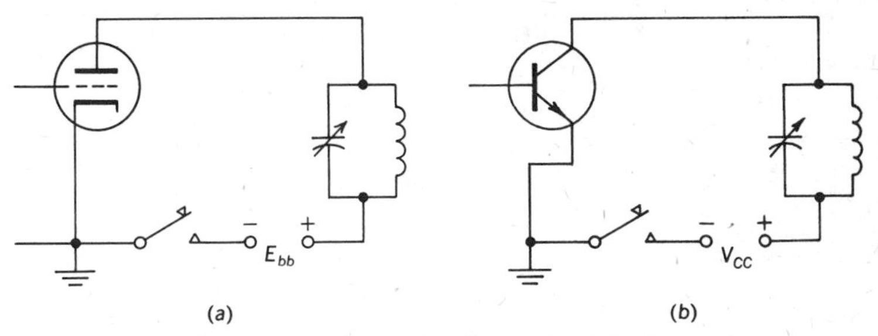

Figure 6-25 Plate and collector-circuit keying.

SCREEN-GRID KEYING. With multigrid tubes, the key could be placed in series with the dc supply to the screen grid. The current and voltage interrupted by this method is generally considerably lower, reducing the possibility of arcing, and improving safety.

BLOCKED-GRID OR BLOCKED-BASE KEYING. A very effective way of controlling CW transmission is to bias an element so that the tube or transistor will remain in cutoff even at the peak of the input carrier signal. Many circuit variations are possible. Three such cases are shown

*For simplicity, keying filters are not shown, and except for Fig. 6–24(a), direct keying is shown even when relay operation should be indicated.

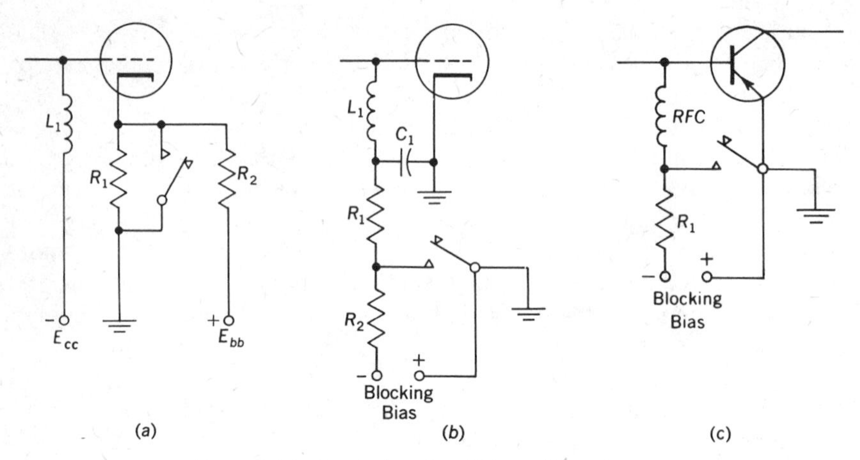

Figure 6-26 Blocked-grid and blocked-base keying.

in the partial schematics of Fig. 6–26. In the first of these, the operating fixed-bias voltage is applied to the control grid. However, the cathode is connected through a high-resistance voltage divider to the plate supply voltage. When the key is open, the cathode is at some positive potential. The combination of E_{cc} and the positive cathode is sufficient to cut off the tube even at the peak of the RF input. Meanwhile, the plate-cathode voltage is reduced, so that cutoff occurs at a lower bias value. In the key-down position, the cathode is grounded, and normal operating voltages are restored.

The circuit of Fig. 6–26(b) requires an additional bias voltage—the blocking bias—sufficient to hold the tube in cutoff at the peak of the RF input signal. With the key down, this blocking voltage is removed, and the circuit functions as a normal grid-leak biased amplifier. Resistor R_1, in combination with C_1, also serves as a time delay filter to remove key clicks. A similar action can be obtained with the blocking bias applied to the suppressor grid.

Figure 6–26(c) shows the same principle applied to a transistor. With the key up, a positive bias voltage is applied to the base, and this bias is of higher value than the peak excitation voltage. The PNP transistor is kept in cutoff for the full excitation cycle. Conversely, in the key-down position, the base voltage is shorted, and normal class-C action is restored.

ELECTRON-TUBE OR SWITCHING-TRANSISTOR KEYING. A circuit that provides excellent safety, has negligible sparking, and is easy to filter, uses a switching tube or transistor to control the transmitter. In turn, the tube or semiconductor is switched between cutoff and heavy conduction by the action of the key. When in heavy conduction, the resistance of the switching device should be as low as possible. Such a circuit is shown in Fig. 6–27. Notice that it uses a combination of emitter

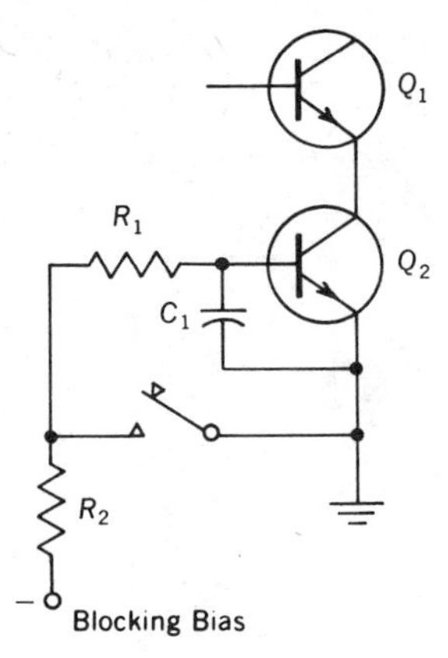

Figure 6-27 Switching transistor keying.

keying and blocked-base keying, with Q_2 acting as the "relay" in the emitter leg of the RF amplifier Q_1. The switching transistor Q_2 is "turned on or off" by the blocked-base keying circuit.

Tone-Modulated Keying

To improve long-distance reception of code signals, the RF carrier can be modulated by a low-frequency signal (400–1000 Hz), in addition to being keyed. This produces additional side-frequency components. Whereas a keyed signal may have a total frequency spread of approximately 100 Hz ($\pm$50 Hz), a tone-modulated signal might have components as far away from the carrier frequency as $\pm$1000 Hz. This is of advantage in reducing signal fading at the receiver due to ionospheric conditions. It has been found that this fading effect varies for different frequencies. Because of the additional frequency components contained in a tone-modulated wave, there is less chance that the entire signal will fade simultaneously.

Frequency-Shift Keying

Receivers generally employ automatic gain control (AGC) or automatic volume control (AVC) circuits that vary the gain of the RF amplifiers to counteract changes in the received signal strength. Such circuits are invaluable in compensating for fading and in mobile equipment where the signal level can vary with location. Yet, when used with the above keying circuits, AGC can create problems. During the space intervals, there is no transmission. The receiver therefore operates at maximum gain, with the result that atmospheric disturbances, man-made static, and inherent receiver noises are heard with appreciable volume. This will interfere with the intelligibility of the mark signals. (These noises could also key an automatic printer or control circuit.) One simple remedy is to disable the AGC circuit, but this would take away the control over varying signal-strength effects.

Another remedy is to maintain transmission at all times. Instead of turning the carrier on and off, its frequency is shifted so that one frequency corresponds to a space, the other to a mark. This is known as *frequency-shift keying* (*FSK*). Since the carrier is not cut off, full quieting action is maintained, improving the readability of the coded transmission. The amount of frequency shift must be kept small so that both signals (mark and space) are within the pass band of the receiver tuning circuits. Frequency shifts of 400–2000 Hz are generally used. Frequency-shift keying has gained in popularity and is used extensively for long-distance telegraphy and particularly with high-speed automatic printing equipment.

Frequency shift can be obtained quite simply by shunting a capacitor (or coil) across a portion of the oscillator tank circuit or across the crystal holder. A relay, or better yet a switching tube or semiconductor, can be used to connect and disconnect this component. The key in turn will activate the switching device from a remote location. However, the most commonly used circuits employ a *reactance tube or transistor* (instead of a coil or capacitor), and instead of changing the frequency of the oscillator circuit, *frequency conversion* techniques are used to effect the shift. Since each of these circuits are treated in later chapters, they will not be discussed at this time.

REVIEW QUESTIONS

1. (*a*) How many separate channels of communication can be established when intelligence is transmitted directly? Explain. (*b*) State another disadvantage of direct transmission. Explain.

2. Refer to Fig. 6–1. (*a*) What does waveshape (a) represent? (*b*) What do the waveshapes in diagram (b) represent? (*c*) How do they differ? (*d*) What type of wave is shown in diagram (c)? (*e*) How is this waveshape obtained? (*f*) What does ΔE_{cm} represent? (*g*) How does the waveshape in diagram (d) differ from that in diagram (c)? (*h*) How is the waveshape in (d) obtained? (*i*) What does the value of ΔE_{cm} depend on?

3. Refer again to Fig. 6–1. (*a*) Compare the amplitudes of E_{s2}, E_{s3}, and E_{s4}. (*b*) Which modulating signal produces the waveshape in (f)? in (g)? (*c*) Compare the ΔE_{cm} values for the waveshapes in (d), (f), and (g). (*d*) Why is this so? (*e*) How do the waveshapes in (f) and (g) differ? (*f*) Why is this so? (*g*) What determines the rate of change in the amplitude of an AM wave?

4. (*a*) In Fig. 6–1(c), (d), (f), and (g), what is the dashed line on top and bottom called? (*b*) Is this another component of the wave? Explain.

5. (*a*) What is the term *modulation factor* a measure of? (*b*) What is the minimum value this factor can have? (*c*) What is its maximum value? (*d*) Give another term used to indicate degree of modulation. (*e*) How are these two terms related?

6. In Fig. 6–2, what do each of the following symbols represent: (*a*) E_{cm}? (*b*) E_{sm}? (*c*) $E_{\min}$? (*d*) $E_{\max}$?

7. Using the symbols listed in Question 6 and assuming linear (symmetrical) modulation, give two equations for calculating the modulation factor.

8. If the modulation process is distorted, rather than linear, how is the degree of modulation evaluated?

9. In Fig. 6–1, approximately what level of modulation is shown in (*a*) diagram (a)? (*b*) diagram (b)? (*c*) diagram (c)? (*d*) diagram (d)?

10. (*a*) What level of modulation is indicated in Fig. 6–3? (*b*) How is this condition obtained? (*c*) What is the effect of this condition during time interval t_1?

11. (*a*) Name two devices that could be used for producing a modulated wave. (*b*) What specific characteristic must the device have to be used in this manner? (*c*) Why is this characteristic necessary?

12. A sinusoidal carrier wave is modulated by a sinusoidal intelligence signal of much lower frequency. (*a*) Is the plate current sinusoidal? Explain. (*b*) Name the components present in the plate current. (*c*) Are all of these components present in the output voltage? Explain. (*d*) Name the components present in the output waveshape. (*e*) Give the algebraic value of each component. (*f*) For 100% modulation, what is the relation between the carrier amplitude and the amplitude of the other components? (*g*) How is the relative amplitude affected by the degree of modulation?

13. (*a*) Give the composite equation showing the three components of a wave that is sine-wave modulated. (*b*) What is the phase relation among these components at $t = 0$?

14. Refer to Fig. 6–4. (*a*) Which condition refers to the above equation at $t = 0$? (*b*) Will the instantaneous condition always remain so? Explain. (*c*) In diagram (a) why is E_l shown with a clockwise arrow? (*d*) Why is the arrow on E_u counterclockwise? (*e*) What is the resultant of these phasors? (*f*) In diagram (b), E_l is shown at a 45° angle to E_c. Why is E_u also at the same angle? (*g*) Why has the resultant value increased? (*h*) When would the resultant have maximum value? (*i*) What would its value be? (*j*) What modulation level is this? (*k*) In diagram (c), E_u and E_l are shown with downward rotating arrows. How is this obtained? What was changed? (*l*) What has happened to the resultant in diagram (d)? (*m*) When will it be a minimum?

15. (*a*) Under what condition will a modulated wave contain sidebands instead of side frequencies? (*b*) Which condition is more commonly encountered in operation? (*c*) How many sidebands are there? (*d*) What determines the maximum extent of the sidebands?

16. (*a*) When a carrier is modulated, what happens to the power level of the carrier itself? (*b*) How much of the intelligence is contained in the carrier component of the modulated wave? (*c*) Where is the intelligence found? (*d*) How much in each of these components?

17. (*a*) At 100% modulation, why does the power in each side frequency equal one quarter of the carrier power level? (*b*) How does the degree of modulation affect the total sideband power? (*c*) Give an equation to show this relation.

18. (*a*) As the modulation level is increased, what happens to the total power in the modulated wave? (*b*) Compared to the carrier power, what is the total power at 100% modulation? (*c*) In an AM transmitter, why is it preferable to maintain as high a modulation level as possible? (*d*) Is there any difference in the power ratings of a given transistor when used as a class-C amplifier modulated or unmodulated? Explain.

19. Refer to Fig. 6–6. (*a*) What value of E_{cc} is generally used? (*b*) What class of operation is this? (*c*) In the absence of modulation, what is the effective value of plate potential? (*d*) Is it constant? (*e*) When modulation is applied what happens to this value? (*f*) What is its maximum value (algebraically)? (*g*) What is its minimum value? (*h*) Compared to E_{bb}, how high and how low may these values go? (*i*) To what condition would these limit values apply?

20. Refer to Fig. 6–7. (*a*) Do these curves represent 100% modulation? (*b*) What factor is responsible for this? (*c*) When modulation is applied, does the bias remain constant? (*d*) Does the excitation remain constant? (*e*) Does the plate supply voltage remain constant? (*f*) Why do the plate current pulses vary in amplitude? (*g*) Why does the cutoff bias value become more negative during the positive half of the modulation cycle? (*h*) Why does the RF waveform in diagram (d) differ from that in diagram (a)? (*i*) Is the plate efficiency high before modulation? Explain. (*j*) Is the plate efficiency high during modulation? Explain from the waveshapes of (a), (b), and (c).

21. Refer to Fig. 6–8. (*a*) Which tube is the modulator? (*b*) What is the function of the other tube? (*c*) What is the function of C_1? (*d*) What is the function of C_3? (*e*) Is any other component associated with this function? (*f*) What two factors govern the capacitance value of C_3? (*g*) What is the function of R_1? (*h*) What is the function of C_4? (*i*) Why are the components C_4R_1 necessary? (*j*) State two disadvantages of the Heising circuit.

22. Refer to Fig. 6–9. (*a*) What is the function of V_1? (*b*) What is the function of V_2 and V_3? (*c*) What type of circuit is this? (*d*) Does plate modulation require such a dual-tube circuit? (*e*) What is T_1 called? (*f*) What acts as the load across the secondary of this transformer? (*g*) What determines the value of this load? (*h*) What is the function of this transformer? (*i*) Why isn't a voltage-dropping network, such as C_4R_1 of Fig. 6–8, used here?

23. When using tetrodes in plate-modulated circuits, what additional requirement must be met to obtain 100% modulation, as compared to triode circuits?

24. Refer to Fig. 6–10. (*a*) State two functions of resistor R_1 in diagram (a). (*b*) Compare the capacitance values needed for C_2 and C_3. (*c*) Why is this so? (*d*) What is the function of L_2? (*e*) What is the function of L_5? (*f*) What is the function of L_6 in diagram (b)? (*g*) In this same diagram, what is the probable reason for the center tap on winding L_5? (*h*) Which circuit, (a) or (b), requires more modulating power? Why? (*i*) Is the screen grid in diagram (c) modulated? Explain. (*j*) What supplies the screen modulating power in this diagram? (*k*) How could this power requirement be reduced?

25. Refer to Fig. 6–11. (*a*) What is the function of V_1? (*b*) V_2? (*c*) V_3? (*d*) Why is this called grid-bias modulation? (*e*) What value of E_{cc} is generally used? (*f*) What determines the capacitance value C_3?

26. Refer to Fig. 6–13. (*a*) Is the RF waveshape in diagram (a) a modulated wave? How can you tell? (*b*) Is the RF waveshape in diagram (b) modulated?

How can you tell? (*c*) During what time interval is there no modulation? (*d*) How do the operating conditions during this interval compare with typical class-C operation? (*e*) How does the plate efficiency during this interval compare with class-C? Why? (*f*) When is optimum class-C plate efficiency obtained? (*g*) Why is this so?

27. Compare grid-bias and plate modulation circuits with respect to (*a*) power output for a given tube type (explain), (*b*) plate efficiency (explain), (*c*) modulation linearity (explain), (*d*) ease of obtaining optimum operation (explain), and (*e*) modulating power requirement (explain).

28. Refer to Fig. 6–14. (*a*) Why does one circuit show a negative E_{cc} supply, while the other is positive? (*b*) What is the normal suppressor-grid potential in a class-C amplifier? (*c*) What value of suppressor-grid bias is used in diagram (a)? (*d*) What value of modulating signal voltage is required (compared to this bias) for 100% modulation? (*e*) Is typical class-C plate efficiency obtained throughout the modulation cycle? Explain.

29. Compare modulation by suppressor-grid injection to modulation by control-grid injection, with regard to (*a*) power output with a given tube, (*b*) plate efficiency, (*c*) ease of adjustment, (*d*) linearity of modulation, (*e*) modulating voltage required, and (*f*) modulating power required.

30. Refer to Fig. 6–14(b). (*a*) What value of E_{cc2} is generally used? (*b*) Why is this change necessary? (*c*) How does the power output and plate efficiency compare with normal class-C values? (*d*) Compare the modulation power requirement of this circuit and the grid-bias circuit. (*e*) Why is this so?

31. (*a*) Can the modulation signal be applied to the cathode circuit? (*b*) How could this be done? (*c*) What is such a circuit called? (*d*) With regard to plate efficiency and power output, how would this circuit compare with grid-bias modulation? (*e*) With plate modulation? (*f*) Why is this so?

32. (*a*) Name three modulation circuits that can be used with transistors. (*b*) Which of these is most commonly used? (*c*) Which has the highest collector efficiency? (*d*) Which requires the least modulating signal power?

33. Refer to Fig. 6–15. (*a*) What type of bias would normally be used in this circuit? (*b*) State two ways by which this bias is obtained. (*c*) What is the function of L_4? (*d*) What is the function of C_8? (*e*) Are there any limitations on the value of C_8? Explain. (*f*) What is the function of T_1? (*g*) Where does the input to T_1 come from?

34. In the circuit of Fig. 6–15, can 100% modulation be achieved? Explain.

35. Refer to Fig. 6–16. (*a*) How is bias for transistor Q_2 obtained? (*b*) Why is the collector connected to a tap on L_4? (*c*) What is the function of L_5? (*d*) What is the function of C_5? (*e*) What limits the maximum value of this capacitor? (*f*) What is the function of T_1? (*g*) Why is the collector circuit of transistor Q_1 returned to this transformer? (*h*) Why is this necessary?

36. Refer to Fig. 6–17(a). (*a*) To what would the two terminals of L_1 be connected? (*b*) To what would the terminal of C_1 be connected? (*c*) Are there any limits to the value of C_1? Explain. (*d*) Are there any limits to the value of C_2? Explain. (*e*) What is the function of R_1? (*f*) Is its value in any way critical? (*g*) What is the function of R_2C_4? (*h*) To what would the collector circuit be connected? (*i*) What is the vacuum-tube equivalent of this circuit called?

37. Refer to Fig. 6–17. (*a*) Compare the tank circuits used in diagrams (a) and (b). (*b*) Why is the base of Q_1 fed to a tap on L_2? (*c*) Why isn't the base of Q_2 similarly fed to a tap on its coil? (*d*) Compare the type of feed used with each modulating signal. (*e*) Compare the base-emitter bias system used in each circuit. (*f*) What advantage does transformer T_1 of diagram (b) provide? (*g*) What is the function of L_3 in this diagram? (*h*) Are there any limiting values to this choke? (*i*) Are there any limitations on the value of capacitor C_2?

38. Refer to Fig. 6–18(a). (*a*) What is the function of C_1R_1? (*b*) Does R_1 have any other function? Explain. (*c*) Are there any limitations on the value of capacitor C_2? (*d*) What is the function of L_2C_4? (*e*) Are there any limitations on the value of C_4? Explain.

39. Refer to Fig. 6–18(b). (*a*) State two advantages of this modulation injecting technique over diagram (a). (*b*) Will the modulating signal affect the input voltage relations? Explain. (*c*) Will it affect output relationships? Explain. (*d*) Name the vacuum-tube equivalent of this circuit.

40. Refer to Fig. 6–19(a). (*a*) What use is made of the RF pickup coil? (*b*) Which type of pattern is obtained using the connections shown? (*c*) What is the function of potentiometer *P*? (*d*) Is any other component associated with this function? (*e*) How can you tell when proper phasing is obtained? (*f*) Will the patterns obtained be stationary? If not, what controls must be adjusted to make them so? (*g*) Why is this so? (*h*) What changes must be made to this circuit to obtain the modulated wave patterns? (*i*) Will these be stationary? If not, how are they made so?

41. Refer to Fig. 6–19(c). (*a*) What transmitting conditions produce the pattern shown in case (1)? Why? (*b*) in case (2)? Why? (*c*) How is percent modulation obtained from the pattern in (c3)? (*d*) What is the degree of modulation shown in (c4)? (*e*) What condition produces the pattern in (c5)?

42. Refer to Fig. 6–20. (*a*) In diagram (a), what is the meaning of the inward curvature of the trapezoid sides? (*b*) State two possible causes for this effect. (*c*) What is the meaning of the curvature shown in diagram (b)? (*d*) State two possible causes for this effect.

43. With linear modulation, as the modulation level is increased to 100%, what happens to (*a*) the plate current reading in the modulated RF amplifier? Why? (*b*) the antenna current reading? Why?

44. In a solid-state transmitter, as the modulation level is increased beyond 100%, what happens to (*a*) the collector current reading? Why? (*b*) the antenna current reading? Why?

45. (*a*) Can the plate current reading decrease when modulation is applied? (*b*) Is this a normal reaction? (*c*) What is this effect called? (*d*) What happens to the power output? (*e*) What is the most common cause for this effect? (*f*) What happens to the antenna current in such cases?

46. Refer to Fig. 6–21. (*a*) Why is this called high-level modulation? (*b*) What change (or changes) would be necessary to convert this to low-level modulation? (*c*) If this unit were to be used on a fixed operating frequency, what would be the most desirable circuit for the *master oscillator* block? Why? (*d*) Is a buffer amplifier always used? (*e*) What determines this choice? (*f*) Is a buffer amplifier operated in class C? Explain. (*g*) How much and what type of bias is used in these stages? Why? (*h*) What class of operation is used for the intermediate power amplifiers? Why? (*i*) How many stages are normally included in this block? (*j*) If this block diagram represented a television transmitter, what would each of the three lower blocks represent? (*k*) Which block, if any, might contain frequency multipliers? (*l*) When and why would such stages be used?

47. (*a*) What advantage is gained by using low-level modulation? (*b*) Compared to a high-level system using the same type of tubes in the RF stages, state two disadvantages of the low-level system. (*c*) Why is this so?

48. (*a*) What type of output wave is obtained from a CW transmitter? (*b*) What is the process for obtaining such an output called? (*c*) Name three applications for CW transmitters.

49. (*a*) In CW transmission, what does the term *mark* mean? (*b*) What is the opposite of a mark called? (*c*) How are these two effects produced? (*d*) What is a back wave? (*e*) What is the effect of a back wave?

50. Refer to Fig. 6–22. (*a*) Which of these waveshapes is a keyed waveform? (*b*) What is the relationship between waveshapes (a) and (b)? (*c*) Which of the keyed waveshapes is the preferable waveform? (*d*) What makes it so? (*e*) How can this preferable waveform be obtained?

51. Refer to Fig. 6–23. (*a*) What is the specific function of section (a) of this filter? (*b*) What is section (b) often called? (*c*) What is its function? (*d*) Is its effectiveness improved by increasing the component values? Explain. (*e*) Can keying affect the carrier frequency? When and why? (*f*) Is this desirable? Explain.

52. (*a*) In hand-keyed circuits, why is there a possibility of danger to the operator? (*b*) How is this possibility avoided?

53. (*a*) When keying is applied to a low-level stage of a vacuum-tube transmitter, how does this affect the choice of circuitry for the subsequent stages? (*b*) Why is this necessary?

54. In Fig. 6–24(a), since the relay armature is at ground potential, could direct keying be used with complete safety? Explain.

55. Refer to Fig. 6–24(b). (*a*) Why is transmission interrupted when the key is up? (*b*) Does this circuit present any high-voltage hazard? Explain.

56. (*a*) Where is the key located when "primary" keying is employed? (*b*) Is this technique suitable for high-speed keying? Explain. (*c*) Is it suitable with high-power transmitters? Why?

57. Does any voltage hazard exist with the keying circuits of Fig. 6–25? Explain.

58. Is there any advantage in screen-grid keying over plate circuit keying?

59. Refer to Fig. 6–26(a). (*a*) What type of bias is used in this circuit? (*b*) How does the value of E_{cc} compare to the value normally used for class-C service? (*c*) Why is transmission interrupted when the key is up? (*d*) Why must R_1-R_2 be a high-resistance voltage divider?

60. Refer to Fig. 6–26(b). (*a*) What type of "operating" bias does this circuit use? (*b*) How does the blocking-bias voltage compare to the E_{cc} value of diagram (a)? (*c*) Why is this necessary? (*d*) Why is resistor R_2 necessary? (*e*) What is the primary function of resistor R_1? (*f*) Does it serve any other function? (*g*) Is any other component associated with this second function?

61. Refer to Fig. 6–26(c). (*a*) Why is this circuit fairly immune to key sparking interference? (*b*) How does the voltage hazard compare with cathode, plate, or screen-grid keying?

62. Refer to Fig. 6–27. (*a*) What is the function of transistor Q_1? (*b*) What is the function of transistor Q_2? (*c*) What two basic keying circuits are combined in this schematic? (*d*) Why is this circuit safer for the operator than any of the previous circuits? Explain.

63. (*a*) What is meant by "tone-modulated" keying? (*b*) What advantage does this system have over simple on-off keying? (*c*) What is a disadvantage of on-off keying whether tone-modulated or not? (*d*) Why is this so?

64. (*a*) What other type of keying is used to remedy the above situation? (*b*) In this system how is a mark differentiated from a space? (*c*) What range of differentiation is commonly used? (*d*) Why is the change kept small? (*e*) Describe a simple means for obtaining this effect.

PROBLEMS AND DIAGRAMS

1. Sketch a modulated wave showing 50% modulation. Indicate key values to support this degree of modulation.

2. A carrier wave of 160 V is modulated by a sine-wave intelligence signal of 100 V. Assuming linear modulation, find (*a*) the highest peak value of the modulated wave, (*b*) the minimum amplitude of the modulated wave, and (*c*) the modulation factor. (Save these answers for Problem 8.)

3. A modulated wave has a crest amplitude of 60 V and a trough amplitude of 15 V. Assuming linear modulation, find (*a*) the percent modulation, (*b*) the amplitude of the carrier signal, and (*c*) the amplitude of the modulating signal. (Save these answers for Problem 9.)

4. A 1000-kHz carrier is modulated by a 4000-Hz sine wave. (*a*) What frequencies are present in the collector current of the modulated stage? (*b*) What frequencies are present in the output modulated wave?

5. A 3.5-MHz carrier is modulated by a complex wave containing a 2500-Hz and 4-kHz component. List the side frequencies produced in the modulated wave.

6. Using engineering graph paper (20 × 20 per inch), draw the following waveshapes: (*a*) On an axis 2 in. from the top of the paper, a sine wave of amplitude 1 in., 24 div per cycle, starting at zero and going positive. (*b*) On an axis $4\frac{1}{2}$ in. from the top, a cosine wave, amplitude $\frac{1}{2}$ in., 30 div per cycle, starting at positive maximum. (*c*) On an axis 6 in. from the top, another cosine wave amplitude $\frac{1}{2}$ in., 20 div per cycle, starting at negative maximum. (*d*) On the same axis as curve *a*, draw the resultant of waves *a*, *b*, and *c*. In obtaining resultant points, use all zero and maximum points for *each* wave. (*e*) Using a time base of 180 div = 0.001 second, find the frequency of curves *a*, *b*, and *c*. (*f*) Using a scale of 1 in. amplitude = 100 V, write the equation for each curve. (*g*) In terms of modulation theory, what does each curve represent? (*h*) Find the frequency of the modulating signal corresponding to these effects.

7. A carrier has an amplitude of 100 V. What is the amplitude of the upper side frequency at (*a*) 100% modulation? (*b*) 40% modulation? (*c*) What is the amplitude of the lower side frequency at 70% modulation?

8. Find the amplitude of the carrier and each side frequency in Problem 2.

9. Find the amplitude of each side frequency in Problem 3.

10. A television transmitter with a carrier frequency of 54 MHz is modulated to one side only by a camera signal with components ranging from 60 Hz to 3.5 MHz. What is the frequency span of the upper sideband?

11. A modulator puts out a carrier of 150 W and 1000 V at a frequency of 40 MHz before modulation. It is now modulated at 80% with a 4-kHz audio signal. Find (*a*) the carrier power after modulation, (*b*) the amplitude of each side frequency, (*c*) the power level of each side frequency, and (*d*) the total power output. (*e*) If the stage has a plate efficiency of 70%, how much power is required from the audio signal source?

12. How much intelligence power is required to fully modulate a 125-kW carrier if the plate efficiency of the modulated RF amplifier stage is 73%?

13. At 100% modulation, the total power in a modulated wave is 12 kW. (*a*) How is this power distributed among its components? (*b*) With the modulation level reduced to 30%, find the power level in each component.

14. The output wave from a transmitter with single-frequency modulation varies from a maximum amplitude of 540 V to a minimum of 60 V. This wave is produced across an antenna load (effective resistance) of 25 Ω. The plate efficiency of the modulated RF amplifier stage is 67%. Find (*a*) the percent of modulation, (*b*) the power in the carrier, (*c*) the amplitude of each side frequency, and (*d*) the modulating power required.

15. Draw the circuit diagram for a modulated RF amplifier, using (*a*) impedance-coupled RF input, (*b*) shunt-fed grid-leak bias, (*c*) a shunt-fed plate circuit, (*d*) neutralization, if needed, (*e*) plate modulation, transformer coupled, and (*f*) inductive coupling to a high-impedance antenna load.

16. Draw the circuit diagram of a plate-modulated class-C amplifier using a pentode and featuring (*a*) double-tuned inductively coupled RF input, (*b*) fixed bias, (*c*) a shunt-fed plate circuit, and (*d*) coupling to a low-impedance antenna load.

17. A plate-modulated stage has a dc plate input of 12.5 kV and 1.0 A. It has a plate efficiency of 70%. At a modulation factor of 0.8, find (*a*) the carrier power, (*b*) the upper sideband power, (*c*) the modulating power requirement, and (*d*) the modulation transformer turns ratio to make this stage "look like" a 250-Ω load.

18. Draw the circuit diagram for a grid-bias modulated stage, featuring (*a*) single-tuned inductively coupled RF input, (*b*) shunt-fed fixed bias, (*c*) a series-fed plate circuit, (*d*) coupling to a low-impedance antenna load, and (*e*) neutralization, if necessary.

19. Draw the circuit diagram for a suppressor-grid modulated stage, featuring (*a*) shunt-fed grid-leak bias and (*b*) pi-network output coupling to the antenna.

20. Draw the circuit diagram for a screen-grid modulated stage, featuring (*a*) impedance-coupled input, (*b*) grid-leak bias, (*c*) a series-fed plate circuit, and (*d*) coupling to a low-impedance antenna load.

21. Draw the circuit diagram for a collector-modulated NPN transistor stage, using (*a*) double-tuned inductively coupled input, (*b*) internal base-emitter bias, (*c*) a series-fed output tank, and (*d*) inductive coupling to the load.

22. Draw the circuit diagram for a base-modulated PNP transistor stage, using (*a*) impedance-coupled RF input, (*b*) transformer-coupled, series-fed modulating signal, and (*c*) pi-network output coupling.

23. Draw the circuit diagram for a PNP emitter-coupled modulator, using (*a*) inductive-coupled series-tuned RF input, (*b*) internal base-emitter bias, (*c*) transformer-coupled modulating signal, (*d*) a shunt-fed collector tank, and (*e*) inductive coupling to a low-impedance load.

24. Draw a diagram showing how an oscilloscope can be used to check modulation from trapezoidal patterns. Specify the connection points to the transmitter.

25. Draw the trapezoidal patterns that would be obtained in each of the following cases: (*a*) Modulation level of approximately 60%. (*b*) Overmodulation. (*c*) No modulation. (*d*) Poor neutralization and 100% modulation. (*e*) Insufficient excitation and approximately 60% modulation.

26. In Fig. 6–2, the E_{max} and E_{min} values are, respectively, 300 V and 100 V. (*a*) Draw the trapezoidal pattern for this waveshape. (*b*) Indicate the corresponding values on this pattern and calculate the modulation factor.

27. Before modulation, the antenna current is 3.0 A. What should the antenna current be for linear modulation at (*a*) 100% modulation? (*b*) 60% modulation?

28. A transmitter's antenna current rises from 8.0 A to 9.5 A when modulation is applied. Find the modulation factor.

29. Draw the block diagram for a high-level modulation voice transmitter, transmitting at 36 MHz. (*a*) Label each block and specify the class of operation for each block. (*b*) Show the direction of energy flow between all blocks. (*c*) Select typical frequencies and specify the frequency or frequencies present at the output of each block.

30. Repeat Problem 29 for low-level modulation.

31. Draw the circuit diagram for a class-C triode amplifier with plate circuit keying, using a remote-controlled relay, and featuring (*a*) grid neutralization, (*b*) series-fed fixed bias, and (*c*) impedance coupling to the next stage.

32. Draw the circuit diagram for a tetrode class-C amplifier with a filamentary cathode and the keying circuit in the cathode-return. Show (*a*) impedance-coupled RF input, (*b*) shunt-fed grid-leak bias, (*c*) a series-fed plate circuit, (*d*) inductive coupling to the next stage, and (*e*) neutralization, if necessary.

33. Repeat Problem 32 for an indirectly heated triode.

34. Draw the complete circuit diagram for a class-C RF power amplifier using blocked-base keying.

35. Draw the circuit diagram for a beam power class-C RF amplifier with (*a*) a double-tuned RF input circuit, (*b*) shunt-fed fixed bias, (*c*) a series-fed plate circuit, (*d*) single-tuned inductive coupling to the next stage, and (*e*) switching transistor keying.

36. Draw the circuit diagram for a transmitter, including (*a*) a crystal oscillator, (*b*) a buffer amplifier, (*c*) a frequency doubler, (*d*) an intermediate power amplifier, and (*e*) a final power amplifier, plate-modulated and coupled to a low-impedance antenna load.

37. Repeat Problem 36 for a solid-state transmitter, using collector modulation, and capable of 100% modulation.

TYPICAL FCC QUESTIONS

3.294 If a 1500 kilohertz radio wave is modulated by a 2000 hertz sine-wave tone, what frequencies are contained in the modulated wave?

3.298 What is the purpose of a buffer amplifier stage in a transmitter?

3.327 Draw a simple schematic diagram showing a method of coupling a modulator tube to a radio-frequency power amplifier tube to produce plate modulation of the amplified radio-frequency energy.

3.328 Draw a diagram of a carrier wave envelope when modulated 50% by a sinusoidal wave. Indicate on the diagram the dimensions from which percentage of modulation is determined.

3.330 Draw a simple schematic diagram showing a Heising modulation system capable of producing 100% modulation. Indicate power-supply polarity where necessary.

3.331 Draw a simple schematic diagram showing a method of suppressor-grid modulation of a pentode-type vacuum tube.

3.332 Draw a simple schematic diagram showing a method of coupling a modulator tube to a radio-frequency power amplifier tube to produce grid modulation of the amplified radio-frequency energy.

3.333 What is meant by "frequency shift" or "dynamic instability" with reference to a modulated radio-frequency emission?

3.334 What is meant by "high-level" modulation?

3.335 What is meant by "grid modulation"? by "plate modulation"?

3.336 What is meant by "low-level" modulation?

3.342 Why is a high percentage of modulation desirable?

3.343 What are some of the possible results of overmodulation?

3.345 What percentage of antenna current increase should be expected between unmodulated conditions and 100% sinusoidal modulation?

3.346 What might be the cause of a decrease in antenna current of a high-level amplitude-modulated radiotelephone transmitter when modulation is applied?

3.347 Why is it necessary to use an oscillating detector for reception of an unmodulated carrier?

3.363 Draw a simple schematic diagram showing the proper method of obtaining dc screen-grid voltage from the plate supply in the case of a modulated pentode, class-C amplifier.

3.364 What is the purpose of a "buffer" amplifier?

3.373 What is the purpose of a "Faraday" screen between the final tank inductance of a transmitter and the antenna inductance?

3.375 What is the effect of carrier shift in a plate-modulated class-C amplifier?

3.379 In a modulated class-C radio-frequency amplifier, what is the effect of insufficient excitation?

3.384 Should the plate current of a modulated class-C amplifier stage vary or remain constant under modulation conditions? Why?

3.468 Why is a high percentage of modulation desirable in amplitude-modulated transmitters?

3.482 What are causes of downward fluctuation of the antenna current of an amplitude-modulated transmitter when the transmitter is modulated?

3.483 What may cause upward fluctuation of the antenna current of an amplitude-modulated transmitter when the transmitter is modulated?

4.040 Draw a simple schematic diagram of a grid-bias modulation system, including the modulated radio-frequency stage.

4.041 Draw a simple schematic diagram of a class-B audio high-level modulation system, including the modulated radio-frequency stage.

4.042 Draw a sample sketch of the trapezoidal pattern on a cathode ray oscilloscope screen indicating low percent modulation without distortion.

4.043 During 100% modulation, what percentage of the average output power is in the side-bands?

4.045 What are the advantages and disadvantages of class-B modulators?

4.047 What is meant by "low-level" modulation?

4.050 In a modulated amplifier, under what circumstances will the plate current vary as read on a dc meter?

4.051 What could cause downward deflection of the antenna current ammeter of a transmitter when modulation is applied?

4.052 If tests indicate that the positive modulation peaks are greater than the negative peaks in a transmitter employing a class-B audio modulator, what steps should be taken to determine the cause?

4.053 In a properly adjusted grid-bias modulated radio-frequency amplifier, under what circumstances will the plate current vary as read on a dc meter?

4.054 What percentage increase in average output power is obtained under 100% sinusoidal modulation as compared with average unmodulated carrier power?

4.055 In a class-C radio-frequency amplifier stage feeding an antenna system, if there is a positive shift in carrier amplitude under modulation conditions, what may be the trouble?

4.056 Name four causes of distortion in a modulated amplifier stage output.

4.057 If you decrease the percentage of modulation from 100 to 50%, by what percentage have you decreased the power in the sidebands?

4.064 If the daytime transmission-line current of a 10-kilowatt transmitter is 12 amperes, and the transmitter is required to reduce to 5 kilowatts at sunset, what is the new value of transmission-line current?

4.065 If the antenna current of a station is 9.7 amperes for 5 kilowatts, what is the current necessary for a power of 1 kilowatt?

4.066 What is the antenna current when a transmitter is delivering 900 watts into an antenna having a resistance of 16 ohms?

4.067 If the day input power to a certain broadcast station antenna having a resistance of 20 ohms is 2000 watts, what would be the night input power if the antenna current were cut in half?

4.068 The dc input power to the final amplifier stage is exactly 1500 volts and 700 milliamperes. The antenna resistance is 8.2 ohms and the antenna current is 9 amperes. What is the plate efficiency of the final amplifier?

4.072 If a transmitter is modulated 100% by a sinusoidal tone, what percentage increase in antenna current will occur?

4.082 Draw a diagram of a complete class-B modulation system, including the modulated radio-frequency amplifier stage. Indicate points where the various voltages will be connected.

4.083 A certain transmitter has an output of 100 watts. The efficiency of the final modulated amplifier stage is 50%. Assuming that the modulator has an efficiency of 66%, what plate input to the modulator is necessary for 100% modulation of this transmitter? Assume that the modulator output is sinusoidal.

4.085 What undesirable effects result from overmodulation of a broadcast transmitter?

4.086 What do variations in the final amplifier plate current of a transmitter employing low-level modulation usually indicate?

4.111 If a frequency of 500 hertz is beat with a frequency of 550 kilohertz, what will be the resultant frequencies?

4.118 How is the load on a modulator, which modulates the plate circuit of a class-C radio-frequency stage, determined?

4.119 Given a class-C amplifier with a plate voltage of 1000 volts and a plate current of 150 milliamperes which is to be modulated by a class-A amplifier with a plate voltage of 2000 volts, a plate current of 200 milliamperes, and a plate impedance of 15,000 ohms. What is the proper turns ratio for the coupling transformer?

4.122 With respect to the unmodulated values, doubling the excitation voltage of a class-B "linear" radio-frequency amplifier will result in what increase in RF power output?

4.123 What may be the cause of a decrease in antenna current during modulation of a class-B linear RF amplifier?

4.149 Define high-level and low-level modulation.

4.156 Define amplifier gain, stage amplification, and percentage of modulation. Explain how each is determined.

4.159 Discuss the characteristics of a modulated class-C amplifier.

4.204 What pattern on a cathode-ray oscilloscope indicates overmodulation of a standard broadcast station?

4.208 What is the ratio of unmodulated carrier power to instantaneous peak power, at 100% modulation at a standard broadcast station?

4.210 What may cause unsymmetrical modulation of a standard broadcast transmitter?

6.417 What is the purpose of a buffer amplifier stage in a transmitter?

6.418 Draw a simple schematic diagram showing a method of coupling the radio-frequency output of the final power amplifier stage of a transmitter to an antenna.

6.423 Discuss the effects of insufficient radio-frequency excitation on a class-C modulated radio-frequency amplifier insofar as the output signal waveform is concerned.

6.428 Why is a "speech amplifier" used in connection with the modulator of a radiotelephone transmitter?

6.430 How should the bias of a grid-modulated radio-frequency stage be adjusted?

6.431 Compare the characteristics of plate and grid-bias modulation.

6.432 What is meant by "low-level" modulation?

6.433 Should the efficiency of a grid-bias modulated stage be maximum at complete modulation or zero modulation? Explain.

6.434 Does grid current flow in the conventional grid-bias modulated stage of a radiotelephone transmitter under modulated conditions?

6.435 What might be the causes of a positive shift in carrier amplitude during modulation?

6.436 What is the ratio between the dc power input of the plate circuit of the stage being plate modulated and the output audio power of the modulator for 100% sinusoidal modulation?

6.437 What increase in antenna current will be observed when a radiotelephone transmitter is modulated 100% by a sinusoidal waveform?

6.438 Why is a series resistor used in the dc plate supply of a modulated radio-frequency amplifier, between the amplifier and the modulator, in a Heising modulation system?

6.439 What is the purpose of the plate choke used in Heising modulation?

6.440 The dc plate input to a modulated class-C amplifier, with an efficiency of 60%, is 200 watts. What value of sinusoidal audio power is required in order to ensure 100% modulation? 50% modulation?

6.441 A ship's transmitter has an antenna current of 8 amperes using A-1 emission. What would the antenna current be when this transmitter is 100% modulated by sinusoidal modulation?

6.442 If a transmitter is adjusted for minimum power output for telegraph operation, why must the plate voltage be reduced if the transmitter is to be amplitude modulated?

6.445 What is the total bandwidth of a transmitter using A-2 emission with a modulating frequency of 800 hertz and a carrier frequency of 500 kilohertz?

6.450 Draw a block diagram of a MOPA radiotelegraph transmitter with the master oscillator operating on 2017.5 kilohertz and the transmitter output on 8070 kilohertz.

6.452 If a 1500-kilohertz radio wave is modulated by a 2000-hertz sine-wave tone, what frequencies are contained in the modulated wave?

6.456 Draw a simple schematic diagram of a system of keying in the primary of the transformer supplying high voltage to a vacuum-tube transmitter. Indicate any values of inductance, resistance, capacitance which may be deemed necessary to fully understand the correct operation of this type of keying.

6.479 Show by a diagram how a radiotelegraph transmitter can be keyed by the use of a keying relay.

6.480 List the various points in a radiotelegraph transmitter where keying can be accomplished.

6.482 What is meant by frequency shift keying and how is it accomplished?

6.486 If the plate current of the final radio-frequency amplifier in a transmitter suddenly increased and radiation decreased, although the antenna circuit is in good order, what would be the possible causes?

6.492 A master-oscillator, power-amplifier type of transmitter has been operating normally. Suddenly the antenna ammeter reads zero, although all filaments are burning and plate and grid meters are indicating normal voltages and currents. What would be the possible cause(s)?

6.494 Describe a means of reducing sparking at the contacts of a key used with a radiotelegraph transmitter.

6.503 Draw a simple schematic diagram of a "key click filter" suitable for use when a vacuum-tube transmitter is keyed in the negative high-voltage circuit.

6.505 Draw a simple schematic diagram showing how a radiotelegraph transmitter may be keyed by the "grid-blocking" method.

6.531 What is the relationship between the antenna current and radiated power of an antenna?

6.532 What is the purpose of the iron compound cylinders which are found in the inductances of certain marine radiotelegraph transmitters? The position of these cylinders, with respect to the inductances, is adjustable for what purpose?

6.533 What is the meaning of "high-level" modulation?

S-3.138 What is the meaning of the term "carrier frequency"?

S-3.139 If a carrier is amplitude modulated, what causes the sideband frequencies?

S-3.140 What determines the bandwidth of emission for an AM transmission?

S-3.141 Why does exceeding 100% modulation in an AM transmission cause excessive bandwidth of emission?

S-3.142 What is the relationship between percent modulation and the shape of the waveform "envelope" relative to carrier amplitude?

S-3.143 Draw a simplified circuit diagram of the final stages (modulator, modulated amplifier, etc.) of a type of low-level plate-modulated transmitter utilizing a pentode tube in the modulated stage. Explain the principles of operation. Repeat using a tetrode to provide high-level modulation.

S-3.144 How does a linear power amplifier differ from other types?

S-3.145 Draw a simple schematic diagram showing a method of coupling a modulator tube to a radio-frequency power amplifier tube to produce grid modulation of the amplified RF energy. Compare some advantages or disadvantages of this system of modulation with those of plate modulation.

S-3.146 What is meant by "frequency shift" or "dynamic instability" with reference to a modulated RF emission?

S-3.147 What would cause a dip in the antenna current when AM is applied? What are the causes of carrier shift?

S-3.148 What is the relationship between the average power output of the modulator and the plate circuit input of the modulated amplifier under 100% sinusoidal plate modulation? How does this differ when normal voice modulation is employed?

S-3.149 What is the relationship between the amount of power in the sidebands and the intelligibility of the signal at the receiver?

S-3.151 Draw a block diagram of an AM transmitter.

S-4.002 List the fundamental frequency and the first ten harmonic frequencies of a broadcast station licensed to operate at 790 kHz.

S-4.047 Explain in a general way how radio signals are transmitted and received through the use of amplitude modulation.

S-4.048 In amplitude modulation, what is the relationship of sideband power, output carrier power, and percent modulation. Give an example of a problem to determine sideband power if other necessary information is given.

S-4.054 Draw a circuit diagram of a complete radiotelephone transmitter composed of the following stages: (*a*) microphone input connection, (*b*) preamplifier, (*c*) speech amplifier, (*d*) class-B modulator, (*e*) crystal oscillator, (*f*) buffer amplifier, (*g*) class-C modulated amplifier, and (*h*) antenna output connection. Insert meters in the circuit where necessary and explain, step-by-step, how the transmitter is tuned.

S-4.055 Show by a circuit diagram two methods of coupling a standard broadcast transmitter output to an antenna. Include a provision for impedance matching, attenuating harmonics, and guarding against lightning damage.

S-4.058 Cathode-ray oscilloscopes are frequently used to register percentage modulation. Sketch the visual displays of (*a*) 0% modulation, (*b*) 50% modulation, (*c*) 100% modulation, and (*d*) 120% modulation.

S-4.085 Draw circuit diagrams of (*a*) triode class-C amplifier properly coupled to a push-pull power amplifier (modulator) and (*b*) a beam power class-C amplifier coupled to a push-pull class-B power amplifier. For both cases, show the modulating signal input, the RF exciting voltage input, and the modulated output. Include neutralization for the triode case. Explain the operation of both types of class-C plate-modulated amplifiers.

S-4.086 Why is plate modulation more desirable than grid modulation for use in standard broadcast transmitters? Why is grid modulation more desirable in television video transmitters?

S-4.089 Draw a block diagram of a standard (AM) broadcast transmitter complete from the microphone (and/or camera) inputs to the antenna outputs. State the purpose of each stage and explain briefly the overall operation of the transmitter.

7

Demodulation of AM Waves

When a modulated RF wave is picked up at a remote receiving location, the signal strength is very low—in the order of microvolts. RF amplifiers are used to build up the signal level. The next step is to extract the intelligence from the wave. This process is technically called *demodulation.* More commonly, it is also known as *detection,* and the circuits used to achieve this effect are called *detectors.*

BASIC PRINCIPLES

The action in a demodulation circuit is essentially the same as in a modulation circuit. That is, if two (or more) sine waves are applied as inputs to a nonlinear device, the *current* waveform will be a complex wave, containing (1) a dc component, (2) components at each of the original input frequencies, (3) harmonics of each of the input frequencies, and (4) sum and difference frequencies. This same effect occurs in a nonlinear detector circuit because the modulated wave input is actually three signals—a carrier, an upper side frequency (or sideband), and a lower side frequency (or sideband). The desired output from the detector circuit is *the difference frequency,* because this is the original modulating frequency, or the intelligence signal.

As an illustration, consider a transmitter in which a carrier of 1000 Hz is modulated by a 5-kHz audio tone. The components in the modulated wave are 1000 kHz, 1005 kHz, and 995 kHz. When this modulated wave is applied to the nonlinear detector circuit, the difference-frequency component is 5 kHz. Notice that this is the original audio tone that produced the modulated wave.

The process of demodulation can be divided into three simple steps:

1. *Distortion of the Modulated Wave.* This can be readily achieved by feeding the wave to a semiconductor or vacuum tube that is deliberately biased to operate on the nonlinear portion of its dynamic characteristic. The resulting distorted *current* waveform will contain the various components noted above.
2. *Filtering the Low-Frequency Intelligence Signal from the RF Components.* This step requires some form of frequency discriminating circuit that will develop an output *voltage* from the low-frequency components of current, but will bypass, or offer very low impedance (short circuit) to, the high-frequency current components.
3. *Separation of the Intelligence Signal from the DC Component.* The low-pass filter action of the step above will allow the dc component to appear in the output. The final step is to pass the ac (intelligence) signal, but to block out this dc component. Transformer or *R-C* coupling will achieve this result.

Various types of detector circuits have been used. Each has its peculiarities and advantages, but the above steps are common to all of these circuits.

THE DIODE DETECTOR

The most commonly used detector circuit, by far, is the *diode detector.* This circuit gets its name from the nonlinear device that is used—a diode. The diode itself can be a semiconductor or vacuum tube.

Basic Circuit Action

Figure 7–1 shows how the three basic steps in the demodulation process are achieved in the diode detector. The output from the previous RF amplifier is inductively coupled to the detector circuit. The tank L_2C_1 is tuned to resonance with the carrier component of this input signal, and it must have sufficient bandwidth to pass the sideband frequencies. Therefore the voltage e_i across this tank circuit is the modulated wave of Fig. 7–1(b). Since the diode conducts only when its anode is positive compared to cathode, current can flow only during some portion of the positive half-cycle of the RF input. The direction of current flow (electrons) is from point 1 through the diode to 2, down through the R_1C_2

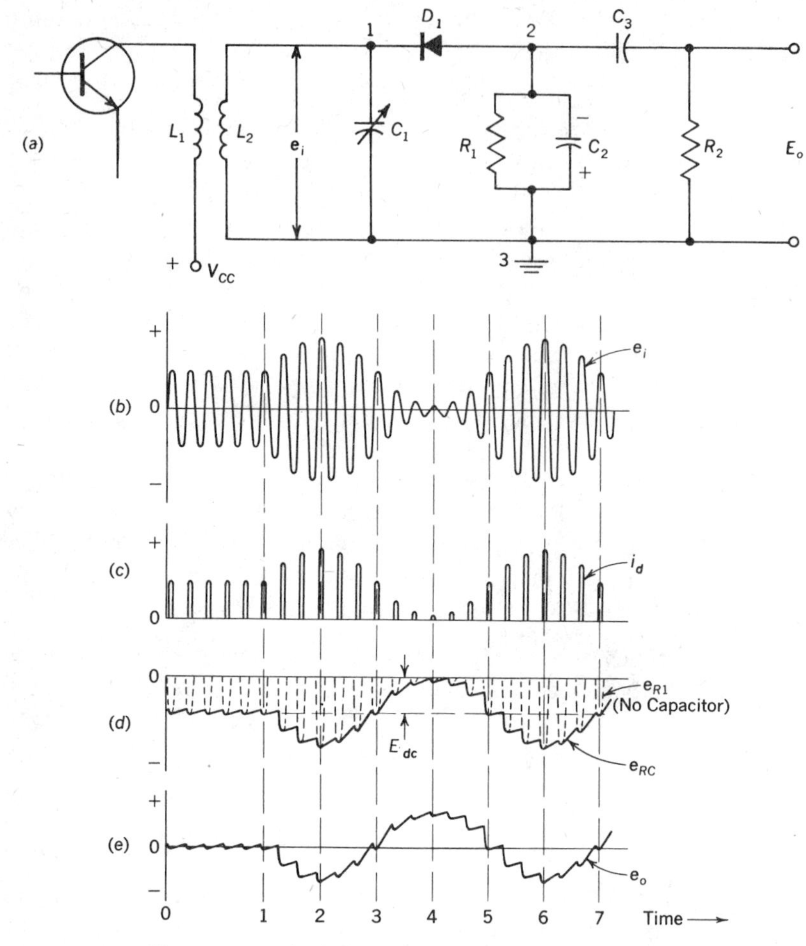

Figure 7-1 A diode detector circuit and its waveshapes.

circuit, and back to the bottom of the tank circuit. This unidirectional current flow will tend to charge capacitor C_2 towards the peak value of the RF wave, with the polarity as shown. This makes the anode negative and will tend to cut off any further current flow. However, as the RF voltage drops and reverses, the capacitor discharges through R_1, so that a current pulse will flow again at the peak of each RF cycle as the input voltage rises above the capacitor voltage. These short-duration current pulses are shown in Fig. 7–1(c).

If there were no capacitor across R_1, the voltage across this resistor would rise and fall following each positive half-cycle of the input signal.

This is shown by the dashed waveshape e_{R1} in Fig. 7–1(d).* At this point before we continue analysis, we must consider some pertinent circuit values. The resistance of the diode *when it conducts* is low—in the order of 1000 Ω. The resistance of R_1 is made appreciably higher, for example 100,000 Ω, and the capacitance value is selected so that the discharge time constant R_1C_2 is long compared to the period of the RF cycles. On the other hand, the charging time constant is quite short because of the low diode resistance. Now we can consider the effect of capacitor C_2. When the diode conducts, the capacitor charges (approximately) to the peak value of the RF input. Between pulses, the capacitor discharges slightly. The result is the waveshape e_{RC} of Fig. 7–1(d). Notice that this waveshape contains a dc component E_{dc}, plus an ac component that essentially follows the modulation envelope.† This ac component therefore represents the intelligence signal. The final step, removal of the dc component, is effectively handled by the *R-C* coupling network R_2C_3. The resulting output, the intelligence signal, is shown in Fig. 7–1(e).

The filtering action of R_1C_2 can also be explained from a complex-wave analysis. This combination, R_1C_2, forms the diode load impedance across which the output voltage is developed. The current through this load impedance is a complex wave (see Fig. 7–1(c)) containing RF components, intelligence components, and a dc component. If the load impedance can be made very low at the radio frequencies but high at the intelligence frequencies, proper filtering action is obtained. Obviously, the reactance of capacitor C_2 must be low (compared to the R_1 value) at RF, but high for all intelligence frequencies.

Design Considerations‡

An ideal detector should have high rectification efficiency, negligible loading effect on the previous stage, and low distortion. The ability of the above diode circuit to fill these requirements depends on the circuit component values—particularly the choice of R_1 and C_2 values.

Detection Efficiency. In the above discussion, we considered the capacitor C_2 to be charged (almost) to the peak value of the RF input, thereby producing the highest possible voltage for e_{RC}. See Fig. 7–1(d). However, because of the internal resistance r_d of the diode, this

*This voltage is shown with negative polarity, because it represents the potential of point 2 with respect to point 3, which is grounded.

†The serrations in this waveshape are caused by RF ripple content that was not removed by the capacitor filtering action. The effect is exaggerated in this diagram because there are only 12 RF pulses per cycle of intelligence. With pulses closer together, the ripple content would become negligible.

‡J. D. Ryder, *Electronic Fundamentals and Applications,* 4th ed. (Englewood Cliffs, N.J., Prentice-Hall, Inc., 1970), Chap. 17.

condition would be realized only if the load resistance R_1 were infinite. The *efficiency of rectification* is a measure of how closely the output voltage approaches its maximum possible value, and it is defined as the ratio of the average value of the diode load voltage to the peak value of the RF input. Unfortunately, this efficiency is not just a simple function of the R_1/r_d, resistance ratio, but it depends also on the capacitance value and on the conduction angle.* It should be obvious, however, that the higher the load resistance value R_1, the higher the detection efficiency will be. For load resistance values of 20 to 100 r_d, the efficiencies will range from 80 to 95%.

LOADING EFFECT. In discussing the basic circuit action, we saw that current flows at the positive peak of each RF cycle. This of course means that a power loss will occur within the diode and in the diode load resistor. Since there is no external power source, the current and power loss must be supplied by the RF source. If the previous stage does not have adequate regulation, this could cause distortion of the input signal waveform.† In addition, because of this current flow, the diode and its load act as an equivalent resistance R_e across the tank circuit. Their shunting effect will lower the Q of the tank, thereby reducing the gain (Q rise) and the selectivity of the tuned circuit. This loading effect is a function not only of the actual load resistance R_1, but also of the capacitance value C_2, and of the conduction angle. If the ratio R_1/r_d is large, the equivalent shunting resistance R_e is approximately $0.5R_1$. Obviously, the shunting effect can be minimized by using high values of diode load resistance.

AMPLITUDE DISTORTION. To prevent amplitude distortion, the dynamic characteristic of the diode should be linear, *when the diode conducts*. Such an ideal characteristic is shown in Fig. 7–2(a). This does not contradict the original statement that nonlinear devices are required for demodulation. Notice the abrupt change in the characteristic curve for negative as compared to positive signals. This is the desired nonlinearity. Meanwhile, in the conducting region—the positive area—the characteristic is perfectly straight. Therefore, regardless of the input signal level, the current flow and output voltage will always be in direct proportion to the input. There will be no amplitude distortion.

Unfortunately, the *static* characteristic of a typical diode does have curvature. By using large values of load resistance, the *dynamic* characteristic can be made fairly straight—except at low signal levels. A typical dynamic characteristic is shown in Fig. 7–2(b). Again we have extreme nonlinearity between the positive and negative portions and excellent

*$\eta_d = (R/\pi r_d)(\sin\theta_1 - \theta_1\cos\theta_1)$, where θ = one-half the conduction angle.

†Notice the similarity to the action of a gate-leak bias circuit.

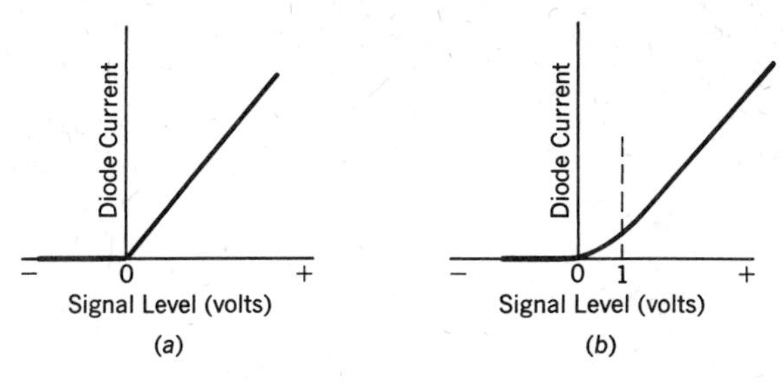

Figure 7-2 Diode dynamic characteristics.

linearity for most of the positive region. However, notice the curvature that still exists in the region 0 to 1. Operation in this region would produce amplitude distortion. This effect is minimized by operating with high signal levels—in the order of 10 V. Yet, regardless of the average signal strength, if the signal is a modulated wave approaching 100% modulation, its level during the modulation trough can readily fall below 1 V. Amplitude distortion cannot be avoided. With the proper choice of *R-C* values, this distortion can be limited to about 2% even with modulation levels of 100%.

Another cause of amplitude distortion is the additional loading effect of the coupling network R_2C_3 of Fig. 7–1(a). These components will cause the *ac* diode load impedance to be less than the *dc* load resistance. The dc load resistance is dependent solely on the value of resistor R_1. At the modulating frequency, X_{C2} can be considered (ideally) as infinite and X_{C3} as zero. This places resistor R_2 in parallel with R_1, so that the ac load impedance, R_{ac}—at the modulating frequency—is resistive and equal to the combined resistance of R_1 and R_2 in parallel. Because of this reduced load resistance to ac, it is therefore possible for the *peak* value of the ac component of plate current to exceed the dc value. Consequently, during some portion of the negative half-cycle of modulation, anode current flow is cut off and distortion results. (This condition is aggravated when automatic volume control circuitry* is added to a diode detector.)

The distortion depends not only on the relative values of R_{ac} and R_1, but also on the modulation level of the input signal. A high percent modulation will increase the ac component of anode current and the possibility of distortion. Depending on the values of R_1 and R_2, there is a limit to the modulation level that the circuit can handle without distortion. This can be explained by using the voltage waveshapes in Fig. 7–1(d). The dc component of the voltage E_{dc} across R_1 is $I_{dc}R_1$. Similarly, the peak value of the ac component (e_{RC}) is (peak I_{ac})R_{ac}. But the ratio of

* These circuits are covered in the next chapter.

these two voltages is dependent on the percent modulation of the input signal, or

$$m = \frac{(\text{peak } I_{\text{ac}})R_{\text{ac}}}{I_{\text{dc}}R_1} \tag{7-1}$$

Since the ratio (peak I_{ac})/I_{dc} should never exceed unity, the maximum percent modulation that the circuit can handle without distortion is limited to

$$m_{\text{max}} = \frac{R_{\text{ac}}}{R_{\text{dc}}} \tag{7-2}$$

For low distortion with high modulation levels, the dc load resistance R_1 should be made as small as possible. But this, in turn, would reduce the rectification efficiency and would cause amplitude distortion because of nonlinearity of the dynamic characteristic. Obviously, a compromise value must be used.

EXAMPLE 7–1

In Fig. 7–1(a), if R_1 is 100,000 Ω and R_2 is 0.27 MΩ, what is the maximum modulation level that can be handled without distortion?

Solution

$$m = \frac{R_{\text{ac}}}{R_{\text{dc}}} = \frac{R_2}{R_1 + R_2} = \frac{0.27}{0.1 + 0.27} = 0.73 \text{ or } 73\%$$

Another technique for reducing the distortion at high modulation levels is shown in Fig. 7–3. Notice the added filter components R_3 and C_4. Capacitor C_4, similarly to C_2, should have a low impedance to RF but a high impedance to all intelligence frequencies. The dc load resistance is now $R_3 + R_1$, and the ac load is $R_3 + R_e$ (where R_e is the resistance of R_1

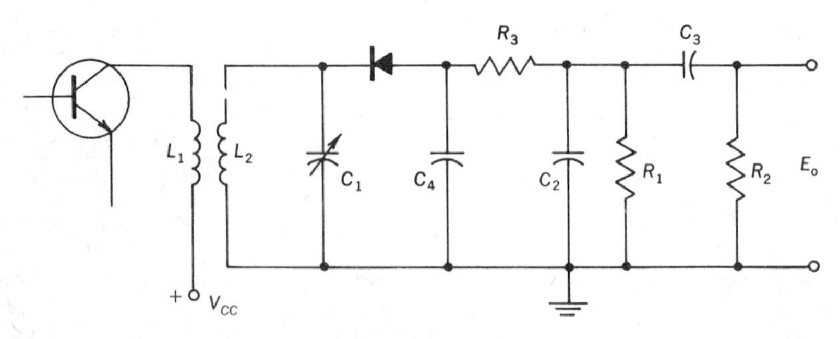

Figure 7-3 A diode detector suitable for high modulation levels.

in parallel with R_2). Because R_3 is common to both loads, the loading effect is reduced. In fact, if R_3 is large, the change in load impedance is negligible. Unfortunately the addition of R_3 causes a reduction in output voltage, but additional gain in other stages can readily compensate for this loss.

FREQUENCY DISTORTION. In examining basic circuit action, we find that, because of the charge and discharge action of capacitor C_2, the waveshape e_{RC} in Fig. 7–1(d) essentially follows the modulation envelope. However, this may not necessarily be true, as is shown in Fig. 7–4(a). As explained earlier, capacitor C_2 charges quickly to the peak of the RF pulse, but between pulses its discharge is relatively slow. This discharge rate is determined by the discharge time constant R_1C_2. During time interval 0 to 1 (no modulation) the action is normal. Also during time interval 1 to 2, as the modulated wave rises to its crest value, since the capacitor charges quickly, voltage e_{RC} follows the modulation envelope. On the other hand, notice that, as the amplitude of the modulated wave falls to its minimum, the capacitor does not *discharge* rapidly enough. The intelligence signal is lost until, at time instant 3, the rising signal amplitude meets the discharge curve. This effect is known as *diagonal clipping*. Obviously the discharge time constant R_1C_2 is too long.

Let us assume that the circuit constants are changed so that the discharge of C_2 can just follow the modulation envelope of Fig. 7–4(a). But we will now double the frequency of the modulating signal. Such a modulation *envelope* is shown as the solid line (curve 1) of Fig. 7–4(b). Notice that the slope of this envelope is steeper than in Fig. 7–4(a). Again diagonal clipping will result. In other words, if the modulating signal were a complex wave, the higher-frequency components could be clipped. This is frequency distortion.

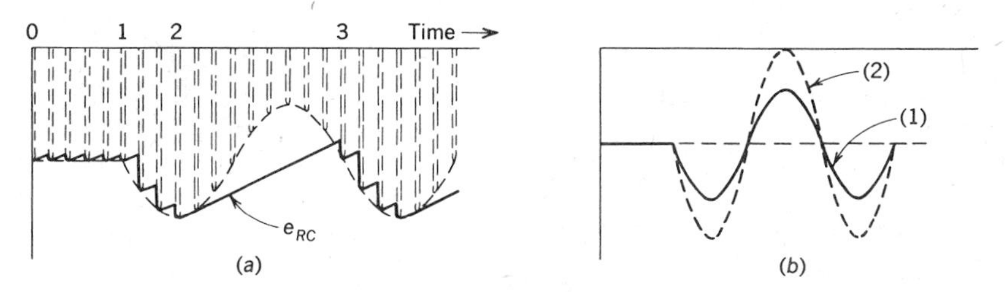

Figure 7-4 Diagonal clipping.

Notice also the modulation envelope shown by the dashed line (curve 2) in Fig. 7–4(b). The frequency is the same as for the solid envelope. This time the modulation *level* has been increased (from approximately 60 to 100%). Again, the slope of the envelope is steeper.

Therefore, the higher the modulation level, the greater the danger of diagonal clipping. Obviously then, the maximum value of the time constant R_1C_2 is limited not only by the frequency of the modulating signal, but also by the modulation factor. More specifically,

$$RC_{\text{max}} = \frac{\sqrt{1 - m^2}}{\omega_s m} \tag{7-3}$$

or

$$\frac{X_C}{R} \geqslant \frac{m}{\sqrt{1 - m^2}} \tag{7-4}$$

where X_C is the reactance of C_2 at the *highest* modulating frequency. With X/R ratios of 2/3, the circuit can handle modulation levels up to 90–95%.

EXAMPLE 7–2

In the diode detector circuit of Example 7–1 ($R_1 = 100{,}000\ \Omega$ and $R_2 = 0.25\ \text{M}\Omega$), what value of capacitor C_2 is needed to prevent diagonal clipping at 73% modulation if the highest modulating frequency is 5000 Hz?

Solution

1. $X_C = R_1 \dfrac{m}{\sqrt{1 - m^2}} = 100{,}000 \dfrac{0.73}{\sqrt{1 - (0.73)^2}} = 106{,}500\ \Omega$

2. $C = \dfrac{1}{2\pi f X_C} = \dfrac{0.159 \times 10^{12}}{5000 \times 106{,}500} = 299\ \text{pF}$

Improved Filter Circuits

In the circuit of Fig. 7–1(a), separation of the intelligence from the RF components is achieved because of the bypass action of capacitor C_2. But when the RF and intelligence frequencies are close together, this simple filter is inadequate. If the capacitance value is too low, it will cause a high reactance to the RF components, and RF will appear in the output. Notice the ripple content in the output voltage of Fig. 7–1(e). Yet, if a larger capacitor is used, it may have too low a reactance at the high end of the intelligence frequency band, and some intelligence will be lost (diagonal clipping). For many applications, this condition is remedied by using the circuit of Fig. 7–3. The ripple output is generally reduced to a negligible value by the added *R-C* filter (R_3C_4).

Particularly severe filtering problems can occur in radar, television, or telemetry receivers in which the pulse intelligence signals may have frequency components as high as 6 MHz, and the RF carrier frequency is in the 40–60-MHz range. In such cases, resistor R_3 of Fig. 7–3 is replaced by an inductor. This converts the network $C_4R_3C_2$ into a low-pass filter. The component values are selected so as to produce a sharp cutoff at just above the highest intelligence frequency. Furthermore, to prevent diagonal clipping (since the reactance of C_2 is low at these frequencies), the diode load resistance value must be drastically reduced. (Values in the order of 1000 to 10,000 Ω are common.)

OTHER TYPES OF AM DETECTORS

Although the diode detector is the most commonly used circuit, other types have been used or, for special reasons, must be used. Let us review a brief chronological development of such circuits. Surprisingly enough, the earliest type of detector circuit used a point-contact semiconductor diode for rectification. The device was rather crude. The point contact was called a *catwhisker*, and it was movable, so that the operator could find a sensitive spot on the semiconductor material. With the development of the vacuum-tube triode, this early crystal detector was dropped. The first type of tube detector circuit was the *grid-leak detector*. The grid and cathode acted as a diode detector, and the recovered audio signal was amplified by the triode action. Since good RF amplifiers were either not available or too costly, it was desirable to obtain as much gain as possible from these detectors. The tube was therefore operated at the low, curved portion of its dynamic characteristic. (In this region the plate current varies as the square of the input signal level. This results in *square-law detection* and high sensitivity.) Unfortunately, however, the distortion level was also increased. But in those days fidelity was not a criterion for performance.

For even higher sensitivity, positive feedback (similar to the Armstrong oscillator circuit) was added. The circuit was now called a *regenerative detector*. Of course, the positive feedback added to the distortion level, but remember, the prime objective was gain. The gain or sensitivity is improved as the amount of positive feedback is increased. Optimum operation is obtained when the circuit is just on the verge of breaking into oscillation. Actually, maximum "gain" would be obtained at even greater feedback levels (when the circuit is driven deep into oscillation) and is limited only by saturation effects. Unfortunately, the circuit then acts as a class-C oscillator; the plate current flows in pulses of constant amplitude and cannot reflect the changes in input signal amplitude. The intelligence is lost.

This difficulty was overcome by *quenching* the oscillations of the detector at many instants during the modulation cycle. The resulting circuit—the *superregenerative detector*—was the most sensitive detector ever designed. However, its use was limited because the circuit is inherently noisy, and the output has a high distortion level. During the 1940s the circuit was revived in a number of UHF applications such as radar beacon receivers—again because good RF amplifiers were not available. Yet, once more, as the state of the art improved, use of this circuitry diminished

As RF amplifier know-how increased, the grid-leak and regenerative detectors were superseded by the *plate detector*. Again a triode is used, but it is cathode-biased so that grid-leak detector action does not occur. Instead, demodulation takes place in the plate circuit. Compared to the grid-leak detector, this circuit has somewhat less gain, but on the other hand, it has less distortion and can handle stronger input signals.

When high fidelity became a design objective, the *infinite-impedance* detector circuit was introduced. This circuit also uses a triode, but with the output taken off the cathode—cathode-follower style. Because of the high inverse feedback, distortion is very low. However, the circuit does not provide for automatic volume control, and its gain is less than one. Since the improvement (lower distortion level) over a well-designed diode circuit is not too great, the infinite-impedance detector never became very popular.

Transistor Detectors

Many transistor receivers use a crystal diode as the detector, in a circuit similar to Fig. 7–1 or Fig. 7–3. However, because of the relatively low gain of transistor RF amplifiers (compared to vacuum-tube circuits), a transistor is sometimes used as the demodulator to provide additional gain. The circuit action of a transistor detector is similar to the vacuum-tube plate detector. The transistor is biased to near cutoff so that collector current flows for less than a full cycle of the input signal. This distorted current contains a dc component; components at each of the original frequencies, at the harmonics of these frequencies, and at the sum-frequencies; and the intelligence signal or difference-frequency component. The first step of the demodulation process has been accomplished.

A typical transistor detector circuit is shown in Fig. 7–5. The operating point is fixed by the voltage divider R_1R_2. The ac load impedance consists of the combination of C_3, R_3, and the coupling network R_4, C_4, and R_5. Here is where the intelligence signal is separated from the RF components. Capacitor C_3 has a low reactance to RF, bypassing these components to ground. However, at the frequency of the intelligence, its capacitive reactance is high, so that this component of collector current flows through R_3 and R_4. An output voltage—at the

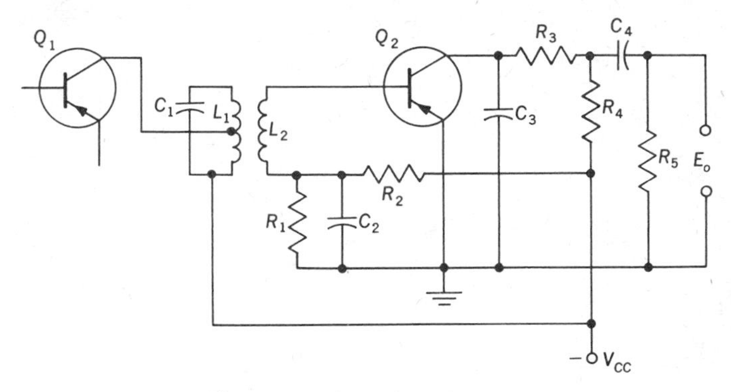

Figure 7-5 A transistor detector.

intelligence frequency—is developed across R_4 and is R-C coupled to the next stage. Besides providing gain, the transistor detector has an added advantage over the diode detector in that it supplies more power for automatic control of the output level. This feature is discussed in the next chapter.

Heterodyne Detectors

In the AM demodulator circuits discussed earlier in this chapter, the intelligence signal was recovered by mixing or beating the sidebands against the carrier of the modulated wave. However, there are some communication methods wherein the RF signal either does not have sidebands (as in Morse code, A1, transmission) or does not have a carrier (as in single sideband, A3a, transmissions). Let us analyze each of these cases briefly.

During the transmission of Morse code signals, the carrier is turned on and off, but only a single frequency, the carrier, is transmitted. In a receiver with a "conventional" detector, this type of transmission produces a click whenever the carrier is turned on or off, but there is no audio signal output. To obtain an audible tone, it is necessary to generate a local RF signal and to mix this local signal with the incoming carrier. The frequency of the local oscillator is selected so as to produce a beat (difference) frequency *in the audio range.* The local oscillator used in this application is generally called a *beat frequency oscillator* (*BFO*). The circuit itself could be any conventional oscillator, such as the Hartley, Armstrong, or Colpitts. A typical circuit for a beat frequency oscillator used with a diode detector is shown in Fig. 7–6.* Let us assume that the pure carrier applied to D_1 through the tank circuit L_2C_1 is at a frequency

*More technically, this circuit is a frequency conversion stage with a transistor oscillator and diode mixer. Frequency conversion is covered in more detail in Chapter 8.

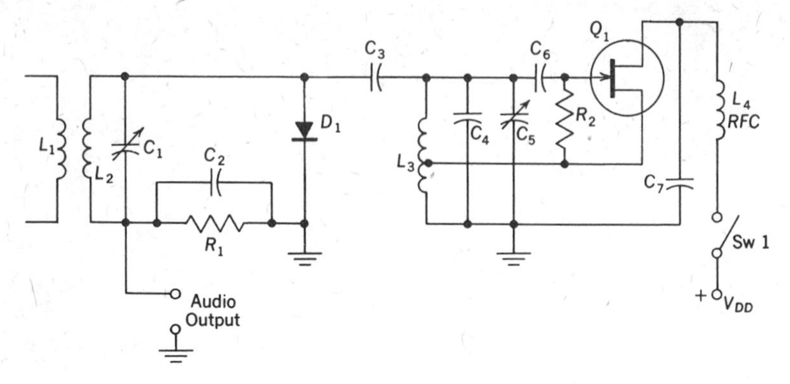

Figure 7-6 A detector circuit suitable for A1 transmission.

of 455 kHz. Simultaneously, the local oscillator signal is capacitively coupled to the diode through C_3. If the frequency of the BFO is set for 453 (or 457) kHz, an audio output voltage at 2 kHz is developed across the diode load R_1C_2. When the carrier is turned off (space), there is no mixing, and no audio output is produced. In Fig. 7–6, the frequency of the BFO is mainly dependent on the tank circuit values L_3C_4, and this LC product is selected to match (approximately) the detector tank circuit. A small variable capacitor C_5 is then used to set the tone of the audio output to the desired pitch.

In A3a transmissions, the carrier and one sideband are suppressed at the transmitter, and only one sideband is transmitted. Because of this lack of a carrier, there will be no mixing in a conventional detector, and no intelligence output will be obtained. It is necessary to supply a locally generated RF signal to replace the missing carrier. Obviously, the beat frequency oscillator circuit of Fig. 7–6 could be used for demodulation of single-sideband signals. However, in this application, frequency stability is very important if distortion of the intelligence signal is to be avoided.* For this reason, a crystal oscillator is preferable to the simple Hartley circuit. Another technique used to ensure stability is to transmit a very small amount of carrier-signal energy (vestigial carrier). At the receiver, this is used to synchronize the local oscillator frequency. (See Chapter 12.)

REVIEW QUESTIONS

1. (*a*) What value of signal voltage would you expect to obtain at a receiving antenna? (*b*) What is generally done before any attempt is made to extract intelligence from this signal?

*With A1 code signals, a change in BFO frequency will only alter the tone of the dot or dash.

2. (*a*) What is the process of extracting intelligence from a modulated wave called? (*b*) Give another name for this process. (*c*) Name the circuit wherein this process is accomplished.

3. (*a*) How many components are contained in an amplitude-modulated wave? (*b*) Name these components. (*c*) Is the intelligence signal one of the components? Explain. (*d*) What combination of the components would produce the intelligence signal?

4. (*a*) What is the first step in the demodulation process? (*b*) Name two devices that could be used to obtain this effect. (*c*) How is the effect produced?

5. The input signal to a detector circuit contains components at 752, 760, and 768 kHz. (*a*) Classify these components by name. (*b*) What intelligence frequency produced the "new" components? (*c*) Which of the three given frequencies are present in the *current* waveshape of the detector? (*d*) Will the current waveshape also contain a component at 1520 kHz? Why? (Give a general answer.) (*e*) Repeat (*d*) for 1528 kHz. (*f*) Repeat (*d*) for 8 kHz. (*g*) Will any other "type" of component be present in the current waveshape? (*h*) Which of the above current components can be grouped together and classified as the RF components?

6. (*a*) What is the second step in the demodulation process? (*b*) In general terms, how is this accomplished?

7. (*a*) What is the third step in the demodulation process? (*b*) State two ways by which this can be accomplished.

8. Refer to Fig. 7–1(a). (*a*) To what frequency is the tank L_2C_1 tuned? (*b*) What bandwidth should this circuit have? Why? (*c*) What circuit component is primarily responsible for the distortion in the anode current waveform? (*d*) Why does capacitor C_2 charge with the polarity as shown? (*e*) To what value does it charge? (*f*) When does the capacitor discharge? (*g*) How does the rate of discharge compare with the rate of charge? Explain. (*h*) What is the function of R_2 and C_3?

9. Refer to the waveshapes in Fig. 7–1. (*a*) What does e_i represent? (*b*) How and where is this voltage obtained? (*c*) What does the waveshape e_{R1} represent? (*d*) How does it compare with e_i? (*e*) Why is it negative in polarity? (*f*) Does this voltage actually exist? Explain. (*g*) What does the waveshape e_{RC} represent? (*h*) How is it obtained? (*i*) What causes the serrations in this waveshape? (*j*) What does the waveshape i_d represent? (*k*) For what portion of an RF cycle does anode current flow? (*l*) During what portion of the waveshape e_{RC} does anode current flow? Explain. (*m*) What does E_{dc} in diagram (d) represent? (*n*) To what input signal value does this voltage correspond? (*o*) What does the waveshape e_o represent? (*p*) How is it obtained from the previous waveshapes? (*q*) How does it compare to the modulating signal? (*r*) In these diagrams what is the relative frequency of the RF and modulating signals? (*s*) If their ratio were increased appreciably, what effect would it have on the e_o waveshape? Explain.

10. (*a*) In the circuit diagram of Fig. 7–1(a), what relation should exist between R_1 and C_2 to remove RF ripple from the output waveshape? (*b*) Explain, using complex wave analysis.

11. (*a*) What is meant by the rectification efficiency of a diode detector? (*b*) What parameter is responsible for loss of efficiency? (*c*) What other circuit values are involved in this efficiency? (*d*) In a given circuit what is the simplest way to ensure high rectification efficiency?

12. (*a*) What is the effect of the diode detector circuit on the effectiveness of the tank circuit? (*b*) On the requirements of the previous stage? (*c*) For large values of R_1, how can this loading effect be evaluated?

13. Refer to Fig. 7–2. (*a*) What does diagram (a) represent? (*b*) Is this considered a linear or a nonlinear characteristic? (*c*) How is this reconciled with the need for nonlinearity in any demodulator? (*d*) What does Fig. 7–2(b) represent? (*e*) How can the linearity of an actual diode be improved? (*f*) What is the result of operating in the 0–1 region? (*g*) How can such distortion be avoided? (*h*) If the input signal is 100% modulated, can distortion be avoided? Explain. (*i*) What is the minimum distortion level obtainable with 100% modulation?

14. Refer to Fig. 7–1(a). (*a*) What is the dc load resistance in this circuit? (*b*) What is the ac load resistance for the frequencies in the modulating signal? (*c*) Explain why this is so. (*d*) Which is lower, the dc or ac load? (*e*) Can this cause any detrimental effects? What specifically? (*f*) During which portion of the modulation cycle could this occur? (*g*) Explain why this is so. (*h*) How is this distortion affected by a low modulation level? (*i*) by a high percent modulation? Explain.

15. Refer to Fig. 7–3. (*a*) What device is used to accomplish the first step in demodulation? (*b*) Compared to Fig. 7–1, what other change makes this circuit more suitable for high modulation levels? (*c*) How should the value of C_4 compare to C_2? Why? (*d*) What is the dc load resistance? (*e*) What is the ac load resistance? (*f*) How can these two values be made more nearly alike? (*g*) What advantage is obtained? (*h*) What is the disadvantage?

16. Refer to Fig. 7–4. (*a*) What is the effect shown during time interval 2–3 of diagram (a) called? (*b*) What is responsible for this effect? (*c*) Why doesn't it occur during time interval 0–1? 1–2? (*d*) What happens to the intelligence output during time interval 2–3? (*e*) How could this defect be remedied?

17. Refer to Fig. 7–4(b). (*a*) What do waveshapes 1 and 2 represent? (*b*) How does envelope 1 compare with the modulating signal envelope of diagram (a)? (*c*) If the time constant of the diode load were adjusted to just avoid clipping with the waveshape in diagram (a), would there be any distortion when used with the modulating signal of envelope 1? Explain. (*d*) Why is diagonal clipping a form of frequency distortion? (*e*) How does envelope 2 compare with envelope 1 as to frequency? (*f*) In what respect do they differ? (*g*) If the circuit constants were adjusted to just avoid clipping with envelope 1, would there be any distortion with envelope 2? (*h*) How does a high modulation level affect the value that can be used for the diode load resistor? (*i*) the value of the bypass capacitor?

18. (*a*) Three demodulator circuits, *X*, *Y*, and *Z*, are designed to operate at signal frequencies of 60 kHz ± 5 kHz, 455 kHz ± 10 kHz, and 40 MHz ± 2 MHz, respectively. Rate these circuits in the order of most difficult to least difficult filtering problem. (*b*) Explain why this is so.

19. (*a*) Which circuit, Fig. 7–1 or Fig. 7–3, has better RF filtering? Explain. (*b*) How can still better filtering action be obtained? (*c*) How is the value of filter capacitance affected by operation at higher carrier and modulation frequencies? (*d*) How is the value of the diode load resistor affected by operation with video modulating signals? Explain.

20. In a diode detector circuit, what value of signal level should be used to avoid distortion?

21. (*a*) Name four types of vacuum-tube detectors that were in former use. (*b*) Rate these four circuits in the order of highest sensitivity. (*c*) Rate them in the order of least distortion.

22. Refer to Fig. 7–5. (*a*) Does transistor Q_2 operate in class A? (*b*) What determines the class of operation? (*c*) Why is this operating condition needed? (*d*) Where does demodulation occur? Explain. (*e*) What is the function of capacitor C_3? (*f*) What two factors influence the value used for this capacitor? (*g*) What other component values are important in the selection of C_3? (*h*) Give two reasons why this circuit may be used in place of a diode detector.

23. (*a*) What two basic circuits are combined to form a heterodyne detector? (*b*) Name two types of transmissions that require the use of such a detector. (*c*) Explain, for each case, why this is so.

24. Refer to Fig. 7–6. What basic circuit is used with (*a*) diode D_1? (*b*) with Q_1? (*c*) What specific name is given to this second circuit when used in this application? (*d*) To what frequency is L_2C_1 tuned? (*e*) What components constitute the tank circuit for Q_1? (*f*) To what frequency is this tank tuned? Why? (*g*) Why does it use two capacitors in parallel? (*h*) Can this circuit be used for demodulation of standard AM signals? Explain.

25. (*a*) What is A3a transmission? (*b*) When such a signal is fed to a standard diode (AM) detector, what will the output be? Explain.

26. (*a*) Can the circuit of Fig. 7–6 be used for demodulation of SSB transmissions? Explain. (*b*) To what frequency should the BFO be tuned? (*c*) What is a possible disadvantage of this circuit? (*d*) Name two circuit modifications that would reduce or remove this weakness. (*e*) Give another technique that is used to eliminate this possibility.

PROBLEMS AND DIAGRAMS

1. Draw the circuit diagram for a diode detector using a semiconductor diode.

2. In the circuit of Fig. 7–1, R_1 and C_2 are, respectively, 100,000 Ω and 200 pF. Calculate the diode load impedance value for (*a*) a 5-kHz audio modulating signal and (*b*) a 1200-kHz carrier frequency.

3. An unmodulated carrier signal of 8.0 V (rms) is applied to the above circuit. A dc meter indicates a current of 95 μA. Find the efficiency of rectification.

4. The diode detector of Problem 2 has a diode resistance r_d of 1000 Ω. (*a*) What is the equivalent loading effect of this circuit? (*b*) The tank circuit has an inductance of 150 μH and an unloaded Q of 100. Find the Q value when used in this detector circuit.

5. The input signal to the detector circuit of Fig. 7–1 has a peak unmodulated value of 7.0 V and rises to a peak of 10.0 V at the modulation crest. The circuit resistances (R_1 and R_2) are each 270,000 Ω. Find (*a*) the dc load, (*b*) the ac load (neglecting capacitance effects), (*c*) the dc component of current, (*d*) the peak value of the intelligence component of current, (*e*) the maximum instantaneous value of the total current, and (*f*) the minimum amplitude (trough) value of the total current. (*g*) Is this current waveshape distorted?

6. Using the above circuit values ($R_1 = R_2 = 270{,}000\ \Omega$), find the maximum percent of modulation that this circuit can handle without distortion.

7. To handle modulation levels of 85%, what value of R_2 is required if R_1 is still 270,000 Ω?

8. In Fig. 7–3, the circuit resistance values are $R_1 = 270{,}000\ \Omega$, $R_2 = 270{,}000\ \Omega$, and $R_3 = 100{,}000\ \Omega$. Find (*a*) the dc and ac load resistances and (*b*) the maximum percent modulation that this circuit can handle without distortion.

9. Repeat (*a*) and (*b*) of Problem 8 for $R_1 = 220{,}000\ \Omega$, $R_2 = 470{,}000\ \Omega$, and $R_3 = 0$.

10. Repeat (*a*) and (*b*) of Problem 8 for $R_1 = 220{,}000\ \Omega$, $R_2 = 470{,}000\ \Omega$, and $R_3 = 47{,}000\ \Omega$.

11. Repeat Problem 10 with R_3 raised to 270,000 Ω.

12. In the diode-detector circuit of Fig. 7–1, what value of capacitor C_2 is needed for a modulation frequency of 5 kHz and a modulation factor of 0.9, if R_1 is 270,000 Ω?

13. The circuit of Fig. 7–1 is used as a video detector (modulating frequencies up to 4.0 MHz). (*a*) If R_1 is 270,000 Ω, what value of capacitor C_2 would be needed to avoid distortion at 90% modulation? (*b*) Is this a practical solution? (*c*) What value of R_1 should be used if C_2 is 50 pF?

14. What is the highest modulation level that the diode detector circuit can handle if R_1 is 500,000 Ω, C_2 is 150 pF, and the modulating frequency is 6 kHz?

15. What is the highest modulating frequency that the diode detector of Fig. 7–1 can handle without diagonal clipping if the modulation level is 85% and R_1 and C_2 are, respectively, 220,000 Ω and 100 pF?

16. Draw the circuit diagram of a diode detector suitable for demodulation of television or pulse signals.

17. Draw the circuit diagram of an NPN transistor detector.

18. Draw the diagram of a circuit suitable for detection of Morse code signals.

19. Draw the diagram of a circuit suitable for demodulation of single-sideband transmissions.

TYPICAL FCC QUESTIONS

3.086 Draw a simple schematic diagram of a diode vacuum tube connected for diode detection and showing a method of coupling to an audio amplifier.

3.087 Draw a simple schematic diagram of a triode vacuum tube connected for plate or "power" detection.

3.088 Draw a simple schematic diagram of a triode vacuum tube connected for grid-leak capacitor detection.

3.123 What operating conditions determine that a tube is being used as a "power detector"?

3.128 Explain the operation of a "grid-leak"-type detector.

3.129 List and explain the characteristics of a "square-law" type of vacuum-tube detector.

3.130 Explain the operation of a diode type of detector.

3.131 Explain the operation of a "power" or "plate-rectification" type of vacuum-tube detector.

3.132 Is a "grid-leak" type of detector more or less sensitive than a "power" detector (plate rectification)? Why?

3.230 Describe the operation of a crystal detector (rectifier).

3.269 What effect does the reception of modulated signals have on the plate current of a grid-leak, grid-capacitor type of detector? On a grid-bias type of detector?

3.349 Explain what circuit conditions are necessary in a regenerative receiver for maximum response to a modulated signal.

3.350 What feedback conditions must be satisfied in a regenerative detector for most stable operation of the detector circuit in an oscillating condition?

3.352 What feedback conditions must be satisfied in a regenerative detector in order to obtain sustained oscillations?

3.356 What are the characteristics of "plate detection"?

3.480 How does the value of resistance in the grid leak of a regenerative-type detector affect the sensitivity of the detector?

6.181 Explain the operation of a diode type of detector.

6.182 Explain the operation of a "grid-leak"-type detector.

6.183 What effect does an incoming signal have upon the plate current of a triode detector of the grid-leak type?

6.184 List and explain the characteristics of a "square-law" type of vacuum-tube detector.

6.185 Is a "grid-leak" type of detector more or less sensitive than a "power" detector (plate rectification)? Why?

6.186 What is the principal advantage in the use of a diode detector instead of a grid-leak-type triode detector?

6.187 What operating conditions determine that a tube is being used as a "power detector"?

6.245 Draw a simple schematic diagram of a diode vacuum tube connected for diode detection, and show a method of coupling to an audio amplifier.

6.270 Draw a simple schematic circuit diagram of a triode vacuum tube connected for plate or "power" detection.

6.271 Draw a simple schematic circuit diagram of a triode vacuum tube connected for grid-leak capacitor detection.

6.463 Why is it sometimes necessary to provide a radio-frequency filter in the plate circuit of a detector tube?

6.537 Describe the principle of operation of a superregenerative detector.

6.550 Describe the operation of a regenerative-type receiver.

6.560 How may a regenerative-type receiver be adjusted for maximum sensitivity?

6.563 Draw a circuit diagram of a crystal detector receiver and explain its principle of operation. Name two substances that can be used as the crystal in such a receiver.

6.566 In a receiver, what is the purpose of an oscillator operating on a frequency near the intermediate frequency of the receiver?

6.580 Draw a simple schematic circuit of a regenerative detector.

8

AM Receivers

The ultimate goal of any communications system is to extract the intelligence from the transmitted signal, at some remote receiving location, and build up this intelligence signal so it can "operate" its load. Typical loads could be the loudspeaker in a sound system, the cathode ray tube in a television system, the printer in a facsimile or teletype system, or the relay or motor in a control system. These functions are performed by the *receiver*. Even in echo-type devices such as radar, sonar, and distance or depth measuring equipment, the transmitter sends out a signal, but a receiver is needed to pick up and recover the intelligence of the returning wave reflected from the "target." Receivers vary in complexity depending on their intended use. For purposes of comparison, several characteristics are of particular importance.

RECEIVER COMPARISON FACTORS

In preparing (or checking) the full specifications for a communications receiver, many factors may be considered, depending on the application. However, four characteristics are of general importance: *sensitivity, selectivity, fidelity,* and *noise figure*.

Sensitivity

If a receiver is to be used for reception of long-distance transmissions or of any weak signals, sensitivity is a very important factor. It is a measure of the level of input signal needed to produce "the standard output."* A sensitivity of about 50 μV is typical of many broadcast-band (540–1600 kHz) receivers. Commercial communication receivers may have sensitivities as high as 0.1–1.0 μV.

Obviously, sensitivity is a function of the amount of amplification or number of stages used in the receiver. This amplification can be obtained before demodulation of the RF signal using RF voltage amplifiers, and it can also be obtained after demodulation using suitable untuned (audio, video, etc.) amplifiers. However, particularly with diode detection, it has been found that less distortion and more gain per stage can be realized with RF amplification. Consequently, receivers are generally designed with fairly similar gain beyond the detector stage, so that their sensitivity rating is mainly dependent on the RF amplification.

At first thought it might seem that a receiver could be designed for any degree of sensitivity merely by increasing the number of stages. Although the gain would be increased, it may not be usable. When the noise level generated within the receiver (or the atmospheric noise at the receiver location) is stronger than the signal to be received, increased sensitivity is wasted. The receiver output will contain more noise than intelligence. For example, if the noise level generated within the receiver is 10 μV, there is no sense in designing such a unit for a 2-μV sensitivity.

Selectivity

Since many transmissions are "on the air" at the same time, a receiver must be capable of selecting the desired transmission while excluding all others. This is the function of the tuned circuits ahead of the detector, and the ability to accomplish this goal depends on the number of tuned circuits, the Q of these circuits, and whether they are synchronous-tuned or stagger-tuned. Obviously, selectivity is closely related to the response curves and bandwidth calculations discussed in Chapter 3. However, a statement of bandwidth is not sufficient for selectivity comparisons. This can be seen from Fig. 8–1(a). Whereas both curves have identical bandwidths, curve 2 has better selectivity because of its steep skirts.

To prevent such ambiguity, the selectivity of a receiver is given by a curve that shows the increase in signal level needed to maintain the standard output† for various input frequencies to either side of the receiver resonant frequency. Such a selectivity curve is shown in Fig. 8–1(b). It

* For sensitivity measurements, the input signal is modulated 30% at 400 Hz. The quantity used as "standard output" varies, depending on the power rating and application. For AM broadcast receivers with at least 1-W power output, 0.5 W is considered as the standard output. With tuners, 1 V has been used as a standard output.

† Expressed in decibels, using the resonant frequency signal input as reference.

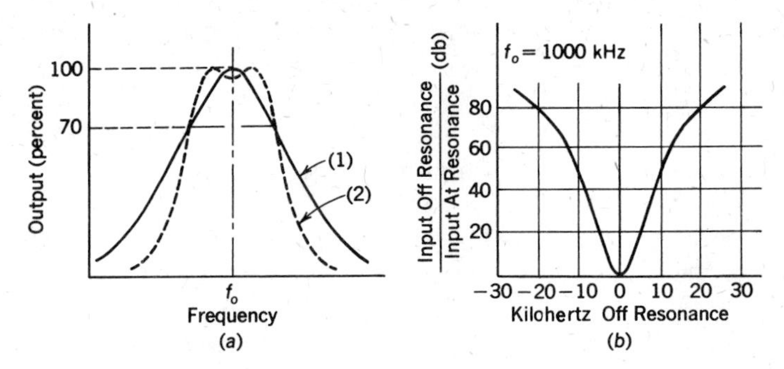

Figure 8-1 Selectivity comparison curves.

should be realized that if the selectivity is too sharp, a receiver will not only discriminate against unwanted signals, but it may also cut off some sideband energy from the desired signal.

Fidelity

This is a measure of how faithfully the receiver reproduces the original intelligence signal. Any distortion of the low-frequency components of the modulating signal is most likely due to the circuitry that follows the detector stage (audio, video, or pulse amplifiers). On the other hand, the high-frequency components of the modulating signal may be affected not only by these post-detection circuits, but also by the selectivity characteristics (bandwidth) of the RF amplifiers preceding the detector stage.

Noise Figure

Although a receiver could be designed for any amount of gain, the *usable* gain is limited by the noise output from the receiver. In a radio receiver, noise is responsible for the crackle and hiss that is heard when the receiver is tuned between stations. In severe cases, noise crashes will override music and speech. In television, noise will appear as snow on the screen (confetti, in a color set). This noise can be a result of external noise fields, antenna system noise, or noise generated within the receiver itself. External noise fields are created by electrical discharges. These could be of natural origin, such as lightning effects, or man-made—by ignition systems, neon lights, machinery with brushes, or other electrical arcing sources. The general term *static* is used to identify such phenomena. The interference from static covers a wide band of frequencies but decreases in strength with an increase in frequency. It is of significant magnitude in the broadcast band, but becomes progressively weaker above 2 MHz, dropping to a negligible value for frequencies above 30 MHz.

Another cause of noise is *thermal agitation.** This can occur inside the receiver or in the antenna system. It is caused by random movement of electrons. This random "current" flowing through a conductor or resistor causes a random noise voltage (IR). As the temperature of the component is increased, the thermal agitation is increased, and the noise voltage is increased. Since this noise effect is distributed uniformly throughout the RF spectrum, the amount of noise in a receiver output is also dependent on the receiver's bandwidth.† Expressed mathematically, the effective value of the noise voltage due to thermal agitation is

$$E_n = \sqrt{4(K)(T)(R)\text{BW}} \tag{8-1}$$

where K = Boltzmann's constant, 1.374×10^{-23} joule per °K
T = the absolute temperature, in °Kelvin (273 + °C)
R = the resistance of the component, in ohms
BW = the receiver bandwidth, in hertz.

A major source of the noise developed within a receiver is the active device (tube or semiconductor) itself. Such noise can be analyzed from several aspects.

1. SHOT EFFECT. Although we normally consider the output current of the device as constant under a no-signal (or dc bias) condition, actually this current is not a continuous and steady flow. Since it consists of many particles, the current depends on the random arrival of electrons (or holes) at the collector, drain, or plate. The resulting noise is called *shot effect* because it sounds like a shower of lead shot falling on a metallic surface. It is convenient, for design purposes, to express this noise in terms of an equivalent resistance shunted across the input of a noiseless active device. This resistor, in turn, following Eq. (8–1), would apply a noise voltage to the input of the ideal device, such that the noise output would be equal to the noise output from the actual device. Equivalent noise resistance values are generally given in the manufacturer's data sheets. Approximate values for FETs and triodes are given by

$$\text{FETs:} \quad R_{\text{equiv}} \cong \frac{1}{g_m}\ \Omega \tag{8-2A}$$

$$\text{Triodes:} \quad R_{\text{equiv}} \cong \frac{2.5}{g_m}\ \Omega \tag{8-2B}$$

*This form of noise is also known as *Johnson noise* or *white noise*.

†It is therefore advisable to restrict receiver bandwidth to the needs of the carrier and sidebands being received.

The FET has a lower shot noise because it operates at a lower temperature. Bipolar transistor noise is more complex and no approximate formula is available.

2. PARTITION NOISE. When a tube has two (or more) positive electrodes, as in a tetrode or pentode, any chance variation in the division of current between the plate and these other positive electrodes will add to the shot-effect noise. As before, this noise can be expressed as an equivalent resistance. Obviously, multielement tubes will have higher noise-equivalent resistances than triodes. Typical values for pentode tubes can range from three to ten times higher than for triodes.

A similar effect also occurs in bipolar transistors. Emitter current divides between base and collector. Due to random recombinations of holes and electrons in the base region, this division is again a chance variation and noise voltages are produced.

3. FLICKER NOISE. This form of noise is caused by imperfections in the surface characteristics, and it affects the diffusion of holes and electrons through the semiconductor material. The effect is more pronounced at low frequencies (below 1000 Hz) and varies inversely with frequency ($1/f$). Since surface imperfections can affect electron emission from a cathode, flicker noise is (to some extent) also found in vacuum tubes.

4. INDUCED GRID OR TRANSIT-TIME NOISE. Any variation in the electron stream as it passes the grid can induce a current in the grid. Since the variation is a random or irregular effect, the resulting voltage is a noise voltage. This effect becomes especially significant at very high frequencies as the period of the input signal approaches the electron transit time. Again this noise can be represented as an equivalent resistance. Due to the lower mobility of electrons (and especially of holes) through a semiconductor material, transit-time noise is an even greater factor in transistors.

From the above discussion, it is obvious that an amplifier stage generates a certain amount of noise. By careful design, particularly of the first stage, the noise output due to thermal agitation and device effects can be minimized. (However, it is possible that, at UHF, an RF amplifier can contribute more noise than gain.) In comparing the performance of receivers with respect to noise, the term *noise figure* (F) (or noise factor) is used. This figure compares the ratio of the signal power to noise power at the input (S_i/N_i), to the signal-to-noise power ratio at the output (S_o/N_o), or

$$F = \frac{(S_i/N_i)}{(S_o/N_o)} \tag{8-3}$$

The noise figure of a receiver can be expressed as a simple ratio as in Eq. (8–3), or in decibels. Obviously, an ideal receiver—one that generates no noise—would have a noise figure of unity or 0 dB.

TUNED RADIO-FREQUENCY (TRF) RECEIVERS

In the early days of radio broadcasting (circa 1920), the most popular type of multitube set was the *tuned radio-frequency* (*TRF*) receiver. These receivers generally consisted of two stages of RF amplification followed by a grid-leak or plate detector, and two stages of audio amplification. Single-tuned inductive coupling was used between the RF stages and to the detector. The "front-end" of such a receiver is shown in Fig. 8–2. For coverage of the broadcast band (540–1600 kHz), the capacitors C_1, C_4, and C_7 are made variable. At one time three separate capacitors were used, and each had to be set independently. This made tuning awkward. Single-dial tuning was achieved by "ganging" the three units into one mechanical assembly, so that all three rotors are on a common shaft. As for the circuit itself, the action of each stage is covered in previous chapters, so no further discussion is necessary.

Although the TRF circuitry shown is direct and simple, it has some inherent weaknesses when used to cover an appreciable frequency span, as is necessary in any communications receiver. If the receiver is designed to give the desired bandwidth at the low-frequency end of its tuning range, the bandwidth becomes much too wide at the high-frequency end of the band.* Adjacent-channel interference results. Conversely, if the set is designed for proper bandwidth at the high end, the selectivity is too sharp at the opposite end, and some of the sidebands are attenuated.

A second disadvantage is the variation in gain with tuning. Three factors combine to increase the sensitivity of the receiver at the high-frequency end of the dial. Since variable-capacitance tuning is used, the tank circuits have minimum capacitance at these frequencies. This produces a higher L/C ratio and higher gain. Simultaneously, capacitance effects between primary and secondary windings in the interstage transformers increase the coupling at the high-frequency end and increase the energy transfer. Finally, because of the increased effect of the interelectrode capacitance C_{gp}, some measure of regeneration is introduced into each stage, and their gain increases. Obviously, if the receiver is designed for good gain at the low-frequency end of its tuning range, it may break

*The Q of the tank circuit remains approximately constant over the tuning band, because as X_L changes with frequency, the effective resistance changes in proportion. Consequently (since $\text{BW} = f_0/Q$), the bandwidth increases directly with the frequency.

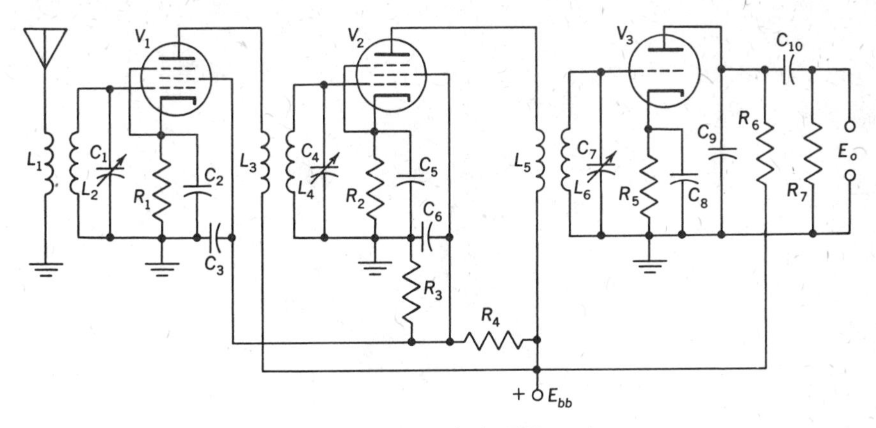

Figure 8-2 Front end of a TRF receiver.

into oscillation when tuned to some higher frequency. Conversely, if designed for stability at its higher frequencies, it will be rather insensitive at the lower end of the dial.

Another drawback with this type of receiver is due to the difficulties involved in multistage variable-frequency operation. For optimum gain, all three tank circuits in Fig. 8–2 should be tuned to identical frequencies throughout the tuning range. This is possible, but awkward, when the tuning capacitors are separate units. With single-dial control, this would require that all coils, all tuning capacitors, each stage's circuit wiring, and all stray capacitances be respectively identical. Such exactness is not practical. Consequently, perfect *tracking** is not possible. The circuits are normally aligned near the high-frequency end of the tuning range by adjusting the *trimmer* capacitors until each stage is tuned to the same frequency. (A typical three-gang tuning capacitor with trimmers is shown in Fig. 8–3.) However, unless the tuning capacitors are identical, small errors in tracking will still occur as the plates are meshed.

If more gain is desired, one (or more) RF stages could be added. This complicates the tracking problem, because now we need a four-gang, five-gang, or six-gang tuning capacitor. In addition, the possibility of oscillation is increased because of the feedback between the capacitor sections. Finally, the bulk of such a multigang tuning capacitor is undesirable. Consequently TRF receivers were generally limited to two RF stages.

In Chapter 2, we saw that a double-tuned coupling network, particularly when slightly (optimum) overcoupled, had a better band-pass

*Simultaneous resonance of the tuned circuits to the same frequency throughout the tuning range.

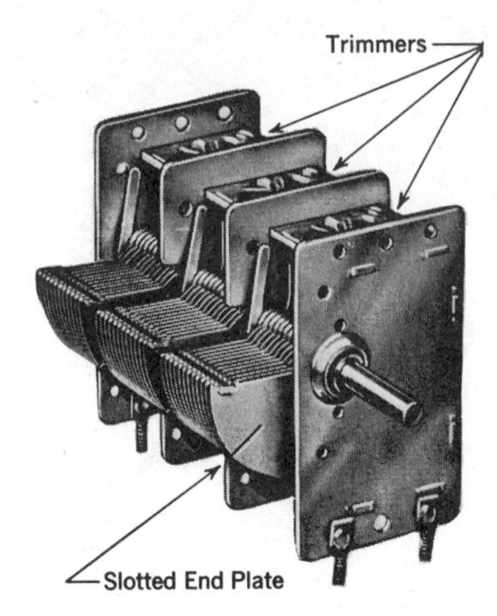

Figure 8-3 A typical three-gang capacitor for the broadcast band. (Courtesy J. W. Miller.)

action. Yet, if we attempted to use this type of coupling in Fig. 8–2, a six-gang capacitor would be needed. Again, because of practical limitations, double-tuned circuits were not used in TRF receivers.

SUPERHETERODYNE RECEIVERS

It should be noted that most of the disadvantages of the TRF circuit would not apply if the receiver were used for fixed-frequency operation. If a selected RF signal, regardless of frequency, could be converted to some predetermined fixed frequency, the advantages of fixed-frequency operation could be realized in a communications receiver. This is the basic principle of the *superheterodyne* receiver, developed in the middle 1920s. Because of its superior gain, selectivity, and sensitivity characteristics, it has become the only receiver circuit in general use.

Block Diagram

Figure 8–4 shows the block diagram of a superheterodyne receiver for use in the AM broadcast band. The first block is an RF amplifier.* This stage must be capable of tuning over the band and could therefore be identical to the RF amplifier in a TRF receiver. Of course modern receivers do not use tubes. Any of the semiconductor small-signal RF amplifier circuits discussed in Chapter 3 (including ICs) could be used here. This stage needs no further discussion.

The two blocks within the dashed rectangle are the key to the super-

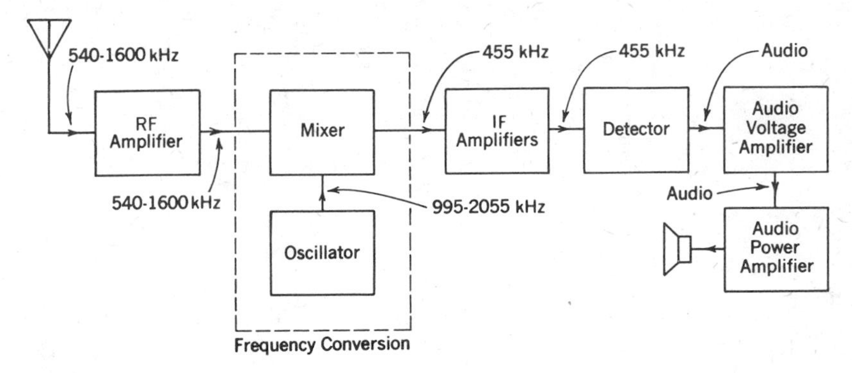

Figure 8-4 Block diagram of a superheterodyne AM broadcast-band receiver.

heterodyne action. Their function is variously known as *frequency conversion, frequency translation,* or *heterodyning.* The mixer circuit is tunable over the entire band, and at any setting of the selector dial, it is tuned to the same frequency as the RF stage. The oscillator section is also a variable-frequency stage, but its frequency is always a fixed amount higher than the resonant frequency of the other two stages. The output from the mixer is the *difference frequency* between the RF signal and the oscillator frequency. Because of the relation between RF and oscillator tuning, the difference will be a constant value, regardless of the setting of the tuning dial. This difference frequency is still a radio frequency. However, to distinguish it from the incoming signal frequency—and also since it lies in between the original carrier and intelligence frequencies—it is called the *intermediate frequency (IF).* In the process, the intelligence contained in the original carrier signal is translated from one place in the RF spectrum, to this lower IF region.

The IF amplifier can be any number of stages. Since it operates at a fixed frequency, it is designed for optimum results at that one frequency. There is no tracking problem, double-tuned circuits present no difficulties, and each stage can be readily shielded to minimize feedback and oscillation. It is this IF section that is mainly responsible for the sensitivity and selectivity characteristics of the superheterodyne receiver.

In the detector circuit, the intelligence is extracted from the IF carrier. The audio signal, in this case, is fed to suitable voltage and power amplifiers and finally to the loudspeaker load. Television (and radar) receivers use video instead of audio amplifiers, and the output from the video power amplifier is fed to the cathode ray tube.

* For reception of signals in the local area, this stage is often omitted, and the antenna signal is applied directly to the mixer stage.

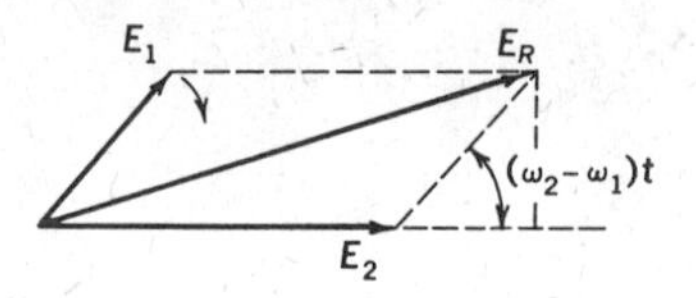

Figure 8-5 Evaluation of the resultant of two signals of unequal frequencies.

Frequency Conversion

The principle of frequency conversion is quite similar to modulation and demodulation in that two signals are fed to a nonlinear device producing a complex current wave containing—among other components—the sum and difference frequencies. In this case the desired output is at the difference frequency. Many applications are made of this principle. For example, we saw it used in Chapter 7 (Fig. 7–6) for the reception of code (CW) transmissions.* When used in the superheterodyne receiver, because of the similarity to demodulation, the frequency conversion circuit is sometimes called the *first detector.*

Let us consider such an application with E_1 and E_2 as the RF signal and the local oscillator signal, respectively. Let us also stipulate, as in the block diagram, that the oscillator signal E_2 be the higher frequency. A phasor representation of these voltages is shown in Fig. 8–5, with E_2 as the reference. Since there is a difference in frequency, E_1 will rotate *clockwise* with respect to E_2, at an angular velocity of $\omega_2 - \omega_1$. At some instant, the conditions will be as shown in Fig. 8–5, with E_1 at an angle of $(\omega_2 - \omega_1)t$. Obviously, the resultant will vary in magnitude depending on the instantaneous angle between the two phasors. This magnitude can be expressed mathematically in terms of the horizontal and vertical components of the phasors:

$$\text{Horizontal components} = E_2 + E_1 \cos (\omega_2 - \omega_1)t$$

$$\text{Vertical components} = E_1 \sin (\omega_2 - \omega_1)t$$

The resultant (the square root of the sum of the squares of these components) consequently varies in some complex manner as a function of E_1. However, if we make E_2 appreciably larger than E_1, the vertical

* This beat frequency idea is also used in the construction of one type of audio oscillator, where the audio output is the difference between two RF oscillator frequencies. It is used again in a variety of frequency-measuring equipment, wherein the frequency of a calibrated variable-frequency source is matched against the unknown signal frequency. When "zero-beat" is obtained, the unknown frequency is the same as the standard frequency. Another application is in the Armstrong FM transmitter, where frequency multiplication is used to obtain a large frequency swing, and frequency conversion is then used to reduce the carrier frequency to the desired value.

component becomes negligible, and the envelope of the resultant will vary in amplitude by an amount equal to E_1 and at a rate (frequency) equal to $\omega_2 - \omega_1$ (or the difference in frequency between the two signals). Therefore, if the incoming RF signal is amplitude-modulated, the resultant will have the same modulation as the original carrier. A numerical example will illustrate this point.

EXAMPLE 8–1

A modulated carrier, 1200 kHz ± 5kHz, is fed to a mixer together with an oscillator frequency of 1400 kHz. What IF frequencies will be produced?

Solution

1. The carrier and side frequencies are 1200, 1195, and 1205 kHz.
2. When mixed with 1400 kHz, the IF carrier is:

$$\text{IF carrier} = 1400 - 1200 = 200 \text{ kHz}$$

3. The side frequencies are

$$\text{Upper S.F.} = 1400 - 1195 = 205 \text{ kHz}$$

$$\text{Lower S.F.} = 1400 - 1205 = 195 \text{ kHz}$$

(Notice that the RF upper side frequency has become the IF *lower* side frequency.)

The action in any frequency-conversion process is essentially twofold—the production of an RF signal by the local oscillator, and the mixing of this local oscillator signal with the incoming modulated carrier. Many circuit variations to accomplish these effects are possible. For example, the oscillator portion could be any of the circuits discussed in Chapter 5. For mixing, a nonlinear device is needed. This could be a diode,* a bipolar transistor, FET, triode, or any multigrid tube. The circuitry can also vary, depending on the method used to couple the two signals into the mixer.

Frequency Conversion Circuits

A two-transistor frequency-conversion circuit using bipolar transistors is shown in Fig. 8–6, where Q_2 is the mixer and Q_3 is the oscillator. The interstage coupling between the RF amplifier Q_1 and the mixer Q_2 uses a tuned primary with the collector of Q_1 connected to a tap, so as to maintain a high-Q tank circuit. The untuned secondary L_2 steps down this impedance to obtain a proper impedance match. The RF signal is fed to

*At UHF where signal levels are quite low, it is common practice to use diodes as mixers. Although they provide no gain, their low noise level is of even greater importance.

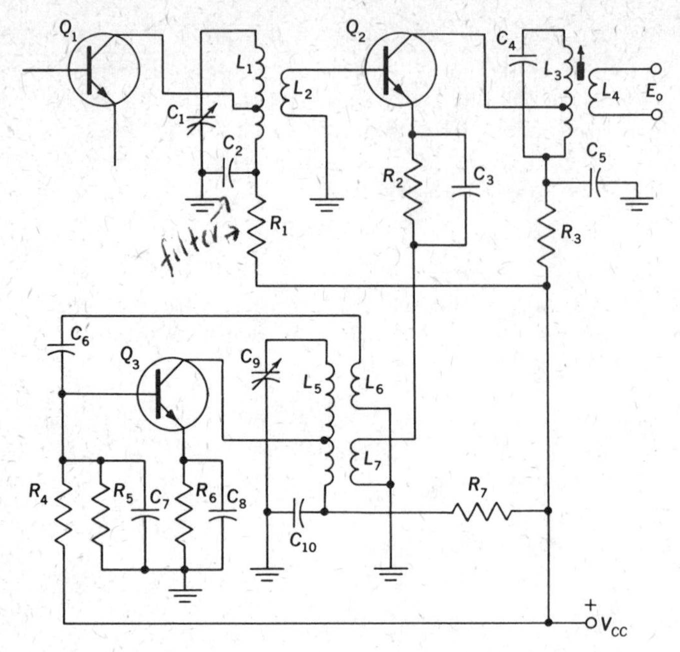

Figure 8-6 A frequency-conversion circuit using bipolar transistors.

the base of Q_2, while the oscillator voltage is inductively coupled into the emitter leg by coil L_7. Bias and bias stabilization are provided by the emitter R_2C_3 network. The collector current of Q_2 flows through L_3C_4, the primary of the IF transformer tuned to the intermediate frequency, and an output voltage is developed at this frequency. The oscillator circuit is of the Armstrong type, with the tuned circuit on the collector side for higher impedance, higher Q, and better stability. The tank C_9L_5 is tuned to the desired oscillator frequency.

Notice that R-C decoupling networks are used in each collector feed. Capacitors C_2 and C_{10} serve a dual function. Not only are they part of their respective decoupling filters, but in addition they serve to complete a low-impedance path between the tank circuit capacitor and inductor. This action is necessary, because capacitors C_1 and C_9 are two sections of the three-gang tuning capacitor* and the rotors are grounded. Meanwhile the inductors cannot be at dc ground because the collector supply voltage is applied through these coils. Some manufacturers get around this problem by grounding the collector side of the supply voltage. This automatically puts the coil and capacitor at ground potential. In other receivers, fixed capacitors are used in the tank circuits, slug tuning is used to vary the resonant frequencies, and either side of the supply source can be grounded.

*The third section tunes the input circuit of the RF stage.

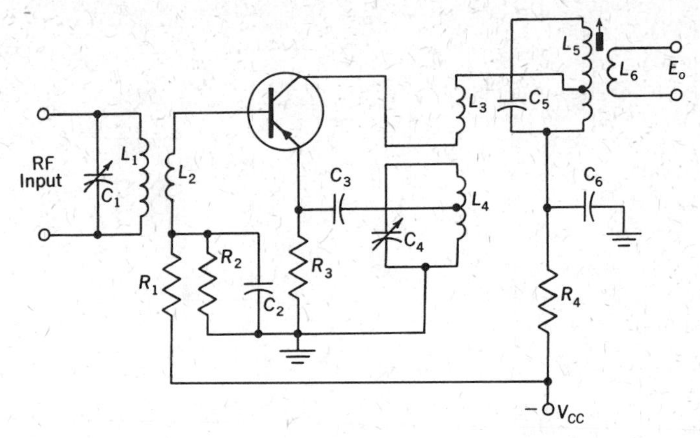

Figure 8-7 A transistor autodyne converter circuit.

Many other variations of the circuit in Fig. 8–6 are possible. PNP transistors can be used. The oscillator circuit can be any of the other types discussed in Chapter 5. (A Hartley version is quite popular.) Finally, the oscillator signal could be inductively or capacitively coupled into the base or emitter leg.

A variation of the above circuit is the *autodyne* converter. This circuit could be described as a self-oscillating mixer, in that the same transistor elements are simultaneously used as an oscillator and a mixer. A typical circuit is shown in Fig. 8–7. The RF input signal is inductively coupled to the base of the transistor by the step-down impedance-matching action of coil L_2. Base bias is obtained via R_1 and R_2. Tank circuit L_4C_4 is tuned to the desired oscillator frequency. Energy from the collector output is coupled into L_4, and the tank develops an RF voltage at the oscillator frequency. The oscillator voltage is injected into the input (emitter) circuit by *R-C* coupling. Feedback from the output, via the tickler winding L_3, keeps the circuit oscillating. Meanwhile, the difference-frequency component of the collector current develops the IF voltage across L_5C_5, the tuned primary winding of the IF transformer. The output from the secondary is fed to the first IF amplifier.

The dual-gate MOSFET makes an excellent mixer circuit. (See Fig. 8–8.) The RF signal is fed to one gate, and the oscillator voltage (from a separate oscillator circuit) is fed to the second gate. This minimizes any chance of interaction (or pulling) between the two input signals. (In this circuit the oscillator voltage is *R-C* coupled to the second gate.) The FET is biased into its nonlinear region using the source-bias R_3C_3 combination. Because of this, the drain current is distorted and will contain many components—one of which is at the difference frequency. The

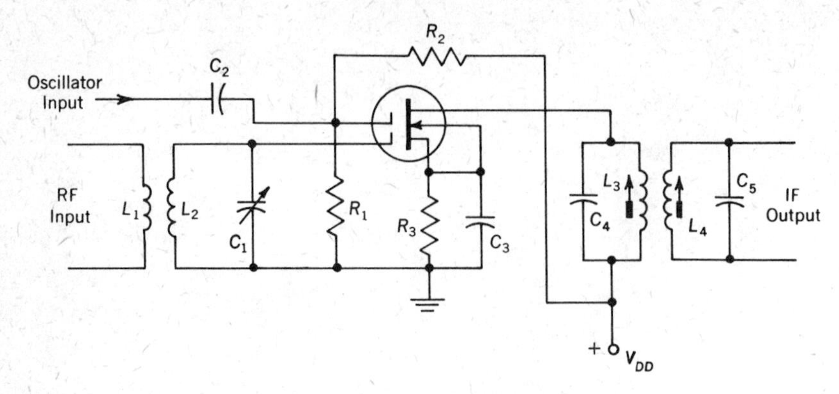

Figure 8-8 A dual-gate MOSFET mixer circuit.

drain tank, L_3C_4, is tuned to this difference frequency and develops an output at this IF value. The output is coupled to the first IF amplifier via the tuned secondary L_4C_5.

Single-gate MOSFETs and JFETs are also used as mixers. However, these do not have two input gates and, generally, both inputs are fed to the one gate. Loose coupling must be used to reduce interaction. One manufacturer, using a JFET mixer, minimizes this interaction by feeding the RF input signal to the source and the oscillator voltage to the gate.

More recently, integrated circuits are also being used for frequency conversion. For example, using the relatively simple ECG 371 IC of Fig. 3–15, the RF signal can be fed to the base of Q_1 through a tank circuit tuned to this frequency. Transistor Q_3, with suitable tank circuit added, can serve as the oscillator. The output is taken from the collector of Q_2 and fed to the primary of the IF transformer.

Finally, since vacuum-tube circuits are still in use, a few words about tube-type frequency converters are in order. Again, basically, a mixer and an oscillator circuit are needed. These could use two completely separate tubes or two tubes in the same physical shell. The oscillator tube (or portion) is a triode, while the mixer tube (or portion) could be a triode or a multigrid tube. Fig. 8–9(a) shows a two-tube circuit. Since the mixer tube is a triode, both signals (RF and oscillator) are fed to the one grid. Diagram (b) shows a dual tube (a *triode-hexode*). The triode section is used for the oscillator, and the hexode, in the mixer circuit. Notice that the oscillator grid is direct-connected to the mixer grid internally. (Obviously no external coupling will be needed between the two circuits.) The RF signal, in this case, is applied to grid G_3.

A second major classification of vacuum-tube frequency conversion circuits is the *pentagrid converter*. Again the dual functions of generating a local RF signal (oscillator action) and mixing are performed. This time, both functions are accomplished by one tube. Do not confuse this with

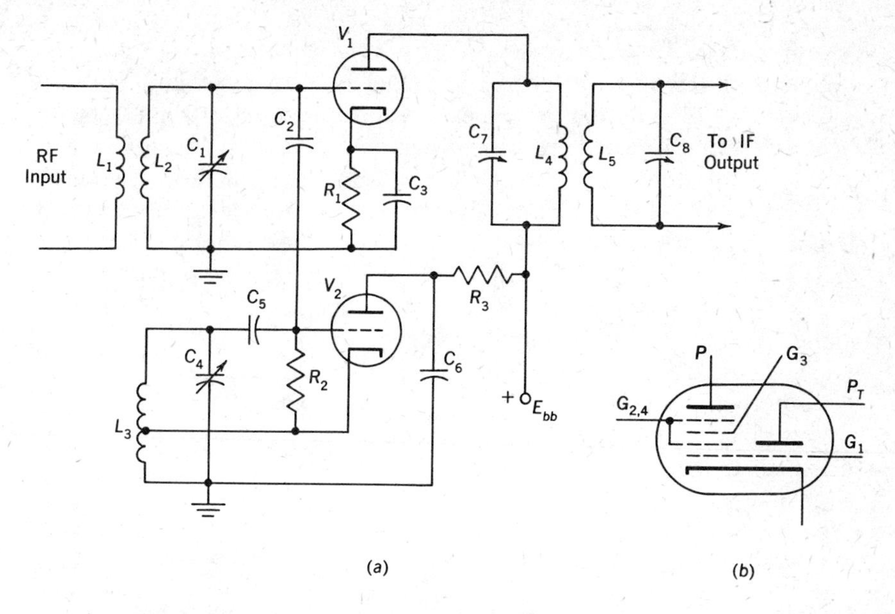

Figure 8-9 A mixer-oscillator frequency-conversion circuit, and a special mixer-oscillator tube.

the triode-hexode action of Fig. 8–9(b). This is a *dual* tube—each with its own electron stream—in one envelope. In a pentagrid converter there is only one electron stream, or truly only one tube. The name pentagrid indicates that these tubes have five grids. However, there are two variations as to grid arrangements, and the circuitry is changed to suit each

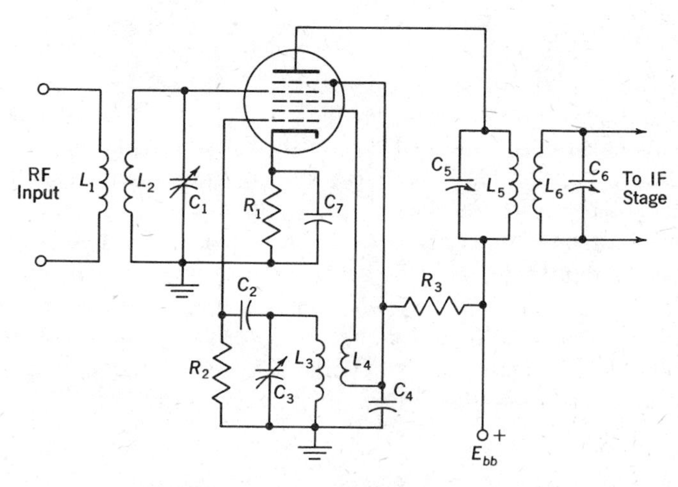

Figure 8-10 A pentagrid converter circuit.

type. Figure 8–10 shows one type of pentagrid converter. In this tube, the incoming RF signal is fed to grid G_4; grid G_1 is the oscillator grid; and grid G_2 acts as the oscillator "plate." (Although the Armstrong or tickler feedback circuit is used here, other types of oscillator circuits could serve equally well.) Because of the oscillator action, the cathode current flows in pulses at the oscillator frequency rate, but since a grid structure is used as the oscillator anode, most of this current passes through G_2 and is accelerated toward the plate by the positive potential on the screen grid $G_{3,5}$. At this point, the current is further controlled by the RF voltage applied to G_4. Obviously, the plate current is a function of the oscillator voltage and the modulated carrier voltage. Therefore mixing takes place, and a current component at the IF value is produced. The tank circuit L_5C_5 is tuned to this frequency, and an output voltage is developed at this intermediate frequency.

Oscillator Tracking

In introducing the superheterodyne principle, it was stated that the oscillator stage is tuned to some fixed amount (the IF value) *above* the incoming RF signal. From a purely theoretical aspect, the oscillator could be tuned either above or below the RF value, and the same difference frequency would result. On the other hand, practical tuning considerations (see Problem 17) make it necessary to use the higher value for AM broadcast-band applications (540–1600 kHz).* Consequently, although the RF and mixer tank circuits should be identical, it is obvious that the oscillator circuit values must differ. For example, assuming a commonly used IF value of 455 kHz, the RF and mixer stages must tune over a band of 540–1600 kHz, but the oscillator range is 955–2055 kHz. Notice that there is only a 28% increase in frequency at the high end of the tuning range, but close to a 100% (84%) change at the low-frequency end. This creates a tracking problem. Using variable-capacitance tuning, the high-frequency end of the band is obtained when the capacitor plates are fully open. The circuit capacitance now consists of the minimum capacitance of the tuning capacitor, stray capacitances of the coil and wiring, and the input capacitance of the transistor (or tube). This total capacitance is approximately the same for each circuit (RF, mixer, and oscillator). To obtain the somewhat higher oscillator frequency, it is therefore necessary to *decrease* the oscillator inductance value by the ratio $(1600/2055)^2$, or to approximately 60% of the RF coil value. Exact tracking is then obtained by adjusting the *trimmer* across each main capacitor section.

*At VHF and higher frequencies—or for fixed frequency operation—the oscillator circuit can be operated below the incoming signal frequency.

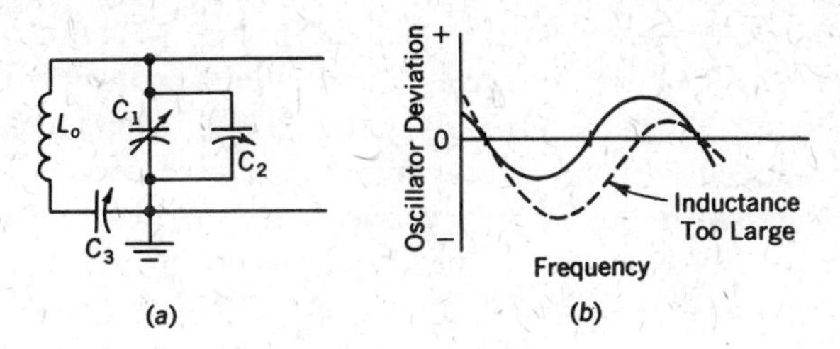

Figure 8-11 Oscillator tracking.

At the low-frequency end of the tuning range, the oscillator frequency is approximately double (1.84) the RF frequency. Therefore, the *LC* product of the oscillator tank circuit should be only 30% of the corresponding RF value. Since the oscillator inductance value was fixed by the requirements at the high-frequency end, any further reduction in the *LC* product must now be obtained by reducing the oscillator capacitance value. This effect is generally obtained by inserting a *padder* capacitor in series with the main tuning capacitor and the coil.* A typical oscillator tank circuit is shown in Fig. 8–11(a), with C_1 as the main tuning capacitor, C_2 as the adjustable trimmer, and C_3 as the adjustable padder. This technique of adjusting the trimmer and padder will give exact tracking at the two ends of the tuning range. Simultaneously, if the oscillator inductance is properly chosen, a third accurate tracking point will be obtained at approximately the center of the tuning range. This is shown in Fig. 8–11(b). With proper design, the deviation will not be excessive at any point in the tuning range. The dashed curve shows the effect of using too large an inductance.

When a receiver is used only on the AM broadcast band, the oscillator section of the tuning capacitor can be specially cut to provide tracking. This section of the gang capacitor can be readily recognized because of its fewer and/or smaller-size plates. With these specially cut plates, the padder capacitor is no longer needed. A disadvantage of this technique is that a "tailored" ganged capacitor can be used only for the one tuning range and the one IF frequency for which it was designed. Obviously, such capacitors cannot be used in multiband communications receivers.

Electronic Tuning

Up to this point, all the variable-frequency tank circuits shown used variable capacitors for tuning the tank over its frequency range. For multistage operation, gang capacitors were used. These units are quite

* For the above broadcast-band illustration, the oscillator maximum capacitance would be approximately one-half the RF value, so that the maximum capacitance of the padder is about the same as the fully closed main capacitor. (See Problem 18.)

bulky. For example, the three-gang capacitor in Fig. 8–3 (vintage 1950) is $3 \times 3 \times 5$ inches, with its plates open. Smaller units are now made for use in the smaller solid-state receivers, but the tuning capacitor is still the largest component in the receiver. Automobile receivers generally use push-button tuning for the driver's convenience. However, the variable capacitors are still there, and the push buttons are merely a mechanical device for stopping the ganged capacitor at several previously selected positions. If anything, they add to the bulk. Some designs use variable-inductance tuning (particularly at the higher frequencies, as in FM receivers). In these circuits, either the number of turns of the coil or the slug core is varied to change the inductance. Although this improves the L/C ratio of the tank circuit, it does not materially reduce the bulk of the tuned circuits. When only a few specific frequencies are to be selected (as in TV channels 2–12), a separate tank is used for each of the desired frequencies, and these channels are then selected by switch rotation. Again, this adds nothing toward miniaturization.

A breakthrough in tuning methods was introduced in 1969. Some manufacturers gave it the dramatic name of *electronic tuning*. Actually, these circuits used a *voltage-variable diode* (VVC diode)* in place of a conventional capacitor. To give a capacitive effect, the diode must be reverse-biased. The capacitance value is changed by varying the dc bias voltage. The higher the reverse-bias voltage, the wider the depletion region on each side of the PN junction, and the lower the capacitance value. Tuning is now accomplished by using a potentiometer to vary the dc bias. The voltage-capacitance curves for four hyper-abrupt junction tuning diodes are shown in Fig. 8–12(a). Notice that a capacitive change of greater than ten times is obtained for a bias change ranging from 2 to 10 V. These diodes are especially designed for tuning over broad frequency ranges—such as the AM broadcast band.

A number of advantages are claimed for VVC diode tuning systems. They are more rugged, more compact, and lighter than conventional capacitor systems. They are not affected by dust or moisture, are obviously ideally suited for remote control, and the bulky mechanical systems for push-button tuning can be replaced by simple switches connected to a voltage divider with adjustable taps. A circuit for VVC diode tuning is shown in Fig. 8–12(b). Notice the symbol used for the VVC diode. As to the other component functions, C_1 is used to prevent short-circuiting of the dc bias voltage through the tank-circuit inductor; R_1C_2 is a decoupling and isolating circuit; and R_2 is the tuning control. For multistage operation, a single potentiometer, through separate decoupling networks, can be used to control the VVC diodes in each of the tank circuits. Obviously, the diodes must have matched characteristics.

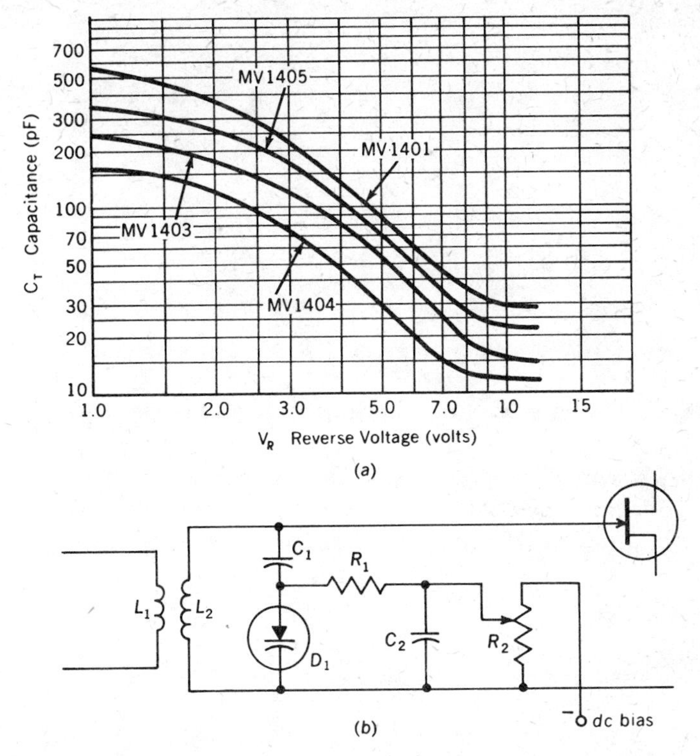

Figure 8-12 VVC diode tuning.

Image Frequency

Let us assume that a communications receiver uses an IF value of 200 kHz and is set to receive a desired signal at 13.8 MHz. Obviously, the oscillator section should be tuned to 13.8 + 0.2, or 14.0 MHz. The difference between the desired RF signal and the local oscillator frequency is equal to the IF value. Unfortunately there is another RF signal which, when mixed with the local oscillator signal, will also produce the correct IF value. This *undesired* RF signal is known as the *image frequency.* Using the above illustration, the image-frequency signal is at 14.0 + 0.2, or 14.2 MHz. Notice that, when this undesired signal at 14.2 MHz mixes with the oscillator frequency at 14.0 MHz, the difference is again 200 kHz.

*This is the same device referred to as a varactor diode in Chapter 4, but when intended for tuning applications, the doping is modified to give better linearity and a larger (min-max) capacitance range.

EXAMPLE 8–2

A broadcast band superheterodyne receiver is tuned to a station at 1010 kHz. The receiver uses an IF of 262 kHz. What is the image frequency corresponding to the above setting?

Solution

1. The oscillator frequency is

$$f_{osc} = f_{RF} + \text{IF} = 1010 + 262 = 1272 \text{ kHz}$$

2. The image frequency is

$$f_{im} = f_{osc} + \text{IF} = 1272 + 262 = 1534 \text{ kHz}$$

In this example, if a station were broadcasting at 1534 kHz, it too might be heard in the output of the receiver. The IF system cannot reject this station. Therefore, prevention of image-frequency reception depends on the combined selectivity of the RF and mixer stages. This is an advantage of using an RF stage ahead of the mixer.

Selection of the IF Value

Several factors should be considered when selecting a value for the intermediate frequency. Sometimes these factors are in direct conflict and a compromise is necessary.

GAIN AND CIRCUIT STABILITY. It is generally easier (and cheaper) to design for high gain at the lower RF frequencies. As the operating frequency is increased, circuit losses (resistive, radiation, and dielectric) tend to increase; stray capacitance and inductive effects are no longer negligible; the cost of suitable circuit components increases; unwanted feedback can cause instability; and transit-time effects can ruin circuit operation. Consequently, high-Q circuits, with high gain and good stability, are more readily obtained at low IF values. A low IF value is desirable.

SELECTIVITY-BANDWIDTH-GAIN. In Chapter 1, we learned that bandwidth is a function of the operating frequency and of the circuit Q. ($\text{BW} = f_0/Q$.) When a high degree of selectivity is desired, low IF values are preferable. However, this can be carried too far. For example, the early superheterodyne receivers were advertised as having "razor-sharp" selectivity—which was true. They used an IF value of 75 or 125 kHz, so that even with a moderate Q value of 50, their bandwidths were only 1.5 or 2.5 kHz. These bandwidths are adequate for code (CW) reception, but much too narrow for audio applications. Even voice frequencies will be distorted. On the other hand, to obtain adequate audio bandwidth (10 kHz) at an intermediate frequency of 75 kHz, the circuit Q must be reduced to 7.5. Such a low-Q value will produce low gain and poor adjacent-channel selectivity (wide-skirt response curve). The obvious solution is to select some higher IF value, commensurate with the bandwidth needed and with reasonably obtainable high-Q values at that frequency. The bandwidth should not be wider than necessary, since that would increase the receiver noise level and impair adjacent-channel selectivity.

For broadcast-band AM transmission, a bandwidth of 10 kHz (maximum) is allowed by the Federal Communications Commission (FCC). Based on a Q of 50, the IF value should not be higher than 500 kHz. Manufacturers standardized at 455 kHz.* With transistorized receivers, because of the low input impedance of the transistor, a Q of 50 is very difficult to obtain. Therefore, a lower IF value is necessary to improve selectivity. A commonly used value is 262 kHz.

Wideband pulse transmissions (radar, telemetry, and television) present another problem. Television signals, for example, require bandwidths up to 4 MHz. If a circuit Q of 50 is desired, an IF value of 200 MHz should be used. But now we run into complications. Many of the transmissions (channels 2–11) are at frequencies below 200 MHz. Furthermore, because of losses, transit-time effects, and stray coupling interactions, it is extremely difficult to achieve any gain at these frequencies. A compromise is necessary between the ideal frequency at which the desired gain and bandwidth is *theoretically* possible, and the actual frequency limits at which good gain can be obtained from practical circuits. At one time, the IF band for television use was in the 21–26-MHz region. As the state of the art improved, this was raised to the 40-MHz region. For similar reasons, a commonly used IF value for radar and other wideband pulse-type receivers is 60 MHz.

IMAGE-FREQUENCY REJECTION. In discussing image frequency earlier in this chapter, it was seen that the oscillator frequency is generally above the desired signal frequency by the IF value, and that the image frequency is, in turn, above the oscillator frequency by an equal amount. This places the image frequency above the desired signal frequency by an amount equal to twice the IF value. Rejection of this image is dependent solely on the selectivity of the tuned circuits in the RF amplifier (if used) and mixer stages. Obviously, the higher the IF value, the further apart the desired and image frequencies will be, and the easier it will be for a tuned circuit to discriminate between these two values.

In the AM broadcast band, image rejection is no problem. Let us assume an RF circuit bandwidth of 20 kHz when the receiver is tuned to

*An advantage of standardizing on one fixed value is that this frequency can be cleared (unassigned), thereby preventing direct pickup of signals into a sensitive IF system.

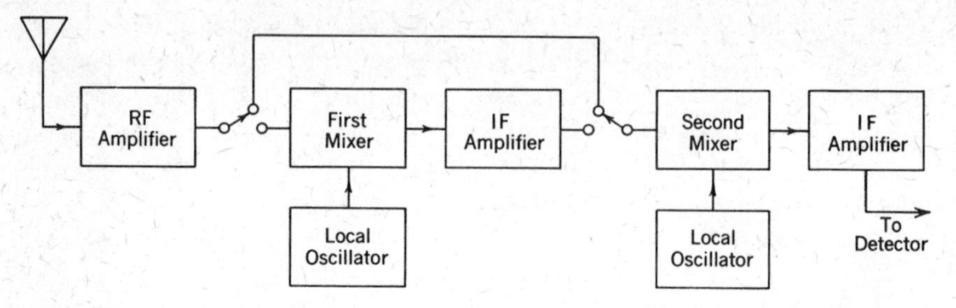

Figure 8-13 A double conversion system.

1500 kHz (Q of 75). With an IF value of 125 kHz, the image frequency is 250 kHz off resonance. On the RF response curve, this image signal is 25 bandwidths away from the resonant peak. The percent response to the image frequency is relatively insignificant. With a 455-kHz IF system, the image rejection is even better.

Now let us consider a communications receiver tuned to a desired station at 22.5 MHz. At this frequency, a Q of 75 would be quite good, but the bandwidth of the RF circuit is now 300 kHz. With an IF of 455 kHz the image-frequency signal is (approximately) 900 kHz, or only 6 bandwidths off resonance. From the general resonance curve data of Chapter 1, it can be seen that the amplitude of the image signal will approach 20% of the desired signal output. This interference is definitely objectionable. To obtain better rejection, the IF value must be raised. Yet a higher IF value will result in too wide an IF bandwidth and create adjacent-channel interference. The solution is to use *double conversion* and two cascaded IF systems. Such a receiver is shown in block diagram form in Fig. 8–13. One manufacturer uses 1720 kHz for the first IF amplifier and 455 kHz for the second conversion. Notice the switching provision for feeding the RF signal directly into the second mixer. This provision is necessary in a multiband receiver since, on the low-frequency bands, the incoming signal frequency is lower than the first IF value. Furthermore, with a 455-kHz IF system image, rejection is good to at least 4.0 MHz.

IF Amplifier Circuits

An IF amplifier is basically no different from an RF amplifier, and so any of the circuits shown in Chapter 3 can be used. However, since IF amplifiers are operated at a fixed frequency, the double-tuned inductively coupled circuit can be used for improved band-pass action. With vacuum-tube circuits there is no loading problem, and the tube is connected across the full tank circuit. (Since tubes are not used in new designs, the schematic will not be shown.) When double-tuned circuits are used with bipolar transistors, the base is always connected to a tap on the IF transformer secondary to reduce loading. The collector (or the supply voltage) is often also connected to a tap on the primary side to further reduce loading (and to provide for neutralization). A variety of double-tuned circuit techniques are shown in Fig. 8–14. The collector of Q_1 is tapped down on L_1 to reduce loading. Neutralization is not necessary here since Q_1 is the mixer stage. (Input and output are at different frequencies, so that even if feedback exists, it cannot cause oscillations.) The base of Q_2 is tapped down on L_2 to reduce loading. R_2 and R_3 are the base-bias resistors, and the bias is series-fed through L_2. The supply voltage is tapped up on L_3 to reduce loading and also to provide a phase-opposing voltage for neutralization. The base of Q_3 is tapped down on its tank circuit using a capacitive voltage divider. R_5 and R_6 are the base-bias resistors, but this time the bias is shunt-fed. (It should be realized that such a variety of circuits would not be used in any one commercial unit; they are shown here only for illustrative purposes.)

Because of the relatively heavy loading due to the low input impedance of the transistors, the higher gain and better selectivity capability of the double-tuned circuit is not fully realized. Consequently, bipolar transistor IF amplifiers often use single-tuned interstage coupling, as shown in Fig. 8–15. This circuit is quite similar to the earlier transistor RF amplifiers. Automatic gain control is now provided by returning resistor R_3 to the AGC bus instead of to the negative supply line. Another variation is the use of decoupling filters R_1C_2 and R_5C_5 in the collector return leads. This would be necessary in multistage operation to prevent unwanted coupling from a common power supply. Note also in this case the use of different types of transistors for Q_1 and Q_2.

When wide bandwidth is necessary, as with radar, television, and other pulse-type applications, multistage IF amplifiers with impedance coupling and stagger tuning are commonly used. Although more gain is obtained from a well-designed overcoupled double-tuned or triple-tuned circuit, it has been found that, due to maintenance (alignment) difficulties, the stagger-tuned circuitry is more practical. A typical circuit is shown in Fig. 8–16. This amplifier has a gain of 53 dB and a bandwidth of 5 MHz with a center frequency of 30 MHz. The input and output impedances are each 50 Ω. The wide bandwidth is obtained by stagger tuning the tank circuits ($L_1C_2 = 28$ MHz; $L_2C_6C_7 = 31$ MHz; $L_3C_{10}C_{11} = 29$ MHz; and $L_4C_{14}C_{15} = 32$ MHz). To reduce loading and provide impedance matching, the base of a transistor should be connected to a tap on the previous tank circuit. Notice that, in this circuit, the "tap" is made in the capacitive branch of the collector tank circuits. The rest of the circuit

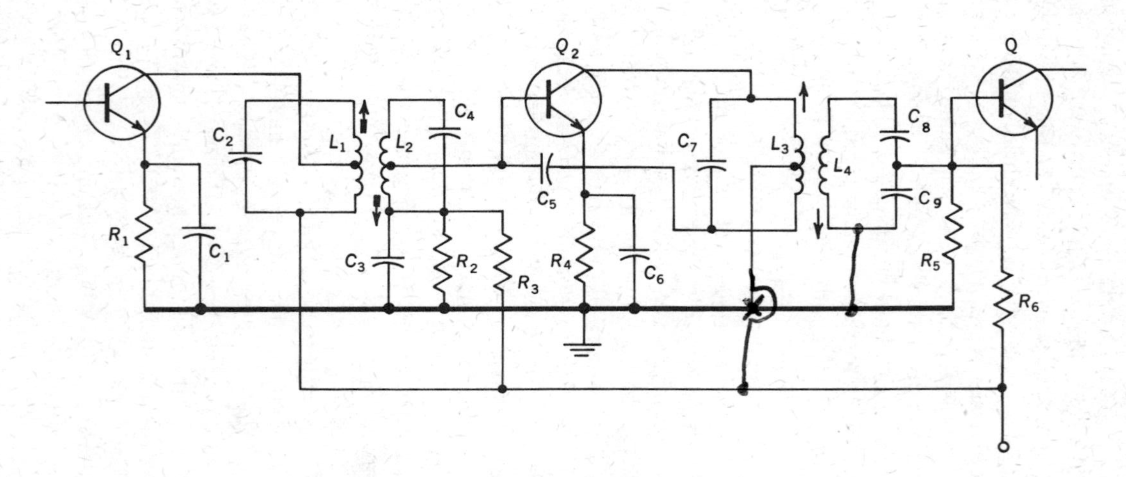

Figure 8-14 IF amplifier techniques.

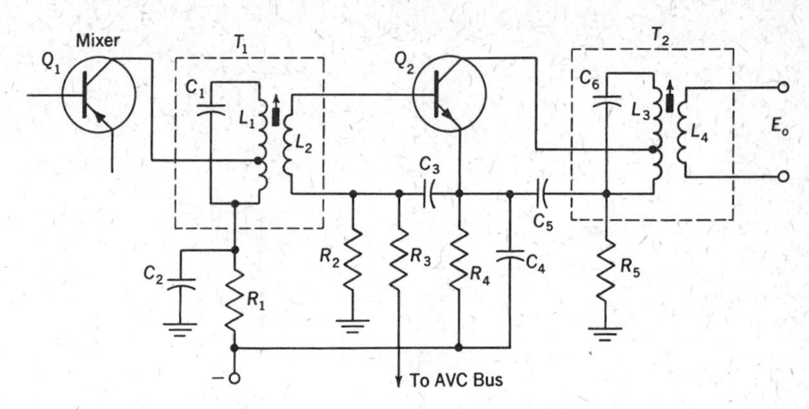

Figure 8-15 A transistor IF amplifier.

follows previously discussed circuitry, so further analysis will be left for discussion. (See Review Question 50.)

The IF circuits shown so far all use *L-C* tank circuits for selectivity. As filters, *L-C* circuits have three drawbacks. Their band-pass action, as discussed in Chapters 1 and 2, falls rather short of the ideal rectangular response curve. (Double-tuned circuits come closer to the ideal characteristic, but still the skirts are not steep enough.) Secondly, with time, the alignment of the IF transformers' resonance may drift, and realignment is necessary. Finally, the inductive component of the tank is not compatible with IC miniaturization techniques. When very narrow bandwidth or very steep skirts for adjacent-channel selectivity is required (as in single-sideband receivers), discrete crystal filters have been used. However, these are too expensive for consumer product applications, and their Q is so high as to make the bandwidth too narrow for broadcast-band receivers.

A study of techniques for inductorless filters led to the development of mechanical filters, monolithic filters, and ceramic filters. Of these, the ceramic filter seems best suited for receiver applications. They are available for operating frequencies as low as 100 kHz, up to the FM IF value of 10.7 MHz, and with Q values of from 30 to 1500. The basic building block for ceramic filters is the ceramic resonator. One type is a simple rounded disc of specially formulated piezoelectric ceramic, with electrodes plated on each face. These resonators operate in their radial mode, and the resonant frequency is determined by the dimensions of the element and the dielectric constant of the material. A typical radial resonator for use at 455 kHz has a diameter of 0.56 cm and a thickness of 0.038 cm (or approximately $\frac{1}{4}$ in. in diameter and $\frac{1}{64}$ in. thick). Several of these

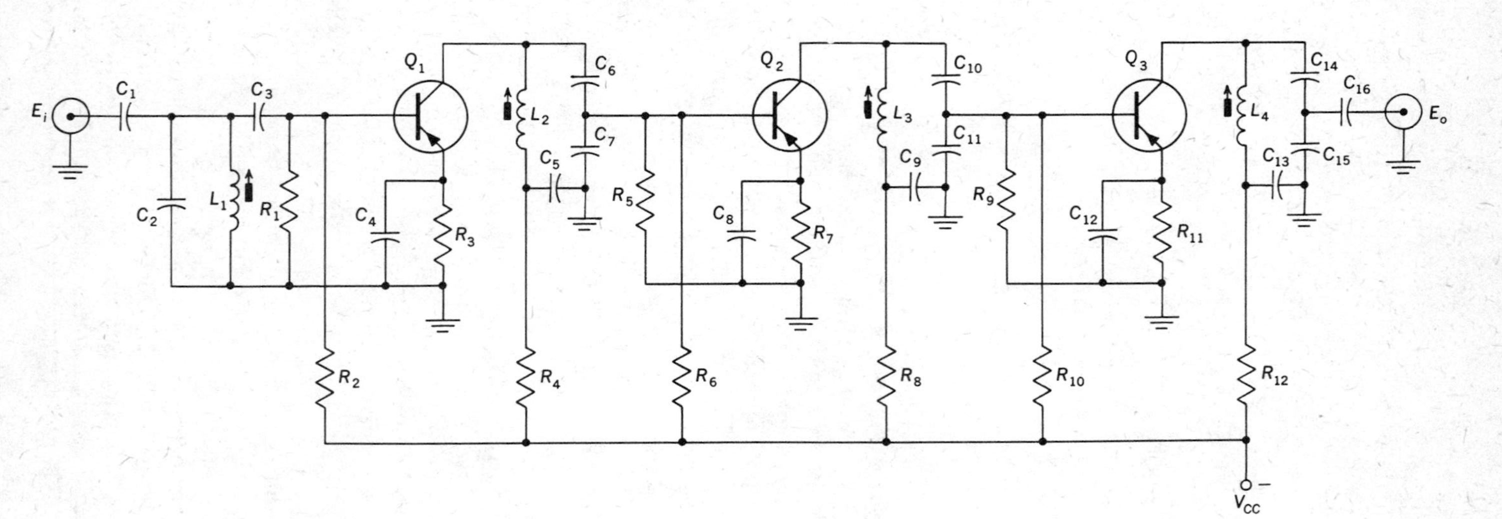

Figure 8-16 A 30-MHz wideband IF amplifier.

resonators can be combined to obtain the desired bandwidth and steepness. Ceramic filters can produce better performance (selectivity, stability, and ruggedness) than conventional transformers; they are cost-competitive with transformers; and their small size lends to miniaturization and IC designs. Figure 8–17 shows a three-terminal ceramic filter used in an AM IF amplifier circuit. Resistor R_3 is necessary because there is no dc path through the filter. In this respect, the circuit is similar to a shunt-fed tank circuit.

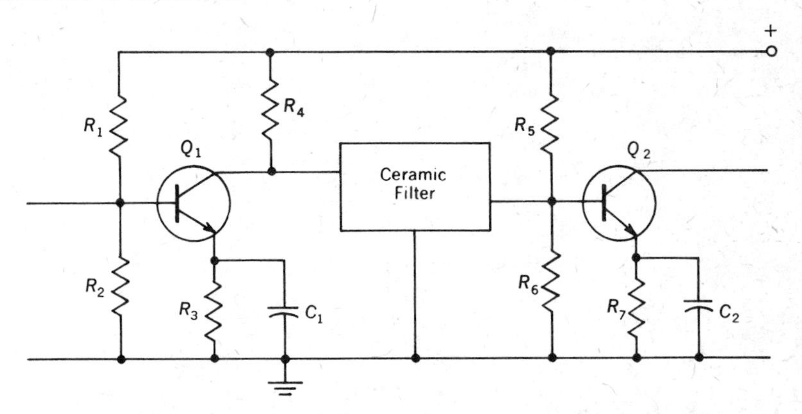

Figure 8-17 IF amplifier with ceramic filter. (Courtesy Sylvania Electric Products, Inc.)

Field-effect transistors (both JFETs and MOSFETs) have also been used in IF amplifier circuits. Actually, they are ideally suited. Their higher input impedance does not load the tank circuit, and they require negligible power for AGC action. In fact, the dual-gate unit simplifies gain-control circuitry because the AGC voltage can be applied to gate 2. Yet, FET IF circuits have not become popular. One reason is the higher cost of the FET and the adequate performance of the bipolar unit. A possibly stronger reason is that the development of linear ICs paralleled the advancement in FET technology, and manufacturers have gone from discrete bipolar transistor stages directly to IC circuitry.

The application of ICs to the IF section of superheterodyne receivers has taken two paths. Some designers use a simple differential-amplifier IC (similar to the ECG 371 shown in Chapter 3) to replace an individual stage transistor. The coupling network between the ICs could be the conventional IF transformer or the newer ceramic (or even quartz crystal) filters. An example of this individual stage design is shown in Fig. 8–18.

Another technique is to use a more sophisticated IC package that will supply the total IF amplification—and in some cases also include the detector circuit. The tuned circuits and all capacitors, however, are still

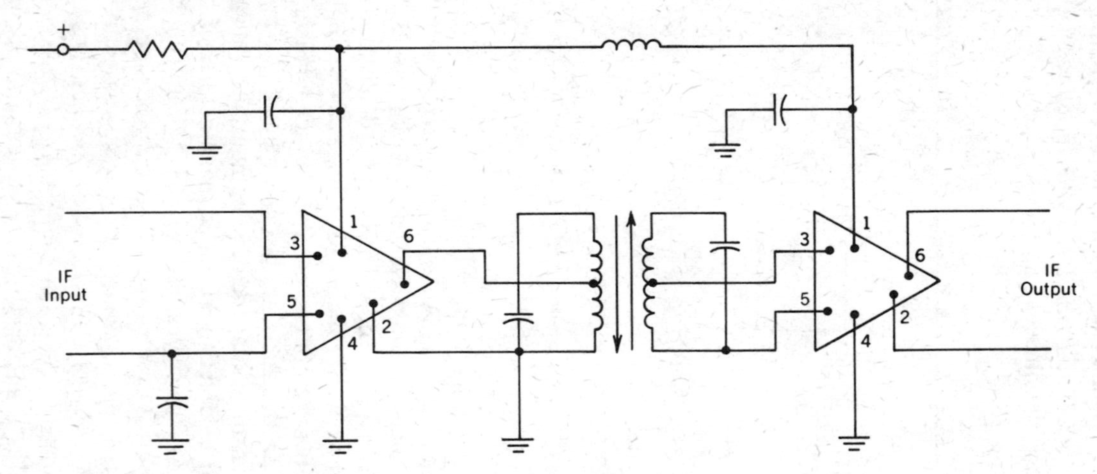

Figure 8-18 IF amplifier using μA 703 ICs. (Courtesy H. H. Scott, Inc.)

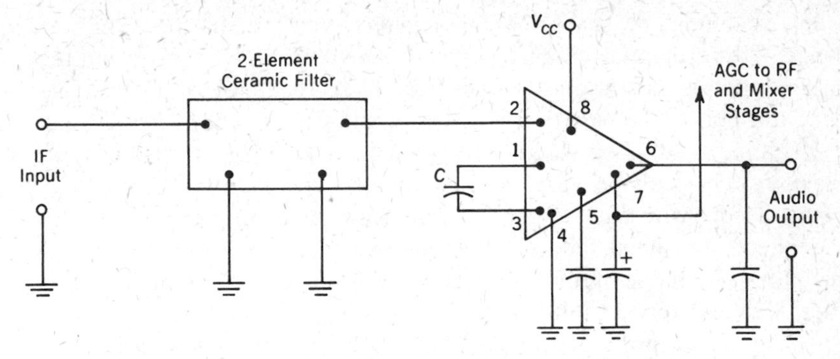

Figure 8-19 Broadband AM receiver IF strip (ECG 372). (Courtesy Sylvania Electric Products, Inc.)

outside the IC unit. Again some use *L-C* tank circuits; others use ceramic filters. An example of this type is shown in Fig. 8–19. This IC consists of a high-gain section, a detector, and an automatic gain amplifier. AGC control is applied internally to the amplifier section and is also available for application to the RF and mixer stages. This unit is intended for use as an IF amplifier at frequencies of from 50 kHz to 2 MHz. Selectivity is obtained by means of an external filter which could be *L-C*, ceramic, crystal, or mechanical, having single or multiple elements. Improved selectivity can be obtained by substituting a second filter between pins 1 and 3, in place of capacitor *C*. The audio output is 0.8 V for an IF input level as low as 50 mV. The total power drain is only 1.4 mA, or 8.4 mW from a +6-V supply. Internal voltage regulators are used, eliminating the need for individual stage decoupling. This IC actually contains 14 transistors, 9 diodes (or diode-connected transistors), and 17 resistors—all in an 8-pin T05 package.

Automatic Gain Control (AGC)

One of the great improvements in radio communications was achieved by providing for automatic control of the signal level at the output of a receiver. This invention is probably the first electronic feedback automation device. Since this feature was originally used with radio receivers to maintain the sound level or volume constant, it was named *automatic volume control* (*AVC*). Actually, this circuit controls the gain (or sensitivity) of the RF and IF sections of the receiver, and it has since been correctly renamed *automatic gain control* (*AGC*). In communications receivers, AGC action will keep the audio volume at the desired level in spite of changes in the signal level. This is of great advantage when scanning, or searching for signals, over the full receiver tuning range. If the manual volume control is adjusted downward when tuned to a strong-signal station—and there is no AGC action—it is possible to tune through weak signals and not know they are missed. Conversely, if the volume is adjusted when monitoring a weak signal, strong signals will blast in. This can be very uncomfortable, particularly if earphones are being used. Automatic control circuits are also important in reducing the effects of signal fading due to ionospheric conditions or multipath reception. Reliable mobile communications and navigational aids owe their very existence to this feature.

The basic principle of any AGC system is to raise the gain of the RF and/or IF amplifiers when the antenna input signal weakens, and conversely to reduce their gain when the input signal strength increases. This can be done by changing the bias or operating point of the active device. A diode detector is an ideal source for such a bias control (AGC) voltage. The output across the diode load (see Fig. 7–1) has a dc component that is negative with respect to ground. Since this dc component is produced by the RF input, its value will vary with the signal level. The action of an AGC circuit can now be shown in block diagram form (Fig. 8–20). The

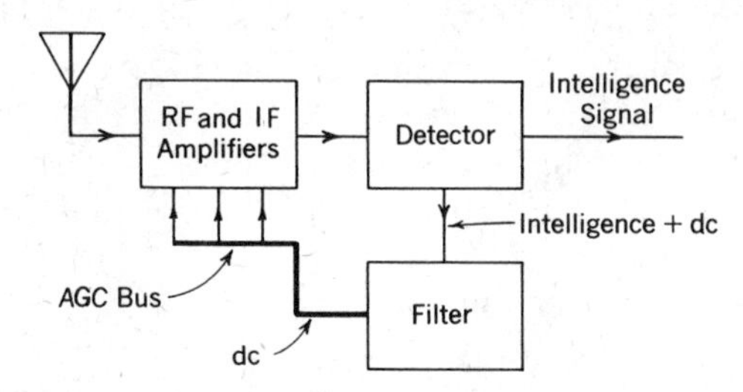

Figure 8-20 The transfer characteristics of a remote-cutoff pentode.

antenna signal is amplified by the RF and IF amplifiers and fed to the diode detector. The voltage developed across the diode load is fed to a filter, and the dc output of the filter is applied as a bias control voltage to each RF and IF amplifier. If the antenna signal strength should decrease, the dc output at the detector will decrease; the bias on the RF and IF amplifiers will decrease, and their gain will increase, compensating for the reduced signal level. The intelligence output *at the detector* tends to remain constant. Conversely, should the signal level at the antenna increase, the resulting increase in the AGC bias level will reduce the gain of the controlled stages, and the detector output will be maintained at a constant level.

The partial schematic of Fig. 8–21 shows the details of how the AGC voltage is obtained from a diode detector and how it can be applied to N-channel FET stages. Notice that the AGC voltage is taken from the

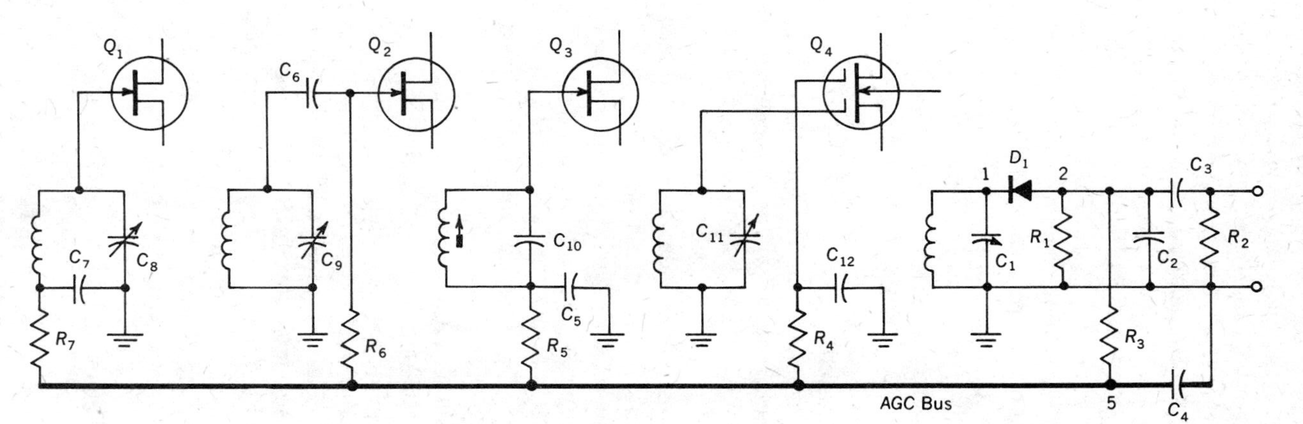

Figure 8-21 A partial schematic of AGC circuits.

diode load resistor, point 2, to ground. Recall from the previous chapter (Fig. 7–1(d)) that this voltage ($e_{R\text{-}C}$) is a complex wave containing the intelligence signal as well as a dc component. If this waveshape were applied directly to the controlled stages, the gain would vary at the modulation rate, causing severe distortion. To remedy this, the R_3C_4 filter is used. Capacitor C_4 must have a low reactance (compared to R_3) at the lowest *modulating* frequency. It therefore bypasses the intelligence signals to ground. Resistor R_3 is needed to isolate the AGC takeoff (point 5) from the intelligence takeoff (point 2).

The AGC voltage can be applied to the controlled stages in various ways. In the Q_1 and Q_3 stages of Fig. 8–21, the dc control bias is in series with the tank circuit signal voltage. These are series-fed circuits. R_7C_7 and R_5C_5 are decoupling filters. Their action is similar to the decoupling filters used in the drain-supply line of multistage amplifiers. Notice that the AGC bias voltage is applied to the gate of Q_2 *in parallel* with the RF input. This is a shunt-fed system. In RF amplifiers where the rotor of the tuning capacitor is grounded, either of the techniques shown with Q_1 or Q_2 can be used. The circuit shown for Q_3 is generally used with IF amplifiers. Q_4 shows how AGC can be applied to gate 2 of a dual-gate MOSFET. Notice that a shunt-fed circuit is used.

Other variations of AGC development and feeding techniques are possible in addition to the four shown here. It should be realized, of course, that no one receiver would use such a variety—and these are grouped here only for illustrative purpose. One other point: If the FETs used were P-channel, a positive AGC control voltage would be required. This could be obtained by reversing the diode connections in the detector circuit.

The addition of an AGC circuit can have an adverse effect on the operation of the diode detector. Notice that R_3 is effectively in parallel with R_1, reducing the *ac* load resistance. As discussed in Chapter 7, this can cause amplitude distortion at high modulation levels. It is therefore important to make R_3 large compared to R_1. On the other hand, too large a value can raise the time constant R_3C_4 excessively, so that the AGC bias cannot respond to rapidly varying signal strengths. Commonly used values for this time constant range 0.1–0.5 s.

The above loading can be eliminated by separating the AGC function from the detector circuit. Figure 8–22 shows a separate AGC circuit. The RF signal output from the last IF stage is also coupled (capacitively) to a second diode—the AGC diode. Assuming the cathode to be grounded, the diode conducts during the first half of the positive half-cycle of the RF input signal, and capacitor C_1 charges to the peak RF value with the polarity indicated. As the signal level drops, the capacitor cannot discharge through the diode, but it tries to discharge through R_2

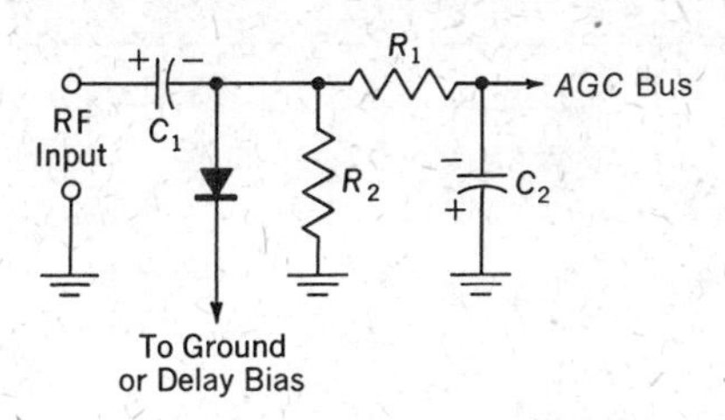

Figure 8-22 A separate AGC Circuit.

(and R_1C_2). This is a relatively long time constant circuit, and only a slight discharge occurs. In this process C_2 picks up a small charge, with the polarity as shown. This latter action is cumulative. After a few cycles, the charge on C_2 will build up to the full value, producing the AGC bias voltage. Because of the even longer time constant of the R_1C_2 filter, RF ripples and audio modulation do not affect this bias value.

A separate AGC diode is also used in *delayed* AGC circuits. Such a feature is particularly desirable when receiving weak signals, as in long-distance communications. With standard AGC action, when a signal is received—no matter how weak it is—AGC action reduces the sensitivity of the receiver. This is not desirable. For weak signals, full sensitivity is needed. The AGC action should be stopped or *delayed* if the signal strength being received is below some minimum value. This effect can be achieved by preventing the AGC diode from conducting on weak signals. For example, the cathode of the diode in Fig. 8–22 can be connected to a positive bias voltage, such as +3 V. Now the diode cannot conduct if the input signal level is less than 3 V at its peak. Meanwhile, capacitor C_2 will discharge through R_1 and R_2, the AGC bias drops to zero, and full sensitivity is maintained. For signal levels above 3 V peak, the AGC action is as before.

Bipolar Transistor AGC Circuits

It has been found that the power gain of a bipolar transistor amplifier varies appreciably with the bias currents. This results not only from a change in the current amplification factor (h_{fe}), but also because of changes in the input and output admittances (y_{ie} and y_{oe}). For example, reduction of the emitter current causes appreciable reduction in power gain. Consequently, control of a transistor's bias *current* can be used for AGC action. Since a current is being controlled, power will be required from the AGC source. In this respect, the transistor power detector is superior to the diode detector.

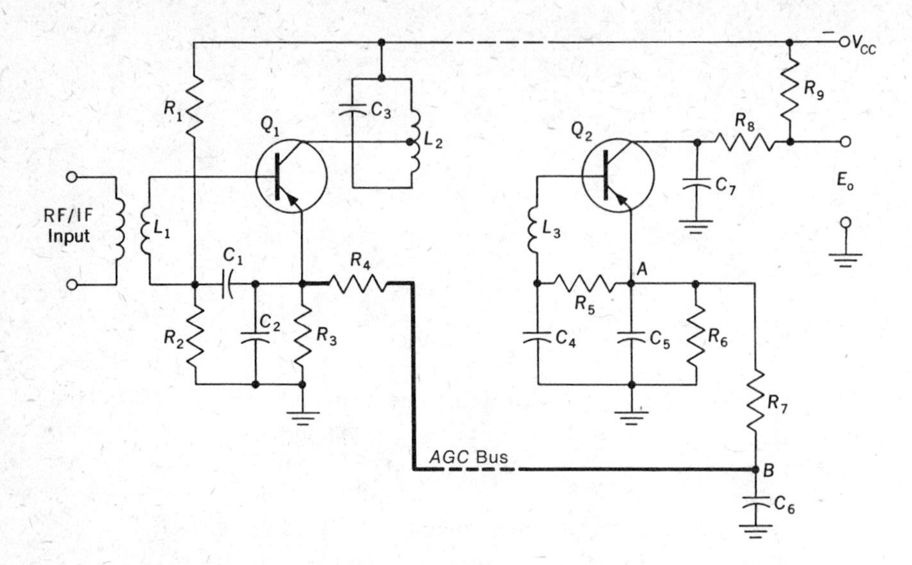

Figure 8-23 A transistor AVC circuit.

Figure 8–23 shows direct control of the emitter bias current of the amplifier stage Q_1. Q_2 is a power detector circuit similar to the previously discussed circuit in Fig. 7–5. This time the emitter, point A, is not at ground potential because of the addition of R_6C_5. If the signal level increases, the emitter current increases and point A becomes more negative. This provides the AGC voltage. R_7C_6 forms the AGC filter to remove any modulation from the control voltage. This voltage is in turn applied to the emitter of Q_1 (or any other RF or IF amplifier) through a resistor R_4 to provide a suitable control current. Should the signal level reaching the detector tend to increase, the emitter current of Q_2 increases. Part of this increase, flowing through R_7,R_4, and down through R_3, makes the emitter of Q_1 more negative. Since this is a PNP transistor, this action opposes the forward bias, reducing the emitter current and the stage gain, thus tending to maintain a constant signal level at the detector. In some applications, resistor R_6 is omitted, so that the full change in the emitter current of Q_2 is available for control.

The above system requires appreciable AGC power and necessitates use of the power detector or an AGC dc amplifier, if a diode detector is used. An appreciable saving in power requirement can be achieved by feeding the AGC signal to the base of the controlled transistors. Because of the amplification of the transistor, a much smaller base-bias current change can be used to obtain a relatively large emitter current change. This circuitry can therefore be used either with a power detector or with a diode detector. It should be realized, however, that when applying the control voltage to the base, the polarity must be reversed. Figure 8–24

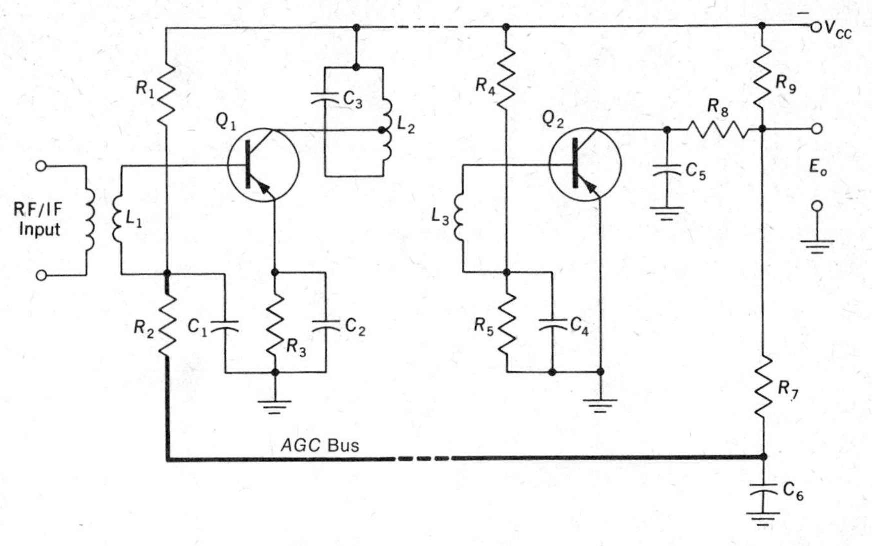

Figure 8-24 A base-bias AVC circuit.

shows how the previous transistor detector circuit can be modified to control base bias. Notice that the control voltage is now taken from the collector circuit. Because of the voltage divider action of R_9 and R_7, the AGC bus voltage is negative. However, when an RF signal is applied to the detector, there is an increase in the collector current flow through R_9, and the increased voltage drop across R_9 makes the AGC bus less negative. The stronger the signal level, the less negative the AGC bus voltage. Meanwhile, the RF (or IF) amplifier Q_1 is forward-biased (base negative compared to emitter) by the combined action of the voltage divider R_1 and R_2 and the AGC bus voltage. If the AGC bus voltage becomes less negative (due to a stronger signal level), the forward bias on the controlled stage is reduced; the emitter current is reduced; the gain of the stage is reduced; and the detector signal level tends to remain constant.

Because of the lower AGC power required with base current control, diode detector circuits are often used to provide the control voltage. Such a circuit is shown in Fig. 8–25 using an NPN controlled stage.

The action of the diode detector circuit produces a negative AGC voltage, with the amplitude of this voltage dependent on the RF signal level reaching the detector. The voltage divider resistors R_1 and R_2 are selected to maintain the base of Q_1 forward-biased (positive with respect to the emitter). The AGC voltage counteracts this forward bias, reducing the stage gain and tending to maintain a constant output from the detector.

The above circuits have one drawback. They are not effective on strong signal levels. Because of leakage current (I_{ceo}), it is impossible to

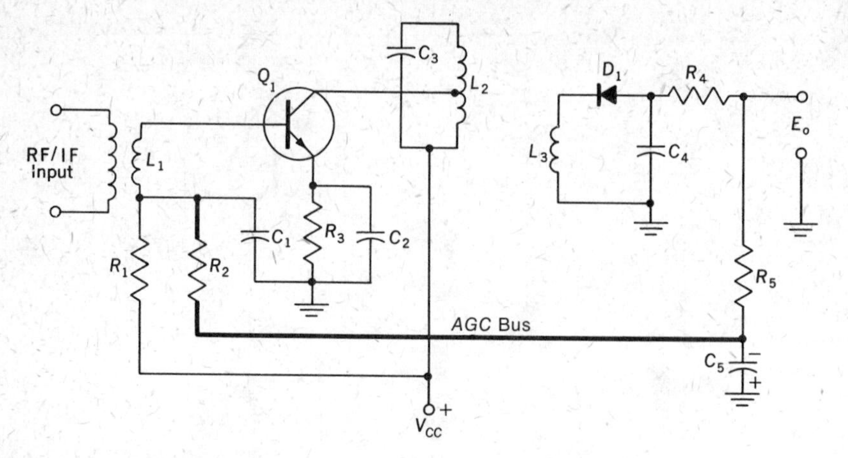

Figure 8-25 An AVC circuit using a diode detector.

get sufficient reduction in gain. It is therefore necessary to use auxiliary circuits that take effect at high signal levels. A commonly used technique is to shunt a load across an interstage tank circuit so as to lower its gain. This technique is shown in Fig. 8-26. Except for the diode between points A and B, the controlled amplifier circuit is identical to Fig. 8-25. For low and moderate RF signal levels, the circuit action is as described above. Resistors R_1 and R_5 are chosen (in combination with the respective collector currents) so that point A is positive with respect to B, and diode D_1 does not conduct. Since its impedance is relatively high, it has no effect on the circuit action.

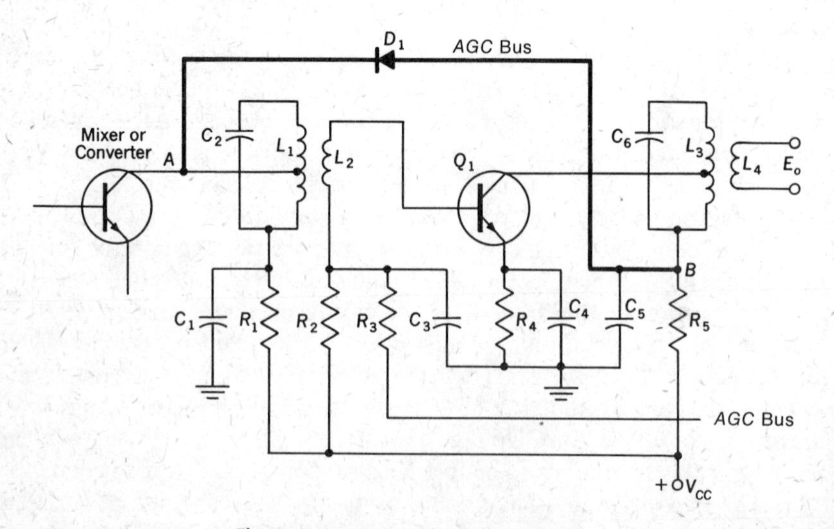

Figure 8-26 An auxiliary AVC diode.

This diode begins to take effect at some predetermined (moderately high) signal level. For proper action, it is important that the converter or mixer stage *not* be controlled from the AGC bus. Consequently, the potential of point A remains essentially constant, whereas at higher signal levels, the collector current of Q_1 will decrease and the potential of point B will rise—approaching the potential of point A. If the RF signal level reaching the detector is sufficiently high, point B can become positive compared to A; the diode conducts; and its resistance decreases, loading the tank circuit primary (L_1C_2) and reducing the voltage coupled into Q_1. This counteracts the increase in signal level and enhances the AGC action.

Superheterodyne Tuner Circuits

Up to this point we have seen a variety of the individual circuits used in superheterodyne receivers. Obviously, many combinations of these functional stages are possible. To give a better appreciation of the RF section of a receiver, together with typical component values, a commercial circuit will be discussed. Figure 8–27 shows the details of a table model receiver. (The power amplifier stage and the power supply circuitry have been omitted.) RF signals are picked up directly by the antenna coil L_1.* The input tank circuit is tuned to the desired frequency by one section (C_{1B}) of a two-gang variable capacitor. The signal is then coupled through the low-impedance winding to the base of Q_1. Simultaneously, the oscillator voltage is coupled into the emitter leg by coil L_3, and an output at the intermediate frequency (455 kHz) is developed across the primary of T_1. Q_2 and Q_3 are IF amplifiers. Notice that these stages are neutralized. Also notice that AGC is applied from the diode detector E_2 to the base of Q_2, and that additional control is obtained by using the auxiliary shunt diode circuit of Fig. 8–26. The demodulated output from the diode detector is fed through the volume control to the first audio amplifier stage Q_4. Its output E_o is then fed (not shown) to another 2SB54 as the driver stage and then to a pair of 2SB56 transistors in push-pull.

RECEIVER ALIGNMENT

In order to obtain optimum sensitivity and proper bandwidth, the tuned circuits of a receiver must be aligned in accordance with design specifications. Two basic techniques are available for this purpose. They may be classified as the *steady-frequency* method and the *sweep-frequency*

*This coil, known as a ferrite loopstick, is wound on a long ferrite (powdered-iron) rod and mounted on top of the chassis to intercept as much RF field as possible. Because of their high Q, high signal pickup, and small size, they are extensively used for urban (and suburban) applications, where strong signals are available.

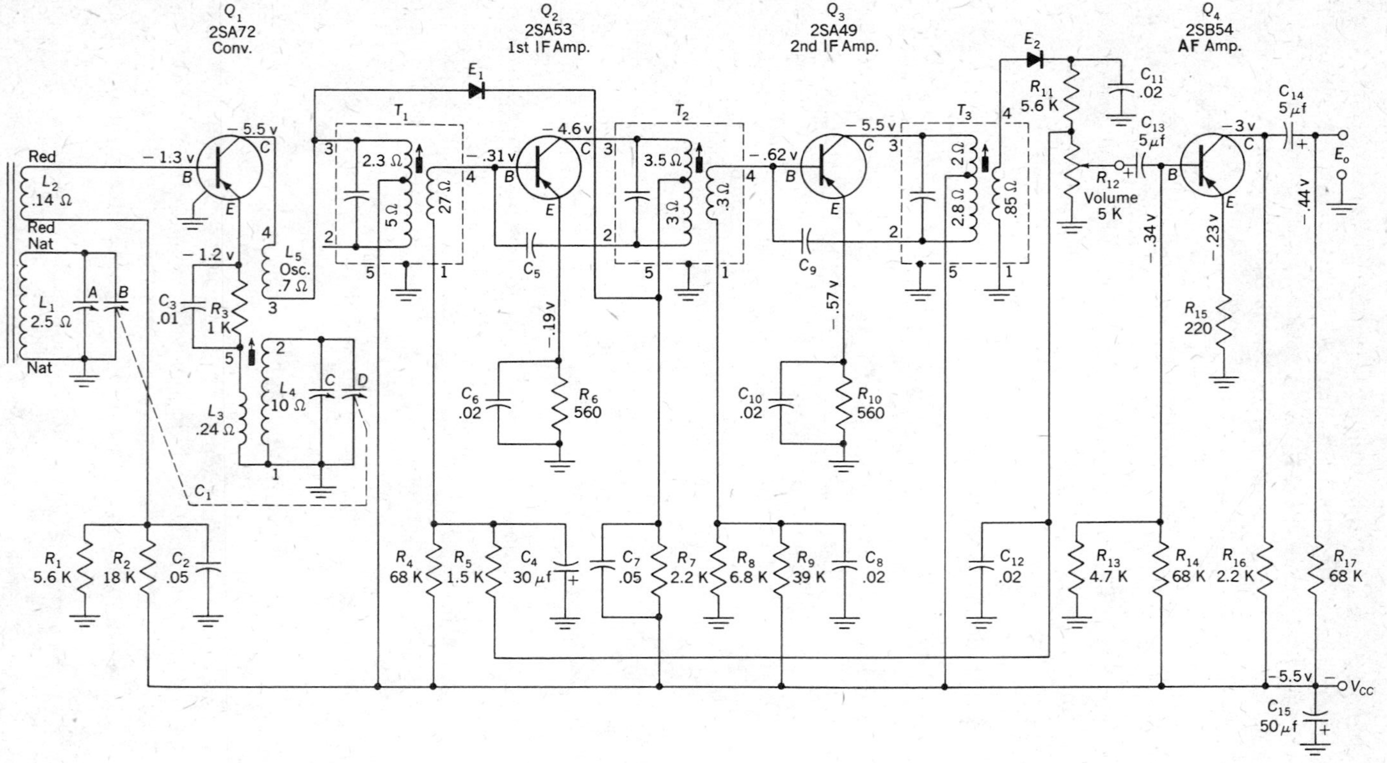

Figure 8-27 A transistor superheterodyne tuner.

method. In either case a suitable RF signal generator is needed to feed a known RF signal into the circuit being adjusted, and an indicator is needed to check for maximum (or in some cases, minimum) response. The accuracy of the alignment will depend, primarily, on the calibration accuracy of the RF signal generator.

Equipment Needed

In the steady-frequency method, the signal generator used should have high accuracy, and its output should remain constant over a wide range of frequencies. Crystal controlled or crystal calibrated units are very desirable. Obviously the generator should cover the IF and RF bands of the receiver under test. For alignment purposes, the RF output can be either a pure sine wave (unmodulated) or it may be amplitude modulated. Most signal generators have provisions for 30% modulation with 400 Hz.

If an unmodulated RF signal is fed to the receiver under test, the output indication must be taken from the detector circuit. Since the signal is unmodulated, the output at the detector is dc, and the indicator used must be a dc voltmeter. The meter should have a high resistance to minimize any loading effect. An electronic voltmeter (VTVM or FET VOM) on its dc range is an excellent indicator. On the other hand, when an AM input signal is used, the output can be taken from any point beyond the detector stage. This time the indicator must be a high-impedance *ac* voltmeter. Again a VTVM (on the ac range) makes an excellent indicator.

For alignment by the sweep-frequency method, a frequency-modulated RF signal generator is necessary. The output of such a *sweep generator* can be made to vary automatically above and below the desired test frequency by some preset amount. The sweep rate used is either 60 or 400 Hz. The amplitude of the output signal does not vary. When such a signal is applied to a receiver, since there is no amplitude modulation, the voltage developed at the detector output will be dc. However, since the signal varies in frequency, the signal level reaching the detector will vary *with the frequency* in accordance with the response curve of the tuned circuits. Furthermore, since the signal frequency varies at a 60 (or 400) Hz rate, the magnitude of the dc output will also vary at this rate. Obviously, a dc voltmeter cannot be used as an indicator.

On the other hand, by using an oscilloscope as the indicator, the receiver response curve will be traced on the screen. This can be explained as follows. Let Fig. 8–28 represent the response curve of the IF system of an AM broadcast-band superheterodyne, plotted in terms of detector output voltage versus frequency. As the frequency of the input

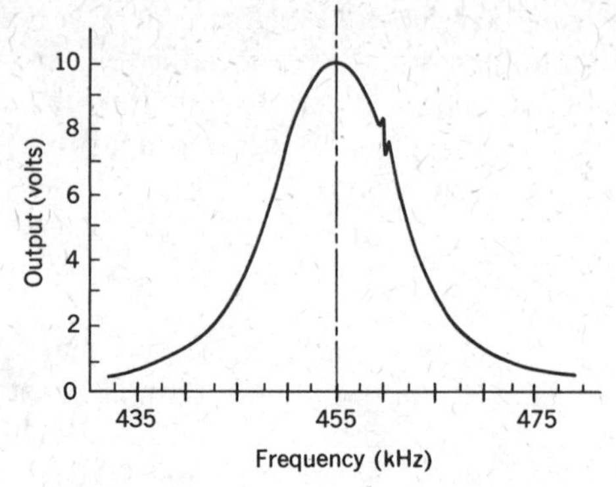

Figure 8-28 A typical IF response curve.

signal is increased from (approximately) 410 to 500 kHz, the output voltage rises from (approximately) 0.5 V to a maximum of 10.0 V (at resonance), and falls to 0.5 V as the resonant frequency is passed. Then, as the input frequency is reduced back to 410 kHz, the output voltage rises to a maximum and falls back to 0.5 V—*retracing the above response curve.* With a 60-Hz frequency sweep rate, this response curve is traced and retraced 60 times per second.

If the detector output is applied to the vertical input terminals of the oscilloscope, a vertical line will be seen on the screen. Actually, this line varies in magnitude from some minimum to a maximum value, 60 times per second, in accordance with the variation of the detector output voltage. Now if we apply the sweeping voltage* from the signal generator to the horizontal input terminals of the oscilloscope (and set the scope for external sweep), the electron beam will also be deflected left and right at the sweep rate. This technique synchronizes the horizontal deflection with the changes in detector output voltage. The screen pattern will be stationary and will duplicate the curve shown in Fig. 8–28. Because the response pattern can be seen, the sweep-alignment method is also known as *visual alignment.*

Since the frequency of a sweep generator must swing above and below a given value by some preset amount, the accuracy of its center frequency is questionable. Also, since the sweep rate may not be linear, we cannot tell from the screen pattern at what specific frequencies the output falls to 70% of its maximum value. Therefore, accurate bandwidth determinations cannot be made by inspection. This ambiguity can be resolved by using an accurate, unmodulated signal generator to indicate specific frequencies directly on the oscilloscope screen. When used for this purpose, the generator is called a *marker generator,* and its effect is to

* This is not the RF signal, but the 60 (or 400) Hz voltage that causes the RF signal to vary in frequency. Frequency modulation is discussed in detail in Chapter 9.

produce a *pip* on the screen pattern, as shown in Fig. 8–28. As the frequency of the marker generator is varied, this pip will ride up and down on the response curve. The specific frequency of any point on the response curve can be found by varying the marker generator frequency until the pip is at the desired point, and then reading the frequency on the marker-generator dial.

General Techniques

Before discussing the alignment procedures for any specific type (or section) of receiver, it is well to review techniques that are common to any application.

CONNECTIONS. It is general practice to connect indicators (voltmeters or oscilloscopes) through *isolating* resistors. This will reduce any resistive and capacitive loading that could widen or detune the response curve. The higher the resistance value, the less the loading. Since this reduces the output indication, choice of resistor value is limited by the output voltage available and the sensitivity of the indicator. The isolating resistor should be located at the receiver end of the test lead. (A VTVM generally has such a resistor at the tip of its dc probe.)

Signal generators should be connected to the receiver in series with a blocking capacitor. This will prevent short-circuiting of bias voltages by the low dc resistance of the generator's output circuit. It also prevents burn-out of the generator output circuit. (An exception to this practice may be made when feeding the signal into the antenna input, in which case—particularly at VHF—an impedance-matching network is preferable.)

SIGNAL LEVEL. Quite often spurious signals, noise, or even oscillations will produce an appreciable output on the indicator, even before a signal is applied from the signal generator. It is therefore important, *before any alignment is attempted,* to make sure that the indication noted results from the desired RF signal. With the signal generator set at the correct frequency, the signal level should be increased until the indicated output rises above its initial value. It is also important that the signal not be too strong or it may overdrive a stage. If a stage is saturated, its output will remain constant even though appreciably detuned, giving an exaggerated flat-top appearance. Obviously, a proper signal level is one which is above the noise level but below the saturation level. The voltmeter indication, or oscilloscope amplitude, should vary in direct proportion to the signal generator output-level control setting. Furthermore, as alignment progresses, if the indicator output tends to go off-scale, the signal level should be reduced rather than the indicator range changed.

AGC CIRCUIT. Faulty alignment can be masked by AGC action. As a circuit is detuned to a point where the output level should decrease, AGC action will increase the gain and hold the output essentially constant. This makes alignment difficult. When the controlled stages obtain their normal bias by some other means (such as emitter or source bias), the AGC action can be defeated by short-circuiting the AGC bus to ground. In some receivers, this bus also supplies the normal bias. It is then necessary to disconnect the AGC voltage and substitute a suitable battery (fixed) voltage.

SPURIOUS SIGNALS. The only signal that should reach the circuit under test is the RF signal from the signal generator. Any other signal would now be a spurious signal. Two possible sources of undesirable signals are the antenna and the local oscillator. Obviously the antenna should be disconnected, and the antenna (or antenna-ground) terminals shorted. If the receiver uses a built-in antenna (as in Fig. 8–27), the tuning capacitor C_{1B} can be shorted. The local oscillator can be disabled by shorting the oscillator section of the tuning capacitor.

IF Alignment – the Steady-Frequency Method

Assuming that the IF strip does not use double-tuned *overcoupled* interstage networks, the signal generator can be connected across the input of the mixer (or converter stage) in series with a blocking capacitor. If the generator is set for unmodulated output, the indicator (a high-impedance dc voltmeter) is connected (through an isolating resistor) across the diode detector load resistor. For synchronous-tuned strips, the generator is set for the proper center frequency, and after the procedures under *General Techniques* are observed, all IF tuning adjustments are reset for a maximum indication on the voltmeter. The adjusting procedure should be repeated because of possible interaction of one tank circuit on another. A complete point-by-point response curve should then be taken to check for center frequency, symmetry, and bandwidth. If the receiver uses stagger tuning, the signal generator is set to one of the key frequencies, but only the corresponding tank circuit is tuned for maximum output. The process is repeated for each of the stagger-tuned frequencies. Again, the adjusting procedure should be repeated. With stagger-tuned systems a complete point-by-point response curve is necessary to get the overall picture.

When the IF strip contains overcoupled double-tuned interstage networks, the response curve will have the typical double peak and center dip. (See curves 3 and 4 of Fig. 2–9.) Obviously, such circuits cannot be aligned by tuning them for maximum output, nor can they be aligned by tuning them for a "little less than maximum." However, any overcoupled circuit can be converted to undercoupled by reducing the Q of its primary and/or secondary. This is readily done by shunting a resistor across one winding. The value used should be the highest value that will eliminate the double peaks and center dip. The alignment should now be done stage-by-stage, starting with the last IF stage. The voltmeter, as before, is connected across the diode load resistor. The signal generator is connected between gate and ground (or base and ground) of the last IF stage. With either side of the interstage transformer loaded, both tanks are tuned for maximum output. Then the load is removed. A point-by-point response check should now be made. The curve should have symmetrical peaks and a slight center dip. The signal generator is now reconnected across the input of the next previous stage, and the interstage transformer between these stages is aligned as before. This process is repeated until all IF stages are aligned. Notice that the indicator is kept across the diode detector load at all times.

IF Alignment—the Sweep-Frequency Method

Both signal generators (sweep and marker) are connected at the input of the mixer (or converter stage) through blocking capacitors. The vertical input terminals of the oscilloscope are connected across the diode detector load resistor; the oscilloscope is set for external sweep; and the sweep voltage from the sweep generator is fed to the horizontal input terminals.

To obtain a response pattern, the sweep generator main frequency dial is set to the center frequency of the desired response curve. The output amplitude, initially, can be set at maximum. The sweep *width* control is preferably set at ten times the anticipated bandwidth (or wider). This will ensure that a pattern will be seen even if the receiver is out of alignment or the sweep generator center frequency is off calibration. Of course, the response curve will occupy only a small portion of the screen width. Let us assume that the pattern appears off-center as shown in Fig. 8–29(a). The pattern is centered by resetting the main frequency dial of the sweep generator—*slowly*. However, the pattern is still too narrow

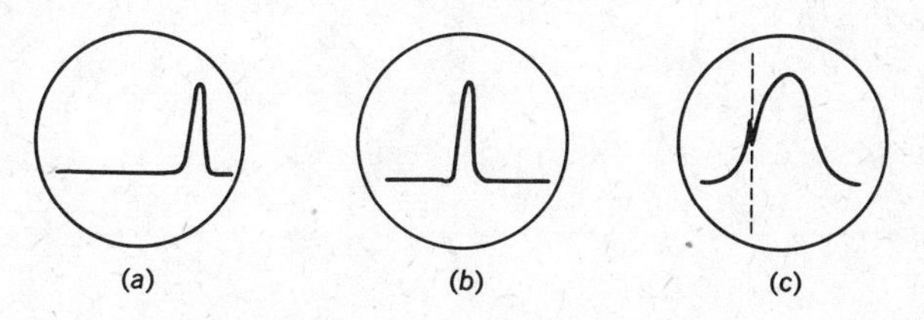

Figure 8-29 Visual alignment patterns.

(Fig. 8–29(b)). This is corrected by reducing the sweep width. The pattern may also tend to go off-center again. A combination of reducing sweep width and adjusting the main frequency will produce the desired effect.

At this point, let us assume we have the seemingly desirable pattern of Fig. 8–29(c) (disregarding the pip). Yet we do not know the exact center frequency and bandwidth of this response curve. Now we can apply the marker generator signal. With the marker frequency set at the center frequency value, the amplitude is raised till a pip appears on the screen. (The marker amplitude should be kept at the minimum possible value, so as not to distort the response curve.) For correct alignment, the pip should appear at the center (peak) of this response curve. A bandwidth check is made by changing the marker frequency until the pip moves down, on one side, to a position at 0.7 of maximum amplitude. Note the frequency. Repeat for a marker pip on the opposite slope of the response curve. These two readings will establish the bandwidth. If center frequency and bandwidth are correct, no alignment is necessary.

However, consider the case wherein the center frequency pip appears to one side. Obviously, the receiver alignment is at fault. Since this is an overall IF response curve, any one (or more) of the IF tank circuits can be off resonance. Assuming that the IF strip is synchronous-tuned, correction is quite simple. Note the vertical line (oscilloscope reticule) with which the pip corresponds. Mark this line, and reduce the marker amplitude to zero. *Now, watching only this line, tune any and all IF adjustments for maximum curve height at this screen location.* Pay no attention to pattern changes at any other part of the screen. When all such adjustments have been made, raise the marker amplitude. The pip should be at the peak and the curve should be symmetrical, but the pattern will be off to one side. Recenter the pattern and check again for center frequency and bandwidth.

If the IF strip employs stagger tuning, the correcting technique is quite similar. Set the marker to one of the key frequencies. Note the scope vertical with which this pip corresponds, and *tune only the per-*

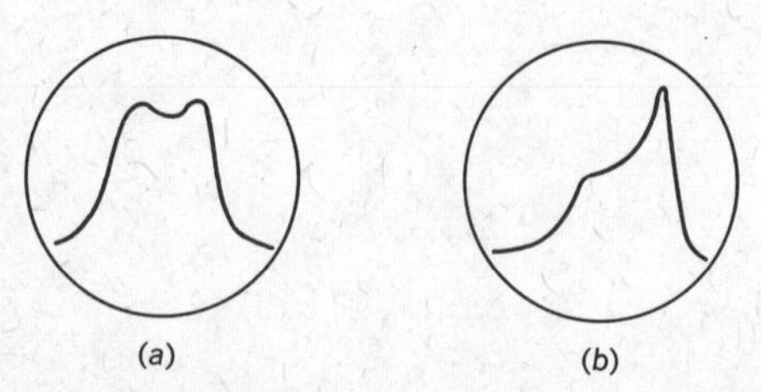

Figure 8-30 Visual patterns created by overcoupled circuits.

tinent tank(s) for maximum amplitude at that vertical. Repeat for each of the key frequencies, center the pattern, and then check center frequency and bandwidth.

When the receiver uses double-tuned overcoupled circuitry, the alignment should be made stage by stage. The oscilloscope connections remain as before, but the sweep generator is now connected to the input of the last IF stage. The procedures described earlier are used to obtain a centered pattern. The pattern should appear as in Fig. 8–30(a). The marker generator can be used to check center frequency and bandwidth. If a distorted pattern, as in Fig. 8–30(b), is obtained, it is readily corrected by touching up the primary or secondary (or both adjustments) for equal-height peaks and symmetry. On the other hand, if serious misalignment is evident, it is preferable as a first approach to note the oscilloscope vertical corresponding to the center-frequency pip, tune for maximum along this vertical (loading one side, if necessary), and then touch up as before. The process is repeated, one IF transformer at a time, working back until the sweep generator is across the mixer (or converter) input.

Front-End Alignment

In a superheterodyne receiver, selectivity is achieved mainly in the IF strip, while the response of the front-end section is relatively broad. Consequently, the steady-frequency method is generally used to check for alignment of the oscillator, mixer, and RF stages. If the receiver has a self-contained antenna (loop or loopstick), the signal generator output is connected to a second loop or to a rod, and the ground terminal of the signal generator is connected to the receiver chassis. The radiation from this loop or rod is then picked up by the receiver. With broadcast-band receivers having an antenna-ground terminal, the signal generator should be connected to these terminals in series with a *dummy antenna* to duplicate the impedance of the antenna. Such a network is shown in Fig. 8–31(a). More often, however, only the 200-pF capacitor is used, since this gives adequate results. In the VHF range, and particularly with receivers requiring balanced input signals, an impedance match must be obtained between the signal generator and the receiver input. A resistive matching network is generally employed. Figure 8–31(b) shows such a network for matching between an unbalanced 50-Ω generator output and a balanced 300-Ω receiver input.

Alignment is generally made at two frequencies, one near the high-frequency end and the other near the low-frequency end of the tuning range. For broadcast-band receivers, 1400 kHz and 600 kHz are the commonly selected check frequencies. With the RF and oscillator stages returned to normal operating condition, the receiver dial is first set to the

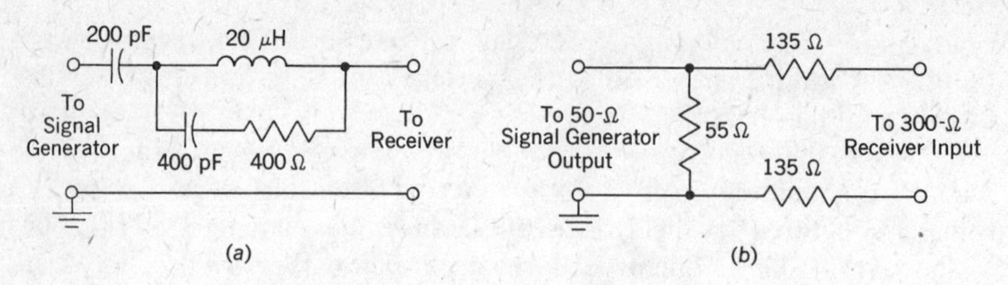

Figure 8-31 Antenna-matching networks.

high-frequency check point (1400 kHz), and the signal generator is set to the same frequency. For use with an unmodulated RF output, the indicator (a dc voltmeter) is connected across the diode detector load.* Once a proper signal level has been set, the trimmer capacitors across the oscillator, mixer, and RF sections of the main tuning capacitor are adjusted for maximum output. The receiver dial and the signal generator are then set to the low-frequency check point (600 kHz). The procedure now varies somewhat, depending on the adjustments incorporated into the tuner:

1. If the oscillator has a padder capacitor, and the RF and mixer coils are slug-tuned, each of these is adjusted for maximum output.
2. If the RF and mixer coils are not adjustable, the padder capacitor is tuned for maximum output, *while rocking the tuning capacitor slowly across the test frequency setting*. It may be found that a higher output can be obtained when the tuning dial is slightly to one side (or the other). This will give optimum alignment, even though the tuning dial calibration will be slightly off.
3. If the oscillator section of the tuning capacitor uses specially cut plates, and all coils (RF, mixer, and oscillator) are adjustable, tune the coils for maximum output. If only the oscillator coil is adjustable, tune this for maximum output while rocking the tuning capacitor as in step 2.

In each of the above cases, the high-frequency alignment should be rechecked and the process repeated until no further changes are necessary. It should be emphasized that, since the signals are passing through the IF amplifiers, alignment of the IF strip must precede the front-end alignment.

REVIEW QUESTIONS

1. Name four characteristics used in a comparison of receiver performance.

2. (*a*) Of what significance is the sensitivity of a receiver? (*b*) What unit is used to specify the sensitivity of a receiver? (*c*) What does the sensitivity rating of a re-

*An ac voltmeter connected across any convenient audio takeoff can be used if the RF signal is amplitude-modulated.

ceiver mean? (*d*) What value of sensitivity may be expected from an average AM broadcast-band home receiver? (*e*) What value may be expected from a communications receiver? (*f*) What section of a receiver is primarily responsible for sensitivity? (*g*) Is there a limit to the sensitivity design of a receiver? Explain.

3. Refer to Fig. 8–1(a). (*a*) Which curve has the wider bandwidth? Explain. (*b*) Which curve has the higher selectivity? Explain. (*c*) Of what importance is the selectivity of a receiver? (*d*) Can a receiver ever have too much selectivity? Explain.

4. Refer to Fig. 8–1(b). (*a*) What characteristic of a receiver is depicted here? (*b*) Why is a graphical representation necessary? (*c*) To what frequency is the receiver tuned? (*d*) Explain briefly how the curve is obtained. (*e*) What ordinate value corresponds to −10 kHz off resonance? (*f*) What does this value mean?

5. (*a*) What does the fidelity of a receiver signify? (*b*) Can the front-end or predetector circuits of a receiver affect the fidelity of its high-frequency intelligence output? Explain.

6. (*a*) State several sources that can create noise fields in the atmosphere. (*b*) What single term can be used to describe such disturbances? (*c*) How does the magnitude of these noise fields vary (if any) with frequency? (*d*) Can its effects be neglected at any frequency?

7. (*a*) Can noise voltages originate in an antenna system? (*b*) What causes such an effect? (*c*) What is this effect called? (*d*) Can this effect also occur within a receiver? Where? (*e*) How does this type of noise vary (if any) with frequency?

8. (*a*) With respect to the noise output due to thermal agitation, does the bandwidth of the receiver have any effect? Explain. (*b*) Name two other factors that affect this noise level. (*c*) How does the value of noise voltage vary with each?

9. (*a*) Where does "shot effect" occur? (*b*) Explain how it occurs. (*c*) What is the result of this action? (*d*) How is this result indicated in an equivalent diagram?

10. (*a*) Compare the FET and the triode as to their possible noise output. (*b*) Why is this so?

11. (*a*) Name three other sources of noise in transistors and tubes. (*b*) Explain each effect.

12. (*a*) What term is used to compare the noise levels of receivers? (*b*) Give the equation for this term. (*c*) What value of this term would indicate that a receiver has exceptionally low noise level? (*d*) Give a value to indicate a higher noise level.

13. Refer to Fig. 8–2. (*a*) What is the function of V_1? V_2? V_3? (*b*) What type of detector does this receiver use? (*c*) What type of coupling is used between V_1 and V_2? (*d*) Why aren't double-tuned coupling networks used? (*e*) What is the function of R_3, R_4? (*f*) What is the function of C_3? (*g*) What is the function of C_9? (*h*) Compare the operating point bias used for V_1, V_2, and V_3. (*i*) Why is this so?

14. (*a*) Is the selectivity of the tuner in Fig. 8–2 constant over its tuning range? Explain why this is so. (*b*) At which end of the dial will the selectivity be sharpest? (*c*) Is this desirable? Explain.

15. (*a*) At which end of its tuning range is the sensitivity of a TRF receiver higher? (*b*) Give three possible reasons for this increase.

16. (*a*) With reference to Fig. 8–2, what is meant by "tracking"? (*b*) Can perfect tracking be obtained at any one frequency? If so, how, and at what frequency? (*c*) What makes perfect tracking over the full tuning range practically unattainable in a commercial receiver?

17. In Fig. 8–3, will the trimmer capacitors have their greatest effect when the main capacitor is fully open or fully meshed? Explain.

18. Refer to Fig. 8–4. (*a*) Which of the blocks shown may not necessarily be used? (*b*) When and why is this so? (*c*) What type of active device would be used in the first block for AM broadcast-band operation? (*d*) What change (or changes) would be necessary to adapt the circuit of Fig. 3–6, to this block? (*e*) For VHF or the lower end of the UHF band, what type of circuit could be used in this first block? (*f*) Why are the "mixer" and "oscillator" blocks grouped together? (*g*) Give three names used for this combined action. (*h*) What relationship exists between the frequencies entering and leaving the mixer block? (*i*) Express this relationship by an equation. (*j*) Why is the output of the mixer called an "intermediate" frequency? (*k*) When the tuning dial of the receiver is varied over its full range (540–1600 kHz), over what range does the tuning of the IF amplifier change? (*l*) Are double-tuned coupling networks practical in IF amplifiers? Explain. (*m*) Give two other advantages of IF amplifiers over the RF amplifiers in a TRF receiver.

19. Give four applications of frequency conversion other than production of the IF in a superheterodyne receiver.

20. (*a*) What is the "first detector" in a superheterodyne receiver? (*b*) Why is such a terminology used?

21. Refer to Fig. 8–5. (*a*) What does phasor E_2 represent? (*b*) What does phasor E_1 represent? (*c*) Why is E_1 shown rotating clockwise? (*d*) What do the symbols ω_2 and ω_1 represent? (*e*) At what velocity does E_1 rotate around E_2? (*f*) At the instant shown, what is the angle between these two phasors? (*g*) What are the maximum and minimum values of the resultant? (*h*) At the instant shown, what is the horizontal component of the resultant? (*i*) What is its vertical component?

22. (*a*) What relation should exist between the oscillator and signal voltages fed to a mixer, so as to preserve the fidelity of the modulation envelope? (*b*) Explain, using the phasors of Fig. 8–5.

23. What two basic actions must occur in any frequency-conversion circuit?

24. (*a*) Compare a diode mixer with a triode or multigrid mixer with respect to gain. (*b*) Why are diode mixers preferable in UHF applications?

25. Refer to Fig. 8–6. (*a*) What is the function of Q_1? Q_2? Q_3? (*b*) Why is the collector of Q_1 connected to a tap on L_1? (*c*) Why is L_2 untuned? (*d*) What is the function of R_2C_3? (*e*) Give two functions of capacitor C_2. (*f*) What specific type of oscillator is used here? (*g*) How is the oscillator signal injected into the mixer circuit? (*h*) To what frequency is L_3C_4 tuned?

26. Refer to Fig. 8–7. (*a*) What basic type of oscillator circuit is used here? (*b*) Why is the emitter connected to a tap on L_4? (*c*) How is the oscillator voltage coupled to the input of the transistor? (*d*) What is the function of R_1 and R_2?

27. Refer to Fig. 8–8. (*a*) What type of active device is this? (*b*) What is the function of this circuit? (*c*) Where is the oscillator voltage developed? (*d*) How is it coupled to this circuit? (*e*) To what frequency is L_2C_1 tuned? (*f*) To what frequency is L_4C_3 tuned?

28. What is the distinction between a mixer-oscillator and a pentagrid converter?

29. Refer to Fig. 8–9. (*a*) What is the function of tube V_2? (*b*) What is the function of tube V_1? (*c*) What specific type of circuit is used with tube V_2? (*d*) How is the oscillator signal coupled into the mixer? (*e*) What determines the frequency of the oscillator voltage? (*f*) What determines the RF input? (*g*) To what frequency is L_4C_7 tuned?

30. Refer to Fig. 8–10. (*a*) What is the function of each of the grids in this tube? (*b*) What type of oscillator circuit is used here?

31. (*a*) In a TRF or superheterodyne receiver using individual stage tuning, is there any tracking problem? Explain. (*b*) Is there any tracking problem in a TRF receiver with single-dial control? Explain. (*c*) Why is a serious tracking problem created with single-dial tuning in a superheterodyne receiver?

32. Refer to Fig. 8–11(a). (*a*) Where would this tank circuit be used? (*b*) How does the value of L_o compare with the inductance value used in the other front-end tank circuits? (*c*) What is C_2 and what is it used for? (*d*) What is C_3 called? (*e*) Why is it necessary? (*f*) At what frequency is it adjusted, and for what effect?

33. Explain the significance of the curves in Fig. 8–11(b).

34. Explain another technique used to obtain oscillator tracking.

35. When a gang capacitor with a specially cut oscillator section is used (*a*) are trimmer capacitors needed? Explain. (*b*) is a padder capacitor needed? Explain. (*c*) Can the same gang capacitor be used for either a 455-kHz or 262-kHz IF system? Explain. (*d*) Can this unit be used in a multiband communications receiver? Explain.

36. Refer to Fig. 8–12. (*a*) Give a "promotional" name for this type of tuning circuit. (*b*) What is D_1? (*c*) How does the series capacitor, C_1, affect the tuning? (*d*) What is the function of R_2? (*e*) Which position of this control will give the lowest frequency? Explain. (*f*) Give three advantages of this tuning method, as compared to the conventional variable capacitor.

37. (*a*) Is reception of an image-frequency signal desirable or undesirable? Explain. (*b*) If the IF system has a wide bandwidth (poor selectivity), how does this affect the possibility of image-frequency reception? Explain. (*c*) If the RF and mixer have poor selectivity, how does this affect the possibility of image-frequency interference? Explain. (*d*) Since a TRF receiver generally has poorer selectivity than a superheterodyne receiver, is it more prone to image-frequency interference? Explain.

38. In the design of an RF amplifier, are gain and stability more readily obtainable at high or low radio frequencies? Explain.

39. Can wide bandwidths and high gain be obtained at low radio frequencies? Explain.

40. Are very narrow bandwidths readily obtainable at high radio frequencies? Explain.

41. FM broadcast receivers should have a bandwidth of 200 kHz. Would it be reasonable to use a 455-kHz IF system in these receivers? Explain.

42. AM broadcast systems use a bandwidth of 10 kHz. For high fidelity this could be raised to 30 kHz. Would it be reasonable to raise the operating frequencies and use an IF system of 30 MHz? Explain.

43. In AM broadcast receivers (*a*) is it *necessary* that all receivers use the same IF value so that they can receive common transmissions? (*b*) What advantage (if any) is gained by standardizing on a common IF value?

44. State two receiver-design factors that would reduce the danger of image-frequency interference.

45. Refer to Fig. 8–13. (*a*) In what type of receivers is this system most likely to be found? (*b*) What is the purpose of double conversion? (*c*) Give typical values for the first and second IF. (*d*) Why is the switching provided?

46. (*a*) What type of interstage coupling is most commonly used with IF amplifiers? (*b*) What is the advantage of this type of coupling? (*c*) Why is it practical in IF rather than RF amplifiers? (*d*) Is this circuitry usable with bipolar transistor IF circuits? (*e*) How is loading minimized?

47. Refer to Fig. 8–14. (*a*) What type of coupling does this circuit use between stages? (*b*) Is Q_1 neutralized? Explain. (*c*) Why is the collector of Q_1 tapped down on L_1? (*d*) What do the arrows between L_1 and L_2 signify? (*e*) Why does one arrow point up and the other down? (*f*) Why is the base of Q_2 tapped down on L_2? (*g*) How is Q_2 biased on the input side? (*h*) What is the function of R_4C_6? (*i*) Give two reasons for connecting the power supply lead to a tap on L_3? (*j*) What is the function of C_5? (*k*) How is loading by the base circuit of Q_3 minimized?

48. Refer to Fig. 8–15. (*a*) What polarity is applied to the collector of Q_1? (*b*) to the collector of Q_2? (*c*) Explain. (*d*) If AGC were not used, what change would be made to this diagram? (*e*) If Q_1 were an NPN transistor, what change would be made to this diagram?

49. (*a*) Can double-tuned overcoupled circuitry be used in wideband IF amplifiers? (*b*) What would be the advantage of such coupling? (*c*) What is the disadvantage of such coupling?

50. Refer to Fig. 8–16. (*a*) What is the center frequency of this amplifier? (*b*) What is its bandwidth? (*c*) How is this bandwidth obtained? (*d*) What are the individual resonant frequencies of each tank? (*e*) What advantage is obtained by not having adjacent tanks at adjacent frequencies? (*f*) What components set the bias for Q_1? Q_2? Q_3? (*g*) Is there any bias stabilization? Explain. (*h*) How is impedance matching obtained between Q_1 and Q_2? (*i*) On this basis, which is the larger capacitor, C_6 or C_7? (*j*) What other function does capacitor C_6 serve? (*k*) What is the function of R_8? (*l*) Is any other component associated with this function? (*m*) Give a second function for capacitor C_9.

51. (*a*) Name three devices that are used to provide selectivity in IF systems in place of *L-C* tank circuits. (*b*) Which of these is most popular in consumer receivers? (*c*) Give three advantages of this device over a standard tank circuit.

52. Refer to Fig. 8–17. (*a*) How is selectivity obtained in this circuit? (*b*) What is the function of R_2? (*c*) What is the function of R_4?

53. (*a*) Can FETs be used in IF amplifier circuits? (*b*) What advantage do FETs have over bipolar transistors? (*c*) What advantage does a dual-gate FET have over all other transistors?

54. (*a*) In Fig. 8–18, what do the triangles represent? (*b*) How many of these would be used in one receiver? (*c*) Can the tank circuit be replaced by a ceramic filter?

55. (*a*) In Fig. 8–19, what does the triangle represent? (*b*) How many of these would be used in one receiver? Explain. (*c*) How could better selectivity be obtained in this strip?

56. A station is transmitting a concert program and the dynamic range of the orchestration varies from pianissimo (very quiet) to fortissimo (very loud). (*a*) What effect will an automatic gain control circuit in the receiver have on this tonal range? Explain. (*b*) What is the purpose of AGC circuits?

57. Refer to Fig. 8–20. (*a*) What type of detector is suited to this purpose? (*b*) What is the polarity of the dc component in the output? (*c*) What is the purpose of the filter? (*d*) To what specific part of the RF and IF amplifiers is the AGC bus voltage applied? (*e*) Explain the effect of the AGC action if the antenna signal level should decrease.

58. Refer to Fig. 8–21. (*a*) What voltages are available between point 2 and ground? (*b*) Can this be used directly for AGC bias? Explain. (*c*) What is the purpose of C_4? (*d*) What determines the value of this capacitor? (*e*) What is the function of resistor R_3? (*f*) What effect would it have on the output voltage E_o if it were omitted? Explain. (*g*) What type of AGC feed is used with Q_1? with Q_2? with Q_3? with Q_4? (*h*) What is the function of R_7? (*i*) Does any other component assist in this function? Explain. (*j*) What other function does C_7 serve? Explain. (*k*) What makes this second function of C_7 necessary? (*l*) What is the function of C_6? (*m*) Which circuit(s) *cannot* be used for an RF amplifier stage? Explain.

59. (*a*) Does an AGC circuit, as in Fig. 8–21, affect the detector function? Explain. (*b*) How can this effect be minimized?

60. (*a*) What values of time constant are generally used for AGC filters? (*b*) What effect would too long a time constant have? (*c*) What effect would too short a time constant have?

61. Refer to Fig. 8–22. (*a*) Is this a diode detector? (*b*) What is the purpose of this circuit? (*c*) Is the polarity shown on C_1 correct? How is this charge obtained, if the cathode of the diode is grounded? (*d*) What is the function of R_1C_2? (*e*) How is the charge on C_2 obtained? (*f*) What is the purpose of applying a delay bias to the cathode? (*g*) What is the advantage of such an action? (*h*) What polarity of bias is required? Explain.

62. Compare and explain the power requirements for an AGC source for FETs and bipolar transistors.

63. Is there any preference between a semiconductor diode detector and a transistor power detector as a source of AGC voltage for transistor circuits? Explain.

64. Refer to Fig. 8–23. (*a*) What is the function of transistor Q_1? (*b*) What is the basic function of transistor Q_2? (*c*) What type of circuit is this? (*d*) What is the polarity of the AGC bus voltage? (*e*) How is this polarity obtained? (*f*) What components, if any, are used to filter the AGC voltage? (*g*) What is the function of R_4? (*h*) If the input signal strength tends to decrease, explain how AGC action is obtained.

65. (*a*) Is any advantage gained by feeding the AGC signal to the base of a transistor instead of to the emitter? Explain. (*b*) In Fig. 8–23 can this AGC bus voltage be applied to the base of Q_1 for suitable control action? Explain.

66. Refer to Fig. 8–24. (*a*) What is the basic function of transistor Q_1? Q_2? (*b*) What is the polarity of the AGC bus voltage? (*c*) How is this polarity obtained? (*d*) Explain how AGC action is obtained if the input signal strength decreases.

67. Refer to Fig. 8–25. (*a*) What is the polarity of the AGC bus voltage? (*b*) How is this polarity obtained? (*c*) What components, if any, are used to filter the AGC voltage? (*d*) What is the normal base-bias polarity for transistor Q_1? (*e*) Explain how AGC action is obtained if the input signal strength increases.

68. Refer to Fig. 8–26. (*a*) What is the function of diode D_1? (*b*) What makes this function necessary? (*c*) What is the polarity of point *A*, compared to point *B*, at low RF signal levels? (*d*) How is this polarity obtained? (*e*) Can a high signal level affect these dc potentials? Explain. (*f*) Explain how auxiliary AGC action is obtained at high signal levels. (*g*) How is this AGC action affected by tying the mixer transistor to the AGC bus?

69. Refer to Fig. 8–27. (*a*) What component picks up the RF signals? (*b*) Why is a second winding used here? (*c*) What is the function of Q_1? (*d*) What specific circuit is this? (*e*) Which is the oscillator section of the tuning capacitor? (*f*) What is C_{1D}? (*g*) Where is the padder capacitor? Explain. (*h*) What is the function of R_1, R_2? (*i*) Why is the collector voltage of Q_1 fed to a tap on T_1 primary? (*j*) What is the function of E_1? (*k*) Is there any other reason for feeding the collector voltage for Q_2 to a tap on its tank circuit? (*l*) What is the function of C_9? (*m*) What device is used for demodulation? (*n*) What type of circuit is this? (*o*) Is any AGC action obtained from this circuit? (*p*) Trace its path.

70. (*a*) Give two reasons for aligning (or checking the alignment of) a receiver. (*b*) Name the two basic alignment techniques.

71. (*a*) Which of the above two methods uses a voltmeter as an indicator? (*b*) Should the voltmeter be dc, ac, or RF? Explain. (*c*) What other attribute should the voltmeter have? (*d*) What type of voltmeter is ideally suited in this respect? (*e*) Can a 20,000 ohms-per-volt voltohmmeter be used?

72. (*a*) What is a sweep generator? (*b*) What is it used for? (*c*) What are typical sweep rates? (*d*) If a signal from such a generator is applied to a receiver, will the output from the detector be zero, dc, or will it vary? Explain. (*e*) What type of indicator is used for alignment with a sweep generator?

73. In sweep alignment, (*a*) to where on the oscilloscope do you connect the detector output? (*b*) What would be seen with this as the only signal applied? (*c*) Why is *external* sweep used? (*d*) What source is used for sweep voltage? (*e*) Where on the oscilloscope is this connected?

74. (*a*) Explain how the curve in Fig. 8–28 is produced on the oscilloscope screen when using the sweep-alignment method. (*b*) What is the jagged discontinuity on the upper right-hand slope of this curve?

75. (*a*) What is a marker generator? (*b*) Why is it necessary to use a marker generator in addition to a sweep generator? (*c*) How is the bandwidth of a response pattern evaluated with this generator?

76. (*a*) What precaution is advisable when connecting an indicator to the circuit? (*b*) Give two reasons for this. (*c*) What precaution is used with signal generator connections? (*d*) Give two reasons for this.

77. (*a*) What constitutes a "proper" signal level for alignment? (*b*) What is the objection to too-strong a signal? (*c*) to too-weak a signal?

78. (*a*) Does a strong AGC action make alignment easier? Explain. (*b*) How is this usually handled? (*c*) Is auxiliary bias ever needed? Explain.

79. (*a*) Name two possible sources for spurious signals. (*b*) How can each be eliminated?

80. The IF system of the receiver in Fig. 8–27 is to be aligned using the fixed-frequency system. (*a*) Where would you connect the signal generator? (*b*) Where would you connect the voltmeter if the signal generator is unmodulated? (*c*) Where else could you connect the indicator if the signal generator is amplitude modulated? (*d*) How would you disable the AGC action? (*e*) How would you prevent spurious signals from antenna pickup? (*f*) from the oscillator? (*g*) Describe the alignment procedure. (*h*) Describe how to obtain the full response curve.

81. Assume that the above circuit uses double-tuned overcoupled IF transformers. (*a*) Where would you first connect the signal generator? (*b*) Where would you first connect the indicator? (*c*) What else must be done *before* aligning? (*d*) What alignment adjustments are now made? (*e*) Is the voltmeter connection now changed? (*f*) Is the signal generator connection changed? To where? (*g*) Complete the alignment procedure.

82. The IF system of the receiver in Fig. 8–27 is to be aligned, using the steady-frequency method. Assume that it is a stagger-tuned strip with T_1 at 258 kHz, T_2 at 262 kHz, and T_3 at 266 kHz. (*a*) The signal generator is unmodulated; where would you connect it? (*b*) What would you use as indicator and where would you connect it? (*c*) Assume that all general techniques have been followed; describe the remainder of the alignment.

83. A sweep alignment is to be performed on the IF system of the receiver in Fig. 8–27. The center frequency is 455 kHz and the bandwidth is 10 kHz. (*a*) Where would you connect the sweep generator? (*b*) Where would you connect the marker generator? (*c*) Where would you connect the vertical oscilloscope input? (*d*) the horizontal oscilloscope input? (*e*) What setting would you use for the main frequency dial of the sweep generator? (*f*) the sweep width dial? (*g*) If the oscilloscope pattern obtained is as in Fig. 8–29(a), state two possible reasons why. (*h*) How is the pattern of Fig. 8–29(b) now obtained? (*i*) Why is the pattern so narrow? (*j*) How is the pattern in Fig. 8–29(c) obtained from pattern (b)? (*k*) When the marker generator is set at 455 kHz, the pip appears as shown in Fig. 8–29(c). Explain how to get proper alignment.

84. A visual alignment check of the IF system of the receiver in Question 82 produces the pattern shown in Fig. 8–30(b). Describe step-by-step how to proceed with realignment.

85. In visual alignment of an overcoupled IF strip, where is the sweep generator connected? Explain.

86. (*a*) In alignment of a receiver front end, which basic technique is generally used? (*b*) Why is this so?

87. The front end of the receiver in Fig. 8–27 is to be aligned. (*a*) How is the RF signal injected into the circuit? (*b*) Where is the indicator connected? (*c*) To what frequency should the receiver be set? (*d*) To what frequency is the signal generator set? (*e*) What adjustments are made? (*f*) To what frequency are the receiver and signal generator now set? (*g*) What adjustments are now made? (*h*) Is it necessary to rock the tuning capacitor during this adjustment? Explain. (*i*) Is it necessary to repeat the adjustments made in step (*e*)? Explain.

88. Refer to Fig. 8–31. (*a*) When is the circuit shown in diagram (a) used? (*b*) What modification is more commonly used? (*c*) When is the circuit of diagram (b) used?

PROBLEMS AND DIAGRAMS

1. Calculate the noise voltage developed by a 270,000-Ω gate resistor in an impedance-coupled RF amplifier, if its operating temperature is 27°C and the bandwidth of the stage is 250 kHz.

2. A FET RF amplifier has the following characteristics: input conductance, g_{is}, 0.027 mmho; output conductance, g_{os}, 0.02 mmho; and forward transconductance, g_{fs}, 7.22 mmho. (*a*) Find the noise-equivalent resistance of this FET. (*b*) What noise voltage would it develop at its output if the stage has a gain of 15 and a bandwidth of 2 MHz? (Assume 27°C.)

3. Draw the circuit diagram of a TRF receiver using one stage of RF and a grid-leak detector transformer coupled to its output.

4. Draw the circuit diagram of a TRF receiver using two RF stages and a diode detector with *R-C* coupling to the next stage.

5. In Fig. 8–2, what change in capacitance (maximum to minimum ratio) is necessary to tune the RF circuits over the broadcast-band range (540–1600 kHz)?

6. A broadcast-band TRF receiver has a bandwidth of 10 kHz at 600 kHz. (*a*) Assuming that the *Q* of its tank circuits remains essentially constant, what is the bandwidth of the receiver at 1600 kHz? (*b*) If the design were altered to produce a bandwidth of 10 kHz at 1600 kHz, what would the bandwidth be at 540 kHz?

7. In Fig. 8–4, if the RF amplifier stage is tuned to a frequency of 630 kHz, (*a*) to what frequency is the mixer input circuit tuned? (*b*) To what frequency is the oscillator tuned? (*c*) What is the output frequency from the mixer?

8. An AM wave has a carrier of 840 kHz and sidebands extending 4 kHz to each side. This signal is fed to a mixer together with a 1290-kHz local oscillator signal. Find (*a*) the IF carrier value and (*b*) the total IF spectrum coverage.

9. An FM wave has a carrier at 99.6 MHz and swings ±75 kHz to either side of this center frequency. In a receiver, this signal is mixed with an oscillator frequency of 110.3 MHz. Find the IF center frequency and the frequency swing.

10. When tuned to channel 4, a TV receiver has an oscillator frequency of 93 MHz, a carrier frequency of 67.25 MHz, an upper sideband width of 4.5 MHz, and a lower sideband width of 1.25 MHz. Find the IF carrier frequency and the frequency limits of each sideband.

11. Draw the circuit diagram of a mixer-oscillator circuit, using a Hartley oscillator inductively coupled into the emitter leg of a bipolar transistor as the mixer.

12. Draw the circuit diagram of a mixer-oscillator using PNP bipolar transistors and grounding the negative side of the power supply.

13. Draw the circuit diagram of an autodyne frequency converter using an NPN bipolar transistor.

14. Draw the circuit diagram of a bipolar transistor mixer-oscillator using a Hartley oscillator and capacitive coupling of the oscillator signal into a dual-gate MOSFET mixer.

15. Draw the circuit diagram of a mixer-oscillator circuit using an Armstrong oscillator and the triode-hexode of Fig. 8–9(b).

16. Redraw the pentagrid converter circuit of Fig. 8–10 using a Hartley oscillator. Show where the two L_1 leads would be connected if an RF stage preceded this stage.

17. A broadcast-band AM receiver (540–1600 kHz) uses an IF value of 455 kHz. (*a*) What capacitor ratio (maximum to minimum) is needed to cover the RF tuning range? (*b*) What capacitance ratio is needed in the oscillator circuit if the oscillator frequency is above the input signal frequency? (*c*) Repeat (*b*) for oscillator tuning below the incoming RF signal. (*d*) Which of these conditions, (*b*) or (*c*), is more practical? Why?

18. An AM broadcast-band superheterodyne receiver uses an IF value of 455 kHz. Show mathematically that (*a*) the oscillator inductance value should be approximately 60% of the RF coil value, (*b*) the *LC* product of the oscillator tank circuit should be only (approximately) 30% of the corresponding RF value, at the low-frequency end of the tuning range (540 kHz), and (*c*) the maximum capacitance in the oscillator tank circuit should be approximately one-half the RF capacitance value.

19. In each of the following cases, what is the frequency of the image signal: (*a*) RF at 1200 kHz with an IF of 455 kHz. (*b*) RF at 650 kHz with an IF of 262 kHz. (*c*) RF at 106 MHz with an IF of 10.7 MHz. (*d*) RF at 3000 MHz with an IF of 60 MHz.

20. Draw the circuit diagram for a PNP transistor IF amplifier with double-tuned input and output coupling networks, but without AGC.

21. Draw the circuit diagram of a transistor IF stage using single-tuned coupling, an NPN transistor, and a grounded negative supply line.

22. Draw a circuit diagram for a two-stage PNP transistor amplifier with a ceramic filter for interstage coupling.

23. Draw a circuit diagram for a FET IF amplifier using double-tuned coupling for input and output. Show the input coming from the mixer and the output going to the next FET IF stage.

24. Draw circuit diagrams showing two methods for applying AGC to an N-channel FET RF stage with the rotor of its tuning capacitor grounded.

25. Repeat Problem 24 using P-channel FETs.

26. In Fig. 8–21, if R_4 is 1.0 MΩ, what range of values would be suitable for C_4?

27. Draw the circuit diagram for a delayed AGC supply.

28. Draw the circuit diagram showing a transistor detector supplying an AGC signal to the emitter of an RF amplifier. Use NPN transistors.

29. Draw the circuit diagram for a transistor detector supplying an AGC signal to the base of a PNP IF stage.

30. Draw the circuit diagram for a diode detector supplying an AGC signal to the base of a PNP IF amplifier. Show also auxiliary AGC diode circuitry.

31. Draw the circuit diagram for a transistor (NPN) superheterodyne tuner, featuring an RF stage, a separate mixer and oscillator, an IF stage, a transistor detector, AGC to the RF and IF stages, and an auxiliary AGC diode.

32. Draw a block diagram showing the equipment needed and their interconnection for sweep alignment of a receiver IF system. Indicate to which points in the equipment the connections are made.

33. Refer to Fig. 8–31(b). (*a*) Calculate the impedance the generator would "look into" when the network is connected to the receiver. (*b*) Calculate the impedance the receiver would "look into."

34. What circuit values would be needed in Fig. 8–31(b) to match a 75-Ω generator into a 300-Ω receiver input?

TYPICAL FCC QUESTIONS

3.212 How may the radio-frequency interference, often caused by sparking at the brushes of a high-voltage generator, be minimized?

3.246 What type of modulation is largely contained in "static" and "lightning" radio waves?

3.263 Compare the selectivity and sensitivity of the following types of receivers: (*a*) Tuned radio-frequency receiver. (*b*) Superregenerative receiver. (*c*) Superheterodyne receiver.

3.264 What type of radio receivers contain intermediate-frequency transformers?

3.265 What type of radio receiver is subject to image interference?

3.266 What type of radiotelephone receiver using vacuum tubes does not require an oscillator?

3.267 Describe the operation of a regenerative receiver.

3.268 How may a regenerative receiver be adjusted for maximum sensitivity?

3.270 What is meant by double detection in a receiver?

3.273 Explain the purpose and operation of the first detector in a superheterodyne receiver.

3.291 Do oscillators operating on adjacent frequencies have a tendency to synchronize oscillation or drift apart in frequency?

3.351 What are the advantages to be obtained from adding a tuned radio-frequency amplifier stage ahead of the first detector (converter) stage of a superheterodyne receiver?

3.353 How is "automatic volume control" accomplished in a radio receiver?

3.354 If a superheterodyne receiver is tuned to a desired signal at 1000 kilohertz, and its conversion oscillator is operating at 1300 kilohertz, what would be the frequency of an incoming signal which would possibly cause "image" reception?

3.355 If a tube in the only radio-frequency stage of your receiver burned out, how could temporary repairs or modifications be made to permit operation of the receiver if no spare tube is available?

3.358 What would be the effect upon a radio receiver if the vacuum-tube plate potential were reversed in polarity?

3.464 Explain the relation between the signal frequency, the oscillator frequency, and the image frequency in a superheterodyne receiver.

3.496 What is the purpose of a squelch circuit in a radio communication receiver?

3.512 Draw a block diagram of a superheterodyne receiver capable of receiving amplitude-modulated signals, and indicate the frequencies present in the various stages when the receiver is tuned to 2450 kilohertz. What is the frequency of a station that might cause image interference to the receiver when tuned to 2450 kilohertz?

3.514 Draw a diagram of a tuned radio-frequency type radio receiver.

4.194 What type of meter is suitable for measuring the AVC voltage in a standard broadcast receiver?

6.534 Draw a block diagram of a superheterodyne receiver capable of receiving continuous-wave radiotelegraph signals.

6.536 Draw a block diagram of a tuned radio-frequency type receiver.

6.538 What is the purpose of a tuned radio-frequency amplifier stage ahead of the mixer stage of a superheterodyne receiver?

6.539 What is the "mixer" tube in the superheterodyne receiver?

6.540 Knowing the intermediate frequency and the signal to which a superheterodyne receiver is tuned, how would you determine the most probable frequency on which "image" reception would occur?

6.541 A superheterodyne-type receiver is adjusted to 2738 kilohertz. The intermediate frequency is 475 kilohertz. What is the frequency to which the input circuit of the second detector must be tuned?

6.542 Explain the reasons why a superheterodyne receiver may not be successfully used for reception of frequencies very near the frequency of the intermediate-frequency amplifier.

6.543 Why do some superheterodyne receivers employ a crystal-controlled oscillator in the first detector?

6.544 How should a superheterodyne communications receiver be adjusted for maximum response to weak CW signals? to strong CW signals?

6.545 Why should a superheterodyne receiver, used for the reception of Al signals, be equipped with at least one stage of radio-frequency amplification ahead of the first detector?

6.546 What is the chief advantage to be gained in the utilization of high intermediate frequencies in a superheterodyne receiver?

6.547 If a superheterodyne receiver is receiving a signal on 1000 kilohertz, and the mixing oscillator is tuned to 1500 kilohertz, what is the intermediate frequency?

6.548 How may "image response" be minimized in a superheterodyne receiver?

6.576 How is "automatic volume control" accomplished in a receiver?

6.583 Explain the purpose and operation of the first detector in a superheterodyne receiver.

6.587 Draw a circuit diagram of a superheterodyne receiver with automatic volume control and explain the principle of operation.

6.605 What is the purpose of a crystal filter in the IF stage of a superheterodyne communications receiver? Under what conditions is this filter used?

6.646 Do oscillators operating on adjacent frequencies have a tendency to synchronize oscillation or drift apart in frequency?

S-3.155 Draw a block diagram of a single-conversion superheterodyne AM receiver. Assume an incident signal and explain briefly what occurs in each stage.

S-3.156 Explain the relation between the signal frequency, the oscillator frequency, and the image frequency in a superheterodyne receiver.

S-3.157 Draw a circuit diagram of an AM second detector and AF amplifier (in one envelope), showing AVC circuitry. Also show coupling to, and identification of, all adjacent stages. (*a*) Explain the principles of operation. (*b*) State some conditions under which readings of AVC voltage would be helpful in troubleshooting a receiver. (*c*) Show how this circuit would be modified to give DAVC.

S-3.158 Draw a BFO circuit diagram and explain its use in detection.

S-3.159 Explain, step by step, how to align an AM receiver, using the following instruments. Also explain briefly what is occurring during each step. (*a*) Signal generator and speaker. (*b*) Signal generator and oscilloscope. (*c*) Signal generator and VTVM.

S-3.160 What would be the advantages and disadvantages of utilizing a band-pass switch on a receiver?

S-3.161 Explain: sensitivity of a receiver; selectivity of a receiver. Why are these important quantities? In what typical units are they usually expressed?

S-4.008 What causes resistance noise in electrical conductors and shot-effect noise in diodes?

S-4.040 What are AGC amplifiers and why are they used?

9

Frequency Modulation

In Chapter 6, it was shown that a sine wave carrier e_c could be modulated by varying its amplitude E_{cm}. This was called amplitude modulation. The amplitude of the intelligence signal determined the amount of change in amplitude, and the frequency of the modulating signal determined the rate of change in amplitude.

FM PRINCIPLES

Modulation can also be obtained by changing some other characteristic of the sine wave. For example, let us use the basic equation (6–1) for the carrier, but this time expressing the angular velocity ω_c of the wave in terms of frequency, or

$$e_c = E_{cm} \sin (2\pi f_c t + \Phi) \tag{9-1}$$

If the frequency, and only the frequency, of this wave is caused to vary in accordance with the intelligence signal, the carrier will be frequency-modulated.

The Effect of the Intelligence Signal

In frequency modulation (FM), the frequency of the carrier swings above and below its *center* or *resting frequency*. The *amount* of change in frequency (ΔF) is dependent on the *amplitude* of the modulating signal, while the *frequency* of the intelligence signal determines the *rate* of change in frequency. These effects of the modulating signal are similar

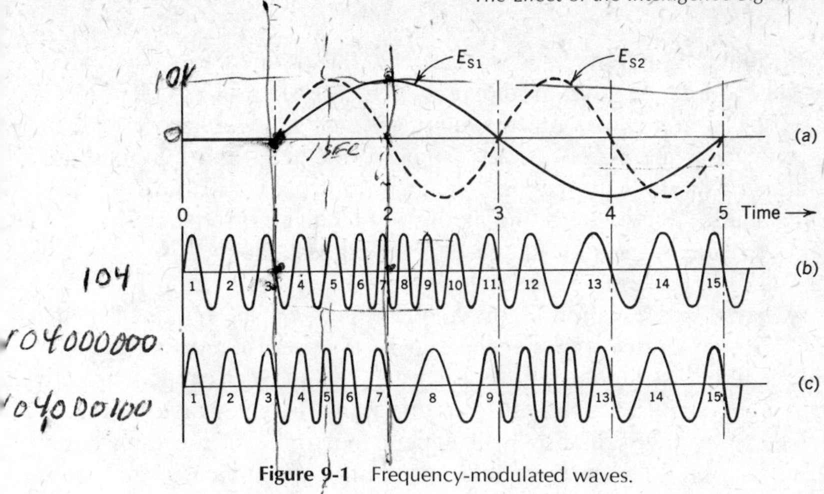

Figure 9-1 Frequency-modulated waves.

to the changes occurring with AM—except that now the change is in the frequency of the output wave.

Two frequency-modulated waves are shown in Fig. 9–1, together with the intelligence signals that produced the modulation. For simplicity of analysis a single modulating frequency (sine wave) is used. The waveshape, Fig. 9–1(b), is produced by the modulating signal E_{s1}. During time interval 0 to 1 there is no modulation, and therefore the first three cycles of the RF waveshape are identical in frequency. (The time base for each half-cycle is the same.) This is the resting frequency. During time interval 1 to 2, the intelligence amplitude rises to a maximum. Notice that the frequency of each RF cycle is increasing until cycle 7, at time instant 2, has the highest frequency. Then, as the amplitude of the modulating signal drops, the frequency of the RF cycles decreases. Cycle 11, corresponding to zero amplitude of the modulating signal, is back at the resting frequency. This effect is repeated during the time interval 3–5. However, since the modulating signal rises to a *negative* maximum, the change in frequency must be in the *opposite* direction. Notice that cycle 13, at time instant 4, has the maximum change in frequency, but is at the *lowest* frequency.*

Now let us examine Fig. 9–1(c) and its modulating signal E_{s2}. Since E_{s2} has the same amplitude as E_{s1}, the amount of change in carrier frequency (ΔF) must be the same as before—or, the lowest and highest instantaneous carrier frequencies must correspond to cycles 13 and 7 of Fig. 9–1(b). On the other hand, notice that the frequency of E_{s2} is

* It is difficult to show an accurate FM wave within the limits of a text page. In Fig. 9–1(b) only 12 RF cycles are shown to one cycle of modulation. A more realistic ratio would be 1000:1. Also, the lowest frequency cycle (13) has been exaggerated to bring out the distinction between it and the resting frequency.

higher (double E_{s1} frequency). Therefore, the frequency of the carrier must vary at a faster rate, as shown in Fig. 9–1(c).

If the amplitude of the modulating signal E_{s1} were increased, the rate of change in frequency would not be affected. The minimum and maximum instantaneous carrier frequencies would still occur at time instants 2 and 4. However, the amount of change in frequency—or *frequency deviation*—would increase. The highest frequency would be higher, and the lowest frequency lower.

A better appreciation of the above effects can be obtained from a simple FM generator. (This technique is not used in transmitters, but it will readily illustrate the point.) Figure 9–2(a) shows a capacitor microphone connected across the tank circuit of an RF (sine-wave) oscillator. The frequency of oscillation will depend not only on the tank circuit values L_1C_1, but also on the additional shunt capacitance of the microphone, and this in turn varies depending on the sound pressure striking the diaphragm. Let us assume that sound wave ① of Fig. 9–2(b) is applied to the microphone. Since the sound wave pressure is zero at time instant 0, it does not affect the microphone diaphragm, and the oscillator is at its resting frequency f_0. During the time interval 0–1, as the sound pressure rises to its maximum value, the microphone diaphragm is pushed in; the capacitance rises to its maximum value; and the oscillator frequency swings to its lowest value. The rest of the analysis involving the frequency swing curve ① should be obvious. Sound wave ②, being of lower amplitude, produces less motion of the microphone diaphragm, less change in capacitance, and a smaller frequency deviation. Sound wave ③ has the same amplitude as sound wave ① and therefore produces the same *amount* of frequency swing; but since the pressure changes more rapidly, the *rate of change* in frequency is increased.

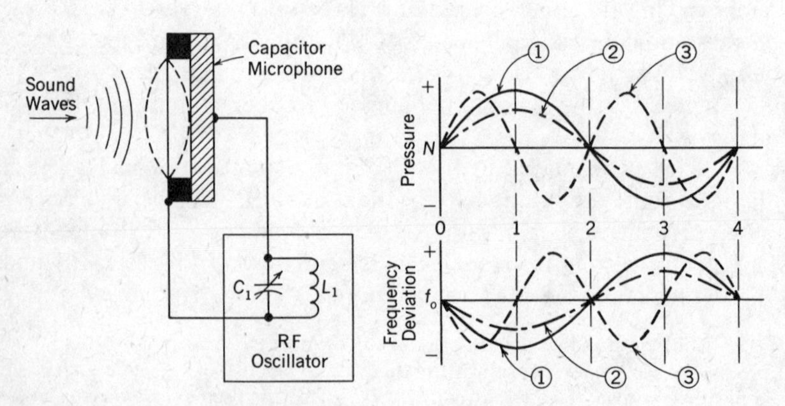

Figure 9-2 Production of FM waves.

To summarize—using a reverse analysis—when an FM wave is demodulated (in the receiver), the amplitude of the recovered intelligence signal depends on the amount of frequency swing, and the frequency of the intelligence signal depends on the rate of change in carrier frequency.

Modulation Level

In an amplitude-modulated wave, the degree or level of modulation is expressed as a percent (or factor), wherein the change (in amplitude) is compared to the original unmodulated amplitude. A modulation limit of 100% is reached when the modulating signal amplitude equals the carrier amplitude, and the modulated wave amplitude varies between zero and double the unmodulated value. Such a comparison factor cannot be used in FM. We certainly do not want the frequency of the modulated wave to vary between zero and twice the resting frequency. A more applicable expression for the modulation level in an FM wave can be obtained as follows. Begin with the basic voltage equation for a sine wave:

$$e = E_m \sin \theta \tag{9-2}$$

The angle θ is the instantaneous angular displacement of the voltage phasor from its zero axis, and at any instant it is equal to ωt. The angular velocity can therefore be written as $\omega = d\theta/dt$ or

$$\theta = \Phi + \int \omega \, dt \tag{9-3}$$

where Φ is the integration constant and represents the initial phase displacement.

In an FM wave, the instantaneous frequency and angular velocity change in accordance with the modulating signal, so that, *for a single-frequency modulating signal (sine wave)*,

$$f_i = f_0 + kf_0 \cos \omega_s t \tag{9-4}$$

where kf_0 is the frequency deviation (ΔF) and is a function of the amplitude of the modulating signal. Also

$$\omega_i = \omega_0 + 2\pi kf_0 \cos \omega_s t \tag{9-5}$$

Using this value of ω_i in Eq. (9–3), integrating and substituting the result into the basic voltage equation (9–2), we get for a frequency-modulated wave*

$$e = E_m \sin \left(\omega_0 t + \frac{2\pi kf_0}{\omega_s} \sin \omega_s t \right) \tag{9-6}$$

The term $(2\pi kf_0/\omega_s)$ is a measure of the change in the angular velocity

*F. E. Terman, *Electronic and Radio Engineering*, Chap. 17.

and, as such, is an indication of the level of modulation. It is therefore called the *modulation index* (m_f). Simplifying this term gives

$$m_f = \frac{\Delta F}{f_s} = \frac{\text{frequency deviation}}{\text{modulating signal frequency}} \tag{9–7}$$

Equation (9–6) can now be rewritten as

$$e = E_m \sin (\omega_0 t + m_f \sin \omega_s t) \tag{9–8}$$

EXAMPLE 9–1

A 100-MHz carrier is modulated by a 12-kHz sine wave so as to cause a frequency swing of ±50 kHz. Find the modulation index.

Solution

$$m_f = \frac{\Delta F}{f_s} = \frac{50}{12} = 4.17$$

Components of an FM Wave

In studying amplitude modulation, we saw that, with a single-frequency modulating signal (sine wave), the modulated wave contains three components—the carrier, the upper side frequency, and the lower side frequency, and the frequency of the new components is $f_c \pm f_s$. With frequency modulation, a more complex product results. Theoretically, there are an infinite number of side frequencies to either side of the carrier and spaced at intervals corresponding to the modulating frequency, or

$$f_c \pm f_s \pm 2f_s \pm 3f_x \cdots \pm nf_s \text{ to infinity}$$

However, beyond a certain spread, the amplitudes of the side frequencies drop to negligible values. Therefore, in a practical situation, the number of *significant* side frequencies is limited and depends on the modulation index. This component structure of an FM wave (with single-frequency modulation) can be analyzed mathematically by expanding Eq. (9–8).*

The result of such expansion yields

$$\begin{aligned} e = E_m\{&J_0(m_f) \sin \omega_0 t \\ &+ J_1(m_f)[\sin (\omega_0 + \omega_s)t - \sin (\omega_0 - \omega_s)t] \\ &+ J_2(m_f)[\sin (\omega_0 + 2\omega_s)t + \sin (\omega_0 - 2\omega_s)t] \\ &+ J_3(m_f)[\sin (\omega_0 + 3\omega_s)t - \sin (\omega_0 - 3\omega_s)t] \\ &+ \cdots \} \end{aligned} \tag{9–9}$$

*The first step is a trigonometric expansion for the sum of two angles. This first result is further expanded by using advanced calculus (Bessel's equations). The end result is of the form $J_n(x)$, a Bessel function of the first kind, where n is the order of the function and x is the argument.

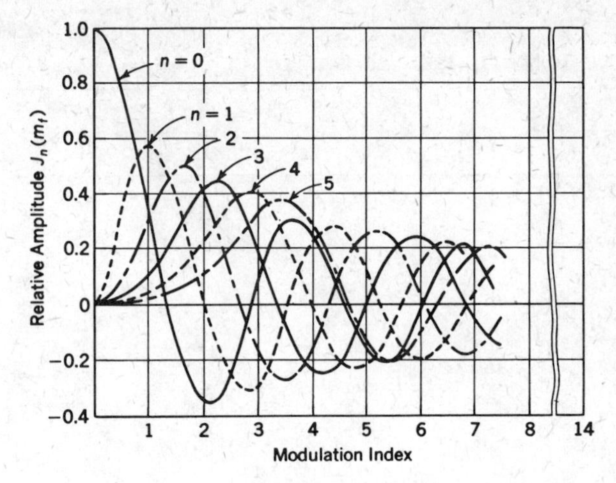

Figure 9-3 Bessel functions of the first kind.

From Eq. (9–9), several conclusions should be immediately obvious. The first component (J_0) is the carrier, and its amplitude will vary depending on the modulation index m_f. The first-order term (J_1) has two components and represents the first-order pair of side frequencies, at frequencies $f_c + f_s$ and $f_c - f_s$. The next term (J_2) is the second-order pair at frequencies $2f_s$ from the carrier (above and below). Theoretically, there are an infinite number of pairs. The amplitude of each component, as given by the general expression ($E_m J_n m_f$), varies with the modulation index and, for any one index, is the product of the unmodulated carrier amplitude E_m and the appropriate Bessel function J_n. The specific amplitudes can be evaluated from a graph of the Bessel functions (see Fig. 9–3) or from a table, such as Table 9–1. Notice from Fig. 9–3 that each component not only varies in amplitude with the modulation index, but also reverses. Also notice that the successive maxima for any one component are lower and lower as the modulation index increases.

Power Relations in an FM Wave

In an amplitude-modulated wave, the amplitude and power level of the carrier component are not affected by the modulation; the intelligence is in the side frequencies; the amplitude and power level of the side frequencies increase with the percent modulation; and consequently the power level of the modulated wave increases with increase in the modulation level. Only two of these conditions apply to a frequency-modulated wave—the intelligence is in the side frequencies, and the side-frequency power increases with the level of modulation. On the other hand, the *total* power in an FM wave does not change, regardless of the degree of modulation. Mathematically, this should be obvious, since

Table 9–1

BESSEL FUNCTIONS OF THE FIRST KIND*

x	*n* OR ORDER																
(m_f)	J_0	J_1	J_2	J_3	J_4	J_5	J_6	J_7	J_8	J_9	J_{10}	J_{11}	J_{12}	J_{13}	J_{14}	J_{15}	J_{16}
0.00	1.00	–	–	–	–	–	–	–	–	–	–	–	–	–	–	–	–
0.25	0.98	0.12	–	–	–	–	–	–	–	–	–	–	–	–	–	–	–
0.5	0.94	0.24	0.03	–	–	–	–	–	–	–	–	–	–	–	–	–	–
1.0	0.77	0.44	0.11	0.02	–	–	–	–	–	–	–	–	–	–	–	–	–
1.5	0.51	0.56	0.23	0.06	0.01	–	–	–	–	–	–	–	–	–	–	–	–
2.0	0.22	0.58	0.35	0.13	0.03	–	–	–	–	–	–	–	–	–	–	–	–
2.5	−0.05	0.50	0.45	0.22	0.07	0.02	–	–	–	–	–	–	–	–	–	–	–
3.0	−0.26	0.34	0.49	0.31	0.13	0.04	0.01	–	–	–	–	–	–	–	–	–	–
4.0	−0.40	−0.07	0.36	0.43	0.28	0.13	0.05	0.02	–	–	–	–	–	–	–	–	–
5.0	−0.18	−0.33	0.05	0.36	0.39	0.26	0.13	0.05	0.02	–	–	–	–	–	–	–	–
6.0	0.15	−0.28	−0.24	0.11	0.36	0.36	0.25	0.13	0.06	0.02	–	–	–	–	–	–	–
7.0	0.30	0.00	−0.30	−0.17	0.16	0.35	0.34	0.23	0.13	0.06	0.02	–	–	–	–	–	–
8.0	0.17	0.23	−0.11	−0.29	−0.10	0.19	0.34	0.32	0.22	0.13	0.06	0.03	–	–	–	–	–
9.0	−0.09	0.24	0.14	−0.18	−0.27	−0.06	0.20	0.33	0.30	0.21	0.12	0.06	0.03	0.01	–	–	–
10.0	−0.25	0.04	0.25	0.06	−0.22	−0.23	−0.01	0.22	0.31	0.29	0.20	0.12	0.06	0.03	0.01	–	–
12.0	0.05	−0.22	−0.08	0.20	0.18	−0.07	−0.24	−0.17	0.05	0.23	0.30	0.27	0.20	0.12	0.07	0.03	0.01
15.0	−0.01	0.21	0.04	−0.19	−0.12	0.13	0.21	0.03	−0.17	−0.22	−0.09	0.10	0.24	0.28	0.25	0.18	0.12

* E. Cambi, *Bessel Functions* (New York, Dover Publications, Inc., 1948).

the amplitude of such a wave is not affected by modulation. A physical appreciation of this effect can be obtained from Fig. 9–2(a). The modulating signal changes the capacitance in the oscillator tank circuit, resulting in a change in frequency (ΔF), but it cannot change the power output of the oscillator. Since the total power level in an FM wave remains constant, any increase in the side-frequency power must be at the expense of the carrier power. This is substantiated by an examination of Table 9–1. At low modulation levels, the amplitude (and power) of the side frequencies is quite low, and the carrier amplitude is high. Notice that, for modulation indices of less than 0.5, only the first-order side frequencies have a significant amplitude, while the carrier amplitude (and power) is almost equal to its unmodulated value. With increasing modulation levels, more and more power is transferred to the side frequencies as their amplitudes increase and/or higher-order components become significant. For example, at a modulation index of 9, the carrier has dropped to less than 10% of its unmodulated value; the amplitude of the fourth-order pair is triple this carrier value; the amplitude of the seventh-order pair is almost four times the carrier amplitude; and now even the twelfth-order pair has a significant value. The energy distribution in an FM wave is shown panoramically in Fig. 9–4, for several values of modulation index.

Bandwidth Requirements

At first impression, it would seem that the bandwidth necessary to transmit or receive an FM signal (without distortion) would be equal

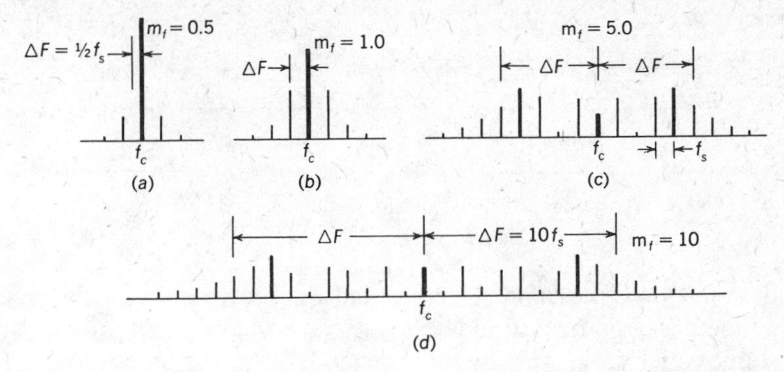

Figure 9-4 Frequency spectra in FM waves.

to twice the frequency deviation ($\pm\Delta F$). An examination of Fig. 9–4 will show that this is not so. In each case there are significant side frequencies that extend beyond the ΔF value. For example, using case (c), and assuming an intelligence signal frequency of 9 kHz, a modulation index of 5.0 would apply only if the deviation is (5×9) or ± 45 kHz. In other words, the signal amplitude is sufficient to cause the carrier to swing ± 45 kHz from its resting frequency. Yet, from the Bessel table, corresponding to $m_f = 5.0$, we notice that there are eight significant side frequencies. Since these are spaced at 9-kHz intervals, the total frequency spectrum or bandwidth needed is $2(8 \times 9)$ or 144 kHz. Notice that this is more than double the deviation.

EXAMPLE 9–2

Find the bandwidth needed for an FM signal if the frequency deviation is 60 kHz and the modulating frequency is 5 kHz.

Solution

1. $m_f = \dfrac{\Delta F}{f_s} = \dfrac{60}{5} = 12.$
2. From the Bessel table, corresponding to $m_f = 12$, find sixteen significant frequencies.
3. Bandwidth $= 2(5 \times 16) = 160$ kHz.

From Example 9–2 we note that a bandwidth of 160 kHz is needed to transmit (or receive) the 5-kHz intelligence signal. On the other hand, in the earlier discussion, a bandwidth of 144 kHz was needed for the 9-kHz modulating signal. Obviously, the frequency of the modulating signal per se has little bearing on the bandwidth required. This is further borne out by the illustrations in Fig. 9–4. In each case the modulating

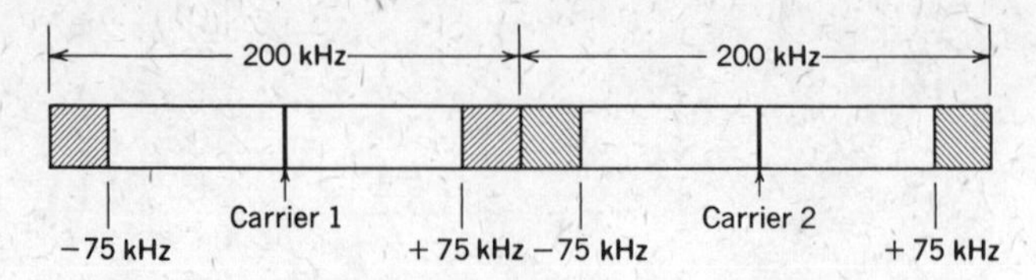

Figure 9-5 Bandwidth allocations for commercial FM broadcasting.

signal frequency is the same. Yet the bandwidth increases with the modulation index. The increased index is due to an increase in the frequency deviation, which is in turn directly dependent on the amplitude of the modulating signal. Therefore, bandwidth is a function of the modulating signal amplitude. However, there is no technical limitation on the amplitude of this signal—as in AM—so that if very high amplitudes are used, the resulting FM wave would occupy a very wide frequency band. This would reduce the number of transmitters that could function (at the same time) in a given frequency spectrum. To prevent interference and to provide for an optimum number of users consistent with good quality, the Federal Communications Commission limits the maximum deviation and allocates the station frequencies.

For commercial FM broadcasting (covering a frequency band of 88 to 108 MHz), the FCC allows a maximum frequency swing of ±75 kHz and a maximum audio modulating frequency of 15 kHz.* This maximum deviation can be considered as corresponding to 100% modulation. On this basis, a carrier swing of ±50 kHz would be equivalent to two-thirds or 67% modulation. When a transmitter utilizes its full deviation, we know that there is additional intelligence beyond the 75-kHz limit. To prevent adjacent-channel interference, the FCC further allocates 25-kHz *guard bands* at each side, so that the total channel allocation is 200 kHz. This is shown in Fig. 9-5.

However, not all FM services use a 75-kHz deviation. To conserve ether space and to allow for many more channels, FM communications stations are limited to a 15-kHz swing. The sound transmitters used in conjunction with television signals have a maximum deviation of 25 kHz. On the other hand, FM transmitters used in radar altimeters (operating at 440 MHz) have deviations of 2 or 20 MHz.

Indirect FM by Phase Shift

If we reexamine Eq. (9-1) for a basic sine wave, we see that there is another factor that can be varied—the starting phase angle Φ. If this angle is made to vary in accordance with the modulating signal, the resul-

*The modulation index in this case is 75/15 or 5. This *special* case, using maximum allowable values, is often called the *deviation ratio*.

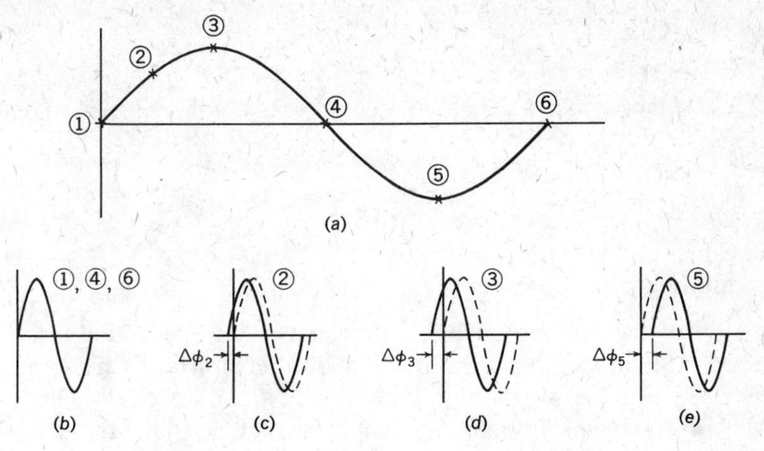

Figure 9-6 Change-of-phase with modulation.

tant wave is *phase-modulated*. The amount of phase shift ($\Delta\Phi$) will depend on the amplitude of the modulating signal, and the rate of phase shift will depend on the frequency of the modulating signal. This effect is shown in Fig. 9–6. At time instant ①, the modulating signal, Fig. 9–6(a), has an amplitude of zero, and the RF cycle (b) is shown starting at zero amplitude. This phasing ($\Phi_1 = 0$) is used as the reference. At time ②, the modulating signal has increased to some positive amplitude, and the RF cycle (c) has advanced in phase by an angle $\Delta\Phi_2$. As the modulation rises to its maximum value, at time ③, the RF cycle (d) increases its lead to Φ_3. On the other hand, when the modulation is negative, the RF cycle lags its reference phase position. Since the modulating signal amplitude varies continuously, the phase of the RF cycles must also vary continuously. But a change in phase can occur only if there is a change in frequency, so that indirectly, phase modulation has resulted in frequency modulation.

The amount of frequency change (ΔF) produced by this *indirect* action depends on two factors—the amount of change in phase $\Delta\Phi$ and the frequency of the modulating signal f_s. The first factor should be fairly obvious: the greater the amount of change in phase (in a given time interval), the faster the rate of change, and therefore the greater the frequency change. Similarly, the higher the frequency of the modulating signal, the less time in which the change $\Delta\Phi$ must be made; the faster the rate of change; and again the greater the change in the carrier frequency. These are both direct proportionalities, or

$$\Delta F_{\text{indirect}} = \Delta\Phi \times f_s \qquad \textbf{(9–10)}$$

where $\Delta\Phi$ is in radians.

EXAMPLE 9–3

A 7.5-kHz modulating signal has sufficient amplitude to cause a phase shift of 50°. Find the amount of indirect FM produced.

Solution

$$\Delta F = \Delta\Phi \times f_s = \frac{50\pi}{180} \times 7.5 = 6.55 \text{ kHz}$$

Interference Suppression

One of the striking qualities of broadcast FM reception (as compared to AM) is the absence of static and other forms of noise or interference. Such noise suppression is characteristic of a well-designed FM system. To explain why this is so, let us first consider the effect of combining two signals (A and B), each of constant amplitude but of slightly different frequencies. We will further consider that signal A, the desired signal, has a higher amplitude than signal B. At some instant, the two signals will be in phase and the resultant amplitude will be the sum of the individual amplitudes. At some other instant, the two signals will be 180° out of phase, and the resultant amplitude will be the difference between the individual amplitudes. At any other time instant, the resultant will vary in amplitude between these two extremes as the limits. Obviously, the resultant is amplitude-modulated, and the rate at which the amplitude varies is the *beat frequency* ($f_1 - f_2$). Now consider the phase relations. When signals A and B are in phase or 180° displaced, the resultant will be in phase with signal A. For any other phase angle θ between signals A and B, the resultant will lead or lag signal A by some angle Φ smaller than $\frac{1}{2}\theta$ (depending on how much signal A exceeds signal B). Therefore the resultant is phase-modulated, and the rate of change of phase is the beat frequency ($f_1 - f_2$). These effects, amplitude and phase modulation of a carrier, will occur whenever a desired signal is mixed with an interfering signal (such as noise or an unwanted carrier).

In a frequency-modulated receiver, elimination of the AM interference effects is quite simple.* One technique is to use one or more *limiter* stages to remove all amplitude variations from the signal. Another technique is to use demodulation circuits (such as the ratio detector or gated beam detector) that do not respond to changes in amplitude. By either means (or combination), none of the interference or noise that caused the amplitude variations in the first place can appear as output.

Unfortunately, the indirect FM effects produced by the unwanted signal cannot be eliminated completely. Any circuit that would nullify the frequency shift due to noise would also cancel the frequency modula-

* The circuitry is discussed in detail later in this chapter.

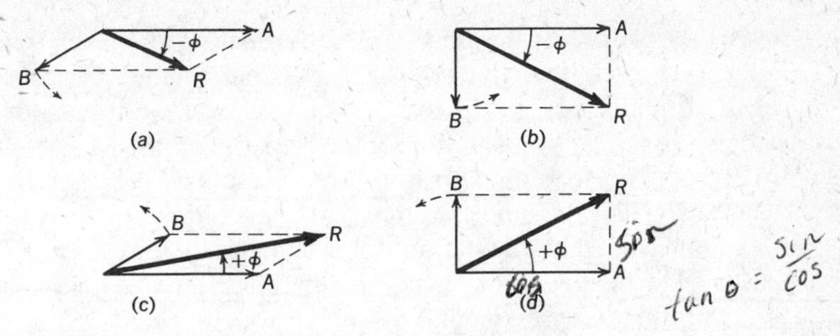

Figure 9-7 Phase shift of resultant caused by interference.

tion of the desired signal. However, the noise effect can be minimized by making the transmitted frequency deviation much larger than the indirect FM product due to interference.

As an illustration, let us assume a very poor signal-to-noise ratio wherein the desired signal is only twice as strong as the interfering signal. Since they are not at the same frequency, the instantaneous phase relation between these two signals will be changing continuously, as explained above. This is shown in Fig. 9–7 for four conditions, using the desired signal A as the reference. It should be readily obvious that the maximum phase shift of the resultant will occur when the original signals are at right angles, as in cases (b) and (d). For a 2:1 signal-to-noise ratio, this $\Delta\Phi$ is 26.6°, or just under 0.5 rad. To evaluate the amount of indirect FM produced, using Eq. (9–9), we must also know f_s, the frequency of the "modulating signal." Since this is the rate at which the phase changes, it corresponds in this case to the beat frequency. The greater the difference between the desired carrier frequency and the interference frequency, the higher the value of f_s, the greater the indirect frequency shift, and the stronger the noise output at the receiver. However, since we are interested in *audible* output, the worst interference case would occur for a difference frequency $(f_1 - f_2)$ of (approximately) 15 kHz. This results in an indirect FM of

$$\Delta F = \Delta\Phi \times f_s = 0.5 \times 15 = 7.5 \text{ kHz}$$

At this point it should be remembered that in an FM receiver the output or volume is proportional to the amount of change in frequency, and in FM broadcasting, a fully modulated carrier – or full volume – corresponds to a deviation of 75 kHz. Therefore, a noise or interference product of 7.5 kHz produces only one-tenth the full volume. We started with a 2:1 signal-to-noise ratio, and we reduced the noise to only one-tenth of the signal level. This is a distinct improvement. This improvement was obtained, not because we are dealing with FM signals, but rather because we used a signal swing that is much larger than the indirect FM that might be produced from noise or interference. Obviously, if wider deviations were premitted by the FCC, even better noise suppression would result. Equally obvious is that the noise reduction on narrow-band FM systems is not as good as for commercial broadcasting.

Since noise reduction is a direct function of the deviation of the desired signal, it is important that FM transmitters operate at close to the maximum allowable deviation, for all modulating signals. Yet, it is well known that in speech or music, the high-frequency sound levels (such as those produced by a piccolo) are much weaker than the output of a bass viol or tuba. Under "normal" modulation conditions, these high-pitched sounds would produce very little frequency deviation, and at the receiver they could readily be drowned by noise signals.* This deficiency can be overcome—at the transmitter—by deliberately overamplifying the high-frequency components of the modulating signal until they are strong enough to produce close to full carrier deviation. This technique is called *pre-emphasis.* Since the higher frequencies require the greatest boost, the response curve of the modulating signal amplifier is made to rise with frequency.

When such a signal is received, because of pre-emphasis, a large frequency deviation is obtained (compared to the indirect ΔF from noise) and good noise suppression is achieved. But these high-pitched sounds are much too loud—also because of pre-emphasis. To counteract this, receivers must include circuitry to reduce the amplitude of the high-frequency sounds so that once again they will be in proper balance with the other voice or music components. Such circuits are known as *de-emphasis* circuits. In order for any receiver to produce the proper tonal response from any transmitter, standardization is necessary. This is achieved by using simple, single-section L filters with rising characteristics at the transmitters, and similar filters with reciprocal drooping characteristics at the receivers. A time constant of 75 μs has been standardized for these circuits. These filters produce an *ultimate* boost or attenuation of 6 dB per octave.

COMMERCIAL RECEIVER CIRCUITS

FM communication channels are found throughout the frequency spectrum. For example, 1600–2900 kHz has been used for police calls; television sound channels for many years have used 54–88 MHz and 174–216 MHz; within this gap, 88–108 MHz is assigned to commercial FM broadcasting, and VHF mobile communications are quite popular in the 144–174 MHz range. In the UHF region, television sound channels are

*Such a condition would also apply to the high-frequency components of pulse signals.

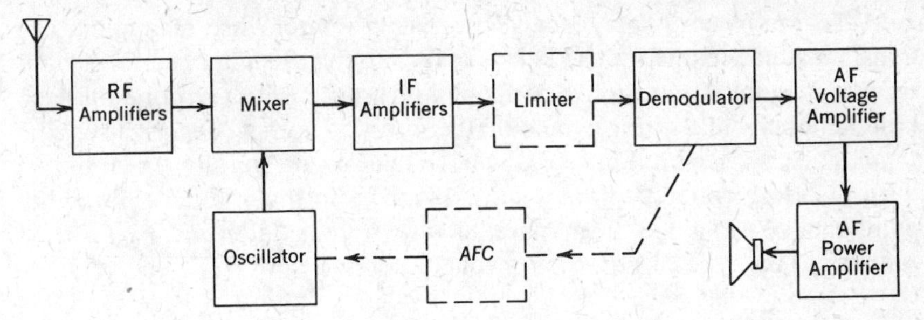

Figure 9-8 Block diagram of an FM receiver.

in the 470–490 MHz range, and many point-to-point communication systems are at frequencies above 4000 MHz. Regardless of the frequency, communications receivers for FM are essentially quite similar to the AM receivers discussed in the previous chapter. The block diagram of a typical FM receiver is shown in Fig. 9–8. It is obviously a superheterodyne receiver.

Types of Tuners

For broadcast FM reception, a receiver should be tunable over the range of 88–108 MHz, and it should have a bandwidth of 200 kHz. This is a function of the tuned circuits in the RF, mixer, and oscillator stages. Many receivers use "conventional" circuitry, that is, fixed inductances and a ganged variable capacitor. Compared to broadcast AM units, the tuning capacitor will have fewer plates (for example, 5 plates instead of 17). This, of itself, will just about double the maximum resonant frequency. Therefore, to tune to the 88–108 MHz region, drastic reduction in the tank circuit inductance is necessary. Consequently, conventional (variable-capacitance) tuning results in low Q and relatively low gain.

Figure 9-9 Inductuner. (Courtesy P. R. Mallory & Co., Inc.)

To improve Q and gain, some manufacturers use a minimum of *fixed* capacitance and tune the circuits by varying the inductance. A common technique is to employ permeability tuning, with the cores of each coil suitably linked for single-dial control. Another design uses the *inductuner* shown in Fig. 9–9, which consists of spirally wound coils with a sliding contact to vary the inductance in the circuit. Still higher gain can be obtained from parallel-wire or coaxial *transmission line tuners*.* This type of tuner is especially common in UHF receivers.

RF Amplifiers

The first block in the "typical" receiver (Fig. 9–8) is an RF amplifier. Use of this stage normally reduces the danger of image-frequency interference, improves the noise figure of the receiver, and reduces the possibility of oscillator radiation from the antenna. However, with an increase in the operating frequency, the gain of this stage decreases and the noise developed increases until at about 1000 MHz, it is actually preferable to omit the RF stage. For the FM broadcast band, the frequency is still low enough to make the use of RF amplifiers desirable. Any of the circuits discussed in Chapter 3 may be used for this stage. However, since their advent, FET amplifiers have been gaining in popularity. The almost ideal square-law characteristics (I_D vs V_G) of these devices greatly reduces cross modulation (CM) and intermodulation (IM) problems.† Dual-gate MOSFETs not only are best in reducing such distortion, but they also provide better AGC action. However, more gain is obtainable from IC stages, either emitter-coupled or cascode.

Bipolar transistors are also used as RF amplifiers in FM receivers. Because of the transistor's low input impedance, untuned or broadband input circuits are the rule. Selectivity is then obtained in the interstage coupling between RF amplifier and the mixer (or converter). At the higher frequencies, common-base circuits are sometimes used. (A transistor has a higher cutoff frequency in the common-base connection f_α as compared to the common-emitter connection f_β.‡)

*These tuners are discussed in more detail in Chapter 10.

†Cross modulation occurs when the signal from a strong undesired station drives the RF amplifier to operate in a steep curved region. The two signals (desired and interfering) could be at widely differing frequencies and completely unrelated. Yet, if the transfer characteristic of the amplifier is beyond square-law curvature, third- and higher-order products are produced. One of these results in transferring the intelligence of the interfering carrier to the desired carrier.

Intermodulation is another product of third- and higher-order nonlinear distortion. It occurs when two undesired signals mix and produce a third frequency equal to the desired station frequency. For example, if f_1 and f_2 are the interfering carriers, then the second harmonic of one ($2f_1$) plus or minus the other (f_2) could equal the desired carrier frequency.

‡$f_\beta \cong f_{\alpha/\beta}$.

At higher frequencies (above 100 MHz), because of transit-time effects, circuits have relatively low input impedance. Since good selectivity and gain is difficult to achieve, some manufacturers use broadband input circuits, even with ICs, adding resistance loading to obtain a sufficiently low Q.

Frequency Conversion

Any of the circuits discussed in Chapter 8 for frequency conversion can be used equally well here. The limitation is not whether the receiver is an AM or FM receiver, but rather depends on the frequency of operation. Since the gain of a transistor decreases at higher frequencies, noise becomes the more important consideration, so that transistor mixers are more common in the VHF range, and diode mixers for UHF applications. Another circuit quite commonly used in broadcast FM receivers is the autodyne converter. Since frequency-conversion circuits were analyzed in Chapter 8, no further discussion is necessary at this time.

IF Amplifiers

Here again the circuitry is essentially the same as that discussed for AM receivers. Transistors, FETs, and ICs (or vacuum tubes) may be used. The interstage coupling will vary, depending on the frequency of operation, the bandwidth requirements, and the maintenance (alignment) problems anticipated. These considerations were covered in Chapter 8 and need no further discussion. For commercial FM broadcast receivers

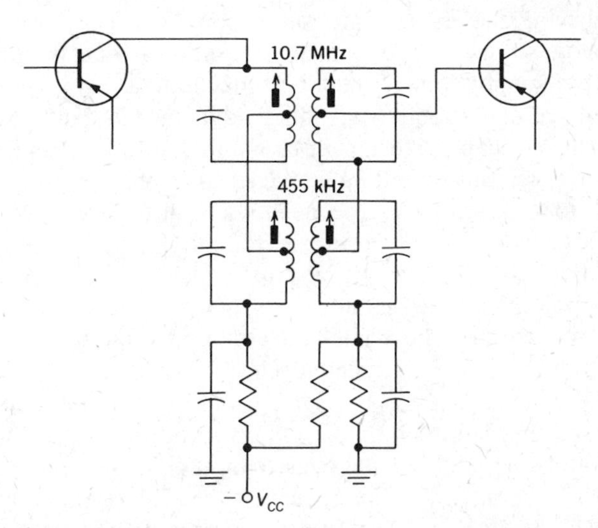

Figure 9-10 Interstage coupling for an AM-FM IF stage.

(bandwidth 200 kHz), the "standard" intermediate frequency is 10.7 MHz.

In combination AM/FM radios, some form of switching is used to change from the AM to the FM circuitry for the front-end and for the demodulation circuits. However, the IF system is left intact as a true combination circuit. A typical illustration is shown in the partial schematic of Fig. 9–10. Notice that two separate IF transformers are used, one tuned to 10.7 MHz and the other to 455 kHz. When receiving FM signals, the lower tank circuits have negligible impedance and do not affect the circuit operation. Therefore, there is no need to remove or short-circuit these components. Similarly, the 10.7-MHz tanks have negligible impedance at 455 kHz and will not affect circuit operation when AM signals are received.

Limiters

The next unit in the block diagram of Fig. 9–8 is a *limiter.* As mentioned earlier in the chapter, such a circuit is needed to remove any amplitude variations (as may be produced by noise or other interference) from the FM wave. Basically a limiter circuit acts as a *slicer, amplitude gate, or clipper-limiter** to cut off both the positive and negative peaks of the input wave at some predetermined level. By proper choice of this level (and with sufficient prior amplification), the output amplitude will be constant.

On the positive half-cycle of input, limiting action is generally obtained by driving the device into saturation. Normally, a transistor is operated below saturation. To ensure saturation within the normal current range, the supply voltage must be reduced to a relatively low value. Reduced voltage operation has an added advantage in that now even low input signal levels can drive the device into saturation. Unfortunately, however, lowering the operating voltages also reduces the stage gain.

Amplitude variations in the negative peaks of the FM wave are readily removed by driving the transistor into cutoff. Here again, low operating potentials will improve this clipping action. The lower the drain or collector potentials, the smaller the signal amplitude needed to drive the device into cutoff. The combined action of a limiter circuit can be studied from Fig. 9–11. A dynamic characteristic for the amplifier is shown in the upper left ($I_D - V_G$ for a FET, or $I_C - I_B$ for a bipolar transistor). Notice the flattening of the upper portion of this curve. This is caused by saturation and by gate or base limiting.* Several cycles of the input signal E_i are shown in the lower left, and the resulting output *current* (drain or collector) is shown in the upper right. Note that, in spite of

*J. J. DeFrance, *General Electronics Circuits,* Chap. 20.

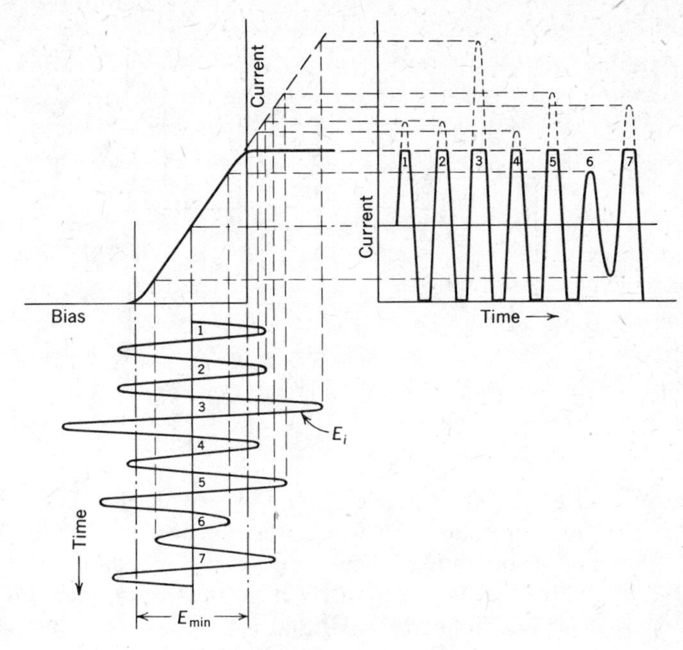

Figure 9-11 Limiter action.

variations in the input signal amplitude, the resulting current pulses are of constant amplitude—as long as the peak-to-peak signal level exceeds E_{min}. Notice also that, because of this limiter action, the current is a complex wave and contains many harmonic frequencies in addition to the fundamental. However, by passing this current through a tank circuit tuned to the input signal frequency, a pure sine-wave output *voltage* will be obtained.

Figure 9–12 shows a JFET limiter circuit. Q_2 should be a "sharp-cutoff" FET. Notice that gate-leak bias is used, in this case, as a shunt-fed circuit. (For series-fed bias, the *R-C* network is preferably inserted at *x*, between the bottom of the tank circuit and ground.) The use of gate-leak bias (as explained in Chapter 5) tends to stabilize the amplitude of the output and improves limiter action. Notice also that the drain voltage is reduced (as compared to other stages) because of resistor R_2. The choice of component values for the gate-leak bias is fairly important. If the time constant is made too long, the bias will not adjust to rapid changes in signal amplitude, and noise can come through. On the other hand, if the *R-C* values are too low, the change in bias may not be sufficient to control the amplification of the stage. Furthermore, since the gate conduction period is increased, the added loading on the tank circuit will impair the selectivity and gain of the stage. A compromise time-constant value of around 2.5 μs is common. With bipolar transistors in

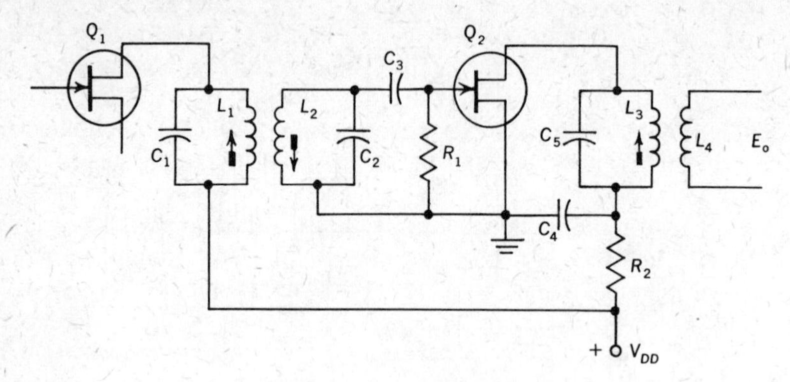

Figure 9-12 A JFET limiter circuit.

the IF strip, limiter action is again obtained by driving the "limiter" stage into cutoff and into saturation. However, there is no significant difference —in circuitry and in operating biases—between a standard IF amplifier and a limiter. This is so because an IF stage is already operated at low bias voltages (base and collector). All that is necessary to obtain limiter action is to ensure that the signal level reaching the "limiter" stage is sufficient to overdrive the stage. Toward this end, FM receivers employ more IF stages than an equivalent AM receiver. (See Fig. 9–20.) One distinct difference may exist between a limiter and a conventional amplifier. A transistor is capable of very low saturation voltages. Therefore, when the transistor is driven into saturation, the collector-base junction can become forward-biased. This places a direct short between input and output. To prevent this, a resistor (200–1000 Ω) is generally connected between the collector and its tank circuit.* Limiter action is also obtained by the use of transistors with AGC characteristics. With such transistors (2N3287), the gain decreases sharply with increase in collector current, tending to maintain the output level constant.

In the above single-stage limiters, limiting action may start at input voltage levels of approximately 0.3 V peak-to peak, but hard limiting does not occur until signal levels closer to 4 V. Furthermore, at still higher signal levels, the gain of the limiter stage will drop and amplitude variations will appear in the output. Far better limiting action can be obtained using two limiter stages in cascade. With FETs, one stage usually has a low time-constant gate-bias circuit (around 1.0 μs) to handle sharp noise bursts, while the other stage uses appreciably larger component values (around 10 μs) to provide higher gain. Dual-limiter circuits also

* Some manufacturers use such resistors in all IF stages, since on strong signals even an earlier stage could be overdriven into saturation.

improve limiter action in another way. In Fig. 9–11 the knee at saturation shows an ideally sharp bend. Usually, the curvature is more gradual, and some amplitude variation in the current pulses will result. These residual variations will be removed by the second limiter. Finally, amplitude variations caused by excessively strong input signal levels can be prevented by using an AGC circuit.

Excellent limiter action can be obtained with an IC emitter-coupled pair such as the ECG 371 or the μA 703 discussed in Chapter 3 and earlier in this chapter. Their dynamic transfer characteristics (see Fig. 3–14(b)) flatten out quite sharply at input signal levels of approximately 60 mV.* Furthermore, the maximum current values are not dependent on saturation of the stage, but rather on the bias setting of the "constant-current" transistor Q_3. With these ICs, the limiting is flat, and no second stage is needed.

Discriminators

One of the frequently used FM demodulator circuits is the Foster-Seeley discriminator. The function of this circuit is to convert the frequency deviations of the FM wave into the intelligence signal. However, since this discriminator also responds to changes in amplitude, it is necessary to use limiters before extracting the intelligence from the input signal. A typical Foster-Seeley discriminator circuit is shown in Fig. 9–13. In analyzing circuit action, several points should be noted. Both tanks (C_2L_1 and L_2C_4) are tuned to the resting frequency of the FM signal. The primary tank must have sufficient bandwidth to maintain essentially constant output for at least the full swing of the FM input. Capacitors C_1, C_3, C_5, and C_6 each have low reactance at the signal frequencies. Because of this, the bottom of the primary tank circuit (point B) is at RF ground potential. Similarly, the junction of C_5 and C_6 (point P) is also at RF ground. Meanwhile points A and N are at the same RF potential. This places the full primary voltage E_p across choke coil L_3. Simultaneously a voltage E_s is *induced* into the secondary winding, developing voltages E_a and E_b. Diode D_1 conducts because of the phasor sum of voltages E_a and E_p. When it conducts, capacitor C_5 charges with the polarity as shown. Conduction of D_2 is dependent on the phasor sum of E_b and E_p. Notice that the resulting charge on C_6 is opposite to the polarity of C_5. Consequently, if the diodes conduct equally, the voltages across R_2 and R_3 are equal and opposite, and the output voltage E_o is zero. On the other hand, the output will swing positive or negative depending on which diode conducts more heavily. The filter R_4C_7 is a de-emphasis circuit.

* Multistage ICs such as the CA 3013 have input limiting voltages (knee) as low as 300 μV.

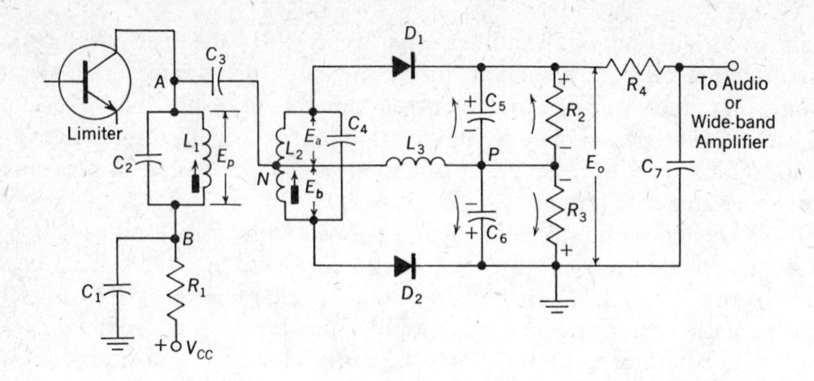

Figure 9-13 A Foster-Seeley discriminator.

Further analysis can now be made with the aid of the phasor diagrams in Fig. 9–14. Diagram (a) shows the conditions that exist when the incoming signal is at its resting frequency value. Starting with the primary voltage E_p as the reference phasor, the induced secondary voltage (by transformer action) is 180° away. Since the secondary circuit is resonant, the circulating tank current I_s is in phase with E_s. This current, flowing through the inductance L_2, produces an IX_L drop $(E_a + E_b)$ that lags the current by 90°. However, *compared to the center tap,* the individual voltages E_a and E_b are 180° apart. The phasor sums E_{D1} and E_{D2} are equal in magnitude; the diodes conduct equally; and the output voltage is zero. This agrees with our FM principles. There is no modulation of the signal when it is at its resting frequency.

Now let us consider a modulation condition that causes the frequency of the incoming signal to swing *above* its resting value. The resulting voltages are shown in Fig. 9–14(b). Voltages E_p and E_s remain as before. But, at this frequency, the induced voltage sees an inductive circuit, and I_s lags by some angle θ. Meanwhile E_a and E_b must still remain at right angles to I_s. Consequently, they swing clockwise with respect to

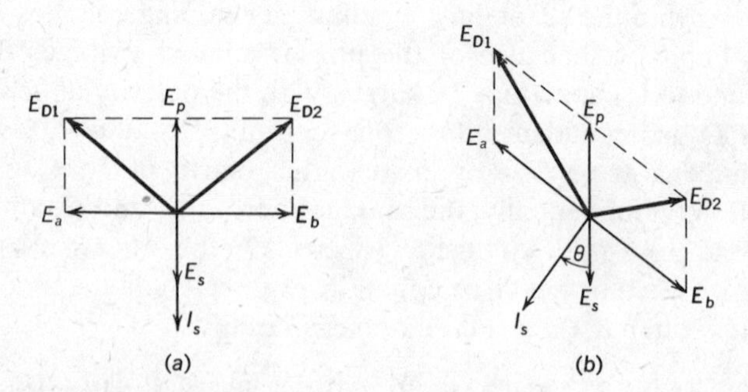

Figure 9-14 Phase relations in a discriminator circuit.

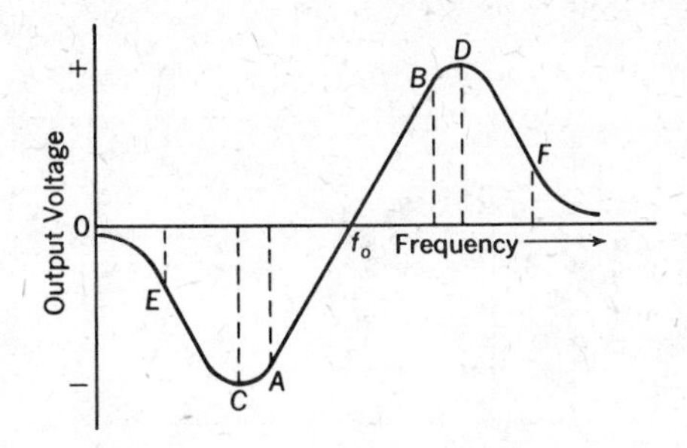

Figure 9-15 Discriminator response curve.

E_p as shown. Notice that E_{D1} is now appreciably greater than E_{D2}. Diode D_1 conducts more heavily, making the output positive. Furthermore, if the frequency deviation increases—up to a point—the angle θ of current lag increases, and the amplitude of E_o increases. However, for frequencies beyond the bandwidth of the primary, the output will fall off, approaching zero. Because of its dependence on phase relations, this circuit is also known as a *phase discriminator.*

It should be obvious that, for incoming signal frequencies *below* resting frequency, the secondary current and voltages E_a and E_b swing *counterclockwise* with respect to E_p, so that E_{D2} will exceed E_{D1}, and the output voltage will be negative. As above, for frequencies beyond the pass band of the primary, the output will drop toward zero. This voltage versus-frequency characteristic of the discriminator circuit is shown in Fig. 9–15. From its general shape, this response curve is often called an S curve. For distortionless demodulation, operation should be restricted to the linear portion (*A* to *B*). This can be achieved by making the bandwidth of the discriminator primary tank greater than the maximum deviation of the FM signal.

Ratio Detectors

Another commonly used FM demodulation circuit is the ratio detector of Fig. 9–16. Notice the strong similarity between this and the previous discriminator circuit. There are, however, three key differences: one of the diodes (D_2) is reversed; there is an added large capacitor ($C_9 \cong 10\ \mu F$) across the load resistors R_2,R_3; and the output is taken from between the junction of the two capacitors (point *P*) and ground.* (The deemphasis circuit has been omitted for simplicity.)

The voltage phasors for the discriminator circuit of Fig. 9–14 apply equally well to the ratio detector, so that at the resting frequency both

*Some circuits ground the bottom of R_3 instead of the junction of R_2 and R_3.

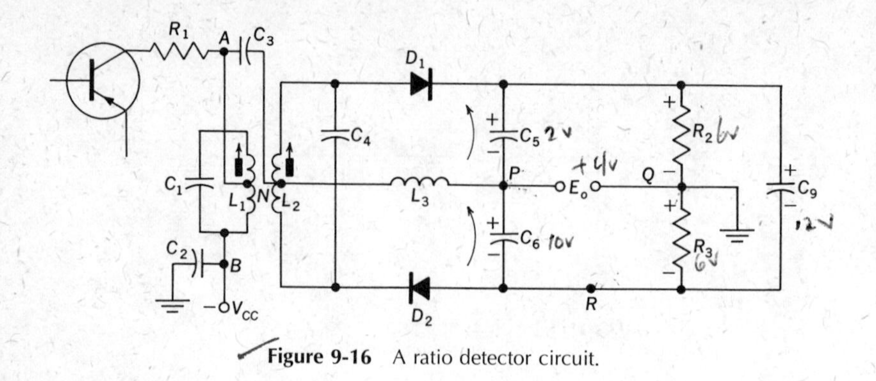

Figure 9-16 A ratio detector circuit.

diodes conduct equally, whereas as the frequency of the incoming signal swings to either side, one or the other diode will conduct more heavily. In addition, because one diode has been reversed, current can now also flow in the overall circuit—from the bottom of L_2 through D_2, through the long time-constant circuit, $R_2R_3C_9$, through D_1, and back to the top of L_2. After several cycles of RF, capacitor C_9 charges to (approximately) the peak value of the voltage across L_2. Herein lies the advantage of the ratio detector. Amplitude variations due to noise or other interference cause very little effect on the charge of capacitor C_9, and the voltage across this long time-constant circuit remains constant.* This in turn means that the voltage across $R_2 + R_3$ and the voltage across $C_5 + C_6$ are held to this value. Obviously, the output voltage must also remain constant. Since the output voltage is not affected by amplitude variations, limiter circuits (theoretically) are not necessary. However, large variations in amplitude may cause sufficient change to produce an audible output. Some manufacturers will therefore include partial limiting even with a ratio detector.

Now let us consider the reaction of this circuit to an FM signal. At the resting frequency, since both diodes conduct equally, the voltages across C_5 and C_6 are equal (at one-half the voltage across C_9). Also, since $R_2 = R_3$, their voltage drops are equal. This places points P and Q at the same potential, and the output voltage is zero. As the input frequency changes so that D_2 conducts more heavily, E_{C6} increases, but E_{C5} decreases (their sum remaining constant), and the output voltage is positive. For example, assuming that E_{C9} is 12 V, E_{C6} could be 10 V and E_{C5}, 2 V. But *compared to point R*, the potential of point P is +10 V. Meanwhile the potential of point Q (also compared to point R) is +6 V. This makes the output voltage +4 V. Conversely, when D_1 conducts more heavily,

*On the other hand, if the receiver is tuned to some other station of higher (or lower) signal level, after a number of RF cycles the time-constant circuit will gradually charge (or discharge) to the new voltage level. (The time constant used is approximately 0.2 s.)

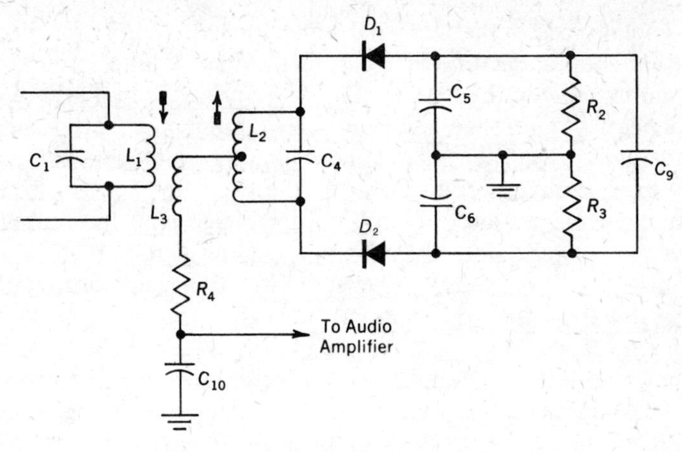

Figure 9-17 A ratio detector variation.

E_{C5} exceeds E_{C6} (their sum still remaining constant), and the output voltage swings negative. Notice that the sum $E_{C5} + E_{C6}$ remains constant—but their ratio changes, depending on the signal frequency. Because of this action, the circuit is called a ratio detector.

A common variation of the ratio detector circuit is shown in Fig. 9-17. The coupling capacitor, C_3, between the primary, L_1, and the reference coil, L_3, has been omitted. Now, L_3 is inductively coupled to L_1 but not to L_2. The RF path for each diode is still from half of L_2, through L_3, now through C_{10}, and either C_5 or C_6 to the respective diode. The circuit action is therefore the same as before.

Other Demodulators

A number of other circuits have been used at one time or another for demodulating FM signals. Among these are the locked-in oscillator (or synchronized oscillator, or Bradley detector), the gated-beam detector, the cycle-counting detector, and the Fremodyne superregenerative detector. In fact, even the AM diode detector can be used on FM signals, if it is detuned slightly. (This is known as slope detection.) However, none of these circuits have achieved any lasting popularity and will therefore not be discussed further.

Automatic Frequency Control

For operation at high frequencies, such as are used in FM transmissions, stability of the local oscillator circuit is very important. For example, a drift of 0.1% at 100 MHz will cause a 100-kHz shift in the IF band, resulting in serious distortion. To prevent such distortion—and even complete loss of the IF signal—many manufacturers resort to automatic frequency control (AFC) circuits. Many circuit variations have been used to achieve this effect. The basic principle in all these circuits is to connect a "controlled" capacitance (or inductance) across the oscillator tank, and to vary the value of this shunting component automatically by an "error" voltage produced when the oscillator goes off frequency. A most convenient source for this error voltage is at the output of the discriminator or ratio detector. If the oscillator is on frequency, the output is zero. But if its center frequency drifts to one side or the other, the output is either positive or negative, while the magnitude of the correcting voltage is dependent on the amount of drift. It should be obvious, however, that before the demodulator output can be used as an error signal, it must be put through a long time-constant circuit (low-pass filter) to prevent cancellation of the FM intelligence. The two devices most frequently used as the control element are reactance transistors (or tubes) and the special VVC diodes.

REACTANCE TRANSISTOR CIRCUIT. Figure 9–18 shows a JFET that is made to act as a variable capacitance. This effect is primarily dependent on the circuit values C_2 and R_1. The reactance of C_2 (at the oscillator frequency) must be very much larger than the resistance value R_1,* so that current I_1 leads the oscillator voltage by 90°. Obviously, E_g is in phase with I_1, and the RF component of drain current I_d, in turn, is in

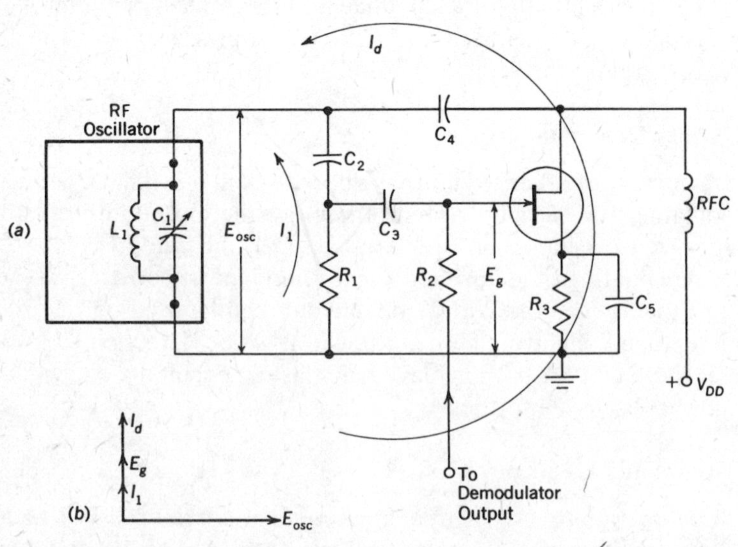

Figure 9-18 A reactance transistor circuit.

*All other capacitors have negligible reactances at this frequency.

phase with E_g and leads E_{osc} by 90°. The magnitude of this current—for given values of E_{osc} and source bias components (R_3C_5)—depends on the voltage fed back from the demodulator circuit, increasing when the demodulator output is positive, and vice versa.

As far as the oscillator is concerned, since the circuit draws a 90° leading current, the FET acts as a capacitive load, and when the current increases, the capacitance effect increases. The phasing of the demodulator output is set so that, should the incoming frequency increase, the output is positive. This increases current I_d, increasing the effective shunt capacitance of the reactance transistor and reducing the oscillator frequency back toward its center-frequency value. The amount of correction for a given error voltage is set by initial design of the FET operating point (source bias and V_{DD}).*

A reactance transistor can be made to act as a variable *inductance* by interchanging the location of components R_1 and C_2, and by making R_1 very much larger than the reactance of capacitor C_2. The analysis of this circuit will be left as a student exercise. (See Problem 29.)

Semiconductor AFC Circuits

Bipolar transistors can be used for automatic frequency control, either in a circuit similar to Fig. 9-18 or by utilizing the collector-to-base capacitance itself in shunt across the oscillator tank circuit. In the latter case, the capacitance C_{CB} is varied by controlling the base bias current. However, a far more popular circuit employs a voltage-variable capacitor (VVC) diode. Silicon diodes are generally used because of their higher breakdown voltage rating. The diodes are reverse-biased to create a depletion layer. The higher the reverse-bias voltage, the greater the width of the layer, and the lower the capacitance. A typical AFC circuit is shown in Fig. 9-19. The VVC diode D_1 is shunted across the oscillator tank circuit L_1C_1 in series with capacitor C_2. The diode is reverse-biased

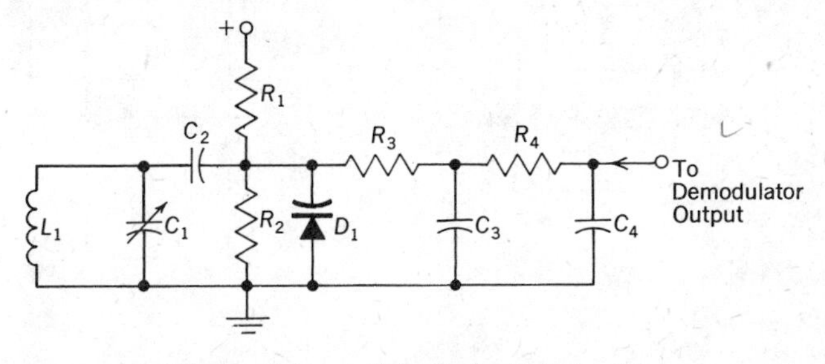

Figure 9-19 A VVC diode AFC circuit.

*From equivalent circuit analyses, it can be shown that the equivalent capacitance of this circuit is $C_{eq} = g_{fs}R_1C_2$.

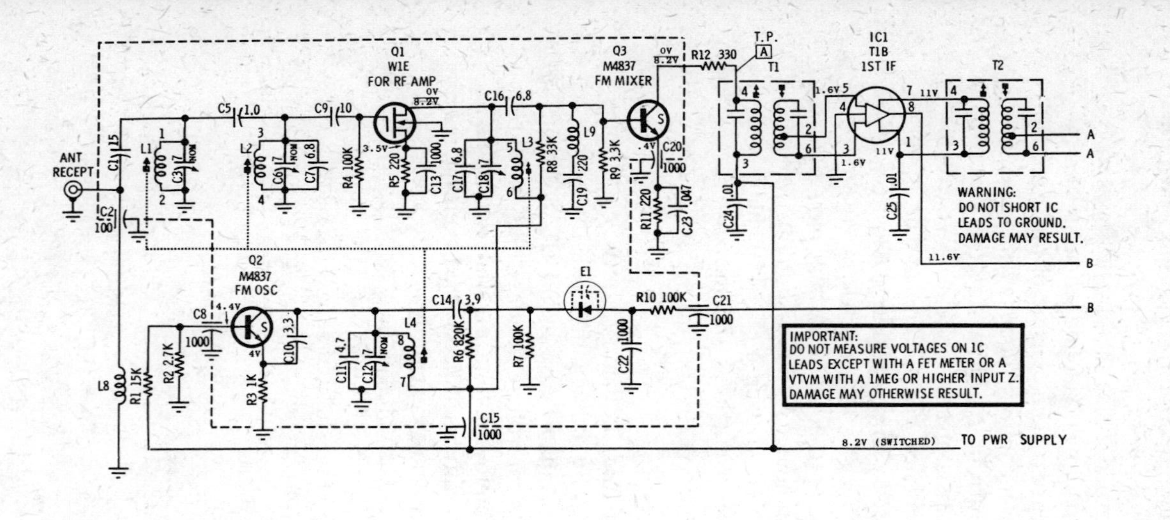

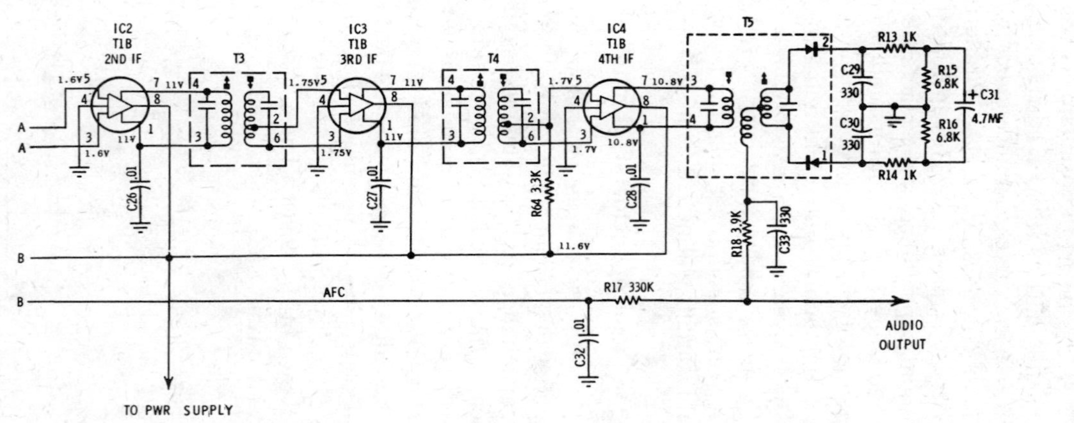

Figure 9-20 Commercial FM tuner. (Courtesy Motorola, Inc.)

by the voltage divider R_1R_2. The filter R_3C_3, R_4C_4 prevents the circuit from reacting to the normal deviation of the FM signal. For even better stability, some circuits will stabilize the diode dc reverse bias with a zener diode.

Complete Tuner Schematics

Figure 9–20 shows the circuit diagram of the tuner portion of a commercial receiver. The complete receiver is an AM-FM automobile radio designed for installation in cars having a 12-V negative ground electrical system. The circuit makes use of a MOSFET in the RF amplifier stage, and a T1B IC in each of the four IF amplifier stages. Except for minor variations, the circuitry of the individual stages has been discussed earlier. Analysis of this diagram will therefore be left to the student. (See Review Question 63.)

ALIGNMENT OF FM RECEIVERS

In most respects, alignment of FM receivers requires the same types of equipment and uses the same general procedures as the alignment of AM receivers.* Since this subject was covered in Chapter 8, it will not be repeated here. However, some differences do exist, and these are discussed below.

IF Alignment—the Steady-Frequency Method

The only difference between AM and FM IF alignment is the location of the indicator. (You will recall that the output is dc, the indicator is a dc voltmeter, and the tank circuits are tuned for maximum output.) In FM receivers, the IF response curve should be centered around the resting frequency. But at this frequency the demodulator output is zero. Some other location must be used for the indicator connection, such that the dc output is proportional to the signal strength and is therefore a measure of the correctness of alignment. In receivers with the Foster-Seeley discriminator, the voltmeter is connected across the gate-leak bias resistor of the limiter stage.† (Depending on the type of bias circuitry, this may or may not coincide with gate-to-ground of the limiter stage.) With ratio detector circuits there may not be a limiter. However, a dc

*Obviously the frequency range and sweep width of the signal generators must be compatible with the needs of the FM receiver under test.

†It is connected across R_1 in Fig. 9–12.

voltage that depends on signal strength is obtained across the long time-constant circuit of the ratio detector. The indicator (voltmeter) is connected here.* In either case, discriminator or ratio detector, all IF tuning adjustments are tuned for maximum indication, following the techniques described in the previous chapter. If the interstage networks are over-coupled, the alignment must be done on a stage-by-stage basis, working backwards toward the mixer, and loading each coupling transformer to obtain a single-peak response while it is being tuned to resonance. These procedures were described earlier and need no further discussion.

IF Alignment – the Sweep-Frequency Method

For visual alignment of the IF stages of an FM receiver, the oscilloscope (vertical input) is connected to the same points as used for the dc voltmeter in the steady-frequency method. The procedure is the same as described for AM receivers. With ratio-detector receivers, one other change is necessary. Since the long time-constant circuit will not respond to amplitude changes at the sweep rate, the output would be a constant dc level, and the oscilloscope screen will show a horizontal line. To obtain the true response curve, the "long time-constant" capacitor (C_9 in Fig. 9–16) must be disconnected.

Discriminator Alignment

When the discriminator stage is aligned, the signal generator(s) is (are) preferably connected from gate (or base)-to-ground of the limiter stage.† Using the steady-frequency method, the indicator (dc voltmeter) is first connected across the lower diode load resistor (R_3 in Fig. 9–13), and the primary of the discriminator transformer is tuned for maximum output. Next, the voltmeter is connected across the total diode load (audio takeoff points), and the secondary is tuned for *zero* output. Notice from the discriminator response curve, Fig. 9–15, that not only is the output zero at the resting frequency, but also that the voltage changes rapidly *and reverses* for a small change in frequency. For this reason, zero-center voltmeters are sometimes used in aligning the secondary. The final step in the discriminator alignment is to take a point-by-point response curve as a check on center frequency and on the linearity for full frequency deviation to either side. Notice that the accuracy of the crossover point (resting frequency) is dependent on the secondary alignment, while symmetry of response to either side is a function of the primary alignment.

*It is connected across $R_2 + R_3$ or C_9 in Fig. 9–16.

†This is especially desirable in visual alignment. The discriminator transformer generally has a wider bandwidth than the cumulative response of the IF stages. A better S-curve pattern is seen with this connection as compared to leaving the signal generators at the mixer gate (or base)-to-ground connection.

For sweep alignment, the oscilloscope vertical input is connected across the total diode load. With proper adjustments, as explained in Chapter 8, a pattern will be seen. The marker generator is used to check the crossover and linearity of the S curve. If alignment is needed, the secondary circuit is trimmed to correct for error in the crossover frequency, and the primary to correct for linearity.

Ratio Detector Alignment

In the steady-frequency method, the signal generator is connected across the input of the last IF stage. The primary of the coupling transformer is tuned for maximum output, with the indicator connected across the long time-constant circuit. Then the voltmeter is connected across the audio takeoff points (between *P* and *Q* in Fig. 9–16), and the secondary is tuned for zero output. (The output polarity should reverse sharply for small changes in trimmer adjustment.) A point-by-point response curve should also be taken to check crossover and linearity. For sweep alignment, the signal generators are connected across the input of the last IF stage, and the oscilloscope vertical is connected across the audio takeoff points. The procedure is the same as that used with the Foster-Seeley discriminator.

Front-End Alignment

Since the bandwidth of the front end is relatively broader than the bandwidth of the IF strip, alignment of the front end is generally done by the steady-frequency method. The signal generator is connected to the antenna terminals through a suitable impedance-matching network (see Fig. 8–31(b)). All of the pertinent precautions or general procedure steps must be performed, and in addition the AFC circuit, if used, must be disabled. As with AM receivers, alignment is adjusted at two points near the ends of the tuning range. The oscillator circuit is aligned first, and for this step, the indicator (dc voltmeter) is connected across the audio takeoff points. The receiver dial is set to the designated high-frequency test point, and the signal generator is set to the same frequency. The oscillator trimmer is adjusted for *zero* output (with sharp reversal). (Obviously, the demodulator circuit must be aligned first.) If the oscillator circuit has a padder or inductance adjustment, the receiver dial and signal generator frequency are set to the designated low-frequency test point, and the padder (or inductance) is adjusted for zero output (with sharp reversal). Both adjustments, high- and low-frequency end, should be repeated till there is no further interaction.

To align the RF and mixer circuits, the indicator voltmeter is connected across the same test points as for IF alignment (that is, across the gate leak of the limiter stage, or across the long time-constant circuit of the ratio detector). With the receiver dial and the signal generator at the high-end check frequency, both the RF and mixer trimmers are adjusted for maximum output. Usually no other adjustments are provided. If we wish to use visual alignment (or to see the response curve) of the RF and mixer circuits, we can connect the sweep and marker generators across the antenna input (through a matching network) and the oscilloscope vertical terminals across the input to the mixer stage. The procedures and adjustments follow the techniques described in the alignment of AM IF systems. Of course, the proper center frequency must be used.

FM TRANSMITTERS

Transmitters for frequency modulation service fall into two broad categories, depending on the technique used to obtain the frequency deviation. One type employs a "direct modulation" process wherein the frequency of the oscillator is made to vary in accordance with the modulating signal. The other type obtains a frequency-modulated wave *indirectly* after phase-modulating the carrier.

Direct Modulation

Many circuits have been used to produce direct modulation of the carrier frequency. In each case some form of reactance (inductance or capacitance) is shunted across the master oscillator circuit, and the effective *L* or *C* value is made to vary in accordance with the modulating signal. One of the most popular circuits is the reactance FET circuit of Fig. 9–18. For this application, however, the demodulator signal fed to the gate in Fig. 9–18 is replaced by the audio or modulating signal. Therefore, as the modulating signal varies in amplitude, the reactance FET bias changes; the drain current changes; the effective capacitance changes; and the oscillator frequency follows the change in the modulating signal. The linearity of modulation depends on the linearity of the transconductance characteristic ($E_G - I_D$) of the FET, while the amount of deviation depends on the slope of the curve. For broadcast-band applications a high-transconductance FET should be used. For optimum linearity, push-pull reactance modulators are sometimes used. This circuit has an added advantage in that the push-pull feature balances out temperature and supply voltage changes, thereby providing better AFC action.

Unfortunately, the amount of frequency deviation that can be obtained from a reactance FET circuit—with good linearity—is limited to approximately 5 kHz. When such circuits are used for broadcast FM transmitters, frequency multiplication is necessary to obtain the full allowable swing of ±75 kHz.

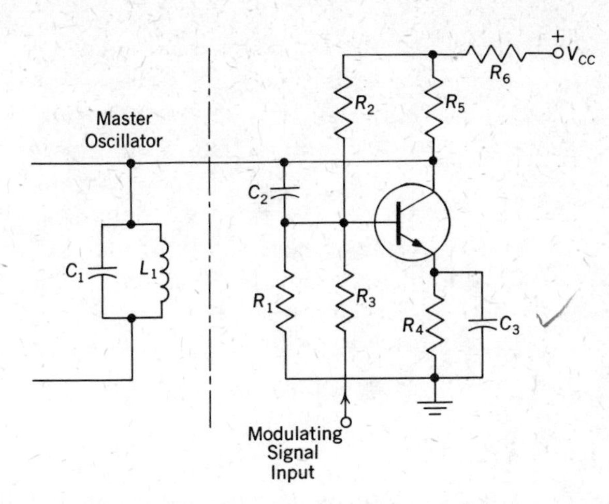

Figure 9-21 A transistor reactance modulator.

Another circuit for direct modulation shunts the *input* of an amplifier stage across the oscillator tank so that the frequency of the oscillator is dependent on the input capacitance C_i of the stage. Because of the Miller effect, this capacitance is primarily determined by the gain of the stage. The transistor is biased to the desired operating point while the modulating signal is applied to the gate, or base, in series with an isolating resistor. The modulating signal will vary the RF operating point, changing the transconductance and gain of the transistor in accordance with the modulating signal. This in turn varies the input capacitance of the stage and the frequency of the oscillator circuit. Since the frequency deviation obtained with this input-capacitance modulator is not as great as from a reactance-type circuit, use of this technique is generally restricted to narrow-band applications.

Bipolar transistors are also used to produce direct frequency modulation. A transistor, in the common-emitter configuration, can be made to act as an equivalent capacitor (or inductor) in much the same manner as a reactance FET. Such a transistor reactance modulator is shown in Fig. 9-21.* The base-emitter forward bias is set by the voltage divider R_2R_1 in combination with the emitter network R_4C_3. When the modulating signal is injected through the isolating resistor R_3, the bias variations will cause the equivalent capacitance of the transistor to vary, and the oscillator frequency is modulated. As in its FET counterpart, the proper phase relations for capacitor action are obtained by making X_{C2} very large compared to R_1. However, in this circuit, X_{C2} must also be low compared to $R_2 + R_5$.

* The no-signal equivalent capacitance is $C_{eq} = (h_{fe}/h_{ie})C_2R_1$.

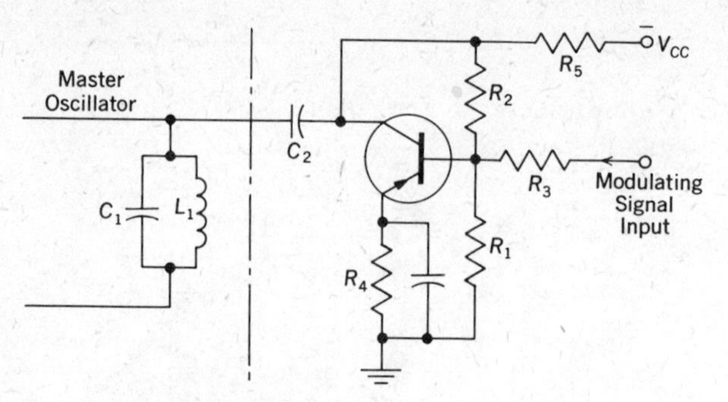

Figure 9-22 A transistor capacitance modulator.

The interelement capacitances of a transistor can also be used to produce direct frequency modulation. Either the emitter-base, collector-base, or collector-emitter capacitance can be used. Since these capacitances are created by the PN junctions, the capacitance values are a function of the width of the depletion layers and will vary with the bias voltages or currents. Figure 9–22 shows how the collector-emitter capacitance is used to frequency-modulate an oscillator. The resting frequency is set by the values of C_1, L_1, and C_{CE}, the latter value being dependent on the no-signal operating biases. When the modulating signal is applied, an increase in emitter (and collector) currents will increase the transistor shunting capacitance, and vice-versa. The oscillator frequency varies accordingly.

Voltage-variable capacitance diodes can also be used for direct frequency modulation. The diode is reverse-biased to produce the capacitance effect. As the modulating signal varies this bias, the capacitance value changes and the oscillator frequency varies accordingly. (See Fig. 9–23.)

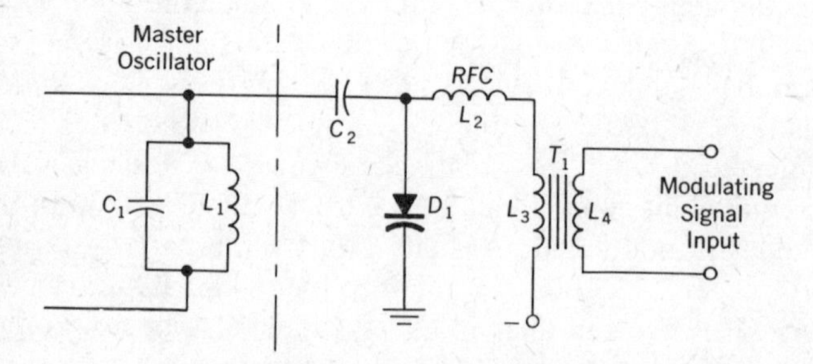

Figure 9-23 A diode modulator.

A Direct FM Transmitter

Any of the modulation methods discussed above can be used in a direct FM transmitter. Figure 9–24 shows the block diagram of the Crosby system as used in some commercial broadcast FM transmitters. Production of the FM wave, as shown in the upper half of this diagram, is quite simple. Since the deviation obtainable from a reactance stage is approximately 5 kHz and a total swing of 75 kHz (to either side) is desired, frequency multiplication must be used. As explained under class-C amplifiers, only doublers and triplers are used. With a total multiplication of 18, the reactance stage circuit is set to produce a maximum deviation of 4.167 kHz, corresponding to full modulation. This produces the desired 75-kHz deviation at the output of the exciter. In order to get the correct "channel" frequency at the exciter output, the master oscillator frequency must be one-eighteenth of the final carrier frequency. For example, a final carrier frequency of 90 MHz requires the use of a 5-MHz oscillator frequency. The power output from an exciter is in the order of 10 W. Depending on the final power output desired, intermediate and final power stages are added as required. An advantage of frequency modulation over amplitude modulation is that all frequency multipliers and power amplifiers can be class C for maximum efficiency, even though the carrier is modulated at its lowest level. Since all of these circuits have been analyzed earlier, no further discussion is needed here.

Direct frequency modulation systems have one inherent drawback. Since the frequency of the master oscillator is varied, this circuit cannot be a crystal oscillator. Therefore the transmitter center frequency is subject to drift. FCC regulations call for a center-frequency stability within ±2 kHz. To meet this requirement, some form of automatic frequency control is necessary. A commonly used AFC system is shown in the lower half of Fig. 9–24. The crystal oscillator supplies the reference frequency against which the carrier signal can be compared. The crystal frequency is multiplied to bring it close to the main transmitter frequency. This reference frequency is fed to a mixer, together with the output of the exciter.* The difference frequency is then fed to a discriminator circuit similar to the Foster-Seeley. The discriminator tank is tuned to the correct difference frequency, so that if the transmitter is on frequency, the discriminator output is zero. On the other hand, if the transmitter center frequency drifts to one side or the other, a plus or minus dc output is developed. This "error" voltage (amplified, if necessary) is fed as an additional bias to the reactance tube, changing the equivalent capacitance

*In some applications the transmitter frequency is "sampled" at some earlier point in the multiplier chain, thereby reducing or eliminating the need for multiplication of the crystal oscillator frequency.

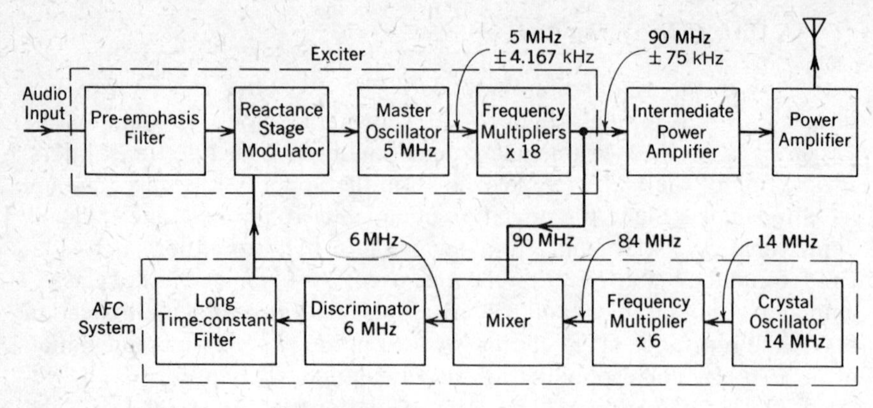

Figure 9-24 The Crosby FM system.

value. By proper phasing of the discriminator circuit, should the transmitter resting frequency drift to a higher value, the mixer output frequency rises, and the discriminator produces a dc output voltage of such polarity as to bias the reactance stage into heavier conduction. This corresponds to an increase in the equivalent capacitance and causes the master oscillator frequency to return toward its assigned frequency. The long time-constant filter between the AFC discriminator and the reactance stage is necessary to prevent cancellation of the transmitter's normal frequency deviation. (See Review Question 85.)

Although the above AFC circuitry can keep the transmitter drift within the FCC limits, it cannot maintain precise control. Stability precisely matching the crystal reference source can be achieved by using an alternate AFC system that establishes and maintains a *phase* lock between the reference crystal and the transmitter carrier. In this system, an RF voltage from the master oscillator is fed to a chain of frequency dividers,* reducing the carrier frequency to a range of 20 to 25 kHz and the frequency deviation to ±20 Hz. A reference crystal oscillator of 100 to 125 kHz is fed to another divider (÷5) and reduced to the same frequency as the master oscillator signal. Both signals (transmitter and reference) are then fed to the phase detector circuit of Fig. 9–25. When the master oscillator is exactly on frequency, the circuit phasing is such that E_2 is 90° out of phase with E_a and E_b. (The latter voltages are obviously 180° apart.) These relations are the same as shown in Fig. 9–14 for the discriminator; diodes D_1 and D_2 conduct equally. The phase detector output (across $R_1 + R_2$) is zero.

*J. J. DeFrance, *General Electronics Circuits*, Chap. 21.

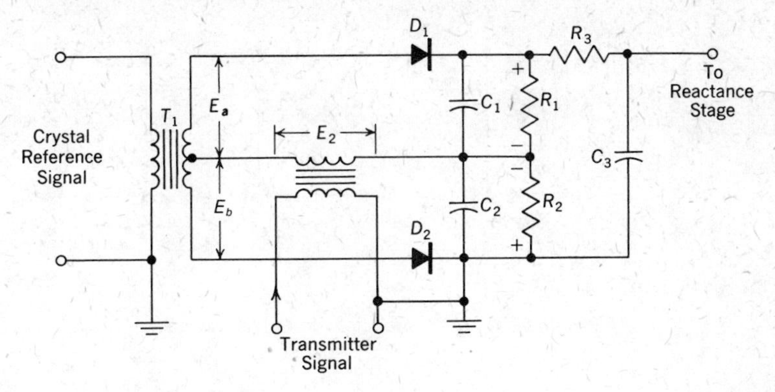

Figure 9-25 An AFC phase detector.

If the frequency of the master oscillator should tend to decrease, the phasing of voltage E_2 will shift toward E_a; diode D_1 will conduct more heavily; and a positive correction voltage will be fed to the reactance stage. A reverse drift in the carrier frequency will cause the phasing of voltage E_2 to shift toward E_b, producing a negative correcting voltage. Thus, any departure from the 90° phasing between the crystal oscillator and transmitter signals is instantaneously corrected by a proper error voltage.

FM from PM

As early as 1931, Hans Roder showed that a frequency-modulated wave* could be produced from an amplitude-modulated wave if (1) the modulating signal were distorted so that the amplitude of each component was inversely proportional to its frequency, (2) the level of modulation (AM) were low, and (3) if the sidebands so produced were shifted 90° with respect to the carrier. To explain why this is so, let us first refer back to Eq. (6–8) for an amplitude-modulated wave

$$e_{\text{mod}} = E_{cm} \sin \omega t - \frac{mE_{cm}}{2} \cos (\omega_c + \omega_s)t + \frac{mE_{cm}}{2} \cos (\omega_c - \omega_s)t$$

and to Fig. 6–4 showing the phasor relations among the three components. Notice that the phasor sum of the side-frequency components is always in phase with the carrier, and therefore there is no phase shift between the resultant (modulated wave) and the unmodulated carrier.

* Hans Roder, "Amplitude, Phase and Frequency Modulation," *Proceedings IRE*, **19**, No. 12 (December, 1931).

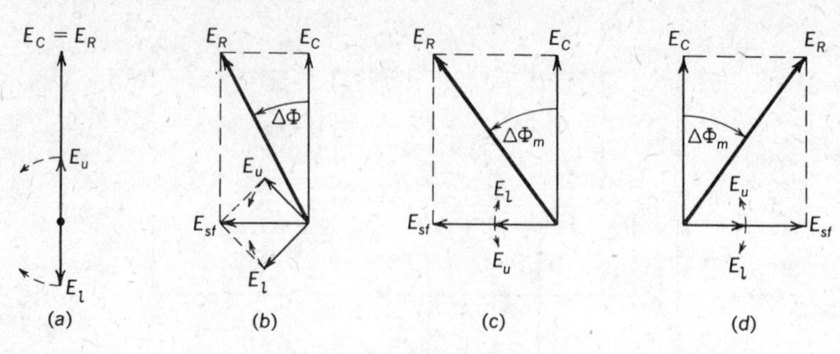

Figure 9-26 Vector relations for a PM wave.

Let us now see what happens when the side-frequency components are separated from the carrier, shifted 90° with respect to the carrier, and then recombined with the carrier. The resulting phasor relations are shown in Fig. 9–26, again using the carrier as the reference phasor. In diagram (a), corresponding to Fig. 6–4(a), the side frequencies have been rotated 90°, so that the upper side frequency E_u is in phase with the carrier. Since the sum of the side-frequency components is zero, the carrier E_c is unaffected at this instant.

At some instant later—diagram (b)—the upper side-frequency component has rotated counterclockwise (downward) by some small angle. Simultaneously, the lower side frequency E_l has rotated clockwise (upward) by the same amount. Notice that the phasor sum of these side-frequency components (E_{sf}) is at right angles to the reference carrier. The resultant modulated wave E_R has shifted away from the unmodulated carrier by an angle $\Delta\phi$. Furthermore, as the side frequencies continue to rotate, their phasor sum increases in magnitude, increasing the phase-shift angle ϕ between the carrier and the resultant modulated wave, until the conditions in Fig. 9–26(c) are reached. At this instant, both side frequencies are in phase and at right angles to the reference carrier. Their sum E_{sf} is a maximum, and the phase-shift angle is a maximum.

It should be obvious that as the side-frequency phasors continue to rotate, the phase-shift angle will decrease to zero, rise to a maximum again when they have rotated through 270° (Fig 9–26(d)), and drop to zero once more as they return to the starting phase positions shown in diagram (a) to complete one cycle of modulation. Notice that because of the 90° phase shift between carrier and side frequencies (Hans Roder's condition 3 above), the resultant modulated wave has swung through a phase-shift angle of $\pm\phi$. However, also notice that the amplitude of the resultant varies in amplitude, from E_c as a minimum to some maximum value as shown in Fig. 9–26(c) and (d). This amplitude modulation is undesirable and can be prevented by keeping the side-frequency components (E_l and E_u) small.* This is the reason for condition 2 above.

So far, we have produced phase modulation. However, we know that indirect FM results from phase modulation, and the amount of frequency swing—*indirect*—*is* $\Delta F_{\text{ind}} = \Delta\phi \times f_s$. This is not true FM, because the amount of change in frequency (ΔF) is now dependent not only on the amplitude of the modulating signal (which determines $\Delta\phi$), but also on the frequency f_s of the modulating signal. Consequently a high-pitched sound will produce a greater deviation than an equally strong (in amplitude) low-pitched sound. This can be corrected by attenuating the modulating signal by a factor inversely proportional to its frequency, so that for a given sound level, the product $\Delta\phi \times f_s$ is constant regardless of the modulating frequency.

The validity of this technique for producing an FM wave can be shown mathematically from Eq. (9–9) and Table 9–1 of the Bessel functions. Since we stipulated earlier that a low level of modulation was prerequisite, let us now further stipulate a modulation index of 0.5 (or less) for the FM wave. From Table 9–1, we see that the carrier component $J_0(m_f)$ is almost at its full unmodulated value (0.94 for $m_f = 0.5$ or 0.98 for $m_f = 0.25$). Also notice that the first-order component $J_1(m_f)$ has a significant value (0.24 for $m_f = 0.5$ or 0.12 for $m_f = 0.25$), whereas the other components are of negligible amplitude. Substituting the amplitude values for $m_f = 0.5$ into Eq. (9–9), we get

$$e = E_m\{0.94 \sin \omega_0 t + 0.24[\sin(\omega_0 + \omega_s)t - \sin(\omega_0 - \omega_s)t]\}$$

To a very close approximation, this can be regrouped into

$$e = E_m \sin \omega_0 t + \frac{m_f E_m}{2} \sin(\omega_0 + \omega_s)t - \frac{m_f E_m}{2} \sin(\omega_0 - \omega_s)t \quad \textbf{(9–11)}$$

Comparing this equation with Eq. (6–8) for an AM wave, it is seen that the only difference is that the FM side frequencies are sine waves, whereas the AM equivalent components are cosine waves. This shows the 90° shift between carrier and side-frequency components.

The Armstrong FM System

A frequency-modulation system based on the above principle was developed by Major Armstrong in 1936,† and the first regularly scheduled broadcast by FM radio was in July, 1939, from his transmitter W2XMN in

* From trigonometric relations, it is obvious that for phase-shift angles of less than 9°, the amplitude variation is less than 1%.

† E. H. Armstrong, "A Method of Reducing Disturbances in Radio Signalling by a System of Frequency Modulation," *Proceedings IRE*, **24**, No. 5 (May, 1936).

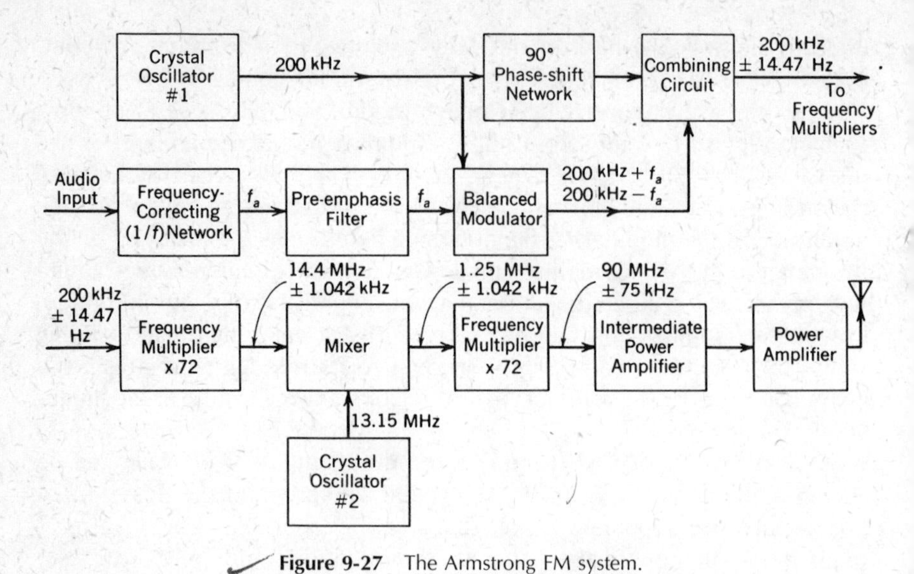

Figure 9-27 The Armstrong FM system.

Alpine, N.J. A block diagram of the Armstrong FM system is shown in Fig. 9–27. The carrier frequency is generated by the first crystal oscillator. The crystal output and the audio signal (suitably corrected) are fed to a *balanced modulator,** wherein the upper and lower AM side frequencies are generated, but the carrier frequency itself is suppressed. Meanwhile the crystal oscillator signal is also fed to a phase-shifting (*R-C*) network to produce a 90° phase shift. The outputs of the balanced modulator and the phase-shift network are fed to a combining circuit,† and the output is a frequency-modulated wave.

Unfortunately, the amount of deviation obtainable by this technique is very low–ranging 10–25 Hz. Consequently, tremendous frequency multiplication is needed to raise this deviation to the full ±75 kHz. For example, if ΔF is 15 Hz, a multiplication of 5000 will be required. However, the oscillator frequency will also be multiplied by this amount, so that if a final channel frequency of 90 MHz is desired, the starting frequency (from the oscillator) should be 18 kHz. Obviously, such a frequency is too low for a crystal oscillator. Instead, most Armstrong transmitter designs use crystal oscillator frequencies of 200 kHz (or 250 kHz).

The total multiplication (which may range 3000–7500, depending on the original frequency deviation) is broken into two sections. The out-

*This circuit is discussed in the next section.

†This could be a linear mixer, as used for microphone mixing, or a common impedance across which both signals are applied.

put of the combining circuit is fed to the first group of frequency multipliers. Then the carrier frequency is reduced by frequency conversion (mixer-oscillator) using a second crystal oscillator. The frequency of this second oscillator is selected so that the mixer output, when multiplied by the second group of frequency multipliers, will produce the desired channel frequency. Additional power amplifiers are then added to develop the required power output. The advantage of the Armstrong system over direct FM systems is the inherent excellent stability obtained with crystal oscillators.

As far as circuitry is concerned—except for the balanced modulator—all of the circuits for this transmitter have been covered, and no further discussion is needed.

The Balanced Modulator

A simple balanced modulator circuit using bipolar transistors is shown in Fig. 9–28. As far as the modulating signal is concerned, this circuit looks like a push-pull audio amplifier. However, notice that the push-pull output circuit uses a tank circuit resonant at the RF (carrier) frequency. The impedance of this load at the modulating signal frequency is negligible. Obviously, no audio voltage is developed in the output. Now disregard the audio signal completely, and consider only the RF

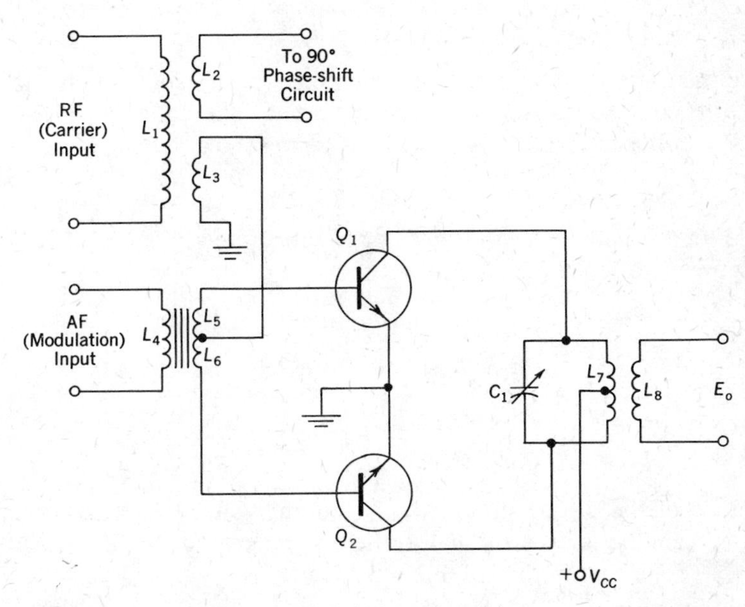

Figure 9-28 A balanced modulator.

input. This is *not* a push-pull input. The RF voltage is applied to each base in series with a coil (L_5 or L_6). Since the base signals to Q_1 and Q_2 are in phase, the RF components of collector current for these transistors are also in phase. These currents flow in opposite directions through the push-pull tank circuit. Their effects will cancel, and the RF output—*at the carrier frequency*—is zero.

We can now consider the combined action. The transistors operate on the nonlinear portion of their characteristics so as to produce modulation products (sum and difference frequencies). Because of the push-pull audio input, the "bias" of the transistors varies at an audio rate—but with opposite phasing. Therefore, as the modulation components (upper and lower side-frequency currents) of one transistor are increasing, the modulation components in the other are decreasing. In the push-pull collector tank, these effects are additive, and since the tank circuit maintains full impedance at these frequencies, an output is developed at the sum and difference frequencies. The circuit has produced an output containing only the side frequencies.

The above action can be verified mathematically. The input to each transistor is the sum of the RF carrier and the modulating signal (with the modulating signals out of phase), or

$$e_{i1} = E_c \sin \omega_c t + E_s \sin \omega_s t \tag{9–12}$$

and

$$e_{i2} = E_c \sin \omega_c t - E_s \sin \omega_s t \tag{9–13}$$

Since the transistors operate on the nonlinear portion of their characteristics, the collector currents will be distorted. The collector current for this square-law characteristic can be represented by a power series:

$$i_c = a_0 + a_1 e_i + a_2 e_i^2 + \cdots \tag{9–14}$$

If the input signal for each transistor is substituted in Eq. (9–14), the ac components of collector current are

$$i_{c1} = a_1 E_c \sin \omega_c t + a_1 E_s \sin \omega_s t + a_2 E_c^2 \sin^2 \omega_c t$$
$$+ 2a_2 E_c E_s \sin \omega_c t \sin \omega_s t + a_2 E_s^2 \sin^2 \omega_s t$$
$$i_{c2} = a_1 E_c \sin \omega_c t - a_1 E_s \sin \omega_s t + a_2 E_c^2 \sin^2 \omega_c t$$
$$- 2a_2 E_c E_s \sin \omega_c t \sin \omega_s t + a_2 E_s^2 \sin^2 \omega_s t$$

Because of the push-pull action, the output voltage E_o across the secondary winding is proportional to the difference between these two currents. Notice that in the subtraction the carrier frequency terms cancel, leaving

$$E_o \propto 2a_1 E_s \sin \omega_s t + 4a_2 E_c E_s \sin \omega_c t \sin \omega_s t \tag{9–15}$$

Expanding Eq. (9–15) by trigonometric identity gives

$$E_o \propto 2a_1E_s \sin \omega_s t + 2a_2E_cE_s \cos (\omega_c - \omega_s)t - 2a_2E_cE_s \cos (\omega_c + \omega_s)t \quad (9\text{–}16)$$

The selectivity of the tuned collector tank will filter out the first component of Eq. (9–16). The remaining components, the cosine functions of the sum and difference frequencies, are obviously the upper and lower side frequencies.

Many variations of the balanced modulator circuit in Fig. 9–28 are possible. For example, the RF signal could be *R-C* coupled and shunt-fed to both bases (still in phase), or a push-push collector circuit could be used with push-pull input—but now the *carrier* is fed as a push-pull signal for cancellation. FETs are also used in a circuit similar to that of Fig. 9–28, and excellent balanced modulator action can be obtained using a dual-independent IC (such as the CA3049) that combines two simple differential amplifiers in one chip.

Tha Phasitron FM System

Another indirect FM system achieved some popularity in the mid 1940s. It used a special tube—the phasitron—that produced a phase shift of approximately 175 Hz from a starting crystal frequency of around 200 kHz. Obviously this required much less frequency multiplication than the Armstrong system. However, the manufacturer has discontinued production of these tubes (except for replacement purposes only), and further discussion is not warranted.*

Other Indirect Modulation Methods

The Crosby and Armstrong FM systems are used in many broadcast FM stations where high fidelity and full frequency deviation are important requirements. However, there are many other applications in voice communication services where an audio-frequency range of 200–3000 Hz is sufficient. Since the lowest frequency is 200 Hz, rather than 20–50 Hz, it is possible to achieve indirect frequency deviations of well over 100 Hz. Furthermore, since these FM systems often operate with maximum deviations of 15 kHz, it is obvious that the number of frequency multiplier stages can be reduced, and the frequency conversion eliminated. The overall transmitter can be simplified appreciably. Still further reduction in the number of frequency multipliers can be obtained (at some sacrifice in linearity) by using phase modulators that will give

* *Electronic Tube Engineering Bulletin ET-820,* General Electric Company (August 1, 1947).

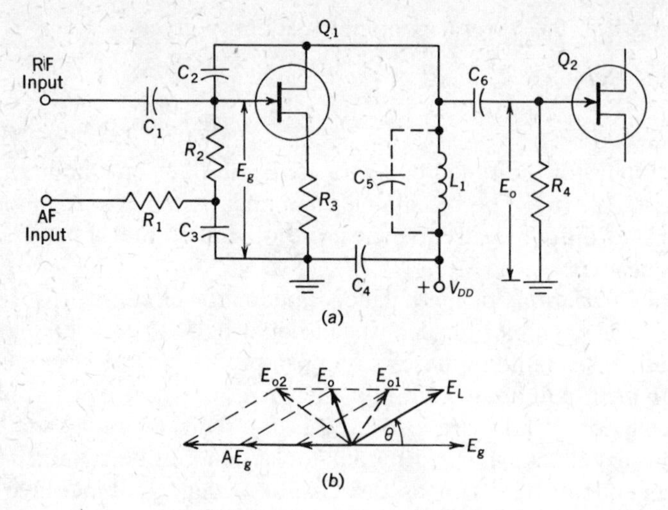

Figure 9-29 The Link phase modulator.

greater frequency deviations. Transmitters using these techniques can be designed with a total frequency multiplication as low as 16 to 48. Two phase modulators commonly used in mobile transmitters are discussed below.

The Link phase modulator is shown in Fig. 9–29. Q_1 is the phase modulator; Q_2 is a frequency multiplier. The oscillator signal (RF) is *R-C* coupled to the gate of Q_1 and simultaneously is also applied across L_1 through capacitor C_2.* All capacitors—except C_2—have negligible impedance at the oscillator frequency, while the drain load L_1 is broadly resonant at this frequency. Obviously, E_g is in phase with the oscillator voltage, while the component E_L of the oscillator voltage across L_1—*as applied through* C_2—is leading by some angle θ. In addition, because of amplifier action, a second component of oscillator voltage, AE_g, is also developed across L_1. Since the FET also acts as a phase inverter, this voltage AE_g leads E_g by 180°. The output voltage E_o is the phasor sum of AE_g and E_L. These phase relations are shown in Fig. 9–29 (b).† When an audio signal is applied to the gate of Q_1, the operating bias swings up and down, varying the stage gain and the magnitude of AE_g above and below the no-modulation value. This, in turn, varies the output voltage to E_{o1} and E_{o2}. Notice that the phase of the output voltage changes with the modulating signal. Components R_1 and C_3 form the frequency correcting ($1/f$) network to convert this PM to FM.

*If the interelectrode capacitance C_{gd} is sufficient, this capacitor may be omitted.

†Resistor R_3, by providing inverse feedback and by biasing the FET to a point of reduced transconductance, creates an approximate equality between E_g, AE_g, and E_L.

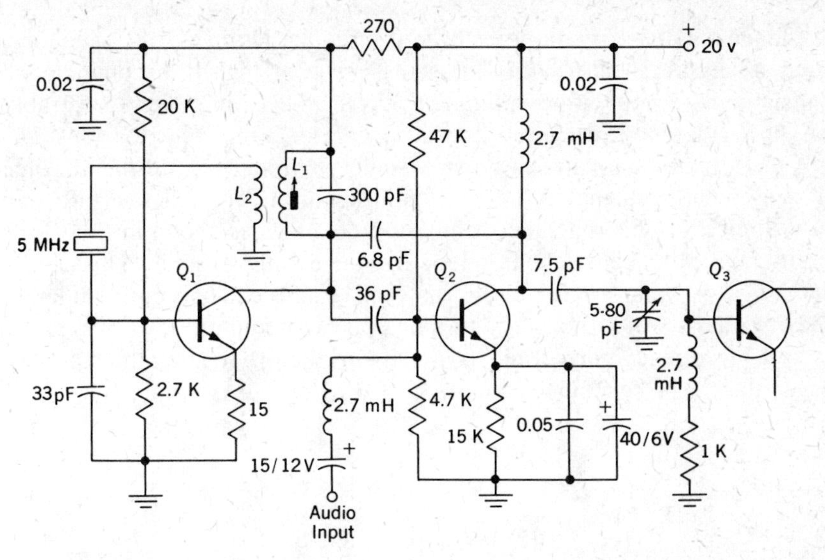

Figure 9-30 A transistor phase modulator. (Courtesy Texas Instruments.)

A portion of a 150-MHz crystal controlled FM transmitter is shown in Fig. 9–30. Q_1 is a common-emitter crystal oscillator operating at (approximately) 5 MHz, with feedback through the crystal to the base. The output of the oscillator is *R-C* coupled to the base of the phase modulator Q_2. The audio modulating signal is fed to the base through a 2.7-mH RF choke. The output of the phase modulator is impedance coupled to a frequency tripler Q_3. The rest of the transmitter (not shown) consists of two more triplers, a driver, and a pair of parallel transistors, for a final output of 250 mW into a 50-Ω load.

A reactance transistor can also be used to produce indirect FM. However, it cannot be used to change the frequency of the crystal oscillator. Instead, it is connected across the tank circuit of *an RF amplifier*

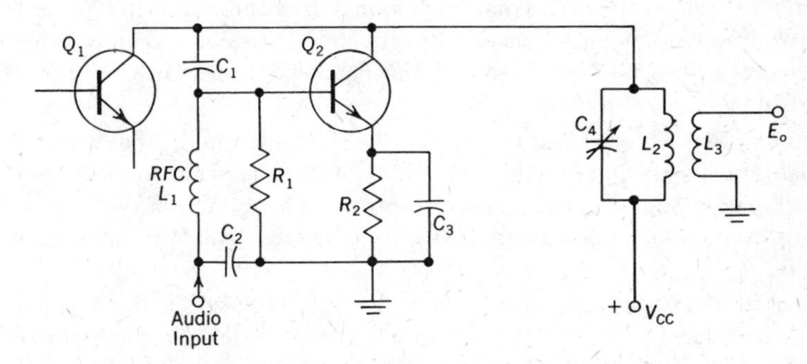

Figure 9-31 A reactance transistor phase-shift modulator.

to change the phase relations as the tank circuit is detuned from resonance. Such a circuit is shown in Fig. 9–31, using a high-transconductance transistor as a capacitive reactance. Q_1 is an RF amplifier (preferably the first stage following the crystal oscillator). Its collector tank circuit C_4L_2 is resonant to the crystal oscillator frequency, in the absence of a modulating signal. When an audio signal is applied, the effective capacitance of the reactance transistor Q_2 changes, detuning the tank circuit above and below the frequency of the incoming signal. This causes the collector tank current to alternately lag and lead the no-modulation phasing. This would normally produce phase modulation. However, by using a "frequency correcting" ($1/f$) network before feeding the modulating signal to the reactance transistor, an FM output is obtained.

Indirect $\Delta F = \Delta\phi F_s$

REVIEW QUESTIONS

1. In a frequency-modulated wave (*a*) what property of the wave is changed by the modulation? (*b*) What is the normal (unmodulated) value of this property called? (*c*) What symbol is used to denote the amount of this change? (*d*) What determines the amount of this change? (*e*) What determines the rate of change of this quantity?

2. Refer to Fig. 9–1. (*a*) What does E_{s1} represent? (*b*) What does E_{s2} represent? (*c*) What does the waveshape in diagram (b) represent? (*d*) What does the waveshape in diagram (c) represent?

3. Refer to Fig. 9–1(b). (*a*) How do RF cycles 1, 2, and 3 compare in frequency? (*b*) What frequency value is this? (*c*) Why is this so? (*d*) Which RF cycle has the highest frequency? (*e*) Why is this so? (*f*) Which RF cycle has the lowest frequency? (*g*) How should the frequency shift of this RF cycle compare with the frequency shift of the highest frequency cycle? Why? (*h*) Which RF cycles, beyond cycle 3, are at the resting frequency? (*i*) Why is this so?

4. Refer to Fig. 9–1(c). (*a*) What modulating signal produces this waveshape? (*b*) How does the amount of change in frequency in this wave compare with the waveshape (b)? (*c*) Why is this so? (*d*) In what respect does this wave differ from waveshape (b)? (*e*) Why is this so?

5. Refer to Fig. 9–2. (*a*) Name two specific oscillator circuits that would *not* be suitable in this application. Why is this so? (*b*) What happens to the output *voltage* from the oscillator when the sound pressure increases? Explain. (*c*) What happens to the *frequency* of the output wave when the sound pressure increases? Explain.

6. Refer to the waveshapes in Fig. 9–2. (*a*) Why is the frequency deviation in curve ② less than in curve ①? (*b*) Which sound curve is responsible for frequency deviation curve ③? (*c*) Why is the deviation in this curve the same as for ①? (*d*) In what respect does this curve differ from curve ①? (*d*) Why is this so?

7. When an FM sound transmission is picked up and demodulated in an FM receiver, what characteristic of the FM wave determines (*a*) the volume of the audio ouptut? (*b*) the pitch of the audio output?

8. In an *amplitude*-modulated wave (*a*) what term is used to denote the level of modulation? (*b*) What is the maximum limit for this level? (*c*) Over what range does the amplitude of the modulated wave vary?

9. (*a*) A 100-MHz RF carrier is to be frequency-modulated. If a maximum modulation limit as in Question 8 were used, how would this affect the modulated output wave? (*b*) Why is this not desirable?

10. (*a*) What term is used to express the modulation level of an FM wave? (*b*) Give the equation for this term.

11. An RF carrier is amplitude-modulated by a pure sine wave. (*a*) From theoretical aspects, how many components does the output wave have? (*b*) What are these components? (*c*) In terms of carrier and modulating-signal frequencies, what is the frequency of each component? (*d*) How are these relations affected in a practical situation?

12. Repeat Question 11(*a*), (*b*), (*c*), and (*d*) for frequency modulation.

13. (*a*) What is meant by the term "significant side frequencies"? (*b*) What determines the number of significant side frequencies? (*c*) What additional information is needed to evaluate the number of significant side frequencies in a given FM wave? (*d*) What other data can be obtained from this source?

14. Refer to Fig. 9–3. (*a*) What does $n = 0$ refer to? (*b*) What does $n = 3$ refer to? (*c*) For a modulation index of 3, find the amplitude of the carrier, the first-order pair, and the fifth-order pair. (*d*) For a modulation index of 7, find the amplitude of the carrier, the second-order pair, and the fifth-order pair.

15. Refer to Table 9–1. (*a*) At a modulation index of 2.0, what is the amplitude of the sixth-order pair? Explain. (*b*) For a modulation index of 3, find the amplitude of the carrier, the first-order pair, and the fifth-order pair. (*c*) Compare your results with those of Question 14. Which technique is easier to use? (*d*) What is an advantage of using the curves in Fig. 9–3? (*e*) Which side-order pair is the *dominant* component (highest amplitude) at modulation indexes of 2.0, 3.0, 5.0, 7.0, and 12.0? (*f*) What is the effect of modulation index on the deviation of the dominant component?

16. As the modulation level is increased, what happens to (*a*) the carrier power in an AM wave? (*b*) the carrier power in an FM wave? (*c*) the sideband power in an AM wave? (*d*) the sideband power in an FM wave? (*e*) the total power in an AM wave? (*f*) the total power in an FM wave?

17. Explain why the total power in an FM wave cannot increase with increase in the modulation level.

18. (*a*) For a given modulating-signal frequency, what determines the deviation of an FM wave? (*b*) Is there any limitation imposed on this deviation? Explain. (*c*) What or who determines this limit? (*d*) What is the maximum deviation allowed in broadcast transmissions? (*e*) What is the deviation ratio for broadcast transmitters? (*f*) How is this value obtained? (*g*) Do all FM transmitters have the same maximum allowable deviation? Explain.

19. Refer to Fig. 9–4. (*a*) In diagram (a), for the modulation index shown, why is ΔF only $0.5f_s$? (*b*) In diagram (c), how many significant side-order pairs are shown? (*c*) How does this compare with the data in Table 9–1? (*d*) In diagram (d), if the intelligence frequency is 6.0 kHz, what is the deviation? (*e*) Are there any significant side frequencies beyond this deviation value? (*f*) What total spectral bandwidth does this signal occupy? (*g*) Does this signal exceed the FCC bandwidth limitations for broadcast FM? Explain.

20. Refer to Fig. 9–5. (*a*) What do the white areas in this spectral diagram signify? (*b*) What do the hatched areas signify? (*c*) Why are these hatched areas necessary? (*d*) What is the total bandwidth for a broadcast transmitter? (*e*) Does a receiver need this full bandwidth? Explain. (*f*) What is the minimum frequency separation between two carriers?

21. If a carrier is phase-modulated (*a*) what characteristic of the carrier is varied? (*b*) What determines the magnitude of this variation? (*c*) What determines the rate of this variation?

22. Refer to Fig. 9–6. (*a*) What does the waveshape in diagram (a) represent? (*b*) What do the solid-line waveshapes in diagrams(b), (c), and (d) represent? (*c*) What do the dashed-line waveshapes represent? (*d*) At what time instant(s) is waveshape (b) obtained? Explain. (*e*) At what time instant is the solid waveshape in (c) obtained? Explain. (*f*) What is the difference between the solid waveshapes in (c) and (d)? Why is this so? (*g*) How does the amount of phase shift in (d) and (e) compare? Why is this so? (*h*) How do these two phase shifts differ? Explain.

23. (*a*) In order to change the starting phase angle of a carrier (with respect to its unmodulated phasing), what other characteristic of the wave must also vary? (*b*) Explain. (*c*) What is this effect called?

24. (*a*) Does the amplitude of the modulating signal have any effect on the amount of indirect FM (ΔF) produced? Explain. (*b*) Does the frequency of the intelligence signal have any effect on the amount of indirect FM produced? Explain. (*c*) Give an equation to show the combined relationships.

25. In the solution to Example 9–3 (page 328), where do the "π" and "180" come from?

26. When two signals of different frequencies are mixed (*a*) describe the effect (if any) on the amplitude of the resultant. (*b*) Describe the effect (if any) on the phase of the resultant with respect to either original signal. (*c*) What is meant by the term "beat frequency"?

27. (*a*) In an FM system, can the amplitude-modulation effects caused by an interfering signal be eliminated *completely*? (*b*) If so, where would it be done? (*c*) State two circuit features used.

28. (*a*) In an FM system, can the indirect FM effects caused by an interfering signal be eliminated *completely*? Explain. (*b*) What is responsible for the reduction of this type of noise or interference? (*c*) Which FM system would have a lower noise level—broadcast FM sound or the FM sound from a TV channel? Explain.

29. Refer to Fig. 9–7. (*a*) What does phasor A represent? (*b*) What does phasor B represent? (*c*) How do their amplitudes compare? (*d*) Why are they shown at a different angular displacement in each of the four diagrams? (*e*) What is the minimum value that angle ϕ can have? (*f*) When would this occur? (*g*) Under what condition(s) would this angle be a maximum? (*h*) For the amplitudes shown, what is the maximum value of this angle in degrees? in radians?

30. Considering the indirect FM products resulting from noise or other interference, which frequency difference (between desired signal and interference) would create the most annoyance from an FM radio: (*a*) 2000 Hz or 10,000 Hz? Why? (*b*) 10,000 Hz or 50,000 Hz? Why? (*c*) What beat frequency would represent the worst possible case?

31. (*a*) In an FM system, where is pre-emphasis used? (*b*) What does this do? (*c*) Why is it applied only to these frequencies? (*d*) Why is it used? (*e*) Describe a simple circuit that would accomplish this effect? (*f*) What other circuit is necessary to complete the action? (*g*) Where is this second circuit located? (*h*) What would be the result if this second circuit were omitted? (*i*) Describe a simple circuit to accomplish the desired action. (*j*) Are the circuit component values in any way critical? Explain.

32. (*a*) What frequency range is used for commercial FM broadcasting? (*b*) Give three frequency ranges wherein the FM sound for television channels is found. (*c*) Specify any other frequency bands used for FM communications.

33. (*a*) What type of receiver is used for the reception of FM signals? (*b*) Name the various sections (blocks) usually found in such a receiver. (*c*) Which of these blocks are involved in the selection of a specific FM signal?

34. (*a*) Describe briefly the type of tank circuit used to cover the 540–1600 kHz AM broadcast band. (*b*) Is this type of tank circuit used for FM bands? (*c*) What is the disadvantage of this type of tuner at FM frequencies? (*d*) Name two types of variable-inductance tuners. (*e*) Why should this type of tuner give better gain? (*f*) Name another type of tuner that is suitable at higher frequencies.

35. (*a*) State three advantages of using an RF stage. (*b*) Are RF stages always an advantage? Explain.

36. What type of active device would be preferable as an RF amplifier: (*a*) Below 100 MHz? Why? (*b*) Above 100 MHz? Why? (*c*) Name two circuits generally used with IC RF stages. (*d*) Why are these preferable to "conventional" circuitry?

37. (*a*) What is a broadband input circuit? (*b*) How does the gain of such a circuit compare with a conventional tank? (*c*) When is this type of input circuit practical? (*d*) How is selectivity obtained in such cases?

38. In bipolar transistor RF amplifiers, (*a*) are broadband inputs practical? Explain. (*b*) Are common-base circuits used? Explain.

39. (*a*) What types of mixers are preferable at the lower FM bands? (*b*) at UHF bands? (*c*) Name a converter circuit that is used at broadcast FM frequencies.

40. For broadcast FM service, what bandwidth is needed for (*a*) the RF amplifier? (*b*) the mixer? (*c*) the oscillator? (*d*) the IF system? (*e*) What is the standard intermediate frequency?

41. Refer to Fig. 9–10. (*a*) What is the effect of the 455-kHz tank circuits if the incoming signal frequency is 10.7 MHz? (*b*) Why is this so? (*c*) What is the impedance of the 10.7-MHz tanks at 455 kHz? Why?

42. (*a*) What is the function of a limiter stage? (*b*) Why may this function be needed? (*c*) Basically, how does a limiter accomplish this function?

43. (*a*) State a reason for using low supply voltages in limiter circuits. (*b*) How is gate limiting accomplished? (*c*) What is the advantage of using gate-leak bias? (*d*) What type of FET is preferable for limiter service?

44. Refer to Fig. 9–11. (*a*) What do the dotted current-waveform peaks represent? (*b*) Are the amplitude variations of this input signal completely eliminated? Give a specific condition. (*c*) What input signal requirement is necessary for successful limiter action? (*d*) Is the current waveshape a sine wave? Explain. (*e*) How can this distortion be eliminated from the output voltage?

45. Refer to Fig. 9–12. (*a*) To what frequency is the tank circuit C_1L_1 tuned? C_2L_2? C_5L_3? (*b*) What type of bias does this stage have? (*c*) What components produce this bias? (*d*) How would you convert this to a series bias circuit? (*e*) What is the function of R_2? (*f*) Why is this function necessary?

46. (*a*) What value of time constant is often used in gate-leak bias circuits for the FM limiter stage? (*b*) What is the advantage of a shorter time constant? (*c*) What is the advantage of a longer time constant? (*d*) How can both these advantages be realized? (*e*) Give another advantage of cascade dual-limiters.

47. (*a*) How does a bipolar transistor limiter stage differ from a conventional transistor IF amplifier? (*b*) Essentially, what determines whether the stage acts as a limiter or normal amplifier? (*c*) Why doesn't limiter action in the last IF stage of an AM receiver remove the intelligence from the IF wave? (*d*) Why are resistors often used in the collector lead of transistor limiters? (*e*) Are such resistors used in transitor IF amplifier stages? Explain.

48. (*a*) Name two types of demodulator circuits commonly used in FM receivers. (*b*) Which of these must be preceded by a limiter stage? Why?

49. Refer to Fig. 9–13. (*a*) To what frequency is the tank circuit C_2L_1 tuned? (*b*) To what frequency is C_4L_2 tuned? (*c*) What bandwidth is necessary for the primary tank? (*d*) What RF voltage is developed across choke coil L_3? (*e*) Explain why this is so. (*f*) What RF voltage(s) is (are) effective in making D_1 conduct? (*g*) What RF voltages cause D_2 to conduct? (*h*) Explain how the polarity shown across C_6 is obtained. (*i*) If the diodes conduct equally, what is the polarity and magnitude of voltage E_o? (*j*) What is the function of components R_4C_7?

50. Refer to Fig. 9–14(a). (*a*) To what condition do these phase relations apply? (*b*) What is E_s? (*c*) Why is it shown 180° from E_p? (*d*) What is I_s, and why is it in phase with E_s? (*e*) What causes voltage E_a? (*f*) Explain its phase relation to any of the previously mentioned quantities. (*g*) Why are E_a and E_b 180° apart? (*h*) What relation does the sum of E_a and E_b have to E_s? (*i*) What is the output voltage for this condition? Explain.

51. Refer to Fig. 9–14(b). (*a*) To what condition do these phase relations apply? (*b*) Why are E_p and E_s not affected by the change in frequency of the input signal? (*c*) Why does I_s now lag E_s? (*d*) Why are phasors E_a and E_b shifted as compared to diagram (a)? (*e*) What is the polarity of the output voltage for this condition? Explain.

52. Refer to Fig. 9–14. (*a*) Under what condition can current I_s be made to lead E_s? (*b*) Why is this so? (*c*) How would this affect the output voltage? Explain.

53. Refer to Fig. 9–15. (*a*) What name is colloquially given to this response curve? (*b*) Over what limits (swing) should the demodulator action extend? Explain. (*c*) To what circuit characteristic do the maximum amplitude points (*C* and *D*) correspond? (*d*) For commercial broadcasts (FM), why should this bandwidth be greater than 200 kHz? (*e*) Why does the output voltage fall off for frequency deviations such as *E* and *F*?

54. How does a ratio detector circuit differ from a Foster-Seeley discriminator with respect to (*a*) the diode connections? (*b*) the output connections? (*c*) Is there any other major difference?

55. (*a*) What determines the relative conduction of the diodes in a ratio detector? (*b*) How does this compare with the action in a discriminator circuit? Why?

56. Refer to Fig. 9–16. (*a*) To what value will capacitor C_9 charge? (*b*) Explain how it charges to this value with the polarity shown. (*c*) What is the effect of the signal-frequency swing on this voltage? (*d*) What is the effect of rapid fluctuations in the signal amplitude on this voltage? Explain. (*e*) At what frequency are the voltages across C_5 and C_6 equal? Why? (*f*) What happens to these voltages as the signal frequency swings? Why? (*g*) Compare the voltages across R_2 and R_3 at the resting frequency. (*h*) What happens to these voltages as the signal frequency swings? Explain. (*i*) Compare the potential of points *P* and *Q* to the common point *R*, at the resting frequency. Explain. (*j*) What is the output voltage at this frequency? (*k*) Why does the output voltage change with carrier frequency swing? (*l*) Why isn't the output voltage affected by fluctuations in the signal amplitude?

57. Name two other types of FM demodulator circuits.

58. (*a*) What does the term "AFC circuit" mean? (*b*) To what stage is this control applied? (*c*) Why is such circuitry desirable in FM receivers as compared to AM broadcast-band receivers? (*d*) Describe in a general way how AFC circuits operate. (*e*) What is a suitable source for the error voltage fed to the AFC circuit? (*f*) Why is this so?

59. Refer to Fig. 9–18. (*a*) Does this circuit act as a variable capacitor or inductor? (*b*) What components determine this action? (*c*) What relation exists between these components values? (*d*) What does current I_1 represent? (*e*) In the phasor diagram, why is this current leading E_{osc} by 90°? (*f*) Why is E_g in phase with I_1? (*g*) What does current I_d represent? (*h*) Why does it lead E_{osc} by 90°? (*i*) Why is the FET circuit considered to act as a capacitor? (*j*) What is applied to this circuit from the terminal marked To Demodulator Output? (*k*) What happens to I_d and the circuit effective capacitance if this demodulator voltage is positive? (*l*) Explain how this can result in AFC action.

60. (*a*) Can a reactance stage be made to act as an inductor? (*b*) What component values must be changed?

61. (*a*) Can transistors be used for AFC circuits? (*b*) Explain briefly one technique for doing this.

62. Refer to Fig. 9–19. (*a*) What is the control element in this circuit? (*b*) What type of diode is used for D_1? (*c*) Why is a special diode used? (*d*) What components determine the operating point bias for this diode? (*e*) Is this a forward or reverse bias? Explain. (*f*) What other factor determines the bias value? (*g*) What is the effect of a negative error-signal from the demodulator? (*h*) What is the function of components R_3, R_4, C_3, C_4? (*i*) What would be the effect if these components were omitted? (*j*) Would such a filter also be needed in the reactance-FET circuit of Fig. 9–18?

63. Refer to Fig. 9–20. (*a*) What type of active device is used in the RF stage? (*b*) How is tuning over the 88–108 MHz range accomplished? (*c*) How many tuned circuits are involved? (*d*) What is the function of diode E_1 in the tuner area? (*e*) From where does it get its "error" voltage? (*f*) What kind of interstage coupling is used in the IF strip? (*g*) What kind of active device does this strip use? (*h*) Which type of demodulator does this tuner use? (*i*) State two circuit features that justify your answer.

64. What equipment is needed to align an FM receiver (*a*) by the steady-frequency method? (*b*) by the sweep-frequency method?

65. What general precautions should be observed when connecting (*a*) signal generators? (*b*) indicators?

66. What is the "proper" signal level to use in aligning FM receivers?

67. In aligning the IF system of the FM receiver of Fig. 9–20 by the steady-frequency method: (*a*) Where would you connect the signal generator? (*b*) Where would you connect the indicator? (*c*) What adjustments would you check? (*d*) For what indication?

68. If the IF transformers of the receiver in Fig. 9–20 were overcoupled, what change is made (*a*) in the indicator connecting point? (*b*) in the generator connecting point? (*c*) in the alignment procedure?

69. In aligning the demodulator of the receiver in Fig. 9–20 (steady-frequency method): (*a*) Where is the signal generator connected? (*b*) What adjustments should be checked? (*c*) Where is the indicator connected for each adjustment? (*d*) How is each adjustment made?

70. For sweep alignment of the discriminator of Fig. 9–20: (*a*) Where are the signal generators connected? (*b*) Where is the oscilloscope vertical input connected? (*c*) Where is the oscilloscope horizontal input connected?

71. For visual alignment of the FM IF system of the receiver in Fig. 9–20: (*a*) Where are the signal generators connected? (*b*) Where is the oscilloscope vertical input connected? (*c*) What circuit change must be made?

72. In aligning the front end of an FM receiver, which of the two basic techniques is preferable? Why?

73. The oscillator of the receiver in Fig. 9–20 is to be aligned using the steady-frequency method. (*a*) Where is the signal generator connected? (*b*. Where is the indicator connected? (*c*) What indication corresponds to correct alignment? (*d*) Is it necessary to align the IF first? Explain. (*e*) Is it necessary to align the demodulator first? Explain. (*f*) Is it necessary to defeat the AFC action? Explain. (*g*) Specifically, how can this be readily done? (*h*) Is C_{18} adjustable? (*i*) At what frequency would you set this adjustment?

74. In aligning the RF and mixer stages of the receiver in Fig. 9–20 (steady-frequency method): (*a*) Where is the indicator connected? (*b*) What indication corresponds to correct alignment? (*c*) At the high-frequency end of the tuning range, what circuit components would be adjusted to improve alignment?

75. (*a*) Name two general classifications of FM transmitters based on the method of producing the deviation. (*b*) Explain (briefly) the difference between these two methods.

76. Explain (briefly) the basic principle used in any direct modulation method.

77. Refer to Fig. 9–18. (*a*) What was the basic purpose of this circuit? (*b*) Can a reactance stage also be used to *vary* the oscillator frequency? Explain. (*c*) What changes (if any) are needed to use this circuit for frequency modulation?

78. (*a*) What characteristic of a transistor can be used as another means of obtaining direct frequency modulation? (*b*) Explain. (*c*) How does this method compare with a reactance stage with regard to the amount of deviation obtainable?

79. Refer to Fig. 9–21. (*a*) What type of master oscillator would *not* be desirable in this application? Explain. (*b*) Does this circuit act as a capacitive or inductive reactance? (*c*) Which components determine this? (*d*) How are their values related for this action? (*e*) What other function does resistor R_1 have? (*f*) Why must the impedance X_{C2} be low compared to $R_2 + R_5$?

80. Refer to Fig. 9–22. (*a*) How (if at all) does this circuit differ from Fig. 9–21 in *principle*? (*b*) What determines the resting frequency of the master oscillator?

81. In Fig. 9–22, as the modulating signal swings positive, explain what happens to (*a*) the emitter-base bias, (*b*) the collector current, and (*c*) the master oscillator frequency.

82. Refer to Fig. 9–23. (*a*) What type of semiconductor diode is this? (*b*) Is the polarity of the supply voltage of any importance? Explain. (*c*) If, for some reason, the negative supply line must be grounded, and the primary of T_1 were connected to a positive polarity, can the circuit be made to work properly? Explain. (*d*) What happens in the given circuit if the modulating signal produces a positive polarity at the top of winding L_3? Explain.

83. Refer to the upper half of Fig. 9–24. (*a*) What name is generally given to the stages contained in the first four blocks of this diagram? (*b*) How much deviation is obtained from the reactance stage modulator? (*c*) Why is frequency multiplication used? (*d*) Give an example of how a multiplication of 18 may be obtained? (*e*) If a final frequency of 98 MHz were desired, how would this be obtained? (*f*) How many stages are used in the Intermediate Power Amplifier block? (*g*) What classification of circuitry would be employed in the last three blocks?

84. Refer to the lower half of Fig. 9–24. (*a*) What is the function of this group of blocks? (*b*) Why is this feature necessary? (*c*) Could this need be eliminated by using crystal control in the master oscillator? Explain. (*d*) To what frequency is the discriminator circuit tuned? (*e*) What is the output of the discriminator for a transmitter frequency of 90 MHz? (*f*) Explain how correction is obtained if the transmitter frequency drifts to a lower value.

85. When the transmitter in Fig. 9–24 is fully modulated: (*a*) Over what frequency range does the final output swing? (*b*) What is the output at the mixer? (*c*) What is the output at the discriminator? (*d*) At what rate does this output vary? (*e*) What would this cause if the discriminator output were fed directly to the reactance stage (*f*) How does the long time-constant circuit prevent this?

86. Refer to Fig. 9–25. (*a*) What is this circuit used for? (*b*) What range of frequency values are generally fed to this circuit? (*c*) How are these low frequencies obtained? (*d*) When the transmitter is "on-frequency," what is the output of this circuit? (*e*) How is this output obtained? (*f*) What is the effect of transmitter drift on the output of this circuit? (*g*) What is the purpose of R_3 and C_2?

87. State three conditions necessary to produce FM from an AM wave.

88. Refer to Fig. 9–26(a). (*a*) What do these three phasors represent? (*b*) How do these phase relations compare with an AM wave? (*c*) Why is E_u rotating counterclockwise and E_l clockwise?

89. In Fig. 9–26(b): (*a*) Through what angle (approximately) has phasor E_u rotated? (*b*) Through what angle has E_l rotated? (*c*) Why are these angles equal in value? (*d*) What is E_{sf}? (*e*) Why is it at 90° to E_c? (*f*) At some instant later, as E_u and E_l rotate further, how will the phase of E_{sf} change? (*g*) What will change? (*h*) What does $\Delta\phi$ represent?

90. In Fig. 9–26(c): (*a*) Through what angle has E_u rotated? (*b*) Through what angle has E_l rotated? (*c*) What is the significance of the subscript *m* on the angle $\Delta\phi_m$?

91. As the side-frequency phasors in Fig. 9–26 rotate from 90° to 180°, what happens to (*a*) the magnitude of E_{sf}? Explain. (*b*) the phase angle $\Delta\phi$? Explain.

92. (*a*) In Fig. 9–26, in addition to the change in phase ($\Delta\phi$), what other change occurs in the modulated wave? (*b*) How can this second change be reduced to negligible values?

93. (*a*) Give the equation for the amount of deviation (ΦF_{ind}) obtained from a PM wave. (*b*) What factor(s) of the modulating signal determine(s) the amount of swing obtained by this indirect method? (*c*) Is this true FM? Explain. (*d*) How can the amount of deviation be made independent of the frequency of the modulating signal?

94. (*a*) To what broad classification does the Crosby transmitter belong? (*b*) To which classification does the Armstrong transmitter belong? (*c*) Does the Armstrong transmitter require AFC circuitry? Explain. (*d*) How much frequency deviation is generally obtainable from the Armstrong system (before multiplication)?

95. Refer to Fig. 9–27. (*a*) Why is the block "Frequency Correcting Network" used? (*b*) If the audio signal entering the "Balanced Modulator" has a frequency of 8 kHz, what frequencies are present in the output of this block? (*c*) What frequencies are present at the output of the "90° Phase-Shift Network" block? (*d*) Still using 8 kHz as the audio signal, what frequencies are present at the output of the "Combining Network"? (*e*) Give an example of how a frequency multiplication of 72 can be obtained. (*f*) Why is a mixer-oscillator combination used in this transmitter? (*g*) What frequency and frequency swing is produced at the output of the "Mixer" block? (*h*) It is desired to change the final transmitter frequency to 98 MHz. What change would be made to either frequency multiplier block?

Explain. (*i*) How could this output frequency be obtained?

96. In Fig. 9–27, instead of using frequency conversion to reduce the carrier frequency, why isn't a lower crystal #1 oscillator frequency used in the first place?

97. Refer to Fig. 9–28. (*a*) If the RF carrier is 220 kHz and the audio signal is 5 kHz, state all the frequencies present in the output E_o. (*b*) Why isn't the audio frequency present in the output? (*c*) Why isn't the carrier frequency present in the output? (*d*) What value of bias (qualitatively) does this circuit use? Why?

98. Why can the amount of indirect FM obtainable from a phase-modulation circuit be increased if the lowest modulation frequency is increased?

99. Refer to Fig. 9–29. (*a*) State two ways by which an RF voltage is developed across L_1. (*b*) By what symbols are these two components represented in the phasor diagram? (*c*) What is the relative magnitude of voltages E_g and AE_g before modulation is applied? (*d*) How is this proportion obtained? (*e*) Which phasor represents the output voltage under a no-modulation condition? (*f*) How is this value obtained? (*g*) Explain the effect of a positive modulating signal on the AE_g value and on the output voltage. (*h*) Is capacitor C_2 always necessary? Explain. (*i*) What is the function of R_1 and C_3?

100. Refer to Fig. 9–30. (*a*) What is the function of the Q_1 stage? (*b*) What type of oscillator circuit is this? (*c*) Why is the tank circuit on the collector side? (*d*) What is the function of the Q_2 stage? (*e*) What is the function of the Q_3 stage?

101. Refer to Fig. 9–31. (*a*) What is the function of the Q_1 stage? (*b*) Is its collector circuit tuned? If so, what determines the frequency to which it is tuned? (*c*) What is the function of the Q_2 stage? (*d*) Does it act as an inductive or capacitive reactance? (*e*) What components determine this effect?

102. In Fig. 9–31, as the audio-modulating signal swings negatively, what happens to (*a*) the reactance effect? (*b*) the resonant frequency of the tank circuit? (*c*) the phasing of the tank current (compared to its no-modulation phase)? (*d*) the output voltage? (*e*) Is this true FM? If so, explain; if not, how can it be made so?

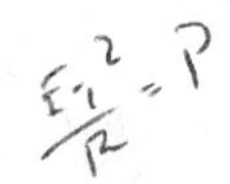

PROBLEMS AND DIAGRAMS

1. An FM transmitter has a frequency swing of ±60 kHz. Find the deviation (ΔF) if the amplitude of the modulating signal is cut in half.

2. An FM broadcast transmitter has a deviation of (±) 20 kHz. What will the deviation be if (*a*) the audio signal amplitude is tripled? (*b*) the audio signal power is doubled? (*c*) the audio signal frequency is halved?

3. Find the modulation index for an FM wave if the frequency swing is ±20 kHz and the modulating signal frequency is 4 kHz.

4. An FM transmitter has a resting frequency of 96 MHz. The carrier amplitude is 40 V. When modulated with an 8-kHz sine-wave signal of 10-V amplitude, the frequency deviation is 40 kHz. Find the modulation index.

5. In Problem 4, find the modulation index when (*a*) the carrier amplitude is raised to 60 V, (*b*) the carrier frequency is raised to 120 MHz, (*c*) the modulating signal frequency is raised to 12 kHz, (*d*) the modulating signal amplitude is raised to 20 V, (*e*) both the (*c*) and (*d*) changes are made.

6. An FM transmitter develops an unmodulated carrier power of 10 W. At a certain modulation level, spectrum analysis shows 4 W of power in the sideband components. (*a*) What is the total power in the modulated wave? (*b*) What is the carrier at this modulation level?

7. An FM wave has a deviation of 60 kHz. The modulating signal producing this deviation is a 12-V, 5-kHz sine wave. Find (*a*) the modulation index, (*b*) the number of significant side frequencies, and (*c*) the total bandwidth required for this signal.

8. Repeat (*a*), (*b*), and (*c*) above for the FM wave of Problem 3.

9. Repeat (*a*), (*b*), and (*c*) above for the FM wave of Problem 4.

10. If the transmitter in Problem 3 is a broadcast FM transmitter, what percent modulation might this be called?

11. What percent modulation would apply to a broadcast transmitter as in Problem 4?

12. Find the amount of indirect FM produced when a carrier is shifted 30° by a 10-V, 5-kHz sine-wave modulating signal.

13. A broadcast FM signal (E_1 = 90-V amplitude at 100 MHz) mixes with an interfering signal (E_2 = 30-V amplitude at 100.012 MHz). (*a*) Over what limits will the resultant amplitude vary? (*b*) What is the percent amplitude modulation in this resultant wave? (*c*) What is the beat frequency? (*d*) Over what limits will the phase of the resultant vary? (*e*) How much indirect FM will this produce? (*f*) Compare the sound level of this interference product against the volume from a fully modulated desired signal.

14. If the signal-to-noise ratio at the receiving location is 4:1, how loud will the noise sound, compared to the desired signal, at the output of (*a*) an FM broadcast receiver? (*b*) an FM communications receiver? Use a beat frequency of 3 kHz for this comparison.

15. Draw the circuit diagram of a FET RF amplifier suitable for use in FM broadcast receivers.

16. Draw the circuit diagram of a bipolar transistor RF amplifier suitable for use in FM broadcast receivers.

17. Draw the circuit diagram for a frequency converter suitable for FM broadcast receivers, using a MOSFET mixer and a bipolar transistor oscillator.

18. Repeat Problem 17, using two bipolar transistors.

19. Repeat Problem 17 for a one-transistor circuit.

20. Draw the circuit diagram for a two-stage transistor IF amplifier suitable for FM broadcast receivers.

21. Repeat Problem 20, using ICs.

22. Draw the circuit diagram for a FET limiter circuit suitable for FM broadcast receivers.

23. Repeat Problem 22, using a bipolar transistor.

24. In Fig. 9–12, if capacitor C_3 is 40 pF, what value of R_1 would be needed for a time constant of 2.5 μs?

25. In Fig. 9–12, if V_{DD} is 15 V and I_D is 4.75 mA, what value of resistor R_2 is needed to produce a drain potential of 7.5 V?

26. Draw the circuit diagram for a discriminator.

27. Draw the circuit diagram of a ratio detector, and include a de-emphasis circuit.

28. Draw a phasor diagram showing the RF voltage and current phase relations that exist in the ratio detector circuit of Fig. 9–14, for an incoming signal frequency slightly below the resting frequency.

29. Draw the circuit diagram of a reactance-FET modulator that acts as an inductive reactance. Illustrate this effect by a phasor diagram of the circuit current and voltage relations.

30. Draw the circuit diagram for a semiconductor AFC circuit.

31. Draw the circuit diagram of a solid-state FM tuner with the following features: an autodyne converter, two IF stages, a ratio detector, and AFC.

32. Draw the circuit diagram for a transistor FM tuner incorporating an RF stage, a mixer-oscillator, two IF stages, a limiter, a discriminator, and AFC.

33. Draw the circuit diagram for a reactance-stage modulator using a FET.

34. Draw the schematic for a direct-modulation circuit using the input capacitance of a FET as the control element.

35. Draw the circuit diagram for a PNP-transistor reactance modulator.

36. Draw the schematic for a transistor direct-modulation circuit using the change in interelement capacitance for modulation.

37. Draw the circuit diagram for a VVC diode modulator using a negatively grounded supply source.

38. Draw the block diagram for Crosby FM transmitter. Show the direction of energy travel between blocks. Specify all frequencies and multiplication values for an output frequency of 103 MHz.

39. In Fig. 9–26(c), the carrier has an amplitude of 10 V and each side frequency is 0.7 V. Find (*a*) the value of the phase-shift angle $\Delta\phi$, (*b*) the magnitude of the resultant, and (*c*) the percent change in magnitude compared to the unmodulated carrier.

40. Draw the phase relations for the carrier and sidebands of Fig. 9–26 corresponding to a 225° rotation of the upper side frequency from its starting position in diagram (a). Using 10 V for the carrier and 3 V for each side frequency, calculate the amplitude of the resultant and the phase-shift angle.

41. A phase-modulation system produces an *indirect* FM of 200 Hz from an audio modulating signal of 10 V at 100 Hz. (*a*) How much frequency deviation would a 10-V, 20-Hz audio signal produce? (*b*) How much deviation would a signal of 10 V at 2000 Hz produce? (*c*) How much reduction in the amplitude of this last signal is needed to make its deviation equal to the original condition?

(*d*) How much reduction is needed to equalize the deviation in case (*b*) to case (*a*)? (*e*) How does the frequency range of the modulating signal affect the deviation obtained from a "corrected" (true) FM indirect system?

42. Draw the schematic diagram for a circuit that can produce a 90° phase shift. Calculate the component values needed to obtain a 90° shift for a signal frequency of 200 kHz.

43. Draw the circuit diagram for a balanced modulator using FETs.

44. Draw the circuit diagram for a balanced modulator using bipolar transistors.

45. Draw the circuit diagram for a balanced modulator using FETs, but with the RF signal *R-C* coupled and shunt fed to the modulators.

46. Draw the circuit diagram for a FET Link phase modulator.

47. Repeat Problem 46 using bipolar transistors.

48. Draw a circuit diagram showing a reactance FET used for phase-shift modulation.

49. Repeat Problem 48 using bipolar transistors.

TYPICAL FCC QUESTIONS

3.117 What is the meaning of the term "plate saturation"?

3.247 What types of radio receivers do not respond to static interference?

3.489 What are the merits of a frequency-modulation communication system compared to an amplitude-modulation communication system?

3.492 Draw a block diagram of a frequency-modulation receiver and explain its principle of operation.

3.493 Draw a block diagram of a frequency-modulated transmitter and indicate the center frequency of the master oscillator and the center frequency radiated by the antenna.

3.494 In a frequency-modulation radio communication system, what is the meaning of modulation index? of deviation ratio? What values of deviation ratio are used in a frequency-modulation radio communication system?

3.495 Why is narrow-band frequency modulation rather than wideband frequency modulation used in radio communication systems?

4.046 Why is frequency modulation undesirable in the standard broadcast band?

4.156 Define: amplifier gain, percentage deviation, stage amplification, and percentage of modulation. Explain how each is determined.

4.217 What is the purpose of a discriminator in an FM broadcast receiver?

4.218 Explain why high-gain antennas are used at FM broadcast stations.

4.219 What is the frequency swing of an FM broadcast transmitter when modulated 60%?

4.220 An FM broadcast transmitter is modulated 50% by a 7000-hertz test tone. When the frequency of the test tone is changed to 5000 hertz and the percentage of modulation is unchanged, what is the transmitter frequency swing?

4.221 What is a common method of obtaining frequency modulation in an FM broadcast transmitter?

4.222 What is meant by pre-emphasis in an FM broadcast transmitter?

4.223 What is the purpose of a de-emphasis circuit in an FM broadcast receiver?

4.224 An FM broadcast transmitter operating on 98.1 megahertz has a reactance-tube-modulated oscillator operating on a frequency of 4905 kilohertz. What is the oscillator frequency swing when the transmitter is modulated 100% by a 2500-hertz tone?

4.225 What characteristic of an audio tone determines the percentage of modulation of an FM broadcast transmitter?

4.226 What determines the rate of frequency swing of an FM broadcast transmitter?

4.227 How wide a frequency band must the intermediate-frequency amplifier of an FM broadcast receiver pass?

4.228 An FM broadcast transmitter is modulated 40% by a 5000-hertz test tone. When the percentage of modulation is doubled, what is the frequency swing of the transmitter?

4.229 If an FM transmitter employs one doubler, one tripler, and one quadrupler, what is the carrier frequency swing when the oscillator frequency swing is 2 kilohertz?

4.230 What is the purpose of a "reactance tube" in an FM broadcast transmitter?

4.231 What is a ratio detector?

4.232 How does the amount of audio power required to modulate a 1000-watt FM broadcast transmitter compare with the amount of audio power required to modulate a 1000-watt standard broadcast transmitter to the same percentage of modulation?

4.233 What is the purpose of a limiter stage in an FM broadcast receiver?

4.234 If the transmission line current of an FM broadcast transmitter is 8.5 amperes without modulation, what is the transmission line current when the percentage of modulation is 90%?

4.235 An FM broadcast transmitter has 370 watts plate power input to the last radio-frequency stage and an antenna field gain of 1.3. The efficiency of the last radio-frequency stage is 65% and the efficiency of the antenna transmission line is 75%. What is the effective radiated power?

4.236 Draw a diagram of an FM broadcast receiver detector circuit.

4.237 Draw a diagram of a means of modulation of an FM broadcast station.

4.238 Draw a diagram of a limiter stage in an FM broadcast receiver.

4.239 How is the operating power of an FM broadcast station determined?

4.243 How wide is an FM broadcast channel?

4.244 What frequency swing is defined as 100% modulation for an FM broadcast station?

4.245 What is the tolerance in operating power of FM broadcast stations?

4.246 What is the meaning of the term "center frequency" in reference to FM broadcasting?

4.250 What is the meaning of the term "frequency swing" in reference to FM broadcast stations?

6.478 Draw the block diagram of an FM transmitter.

6.535 Draw a block diagram of superheterodyne receiver designed for reception of FM signals.

6.561 What types of radio receivers do not respond to static interference?

6.564 Draw a simple schematic circuit diagram of an FM receiver discriminator.

S-3.162 Draw a schematic diagram of a frequency-modulated oscillator using a reactance-tube modulator. Explain its principle of operation.

S-3.163 Discuss the following in reference to frequency modulation. (*a*) The production of sidebands. (*b*) The relationship between the number of sidebands and the modulating frequency. (*c*) The relationship between the number of sidebands and the amplitude of the modulating voltage. (*d*) The relationship between percent modulation and the number of sidebands. (*e*) The relationship between modulation index or deviation ratio and the number of sidebands. (*f*) The relationship between the spacing of the sidebands and the modulating frequency. (*g*) The relationship between the number of sidebands and the bandwidth of emissions. (*h*) The criteria for determining bandwidth of emission. (*i*) Reasons for pre-emphasis.

S-3.164 How is good stability of a reactance-tube modulator achieved?

S-3.165 Draw a circuit diagram of a phase modulator. Explain its operation. Label adjacent stages.

S-3.166 Explain, briefly, what occurs in a waveform if it is phase-modulated.

S-3.167 Explain, in a general way, why an FM deviation meter (modulation meter) would show an indication if coupled to the output of a transmitter which is phase-modulated by a constant amplitude, constant audio frequency. To what would this deviation be proportional?

S-3.168 Draw a circuit diagram of each of the following stages of a phase-modulated FM transmitter. Explain their operation. Label adjacent stages. (*a*) Frequency multiplier (doubler) with capacitive coupling on input and output. (*b*) Power amplifier with variable link coupling to antenna. Include circuit for metering grid and plate currents. (*c*) Speech amplifier with an associated pre-emphasis circuit.

S-3.169 Discuss wideband and narrow-band reception in FM voice communication systems with respect to frequency deviation and bandwidth.

S-3.170 What might be the effect on the transmitted frequency if a tripler stage in an otherwise perfectly aligned FM transmitter were slightly detuned?

S-3.171 Could the harmonic of an FM transmission contain intelligible modulation?

S-3.172 Under what usual conditions of maintenance and/or repair should a transmitter be retuned?

S-3.173 If an indirect FM transmitter without modulation was within carrier frequency tolerance *but,* with modulation, was out of tolerance, what might be some of the possible causes?

S-3.174 In an FM transmitter what would be the effect on antenna current if the grid bias on the final power amplifier were varied?

S-3.175 Explain, briefly, the principles involved in frequency-shift keying (FSK). How is this signal detected?

S-3.176 Assume you have available the following instruments: an ac-dc VTVM, an ammeter, a heterodyne frequency meter (0.0002% accuracy), an absorption wave meter, and an FM modulation meter. Draw and label a block diagram of a voice-modulated (press-to-talk microphone), indirect (phase-modulated) FM transmitter having a crystal multiplication of 12. (*a*) If the desired output frequency were 155.460 MHz, what would be the proper crystal frequency? (*b*) Consider the transmitter strip completely detuned; there are ammeter jacks in the control-grid circuits of the multipliers and the control-grid and cathode circuits of the final circuits of the final amplifier. Explain, in detail, step-by-step, a proper procedure for tuning and aligning all stages except the plate circuit of final power amplifier (P.A.) (*c*) Assume a tunable antenna with adjustable coupling to the plate circuit of the final P.A. With the ammeter in the cathode circuit of the P.A. and with the aid of a tube manual, describe a step-by-step method of obtaining maximum output power without damage to the tube. (*d*) If the P.A. in part (*c*) were a pentode, how would you determine the power input to the stage? (*e*) In part (*c*), how would you determine if the P.A. stage were self-oscillating; if so, what adjustments could be made? (*f*) Assume the transmitter's assigned frequency is 155.460 MHz, with a required tolerance of ±0.0005%. What would be the minimum and maximum frequencies, as read on the frequency meter, which would assure the transmitter being within tolerance? (*g*) Assume the 1-MHz crystal oscillator of the frequency meter has been calibrated with WWV and that the meter is tunable to any frequency between each 1-MHz interval over a range of 20–40 MHz, with usable harmonics up to 640 MHz. Explain in detail what connections and adjustments would be made to measure the signal directly from the transmitter and also by means of a receiver. (*h*) In checking the frequency deviation with the modulation meter, would you expect the greatest deviation by whistling or by speaking in a low voice into the microphone? (*i*) If the transmitter contained a means for limiting and were overmodulating, what measurements and adjustments could be made to determine and remedy the fault.

S-3.177 Draw a schematic diagram of each of the following stages of a superheterodyne FM receiver. Explain the principles of operation. Label adjacent stages. (*a*) Mixer with injected oscillator frequency. (*b*) IF amplifier. (*c*) Limiter. (*d*) Discriminator. (*e*) Differential squelch circuit.

S-3.178 Draw a diagram of a ratio detector and explain its operation.

S-3.179 Explain how spurious signals can be received or created in a receiver. How could this be reduced in sets having sealed untunable filters?

S-3.180 Describe, step-by-step, a proper procedure for aligning an FM double conversion superheterodyne receiver.

S-4.059 Explain in a general way how radio signals are transmitted and received through the use of frequency modulation.

S-4.060 Draw a circuit diagram of a reactance-tube modulator and explain its operation.

S-4.061 What is the difference between frequency and phase modulation.

S-4.062 Describe briefly the operation of the Armstrong and the phasitron methods of obtaining phase modulation.

S-4.063 What is the purpose of pre-emphasis in an FM transmitter? of de-emphasis in an FM receiver? Draw a circuit diagram of a method of obtaining pre-emphasis.

S-4.089 Draw a block diagram of an FM broadcast transmitter complete from the microphone (and/or camera) inputs to the antenna outputs. State the purpose of each stage and explain briefly the overall operation of the transmitter.

10

Transmission Lines

The basic purpose of a transmission line is to transfer energy from a generator, or source, to the load. For example, a transmission line is used to feed RF energy from the final power amplifier of a transmitter at one location to an antenna at some remote location. Lines are also used to interconnect two units a few feet apart, on a rack. When the energy being transmitted is dc or low frequency (power-line or audio), the tie lines present no special problems, but at high frequencies, even a short length of wire seems to take on very peculiar attributes. Yet—if you understand transmission line theory—its action is quite normal. Because of these seemingly peculiar characteristics, transmission lines are extensively used as resonant circuits, as measuring devices, and as impedance-matching devices. With the progress of electronics into higher and higher frequencies, knowledge of the properties of transmission lines is indispensable.

TRAVEL OF ELECTRICAL WAVES DOWN A LONG LINE

It is a well-established fact that electrical waves travel at the speed of light waves—186,000 miles per second (or 300,000,000 meters per second, in the metric system). Let us suppose that we have a line 186,000 miles long, and we apply an alternating voltage with a frequency of 1 Hz at the input end. At intervals of 46,500 miles, we place observers to check voltage readings across the line. (See Fig. 10–1.) Their watches are synchronized. At $t = 0$, just as the input voltage wave reaches its maximum positive value, the main switch is closed. The observer at

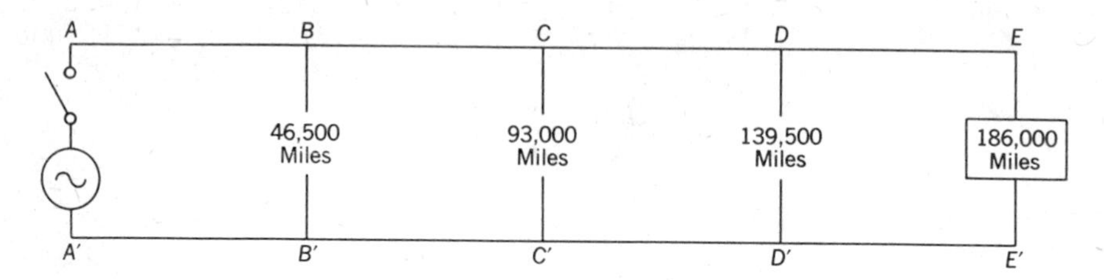

Figure 10-1 A long transmission line.

A–A′ reads "Maximum Positive" on his voltmeter. (All voltages given will be for the potential of the upper line compared to the lower line.) What are the readings at *B–B′*, *C–C′*, *D–D′*, and *E–E′* at this instant of time? These voltages are all zero! Obviously, since it takes time for a wave to travel down the line, the voltage wave could not have reached point *B–B′* or any other location. Now let us consider what the line voltages are at 0.25 s later:

1. At *A–A′*. The frequency of the applied voltage is 1 Hz. It started at its maximum positive value. In 0.25 s, the voltage has gone through one quarter of its cyclic variation and is now zero. The reading of the voltmeter at *A–A′* must be zero.
2. At *B–B′*. Remember that it takes the voltage wave 1 s to travel 186,000 miles. In 0.25 s, it will travel 186,000/4 or exactly 46,500 miles. But that is the exact observation spot *B–B′*. The voltage at *B–B′* is maximum positive. The maximum positive voltage which was sent down the line 0.25 s ago has just reached point *B–B′*.
3. *At C–C′*. The voltage reading must be zero—the voltage wave has not reached there yet.
4. At *D–D′*. Same as 3.
5. At *E–E′*. Same as 3.

Using the same method of analysis, we can now examine the line conditions 0.5 s after closing the main switch.

1. At *A–A′*. Due to cyclic variation of voltage versus time, the input voltage has reached its maximum negative value.
2. At *B–B′*. The maximum positive voltage that was here 0.25 s ago has traveled down the line. Meanwhile the zero potential that was at *A–A′* just 0.25 s ago has now just reached location *B–B′*. Therefore the voltmeter at *B–B′* now reads zero.
3. At *C–C′*. It is now 0.5 s since the line was energized. The original maximum positive voltage passed *B–B′* 0.25 s ago and has

now traveled a total of 186,000/2 or 93,000 miles. Obviously the meter at *C–C′* will register maximum positive.

4. At *D–D′*. The voltmeter reading must still be zero—the voltage wave has not reached here yet.
5. At *E–E′*. Same as 4.

By now, the pattern of voltage variations along the line—with distance and with time—should be obvious. For the frequency and observation points used in this illustration, the condition at any one observation point advances to the next point every 0.25 s. Let us tabulate the above data and also include the line conditions that exist at 0.75, 1.00, and 1.25 s after closing the main switch:

TIME	$E_{A-A'}$	$E_{B-B'}$	$E_{C-C'}$	$E_{D-D'}$	$E_{E-E'}$
0.00	Max positive	0	0	0	0
0.25	0	Max positive	0	0	0
0.50	Max negative	0	Max positive	0	0
0.75	0	Max negative	0	Max positive	0
1.00	Max positive	0	Max negative	0	Max positive
1.25	0	Max positive	0	Max negative	0

Using the step-by-step analysis, let us check the line conditions at $t = 1.25$ s:

1. At *A–A′*. The voltage wave has gone through 1.25 cycles of variation. It has passed through its maximum positive value (at $t = 1.0$ s) and the voltage has dropped to zero again.
2. At *B–B′*. The maximum positive value that was at *A–A′* 0.25 s ago has just reached here.
3. At *C–C′*. It takes 0.5 s for the voltage conditions at *A–A′* to reach *C–C′*. Therefore the zero value that was at *A–A′* at 0.75 s just reaches here at 1.25 s.
4. At *D–D′*. Since it is 139,500 miles away, it takes 139,500/186,000 or 0.75 s for the conditions at *A–A′* to reach *D–D′*. Therefore the maximum negative value that was at *A–A′* at 0.75 s, has just reached here.
5. At *E–E′*. This point is a full 186,000 miles away. The conditions here must duplicate the conditions that were at *A–A′* one full second ago. In this case it is zero.

Notice that in the above explanation we have considered voltage versus *time* conditions *at one point* (such as *A–A′*). We have also considered voltage versus *distance* at some fixed instant of time. In these

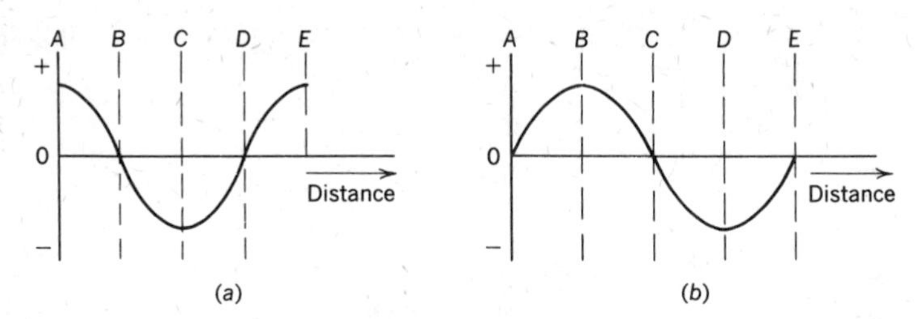

Figure 10-2 Voltage versus distance at a fixed time instant (*a*) at $t = 1.00$ (*b*) at $t = 1.25$ s.

two sentences is found the stumbling block for most students of transmission line theory. Heretofore, we have studied voltage versus time relations—but this is the first exposure to voltage versus *distance*. Since we now see that voltage conditions can vary with distance along a line, we can no longer speak of just voltage versus time—we must specify at what location. Also, when speaking of voltage versus distance, we must specify at what time instant.

As an illustration let us examine the voltage versus distance along the above line at $t = 1.00$ s. The potential of the upper line compared to the lower line varies as follows: maximum positive at A, zero at B, maximum negative at C, zero at D, and maximum positive again at E. If additional voltmeters were placed between these major points, we would find that the voltage variation with distance follows a sine-wave relationship, when the input voltage is a sine wave. Figure 10–2 shows the plot of voltage versus distance for time instants of 1.00 and 1.25 s.

In the above example of transmission line conditions, no voltage was available at the load end of the line for the first second. After this initial "transient" condition, the voltage at E–E' (and at any other point along the line) will go through cyclic variations with time. However, the

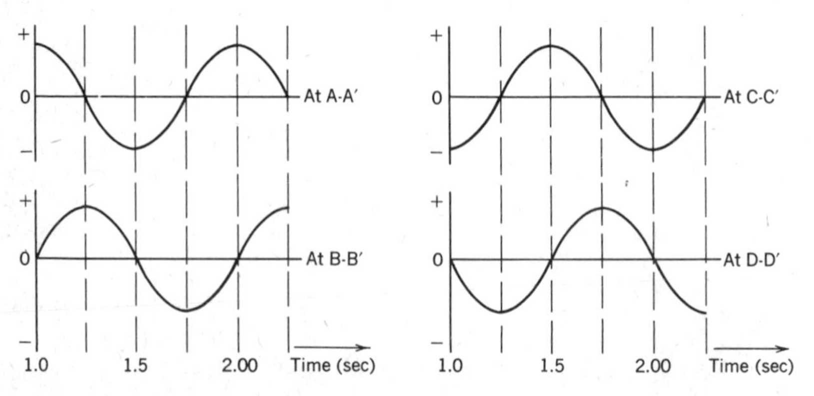

Figure 10-3 Voltage versus time at various points along a line.

cyclic variations will not be the same for any two points, unless they are (in this case) exactly 186,000 miles apart. Figure 10–3 shows the voltage versus time variations at each of the key points in our line. The voltage variation with time at E–E' is a repeat of the curve for location A–A'. Notice that we must pick a particular location before we can show variations with time. Also notice that there is a distinct phase difference between the waves at any two locations.

Had we selected any intermediate locations we would have found an in-between value of phase difference.

Wavelength (λ)

Throughout the above illustration we used an input frequency of 1 Hz. Refer back to Fig. 10–2 for a moment. Remember it was voltage versus *distance*. Points A and E were 186,000 miles apart. Notice that the voltage wave went through 1 cycle of changes (maximum positive to another maximum positive) in a distance of 186,000 miles. Therefore we can say that the *length* of this wave (or its *wavelength*) is 186,000 miles.

Suppose we had used an input voltage with a frequency of 2 Hz. The period, or time for 1 cycle, would have been $1/f$ or 0.5 s. How far would the wave have traveled in this time interval?

$$\text{Distance} = \text{velocity} \times \text{time} = 186{,}000 \times 0.5 = 93{,}000 \text{ miles}$$

In 0.5 s, the wave of the original input condition has traveled 93,000 miles. Meanwhile, in the same interval of time, the source voltage has gone through one complete cycle of variation (with time). But the distance covered by one cycle of variation is the wavelength. By doubling the frequency, we have cut the wavelength in half. In other words, wavelength is inversely proportional to frequency. The exact derivation can be shown by simple mathematics:

$$\text{Distance} = v \times t \tag{10-1}$$

If we replace the general term for time (t) by the time for one cycle (T), we get for the distance covered by one cycle, or wavelength

$$\lambda = v \times T \tag{10-2}$$

but, since the frequency is

$$f = \frac{1}{T} \quad \text{or} \quad T = \frac{1}{f}$$

substituting $1/f$ for T, we get

$$\lambda = \frac{v}{f} \tag{10-3}$$

1. If we wish to express wavelength in meters, we merely use velocity in meters per second:

$$\lambda \text{ (in meters)} = \frac{3 \times 10^8}{f} \quad \text{or} \quad \frac{300}{f \text{ (in MHz)}} \tag{10-4}$$

2. Using velocity in miles per second, we can find the wavelength in miles:

$$\lambda = \frac{186{,}000}{f} \tag{10-5}$$

3. More often we desire wavelength in feet. It is therefore necessary to change the velocity from miles per second to feet per second. In feet per second, this value is 984×10^6. Now, if we also express the frequency in megahertz, we get

$$\lambda \text{ (in feet)} = \frac{984}{f \text{(in MHz)}} \tag{10-6}$$

EXAMPLE 10–1

Find the wavelength in meters, corresponding to a frequency of 1500 kHz.

Solution

$$\lambda = \frac{v}{f} = \frac{3 \times 10^8}{1500 \times 10^3} = 200 \text{ meters}$$

EXAMPLE 10–2

What is the wavelength of the above frequency—expressed in feet?

Solution

This could be done by metric system conversion, but it is much simpler to use the equation for wavelength in feet.

$$\lambda = \frac{v}{f} = \frac{984}{f \text{ (MHz)}} = \frac{984}{1.5} = 656 \text{ ft}$$

EXAMPLE 10–3

Each section of a dipole antenna should be one quarter-wave in length. Find the required length for a television channel of 208 MHz.

Solution

1. $\lambda = \dfrac{984}{f \text{ (MHz)}} = \dfrac{984}{208} = 4.73 \text{ ft}$

2. Each section $= \dfrac{1}{4}\lambda = \dfrac{4.73}{4} = 1.18 \text{ ft}$

Electrical Length of a Line

Transmission line phenomena are apparent only on long lines. So in the introduction we used a line 186,000 miles long. In practice, we will not come across lines of such lengths, but neither will we find frequencies of 1 Hz. Remember that knowledge of transmission lines becomes important as we use higher and higher frequencies. This is illustrated in Example 10–3. One wavelength is only 4.73 ft. A quarter wave is just about 14 in. If we connected two components in "parallel" using 14 in for our leads, one unit might be at maximum voltage while the other is at zero—hardly a parallel circuit such as those we are accustomed to on low frequencies. In this particular case, 14 in at a frequency of 208 MHz cannot be ignored as just a piece of wire—it is a long line and must be treated according to transmission line theory.

What constitutes a "long line"? The answer should be obvious. It is not the physical length in itself, but rather the ratio of physical length to the wavelength corresponding to the frequency used. If the line is so short that all voltages along the line are appreciably constant, it can be considered a short line. On the other hand, if voltages along the line change appreciably, the line must be called long. Obviously a quarter-wave line is a long line. No fixed value can be given as the dividing point between long and short. A line does not change sharply from short to long. As a rough approximation, a line shorter than one-sixteenth of a wavelength may be considered short. At ultrahigh frequencies a few inches may constitute a long line!

Electrical Equivalent of a Line

In its simplest form, a transmission line consists of two parallel wires. Each wire has some resistance per unit length—no matter how small the unit of length. Each wire also has inductance per unit length. These two components must be considered as series elements. In addition there is some capacitance between the two wires—regardless of the spacing between the wires and the dielectric between them. One more factor is that no insulation between the wires can be perfect. Therefore, there is also a leakage resistance between the wires. These last two components are shunt effects. Any transmission line can be replaced by an equivalent electrical circuit containing each of these four elements in proper proportions. Such an equivalent circuit is shown in Fig. 10–4. In the diagram the line constants (L-C-R) are lumped. In the actual line they are distributed over the entire length of the line. Many times when transmission line effects are desired, lumped constants as in Fig. 10–4 are used to form an "artificial line," thereby saving the space that would be required for a real line. The unit length used for evaluating the lumped

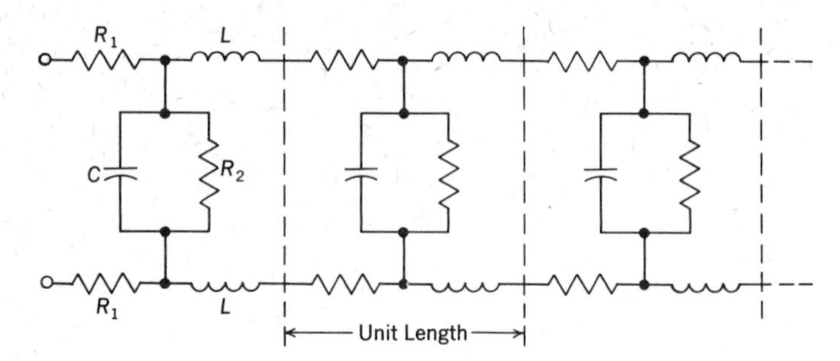

Figure 10-4 Electrical equivalent of a line.

constants of the line could be many feet or fractions of a foot. Since the real line has distributed constants, it should be obvious that more accurate results are obtained by using shorter distances for the "unit length."

Characteristic Impedance (Z_0)

Let us assume that we have a transmission line that is infinite in length. Now let us connect an ammeter at the input end and apply a voltage (E_0) to the line. The ammeter indicates that current is flowing in the line—even though the far end is open. It may seem that we do not have a complete circuit, but we do. (See Fig. 10–4.) The circuit is completed through the distributed constants of the line. The amount of current (I_0) indicated by the ammeter at the input end depends on the input voltage E_0 and on the constants of the line. If we double the applied voltage, the current will double. By Ohm's law, this relationship can be stated as

$$Z_0 = \frac{E_0}{I_0} \tag{10-7}$$

where Z_0 is the *input impedance of the infinite line*. This value Z_0 is known as the *characteristic impedance* of the line.

EXAMPLE 10–4

In estimating a television installation, preliminary calculations indicate that we will need 200 ft of transmission line cable having a characteristic impedance of 300 Ω. When the installation is made, only 150 ft of cable is used. How does this affect the characteristic impedance of our line?

Solution

This is a "trick" question. By definition, the characteristic impedance is the input impedance of a line that is infinite in length. Whether we use 200 ft or 150 ft of cable makes no difference—the input impedance of that line if it were infinite in length has not changed. Therefore Z_0 is independent of the actual line length that we use.

The above solution raises an interesting problem. If the length of the line does not affect the characteristic impedance of a line, how do we get various values of characteristic impedance? The answer is quite simple. The current that flows in the infinite line for any given value of applied voltage depends on the distributed constants of the line. In turn, these constants depend on the size of wire used, the spacing between the wires, and the type of dielectric or insulation between the wires. Types of transmission lines and their characteristic impedances will be discussed in more detail later.

Resistance Equivalent of a Line

From the equivalent circuit of a transmission line, we can consider a line as consisting of an infinite number of "unit sections" added one to the other. Offhand it would seem that because of the shunt elements—capacitance and leakage resistance—the impedance due to an infinite number of sections should drop to zero. And yet the characteristic impedance is not zero, but some definite value. That this is true can be seen by using a resistance analogy. Let us temporarily neglect the inductance and capacitance of the line. (The calculation would be too laborious.) For our illustration we will use 20 Ω for the series resistance and 400 Ω for the leakage resistance. Such a line is shown in Fig. 10–5. To find the input resistance of this line at point J, we will have to work backwards from the right:

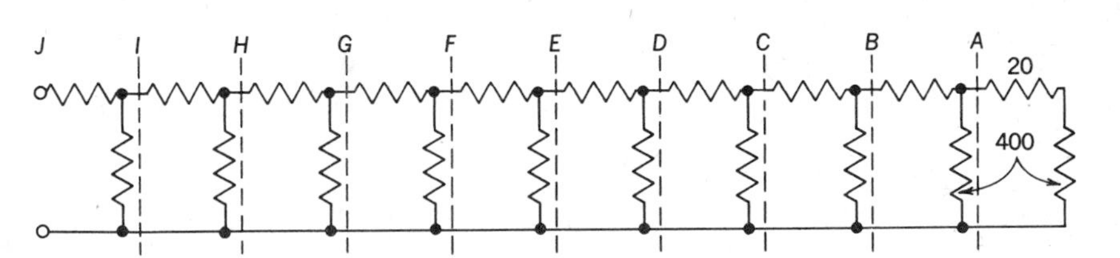

Figure 10-5 Resistance equivalent of a line.

1. At A, the resistance looking to the right consists of the 20- and 400-Ω resistors in series, or 420 Ω.

2. At B, again looking to the right, we have the 420 Ω of section A in parallel with the shunt resistance of 400 Ω. This parallel combination is in series with the 20-Ω resistor, or

$$R_B = \frac{R_1 R_2}{R_1 + R_2} + R_3 = \frac{420 \times 400}{820} + 20 = 205 + 20 = 225\ \Omega$$

3. Using similar reasoning for R_C (looking to the right):

$$R_C = \frac{225 \times 400}{625} + 20 = 144 + 20 = 164\ \Omega$$

4. $$R_D = \frac{164 \times 400}{564} + 20 = 116 + 20 = 136\ \Omega$$

5. $$R_E = \frac{136 \times 400}{536} + 20 = 101.5 + 20 = 121.5\ \Omega$$

6. $$R_F = \frac{121.5 \times 400}{521.5} + 20 = 93.2 + 20 = 113.2\ \Omega$$

7. $$R_G = \frac{113.2 \times 400}{513.2} + 20 = 88.2 + 20 = 108.2\ \Omega$$

8. $$R_H = \frac{108.2 \times 400}{508.2} + 20 = 85.1 + 20 = 105.1\ \Omega$$

9. $$R_I = \frac{105.1 \times 400}{505.1} + 20 = 83.2 + 20 = 103.2\ \Omega$$

10. $$R_J = \frac{103.2 \times 400}{503.2} + 20 = 82.2 + 20 = 102.0\ \Omega$$

By now the characteristic impedance of this line should be obvious. The change from step to step is getting smaller and smaller. The final value will stabilize at 100.0 Ω for an infinite number of sections.

Types of Transmission Lines

Transmission lines are available commercially in various forms. The main types are shown in Fig. 10–6. Each type has its own particular advantages. The choice of type of line for any specific installation would depend on the amount of power to be transmitted, the frequency at which it is to be used, the length of line needed, and the economics of cost versus efficiency. Let us analyze each of these types of lines in more detail.

TWO-WIRE PARALLEL CONDUCTORS. In Fig. 10–6(a) and (b) are shown several types of parallel conductor lines. The characteristic impedance of such lines (with air dielectric)* is given by

$$Z_0 = 276 \log \frac{b}{a} \tag{10-8}$$

where b is the spacing between the wires, and a is the radius of the conductor. From this equation it is obvious that the spacing between the

*For other-than-air dielectric, multiply by the velocity constant of the line. (See p. 392.)

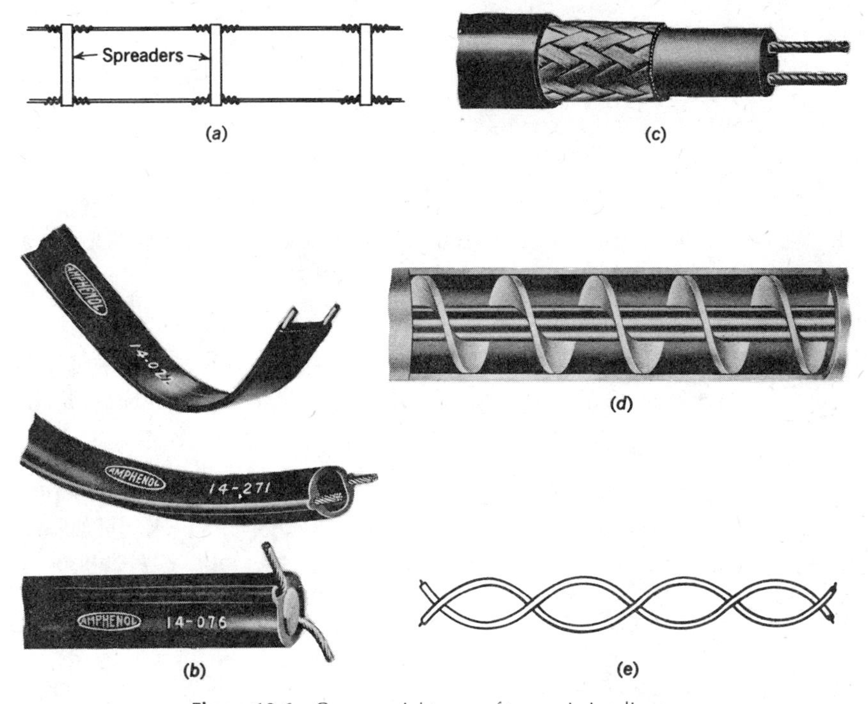

Figure 10-6 Commercial types of transmission lines.

wires must be kept constant; otherwise the characteristic impedance of the line would vary. Due to constructional limitations, parallel-wire lines are available only in characteristic impedances of approximately 75–500 Ω. The type shown in Fig. 10–6(a) is generally used in high-power low-frequency applications. Very often, to reduce stray field pickup by this line, the leads are crisscrossed (transposed). The spacing used between conductors may vary from 2 to 6 in, depending on the voltage between wires and the frequency of the applied voltage. Closer spacing is preferable at higher frequencies to reduce radiation losses.

The transmission lines shown in Fig. 10–6(b) are used extensively in television receiver antenna feed systems. Their low cost, low loss, and simplicity make them ideal for such installations. The spacing used for a 300-Ω twin lead of this type is approximately $\frac{3}{8}$ in. Let us check this using Eq. (10–8).

EXAMPLE 10–5

A twin-lead transmission line uses No. 14 wire and a spacing of $\frac{3}{8}$ in between wires. Find its characteristic impedance.

Solution

1. No. 14 wire has a diameter of 0.064 in

2. $Z_0 = 276 \log \frac{b}{a} = 276 \log \frac{0.375}{0.032} = 298\ \Omega$

In general, parallel-conductor lines have advantages of ease of construction and installation, low cost, and high efficiency due to low losses. The main type of loss and the disadvantage of these lines is high radiation loss, particularly at high frequencies and near metal structures.

SHIELDED-PAIR TRANSMISSION LINES. To reduce the radiation losses of the above lines and even more to reduce the stray field pickup, a parallel two-wire line can be enclosed in a metal braid for shielding. Such a line is shown in Fig. 10–6(c). It is available commercially in characteristic impedances of approximately 40–150 Ω. The insulating material used should be a low-loss dielectric material. But even so, the dielectric and the leakage loss of this line is higher than for the types previously discussed. Due to the outer shielding (which is grounded), this line can be run close to metal structures without serious losses. In addition to the advantages already discussed, the shielded-pair type of line will give best balance between the conductors and ground. It is ideally suited to balanced-line feeds. However, the total losses for this line (depending on the quality of the insulation) will range from just slightly higher, to double the losses of the simple two-wire parallel line.

CONCENTRIC OR COAXIAL TRANSMISSION LINES. Figure 10–6(d) shows one form of concentric or coaxial line. This is the most efficient type of all lines. The total losses are less than one-third of the two-wire parallel line. The construction should be obvious from the diagram. A high quality dielectric should be used for the helical spacer; also, the looser the helix, the less the losses. However, a sufficient number of turns should be used to maintain the inner conductor centered; otherwise the characteristic impedance will change. The outer conductor is usually grounded and acts as a shield. Therefore, in addition to extremely low loss, coaxial lines have the advantages of freedom from stray field pickup and from radiation losses. Its disadvantages are that it is more expensive and that it is basically an unbalanced type of line. (You will see later how such lines can be used for balanced-line feeds.)

The coaxial line as shown in Fig. 10–6(d) is obviously a fairly rigid line. This makes installation difficult. However, coaxial lines are also available in flexible form. In appearance the flexible type resembles the shielded pair of Fig. 10–6(c). The difference is that only one conductor is used; this conductor is centered and the shield acts as the outer conductor. Since the dielectric material is now no longer air, there is a slight sacrifice in increased losses—but it is still much more efficient than the other lines.

Commercially, the characteristic impedance of coaxial lines ranges approximately 10–150 Ω. This range is obtained by changing the diameter of the conductors. The relationship (for air dielectric)* is given by

$$Z_0 = 138 \log \frac{b}{a} \tag{10–9}$$

where b is the inner diameter of the outer conductor, and a is the outer diameter of the inner conductor.

EXAMPLE 10–6

Find the characteristic impedance of a coaxial line if the inner conductor has an outside diameter of $\frac{1}{4}$ in and the outer conductor has an inside diameter of 0.85 in.

Solution

$$Z_0 = 138 \log \frac{b}{a} = 138 \log \frac{0.85}{0.25} = 73.4\ \Omega$$

TWISTED-PAIR TRANSMISSION LINE. The only advantages of the twisted-pair type of line shown in Fig. 10–6(e) are its simplicity, ease of installation, and low cost. Because of high dielectric losses, it is the least efficient of all lines. Its use should be restricted to low frequencies, unless the length of run is quite short, and it should be used only for *nonresonant* line applications. (The distinction between resonant and nonresonant lines will be explained later.) The practical range of characteristic impedance for twisted-pair lines is approximately 70–150 Ω.

Calculation of Characteristic Impedance

In the above discussion, formulas for calculating characteristic impedance from the physical dimensions of a line were given for the two-wire parallel conductor and for the coaxial line. The two formulas are different because of the difference in physical construction. Similar types of equations can be found for the other types of lines. However, since in all cases the characteristic impedance depends on the electrical constants (L, C, and R), one equation based on these constants can be used for any type of line. Neglecting the resistive effects (the error so introduced is usually negligible), this relation is

$$Z_0 = \sqrt{\frac{L}{C}} \tag{10–10}$$

* For other-than-air dielectric, multiply by the velocity constant of the line.

Since the L/C ratio for a given line is constant, any finite length can be used for determination of this value. Using an impedance bridge or Q meter, the inductance can be measured with the far end of the line shorted, and the capacitance can be measured when the far end is open. By substitution in the basic equation, the characteristic impedance can then be calculated.

Velocity Constant of a Line

Earlier in this discussion of transmission lines it was stated that electrical waves travel at the speed of light (186,000 miles, or 300 million meters, per second). This is strictly true only for electromagnetic waves traveling through air. This speed is reduced when the waves travel in lines that use spacing insulators or solid dielectrics. The decrease in velocity is due to the inductance and capacitance of the line: the inductive effect opposes any change in current, and the capacitive effect opposes any change in voltage. The ratio of the actual velocity of the waves through a given line, compared to the velocity through air, is called the *velocity constant* of the line. Velocity constants for common types of lines are listed below:*

1. Parallel line, air dielectric (Fig. 10–6(a)) 0.95–0.975
2. Parallel line, plastic dielectric (Fig. 10–6(b)) 0.80–0.95
3. Shielded pair, rubber insulation (Fig. 10–6(c)) 0.56–0.65
4. Coaxial line, air dielectric (Fig. 10–6(d)) 0.85
5. Twisted pair, rubber insulation (Fig. 10–6(e)) 0.56–0.65

Since the velocity of the wave motion through a line is reduced, so is the wavelength. This fact must be kept in mind when calculating the length of line needed for specific applications. The equation previously given for wavelength must be corrected to

$$\lambda = \frac{984k}{f\,(\text{MHz})} \qquad (10\text{–}11)$$

where k is the velocity constant of the line.

Wave Motion on an Infinite Line

Figure 10–7(a) shows an infinite line connected to an ac supply. Let us stop time at the instant that the input voltage is at its maximum value, with the potential of the upper line positive compared to the lower line. The voltage (line A compared to line B) for any location along the line—at that instant—is shown in Fig. 10–7(b). We also know that, due to the distributed constants of the line, current flows through the line. What is

* With solid dielectrics, the reduction in the speed of propagation is (approximately) inversely proportional to the square root of the dielectric constant of the material.

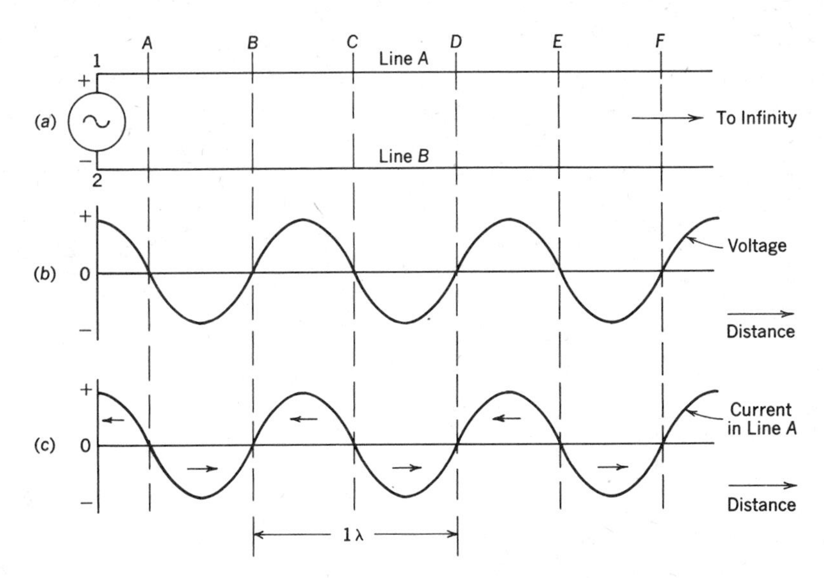

Figure 10-7 Current and voltage relations on an infinite line.

the value and direction of the current at any point along the line? By Ohm's law, $I = E/Z$. But the impedance at any location, from that point toward infinity, is the characteristic impedance Z_0, and it is constant and resistive. Therefore the current versus distance must vary in step with the voltage—or current and voltage are in phase at all points along the line. This can be seen from a comparison of Fig. 10–7(b) and (c). Notice that the potential and current curves for line A reverse every half-wavelength. (If we were to plot the potential of line B with respect to line A, the curves would each be 180° out of phase with the curve shown in Fig. 10–7(b).)

Still keeping time "frozen" at the above instant, let us now look into the sending end of our line at points 1 and 2 (Fig. 10–7(a)). The difference in potential is at a maximum. Therefore a strong electrostatic field must exist between these points. In addition, since the current flowing into or out of the line at these points is also a maximum, a strong magnetic field exists around each wire. The relationship between these two space fields is shown in Fig. 10–8. Notice that the electrostatic field lines are perpendicular to the magnetic lines. The combination of these two fields is known as the *electromagnetic* field.

Now let us release time. As we know from our earlier discussion, the maximum value of voltage at the sending end of the line starts to travel down the line to the right. So does the maximum value of current—and so does the electromagnetic field. Neglecting the velocity constant of the line, these waves (current, voltage, and electromagnetic field) are traveling at the speed of light. At this speed, the energy being fed into the

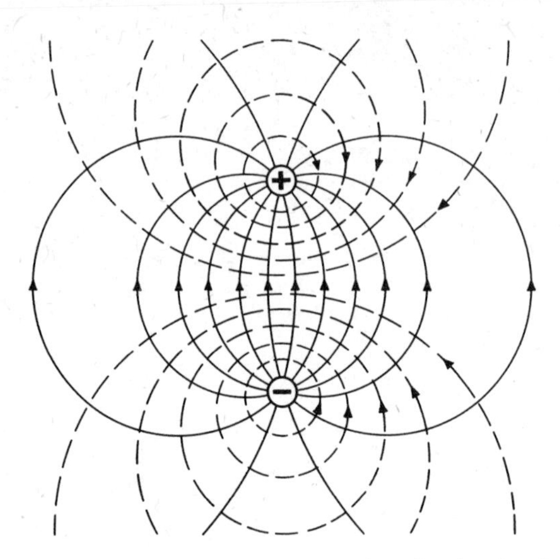

Figure 10-8 Cross-sectional view of a transmission line showing the electromagnetic field.

line is divided 50% in the electric field and 50% in the magnetic field. At reduced speeds, the magnetic field is somewhat weaker than the electric field.

Propagation Constant

In our discussion so far, we have neglected the effect of the resistance of the line. For a short line, this is permissible, particularly if the line is carefully designed for low loss. Yet, no matter how small the loss per foot or per mile, some energy will be lost as the waves travel down the line, and the amplitude of the wave, for each successive wavelength, will be slightly lower. (Since this reduction in amplitude is small, it was not shown in Fig. 10–7.) However, as we approach infinity, the amplitude of the waves approaches zero. In other words, the energy that is put into the line at the sending end is gradually dissipated in the resistance of the line, so that the total loss for an infinite line is 100%. Consequently, when the wave reaches the "infinite end," no energy is left, and no energy returns back to the sending end. Therefore, *there is no reflection of energy in an infinite line.*

In evaluating line loss, it is convenient to express this loss in *decibels per unit length* using an *attenuation constant* α.* Since this value varies with frequency, attenuation data are often given by curves such as loss in dB per 100 ft versus frequency in megahertz.

*Attenuation is also expressed in *nepers,* using the *natural* (Napierian) logarithm. One neper = 8.686 dB.

At the beginning of this chapter, it was shown that as a wave travels down the line, there is a difference in phase between the voltages (and currents) at various points along the line. This effect is more commonly expressed in terms of a *phase constant,* β, in radians per unit length. For example, for two points separated by a distance l, there is a phase shift of βl radians.

Combining these two propagation effects, we get the *propagation constant* γ* of the line:

$$\gamma = \alpha + j\beta = \sqrt{ZY} \tag{10–12}$$

where: Z = series impedance $(R + j\omega L)$ per unit length, and
Y = shunt admittance† $(G + j\omega C)$ per unit length.

At radio frequencies, since $\omega L >> R$ and $\omega C >> G$, the individual components of the propagation constant can be evaluated in terms of the distributed constants of the line, as

$$\alpha = \frac{R}{2Z_0} + \frac{GZ_0}{2} \tag{10–13}$$

and

$$\beta = \omega\sqrt{LC} \tag{10–14}$$

where: R = series resistance per unit length, and
G = shunt conductance per unit length.

Line Terminated by $R = Z_0$

Although it is obvious that an infinite line can never be built, we do have a line that acts as an infinite line. Figure 10–9(a) shows an infinite line starting at point A. The impedance at A, looking into the line, must be the characteristic impedance Z_0. Now suppose we cut off a 1-ft sec-

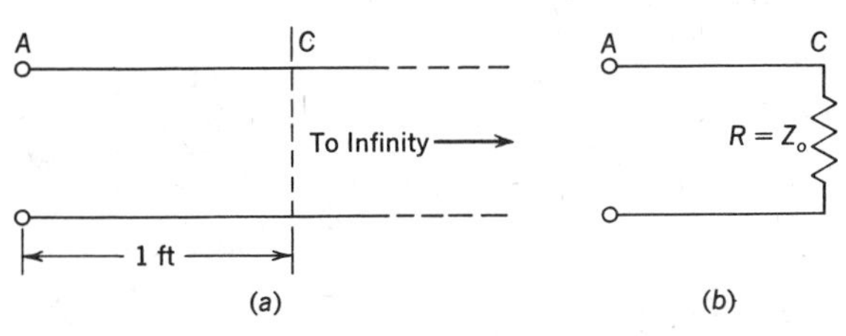

Figure 10-9 Comparison of an infinite line, and a finite line terminated by $R = Z_0$.

*F. E. Terman, *Electronics and Radio Engineering,* Chap. 4.
†Admittance (Y) is the reciprocal of impedance.

tion A–C. The length of the line from C to infinity is still infinite. Therefore, the input impedance at C is still Z_0. Now let us take our 1-ft section A–C and connect a resistive load of $R = Z_0$ at end C. As far as the input end A is concerned, the conditions are as follows:

1. The section A–C is 1 ft long.
2. The impedance at C (due to the load R) is Z_0—just as if the line had extended out to infinity. Therefore, the input impedance at A is still equal to the characteristic impedance Z_0. What is more—this would apply equally well if we had made section A–C 2 ft, 5 ft, or any finite length.

Let us examine the wave motion and energy conditions on any section of line terminated with a load of $R = Z_0$. As in the case of the infinite line, energy travels down the line. This time, however, except for the small amount of energy that is lost in the resistance of the finite line, all the energy is delivered to and dissipated by the load. By this means we can get a maximum of energy transfer between the source and the load. Since all the energy is dissipated in the load, none travels back or returns to the source. In other words, there is no reflection of energy, and as a result no *standing waves* occur on this line. (Reflection of energy and standing waves are discussed later in this chapter.) Again our finite line, terminated by $R = Z_0$, acts as an infinite line.

Earlier in the chapter we showed that the voltage and current values at any one location vary sinusoidally with time. Also, at any one instant of time, voltage and current values vary sinusoidally with distance. But remember that the voltage and current at any one instant and at any one location are always in phase, and the ratio of voltage to current is constant and equal to Z_0. This is true for an infinite line because it has no reflections. But there are no reflections on a properly terminated finite line. Therefore, any length of line terminated by a resistive load of $R = Z_0$ has an input impedance that is resistive and equal in value to the characteristic impedance of the line.

Summary of Infinite-Line Conditions

We have covered many new ideas concerning infinite lines. And so, before we continue with even stranger concepts, let us summarize the salient features of the above discussion:

1. The input impedance of an infinite line is its characteristic impedance (Z_0) and is resistive at radio frequencies.
2. Voltage, current, and electromagnetic waves travel down the line.
3. The ratio of voltage to current at any location throughout the line is constant and equals Z_0.

4. The voltage and current at any location on the line are in time phase with each other.
5. Energy travels down the line—without reflections.
6. Such a line is called *nonresonant*.
7. The line losses on a nonresonant line (per unit length) are a minimum.
8. Any line terminated by a resistive load equal to the characteristic impedance acts as an infinite line:
   (*a*) Input impedance = Z_0 and is resistive.
   (*b*) No reflections.
   (*c*) Current and voltage in phase.
   (*d*) Maximum energy transfer from source to load.

LINES WITH REFLECTIONS

When sound waves traveling through one media (for example, air) hit a discontinuity or change in media—such as striking a wall—some of the sound energy continues through, and another portion of the sound energy is reflected or bounces back toward the original sound source. This effect—the reflected energy—is the echo. The amount of energy reflected, or strength of the echo, will depend on the characteristics of this second media that the original wave encounters. For example, if the wave strikes a hard solid wall, very little energy goes through and practically all of the energy is reflected. On the other hand, some other type of discontinuity, such as a thin wall, might reflect back very little energy. Similar effects are encountered in transmission lines if there is a "discontinuity" in the line. In a transmission line an open circuit is a drastic discontinuity. A short circuit is also a drastic discontinuity. If we connect one line to another line of different characteristic impedance, or to any load other than a resistive load equal to the characteristic impedance, we have introduced a discontinuity in the line. It should be obvious by now that any change in the impedance offered to a wave traveling down a line constitutes a discontinuity.

On the other hand, an infinite line has no discontinuity. As was pointed out, the ratio E/I at any point along the line is constant; the impedance (Z_0) is constant; therefore there is no discontinuity. Similarly, a line terminated by a load $R = Z_0$ has no discontinuity because, with such a termination, there is no change in impedance.

What is the effect of a discontinuity or change in impedance? Just as for sound waves, reflections will occur, and energy will travel back along the line from the point of discontinuity to the original source. The amount of energy that is reflected back will depend upon the extent to which the impedance is disturbed. A few examples are shown below:

1. *Infinite line.* No change in impedance and no reflection.
2. *Open or shorted line.* Drastic change; all the energy is reflected back.
3. *Line terminated by a pure capacitance or pure inductance.* Since pure reactive elements absorb no power, again all the energy is reflected back.
4. *Line terminated by a resistance load but not equal to Z_0.* Some energy is absorbed by the load and some is reflected back. The closer the load resistance matches the characteristic impedance, the less the amount of energy reflected.

An evaluation of the degree of reflection can be obtained by comparing the magnitude (E_r) of the voltage wave reflected back from the discontinuity to the magnitude of the initial voltage wave (E_i) traveling toward the discontinuity. This ratio is called the *reflection coefficient,* Γ (gamma), or

$$\Gamma=\frac{E_r}{E_i} \tag{10-15}$$

Phase of Reflected Waves on an Open Line

If we connect an ac supply to a line that is $1\frac{1}{2}$ wavelength long, the initial wave of current and voltage will start traveling down the line—totally unaware that the line is open. As far as these initial waves are concerned, the line can be considered infinite in length. So, the current and voltage are in phase. Let us assume that the waves reach the open end when they are at maximum negative amplitude. Stop the clock and examine the conditions on the line. One and one-half wavelengths of voltage and current must exist on the line. The input conditions must therefore be at a maximum positive value at this same instant. These voltage and current relations are shown in Fig. 10–10(a) and (b), respectively. Yet a maximum value of current cannot possibly exist at an "open" point!

Let us analyze this effect slowly. A potential exists at a point because of an accumulation of charge (due to an excess or deficiency of electrons) compared to zero potential or ground. The energy associated with a stationary electron (or a charge) is all located in the dielectric or electrostatic field around the charged body. Also electrons in motion constitute current flow. But electrons in motion give rise to a magnetic field. When electrons move at the speed of light, the energy associated with these charges is distributed 50% in the dielectric field and 50% in the magnetic field. This is the situation in any infinite line and also in our open line—up to the point when the current and voltage waves reach the

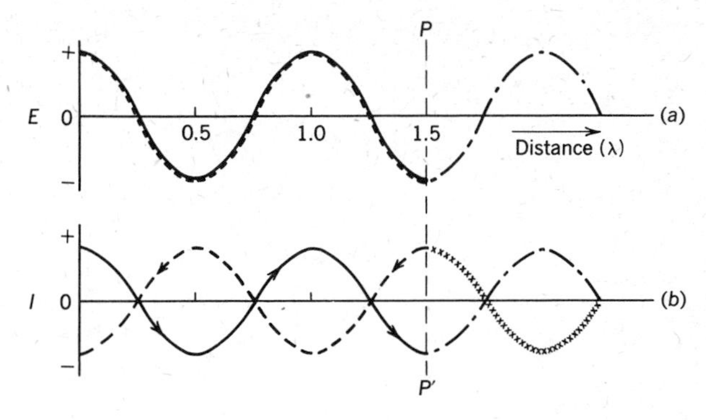

Figure 10-10 Reflection of current and voltage on an open line.

open end of the line. But now the electrons in motion can go no further. They stop—and accumulate at the open end. Since the electrons are no longer in motion, the magnetic field collapses. The sudden accumulation of electrons causes the charge at this point to increase. The potential increases and the dielectric field strength increases. Momentarily all the energy is now in the electrostatic field. This rise of potential at the open end, compared to the potential just behind it on the line, causes a redistribution of electrons. The result is a heavy rush of current in the opposite or *positive* direction. This new current wave traveling back toward the source is called a *reflected wave. The phase of this reflected current wave is reversed 180° from how the original current wave would have continued if the line had extended beyond p–p′*. In Fig. 10–10(b), the dot-dash curve to the right of *p–p′* indicates how the wave would have continued. The x'd curve is drawn 180° reversed. By folding the diagram along *p–p′*, we can see the dashed curve in the reflected direction.

At the same time, the redistribution of electrons changes the potential all along the line. In other words, a reflected wave of voltage also travels back to the source. From another point of view, the collapse of the magnetic field, due to stoppage of current, develops an induced voltage on the line tending to keep the current flowing. This induced voltage is of the same polarity as the initial voltage wave. (This checks with the earlier discussion that the potential increases and the electrostatic field energy increases.) This new voltage wave cannot travel to the right—there is no line there—so it must travel back to the source as a reflected wave of voltage. *The phase of this reflected voltage wave is the same as it would have continued.* In Fig. 10–10(a), the dot-dash line to the right of *p–p′* indicates how the wave would have continued. By folding the diagram along *p–p′*, the dotted curve will be in the reflected direction. For the particular time instant selected, the reflected voltage wave path coincides with the path of the initial wave.

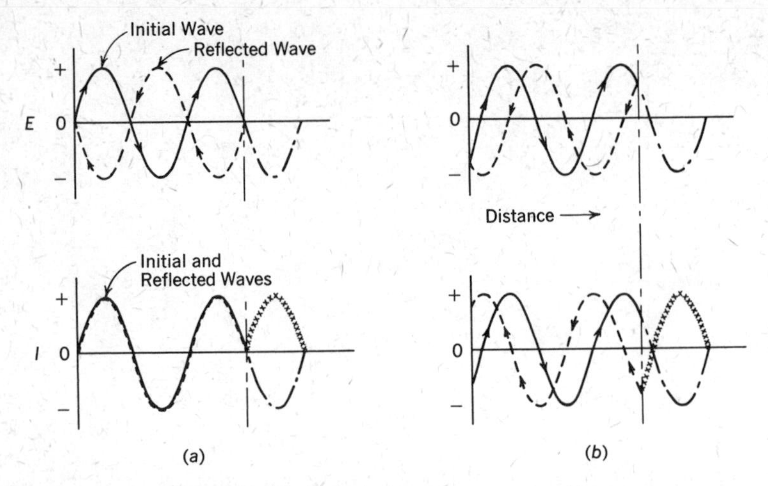

Figure 10-11 Reflected waves on an open line.

From Fig. 10–10, it might seem that the voltage wave reflects back in phase with the initial wave and that the current wave reflects back 180° out of phase with the initial wave. *This is a false conclusion.* It only seems so at this particular time instant. "Phase" applies to voltage and current *time* relations. These plots are of voltage and current versus *distance*. The only true way to arrive at the path of reflected waves for any time instant is as repeated below:

1. In an open line, the voltage wave reflects back as it would have continued.
2. In an open line, the current wave reflects back 180° reversed from how it would have continued.

To further clarify these statements, let us apply this procedure at two other instants of time, as shown in Fig. 10–11. Case (a) is for the initial waves reaching the "open" when they are at their zero value. Case (b) shows the initial waves reaching the end of the line at some intermediate positive value. In each case, notice that the sum of the initial and reflected waves of current at the end of the open line is zero. This checks with our earlier statements—and with common sense—there is no current flow at the open end.

Phase of Reflected Waves on a Shorted Line

Once again the initial waves of current and voltage travel down the line in phase with each other and unaware of the short circuit at the end. Again they reach the end when they are at their maximum negative amplitude. However, a voltage cannot exist across the line at this location, because it is shorted. Following a similar line of reasoning as we used for

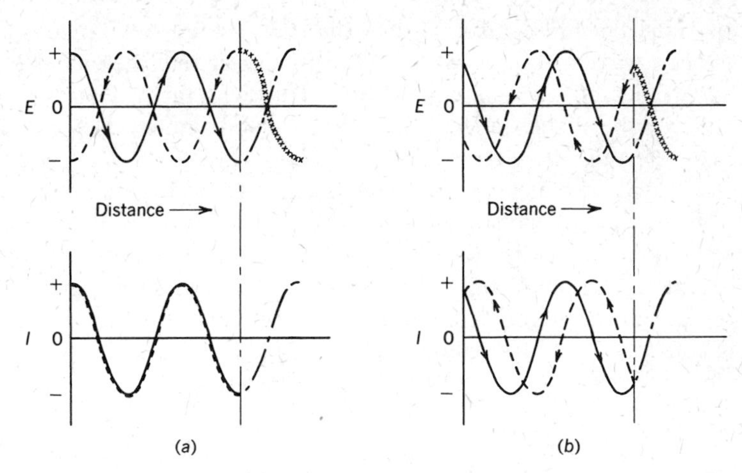

Figure 10-12 Reflected waves on a shorted line.

the "open" line, the only way that the voltage at the shorted end can drop to zero is by having the voltage wave reflect back 180° reversed from how it would have continued. Now remember that the energy in the electromagnetic field was divided equally between the dielectric field and the magnetic field. With the voltage at the shorted end reduced to zero, the electrostatic field collapses. This energy is not lost, nor can it be dissipated in the short circuit. There is only one possible explanation—all the energy must now be in the magnetic field. In order for the magnetic field strength to increase, the current at this point must increase. Therefore, the current wave must reflect back *in the same phase* as it would have continued.

Notice that the phase of the reflected waves for current and voltage in a shorted line are opposite to the reflections we developed for the open line, or:

1. The sum of the initial and reflected current waves for any time instant must always be zero at the end of an *open* line.
2. The sum of the initial and reflected voltage waves for any time instant must always be zero at the end of a *shorted* line.

Figure 10–12 shows the phase of reflected voltage and current waves on a shorted line for two different instants of time.

Standing Wave of Voltage—Open Lines

We have seen how an initial wave of voltage travels down a line, and that if this voltage wave comes to an open line, a reflected wave of voltage is sent back along the line in the same phase as it would have continued. So, if we were to "freeze" time on an open line, the voltage reading at any location must be the resultant due to the sum of the instantaneous voltages of each wave at that location. We saw earlier that the values of the initial and reflected waves at any location vary with time. Therefore, the amplitude of the resultant voltage at any location must also vary with time. To determine how this resultant voltage varies *with time* and with location, it will be necessary to plot voltage versus distance curves for several fixed time instants. These curves are shown in Fig. 10–13. For time reference we will use the instantaneous value of the voltage applied at the input or "sending" end of the line. This is shown as the angle of the rotating phasor at the left of each diagram. In each case the resultant is shown in heavy lines. In this diagram the distance along the line is measured in fractions of a wavelength, *starting from the open end and mea-*

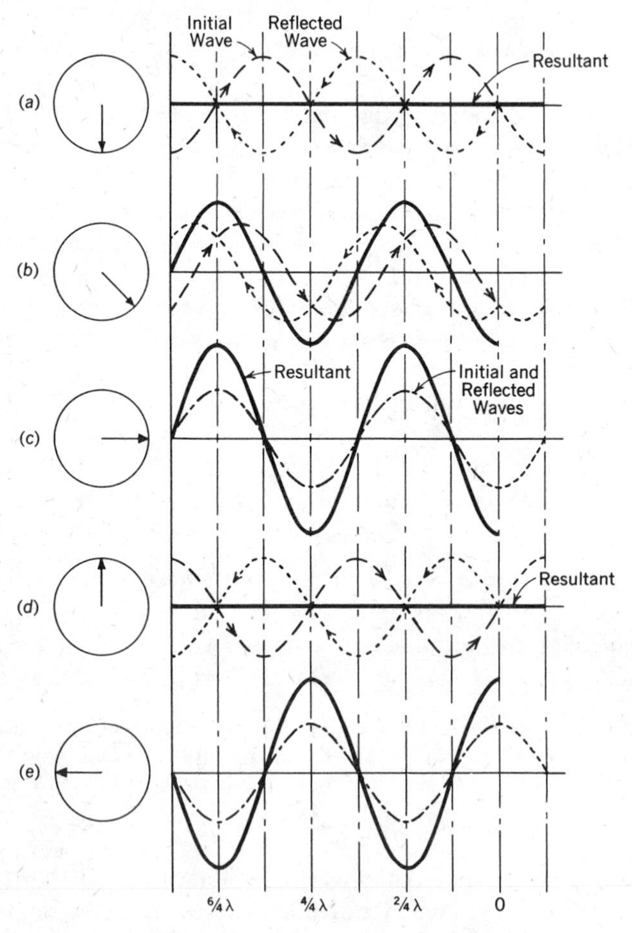

Figure 10-13 Voltage waves on an open line.

suring backwards. The reason for this backwards measurement will become obvious shortly.

In each case (Fig. 10–13), notice that the resultant voltage is zero at every odd quarter-wave point. This would be true no matter what other instants of time we had used for our fixed time reference. If we place voltmeters at each odd quarter-wave point, their readings (rms or peak value) would be zero, regardless of the applied voltage, frequency, or length of the open line. Remember, however, that the distance must be measured from the open end of the line.

Now let us check the value of the resultant voltage at the even quarter-wave points. Examine the half-wave location. In Fig. 10–13(a) the amplitude is zero; at time (b) it is some positive value; at time (c) it reaches a maximum positive value equal to twice the peak value of the initial voltage wave; at time (d) the amplitude is back to zero again; at time (e) it has reached a maximum negative value equal to twice the peak value of the initial voltage wave. Had we used other intermediate instants of time for our fixed time reference, we would find that the value of our resultant would have reached other intermediate values between the above maximum positive and maximum negative values. A little thought should make it obvious that the amplitude of the resultant voltage at this even quarter-wave point will vary sinusoidally with time, from twice the peak value of the initial wave in a positive polarity to an equal amplitude in a negative polarity. If we place a voltmeter at this half-wave location it would indicate (in rms value) twice the rms value of the applied voltage.

Now consider a voltmeter placed at the full-wave location or at the $1\frac{1}{2}$ wavelength location. Notice that here too the amplitude of resultant voltage varies from twice the peak value of the applied voltage in the positive direction to an equal value in the negative direction. The voltmeter readings will therefore be identical at each of the three locations. In fact, regardless of the length of the line, this same voltage would be indicated at all even quarter-wave points—*measured from the open end of the line.*

This time let us examine the condition at some intermediate point, for example, a point halfway between the three-quarter and the full-wave points. At time instant (a) (Fig. 10–13), the value of the resultant is zero; at time (b) it is equal to the maximum value of the initial wave (or one-half of the maximum value of the resultant at the even quarter-wave points); at time (c) it is equal to 86.6% of the peak value of the resultant at the even quarter-wave points. This is the maximum amplitude that the resultant will reach at this location at any time instant. At time instant (d), the value of the resultant is again zero; at time (e) it is the same maximum value as at (c). Here again the amplitude of the resultant varies sinusoidally with time—but now the peak value is only 86.6% of the peak value at the even quarter-wave points. A voltmeter at this location would indicate a lower reading (86.6%) than at the even quarter-wave points.

If a voltmeter is moved from an odd quarter-wave location toward an even quarter-wave location, it would read zero at the odd quarter-wave point, and the reading would increase from zero to twice the applied voltage as we move the meter closer and closer to the even quarter-wave point. Since the voltage along the line varies every quarter-wave between zero and twice the applied voltage, a *standing wave* of voltage is said to exist on the line. The standing wave of voltage for an open line is shown in Fig. 10–14. Remember that the voltage at any location varies sinusoidally with time from some positive maximum to an equal negative maximum value. Yet by convention, the standing wave is drawn in the upward direction only and shows *only the maximum value of the resultant at any location.*

Let us see by a series of questions if you can interpret this standing wave diagram correctly. Refer to Fig. 10–14:

1. Where along this line would a voltmeter reading be highest? If the applied voltage were 100 V (rms), what would this reading be?
   Answer: At any even quarter-wave point, 200 V.
2. What would voltmeters connected across the line at points *A*, *B*, and *C* indicate?
   Answer: At *A* the reading would be the rms value corresponding to a maximum value of *A′*. At *B* the reading would be zero. At *C* the reading would be the rms value corresponding to a maximum amplitude of *C′*.
3. Is the voltage along the line constant at the values shown?
   Answer: No—the diagram shows only the maximum value the voltage will reach. The voltage will vary sinusoidally with time, at the same frequency as the applied voltage.

In the above discussion, we assumed that the line had negligible losses. Now if we introduce the effect of line losses, we will find that the

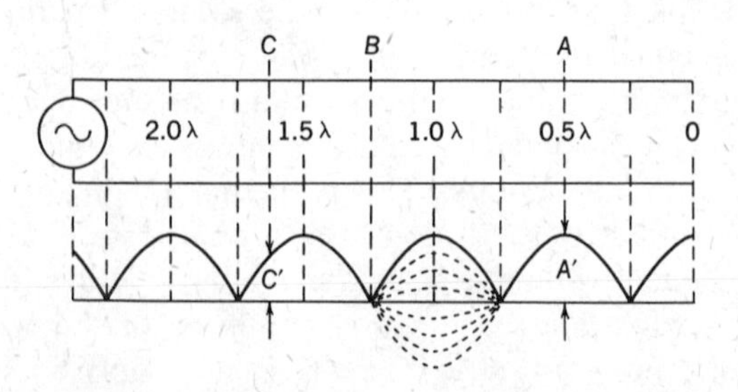

Figure 10-14 Standing wave of voltage on an open line.

initial wave is attenuated as it travels toward the open—its amplitude decreasing. Similarly the reflected wave is attenuated as it travels back to the source. Consequently, the peak value of the resultant will be less than twice the amplitude of the initial wave. Similarly when the initial wave and reflected wave are in phase opposition, the cancellation is not complete, and the resultant value is not zero. The standing wave will be similar to Fig. 10–14, except that the maximum value will be reduced while the minimum value will not be zero.

Standing Wave of Current—Open Lines

When an initial wave of current reaches the open end of a line, a current wave reflects back 180° from how it would have continued. The current at any instant on the line must be the resultant due to the initial

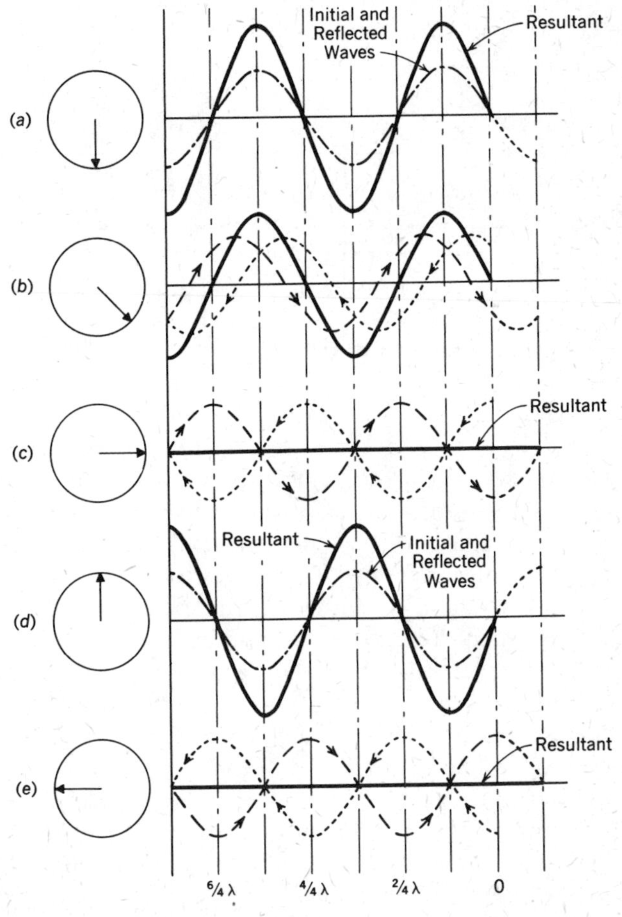

Figure 10-15 Current waves on an open line.

and reflected waves. As in the above cases for voltage waves, the value of the resultant wave of current at any location will depend upon the specific instant at which we chose to analyze the line conditions. Using the same method as we used for voltage waves (Fig. 10–13), let us check the line conditions for five time instants. (These are shown in Fig. 10–15.)

If we examine the resultant current waves in Fig. 10–15, we can see that the resultant is always zero at the open end and at every even quarter-wave point—measured back from the open end. Also, at every odd quarter-wave point, the amplitude of the resultant current varies from zero to twice the peak value of the initial wave, and from positive to negative in direction. Again a standing wave pattern is formed—the current along the line varies from zero at every even quarter-wave point to a maximum at the odd quarter-wave points. This standing wave of current for an open line is shown in Fig. 10–16. Remember that this standing wave shows only the maximum value the current can reach at any location. The current will vary sinusoidally with time to zero and reverse in direction of flow.

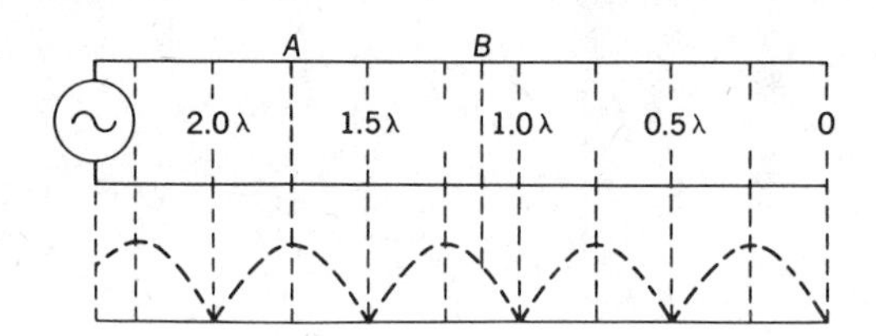

Figure 10-16 Standing wave of current on an open line.

Standing Waves—Shorted Lines

We have just seen that standing waves are caused by reflection of waves. We also know that reflections are caused by discontinuities in a line. Obviously standing waves must also exist on a shorted line. But remember that, on a shorted line, it is the voltage wave that reflects back 180° from how it would have continued, while current reflects back as it would have continued. Therefore the standing wave of voltage will now appear with zero value at the shorted or far end just as we had for current on the open line (Figs. 10–15 and 10–16). Similarly, the standing waves of current on a shorted line will be identical to Figs. 10–13 and 10–14 for voltage on an open line. To help recall the standing wave pattern on a shorted line, remember that at the shorted end the voltage must be zero while the current is a maximum, and that the standing wave pattern reverses every quarter-wave from the shorted end. These relations for a shorted line are shown in Fig. 10–17.

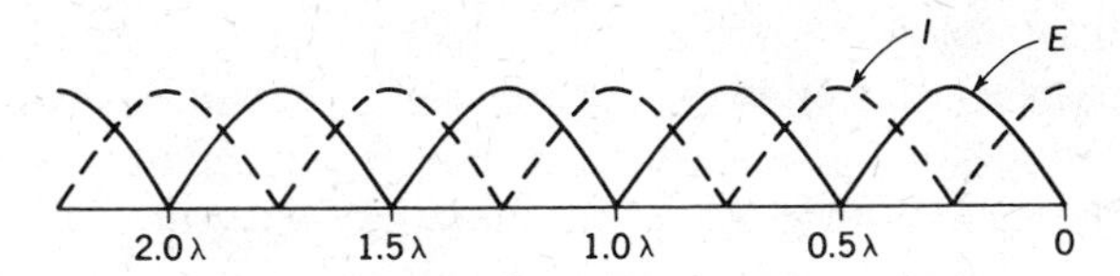

Figure 10-17 Standing waves on a shorted line.

Measurement of Standing Waves on a Line

The existence of standing waves can be readily shown by connecting a neon bulb across the line and sliding it along the line. The brilliancy of the bulb will vary from bright to dim or completely dark as the bulb is moved from the maximum voltage points to the minimum voltage points. If the voltage across the line is high, the bulb can be capacitively coupled to the line merely by holding the leads close to the line. The coupling can be increased by keeping a small portion of the leads parallel to the line. Such a simple means will give qualitative results. If actual voltage values are desired, an RF voltmeter or a dc voltmeter with an RF diode rectifier can be used as the indicator.

To measure the standing wave value of current, an RF ammeter or a dc ammeter with suitable rectifier can be used. Rather than inserting the meter into the line, the meter can be connected to a loop of wire and the wire then magnetically coupled to either line. Such an arrangement will give relative values of current (maximum to minimum). If the actual current value is desired the meter (with coupling loop) must be calibrated against some known value of current.

Standing Wave Ratio (SWR)

If we measured the voltage at numerous places along an open line, we would find that the readings would vary from some maximum value (at the even quarter-wave points) to some minimum value (at the odd quarter-wave points). The minimum voltage would be zero only if the line were a theoretically perfect line with no attenuation. The ratio of the maximum voltage read on the line to the minimum voltage reading is called the *standing wave ratio.* Let us analyze the standing wave ratio for several conditions of operation.

LINE TERMINATED BY $R = Z_0$. In a properly matched line, all the energy is delivered to the load. There are no reflections and no standing waves. The rms value of voltage (or current) along the line is constant. Therefore the ratio maximum-voltage/minimum-voltage is unity. The standing wave ratio for such a line is 1. This gives us a simple method for checking if we have properly matched the line and load.

OPEN LINE OR SHORTED LINE. All the energy that travels down a line will be reflected back to the sending end if the line is open or shorted. If we assume that the line is perfect, there is no attenuation of the waves. Therefore, the maximum value of the standing wave pattern is double the amplitude of the initial wave. The minimum value will be zero, and the standing wave ratio is infinite. Since a practical line has some losses, the amplitudes of the waves decrease as they travel, and the reflected waves will be weaker than the initial waves. The maximum value of the standing wave pattern is reduced, while the minimum value does not fall to zero. Consequently the standing wave ratio is high but not infinite. The greater the attenuation of the line, the lower the standing wave ratio.

LINE TERMINATED BY R GREATER THAN Z_0. Let us consider a line that is terminated with resistive load, but the resistance of the load is much higher than the characteristic impedance of the line. In many respects the situation is somewhat similar to an open line. Some of the energy traveling down the line will be absorbed by the load, but most of the energy travels back as a reflected wave. However, the reflected waves will be weaker than if the line were open. Since there are reflections, standing waves will be formed on the line, but the maximum value of the standing wave pattern is reduced, and similarly the minimum value will not be as low as compared to an open line. Obviously, the standing wave ratio will be less than for an open line. As the resistance of the load is decreased, the standing wave ratio also decreases, reaching unity when $R = Z_0$. If we neglect the attenuation due to line losses, the standing wave ratio of the line is a direct measure of the degree of mismatch. In other words, if $R = 2Z_0$, the standing wave ratio is 2; if $R = 7Z_0$, the standing wave ratio is 7, and so on. Conversely, if the standing wave ratio of the line is 5, this means that the load resistance is five times greater than the characteristic impedance.

LINE TERMINATED BY R LESS THAN Z_0. The last statement in the above paragraph is not necessarily true. A shorted line also has standing waves and a high standing wave ratio. A mismatch, with the load resistance *less* than the characteristic impedance of the line, will also result in reflections and standing waves. Therefore, a standing wave ratio of 5 could also be caused by a load resistance equal in value to one-fifth of the characteristic impedance of the line. Similarly if $R = \frac{1}{3}Z_0$, the standing wave ratio will be 3.

Now we are faced with a new problem. Suppose a transmission line has a standing wave ratio of 4. Obviously, there is a mismatch between line and load. But is the load resistance four times greater or four times smaller than the characteristic impedance of the line? A little thought should make the answer obvious. If the load resistance is lower than the characteristic impedance, the standing wave pattern approaches shorted line conditions. The voltage at the load is a minimum and will rise to a maximum as we move in toward the quarter-wave point. Conversely, if the terminating load resistance is higher than the characteristic impedance, the line approaches an open line, and the load voltage is a maximum. The voltage in this case would decrease as we move in from the load. A simple voltage check will therefore reveal whether the load resistance is too high or too low.

Standing wave ratios can be evaluated if we know the amplitude (or rms value) of the initial and reflected waves, or the reflection coefficient. For example:

$$\text{SWR} = \frac{\text{highest voltage reading}}{\text{lowest voltage reading}} \tag{1}$$

But the highest voltage reading occurs when the initial and reflected voltages are in phase, while the lowest voltage reading occurs when these voltages are 180° out of phase. Therefore

$$\text{SWR} = \frac{E_i + E_r}{E_i - E_r} \tag{2}$$

We also know that

$$\Gamma = \frac{E_r}{E_i} \qquad \text{or} \qquad E_r = \Gamma E_i$$

and replacing E_r in Eq. (2)

$$\text{SWR} = \frac{E_i + \Gamma E_i}{E_i - \Gamma E_i} = \frac{E_i\,(1 + \Gamma)}{E_i\,(1 - \Gamma)}$$

Simplifying,

$$\text{SWR} = \frac{1 + \Gamma}{1 - \Gamma} \tag{10–16A}$$

Obviously, if the SWR is known, Eq. (10–16A) can be solved to find the reflection coefficient:

$$\Gamma = \frac{\text{SWR} - 1}{\text{SWR} + 1} \tag{10–16B}$$

So far, we have seen how the voltage and current on a transmission line vary with distance and with time, for any length of line and for terminations of $R = Z_0$, open or shorted. Very often we also wish to know what the input impedance of a line will be for any length and termination both as to magnitude and type (that is, resistive, capacitive, or inductive). In the following sections, we will analyze various line conditions for impedance.

Input Impedance of a Quarter-Wave Open Line

Any open line will have standing waves of current and voltage. At the open end, the standing wave of voltage is a maximum, and the current wave is zero. One quarter-wave back, the conditions reverse, that is, the standing wave of current is now at its maximum value, while the voltage wave is zero. These standing wave patterns are shown in Fig. 10–18(a). Considering line losses, the above zero values should be replaced by minimum values, since the cancellation of the initial wave by the reflected wave is not complete due to attenuation.

Referring to Fig. 10–18(a), since the current is high (maximum) and the voltage is low (minimum), the ratio E/I is low (minimum). Therefore the input impedance of a quarter-wave line is a minimum. Now let us consider whether this impedance is resistive, inductive, or capacitive. This depends on the phase relation between current and voltage. A quick glance at Fig. 10–18(a) might lead to the conclusion that the current leads the voltage by 90° and that the impedance is capacitive. Such a conclusion is wrong, erroneous, and fallacious! In Fig. 10–18(a) we have plotted a standing wave of voltage and current versus *distance*. Phase relations refer to variations with time. Therefore, before we can draw any such conclusions, we must analyze the conditions existing at the source at one instant of time. This analysis is shown in Fig. 10–18(b).

I_0 and E_0 represent the magnitude and phase relation of the *initial* current and voltage waves at the *open end* of the line at a given instant of time. The input end of the line is one quarter-wave away from the open end. Therefore the initial current and voltage conditions that are *now* at the input must lead the output conditions by 90°. In the diagram they are shown by E_i and I_i. Since these waves are at the input end, they have not been attenuated by line losses and are therefore slightly greater in amplitude than E_0 and I_0. But when the initial waves reach the open end, reflected waves are set up that travel back to the input. The voltage wave

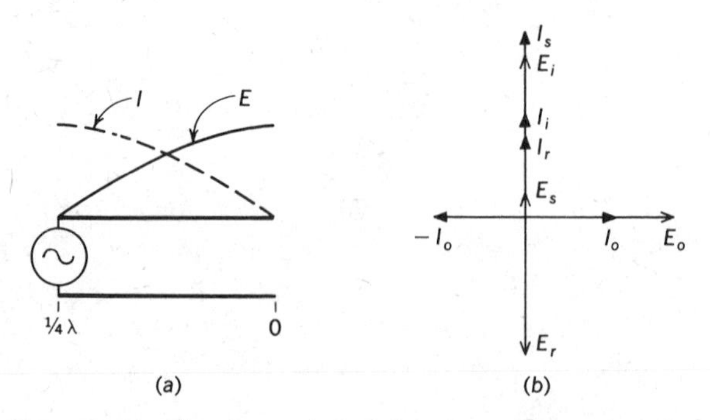

Figure 10-18 The phase relations for a quarter-wave open line.

reflects back as it would have continued. Since it travels back one quarter-wave, any voltage wave that was at the open end some time ago reaches the input end 90° behind the wave *now* existing at the open end. Therefore, at the instant when the initial wave at the input is E_i and the initial wave at the open end is E_0, the reflected wave at the input end is E_r, and the total voltage at the input end (E_s) is the phasor sum of E_i and E_r.

Now let us check the conditions regarding current. We have already seen that when the initial current wave at the input end is I_i, the initial current wave at the open end is I_0, 90° behind I_i. But current is also reflected back, and the reflection is 180° from how it would have continued. So the reflected wave now at the open end is $-I_0$. But a reflected wave (I_r) that was at the open end some time ago now just reaches the input end, and since it traveled one quarter-wave from the open end back to the input end, this reflected wave of current lags $-I_0$ by 90° more. Therefore, for a quarter-wave line, the reflected wave of current reaching the input end is always in phase with the initial wave just starting down the line. The total current at the sending end (I_s) is therefore the phasor sum of I_i and I_r and, neglecting losses, would be double the value of the initial wave. Due to attenuation I_0 is slightly lower in amplitude than I_i; in turn, I_r is slightly lower than I_0, and I_s is less than twice I_i.

Notice that the voltage and current, E_s and I_s, at the input end are *in phase*. Also notice that E_s is at a minimum while I_s is at its maximum value. Therefore the input impedance of a quarter-wave line is a minimum and *resistive*.

Input Impedance of a Half-Wave Open Line

Figure 10–19(a) shows the standing wave pattern for an open line one half-wave long. Since the voltage is a maximum and the current a minimum at the input end, the input impedance must be a maximum. However, the phase relation between the current and voltage at this point cannot be seen from this diagram. This is a distance diagram. As before we must develop the phasor diagram before we can answer this question.

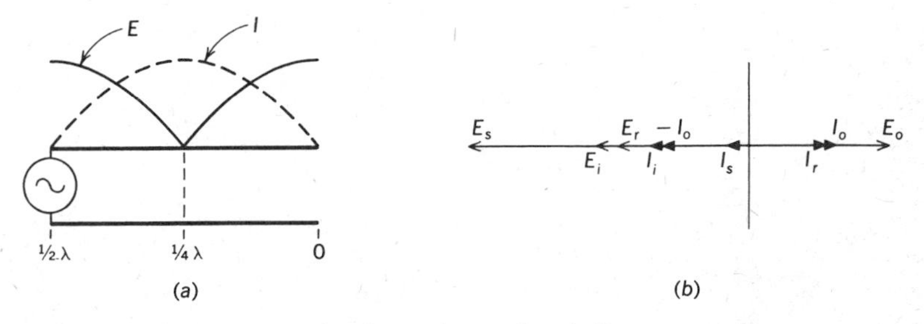

Figure 10-19 The phase relations for a half-wave open line.

Such a diagram is shown in Fig. 10–19(b). Again we will let I_0 and E_0 represent the initial waves of current and voltage at the open end. Since the line is one half-wave long, the initial current and voltage waves that are *now* at the input end (I_i and E_i) must lead the output conditions by 180°. An open line has reflected waves. The voltage wave reflects back as it would have continued, but by the time it reaches the input end a half wave away it is 180° reversed from the initial wave now at the open end. This reflected voltage when it reaches the sending end is therefore always in phase with the initial wave now at the input end. The total voltage at the sending end (E_s) is the phasor sum of E_i and E_r. Its value is therefore almost double the initial voltage.

The reflected wave of current at the open end is $-I_0$ and is 180° reversed from the initial wave at this point (I_0). But the reflected current wave goes through another 180° reversal in its travel back to the input end. Therefore, a reflected wave of current, which was at the open end of the line a short time ago, now reaches the sending end of the line in phase opposition to I_i. The total current at the input end (I_s) is the phasor sum of I_i and I_r. Notice that the resultant current and voltage at the input end (I_s and E_s) are again in phase. However, this time, the voltage is a maximum and the current is a minimum. Therefore the input impedance of a half-wave open line is a maximum and resistive.

Input Impedance of an Open Line less than One-Quarter Wavelength

By now it should be obvious that the only way we can find the impedance of a line with standing waves is to develop the phasor diagram for a given time instant. Let us apply this same system to a line of one-eighth wavelength. E_0 and I_0 in Fig. 10–20 represent the initial waves of voltage and current now at the open end. For an eighth-wavelength line the initial waves of current and voltage now at the sending end (I_i and E_i) must be 45° ahead of the output conditions. The reflected voltage wave

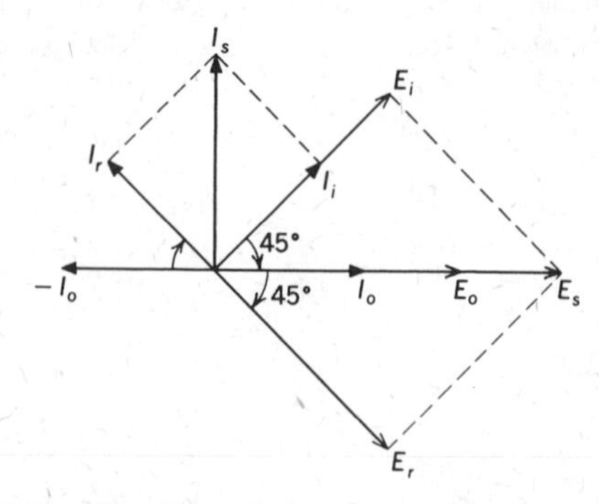

Figure 10-20 The phase relations for an eighth-wave open line.

that reaches the input end of the line (E_r) must lag E_0 by 45°. The resultant voltage at the sending end (E_s) must be the phasor sum of E_i and E_r. The reflected current wave at the open end is $-I_0$. But the reflected current now at the input end (I_r) must lag $-I_0$ by 45°. The total current at the input end (I_s) must be the phasor sum of I_i and I_r. Notice that the current I_s leads the voltage E_s by 90°. Also notice that neither the current nor the voltage is at maximum or minimum value. Therefore the input impedance is intermediate in value and capacitive in nature. A more careful analysis of this diagram will show that the ratio of E_s/I_s is the same as E_i/I_i. But this second ratio, E_i/I_i, is equal to the characteristic impedance of the line! The impedance of an open line exactly one-eighth wavelength long is capacitive and has a reactance X_c equal to Z_0. Had we selected any other length of open line (less than one quarter-wave), we would have found that the input impedance would still be capacitive, but the reactance (X_c) would decrease as we approach one-quarter wavelength.

Input Impedance of an Open Line between One-Quarter and One-Half Wavelength

For convenience let us analyze an open line that is three-eighths wavelength long. It should be obvious that the initial waves at the input end will lead the open-end conditions by 135°. Similarly, the reflected waves at the input end will lag the reflected waves at the output end also by 135°. The resultant voltage and current at the input end are the phasor sums of the initial and reflected values. These relations are shown in Fig. 10–21. This time the current at the input end (I_s) lags the voltage (E_s) by 90°. The input impedance is therefore inductive. For this particular length of line the ratio E_s/I_s is again equal to the ratio E_i/I_i. The magnitude of the input impedance is equal to the characteristic impedance of the line. For any other length of line (between one-quarter and one-half wavelength) the input impedance of the line would still be inductive. The value of the

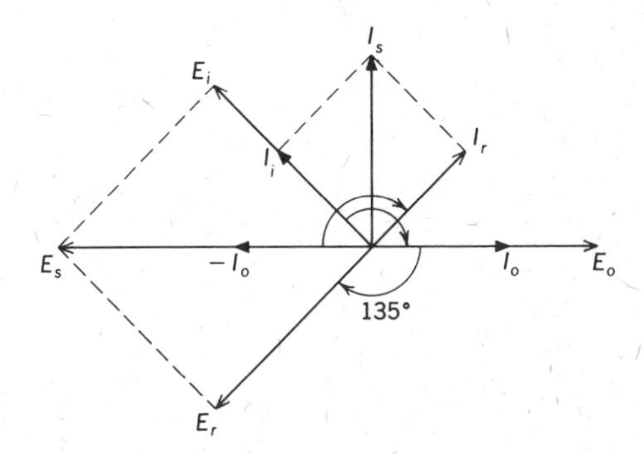

Figure 10-21 The phase relations for a three-eighth-wave open line.

inductive reactance would vary from a minimum for a line just above one-quarter wavelength to a maximum as the length approaches one-half wavelength.

Summary of Input Impedance—Open Lines

From the above discussion we have seen that the impedance of an open line is pure resistance for quarter-wave and half-wave lines, being a minimum at the quarter point and a maximum at the half-wave point. Since all odd quarter-wave points are identical and all even quarter-wave points are also identical as to standing wave patterns, these resistive impedances will be repeated at every quarter-wave point. Similarly, the input impedance for a line between one-half and three-quarter wavelength will repeat the conditions found for lines between zero and quarter-wave in length. The variation of impedance of a line with the length of the line is represented pictorially in Fig. 10–22.

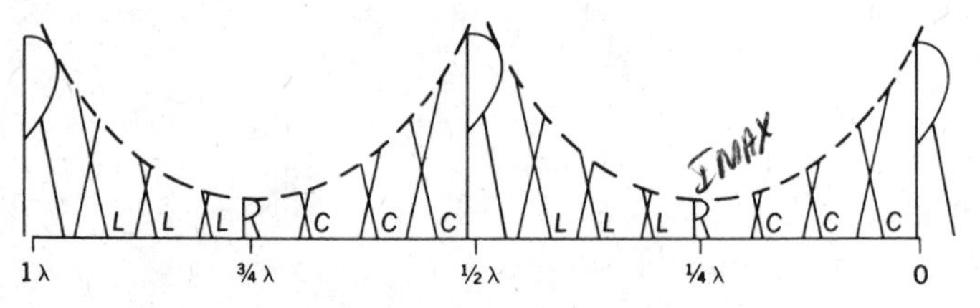

Figure 10-22 Variation of input impedance with length on an open line.

Input Impedance Variations—Shorted Lines

We saw earlier that the standing wave patterns for current and voltage on a shorted line are just the reverse of the patterns for an open line. Therefore, without going through detailed phasor-diagram analysis, it should be obvious that the impedance variation with length for a shorted line will be just the opposite of the results obtained on open lines. To summarize:

1. *Shorted line, one quarter-wave.* The standing wave pattern at the input end is high voltage (maximum) and low current (minimum). The input impedance is a maximum and resistive.
2. *Shorted line, one half-wave.* The input impedance is low and resistive.
3. *Shorted line, less than one quarter-wave.* The input impedance is inductive. The impedance is low for a short line but increases as we approach a quarter wave.
4. *Shorted line, between one quarter-wave and one half-wave.* This section of line acts as a capacitor. The input impedance (X_c) is

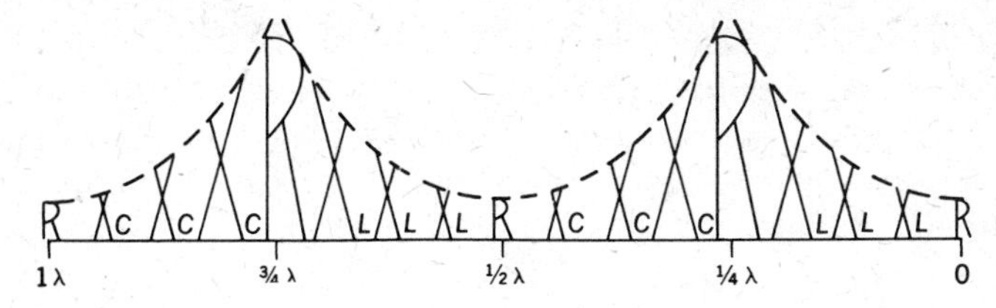

Figure 10-23 Variation of input impedance with length on a shorted line.

high for a length just above a quarter-wave section and decreases as we approach a half-wave section. These impedance variations are shown pictorially in Fig. 10–23.

Evaluation of Impedance—the Smith Chart

From the discussions of input impedance, we have seen, *qualitatively,* how the impedance varies with line length, but—except for the special cases of one-eighth and three-eighth wavelength—no evaluation of the actual input impedance value was made. Furthermore, these discussions applied only to open or shorted lines without losses and did not consider any other load terminations. Yet, transmission lines are basically used for power transfer, and optimum efficiency requires proper impedance match between source, line, and load. It is therefore necessary to evaluate the input impedance of a line, not only for various lengths, but also for various load conditions. Although mathematical solutions can be made,* such solutions are laborious. Instead, it is common practice to solve for such values graphically. The most widely used technique employs the *Smith chart,* shown in Fig. 10–24. The chart consists of two sets of orthogonal circles. One set, tangent at the right-hand center of the chart (the infinity point), represents the ratio R/Z_0, where R is the resistive component of the line impedance and Z_0 is the characteristic impedance. The range of usable values for R/Z_0 extends 0–50. Two such circles—$R/Z_0 = 0.3$ and $R/Z_0 = 2.0$—have been darkened on the chart for easy identification.

The second set of circles—or really arcs—represents the ratio jX/Z_0, where X is the reactive component of the line impedance. These arcs all start from a common point (the infinity point). Those curving downward represent negative or capacitive reactances; those curving upward are for positive or inductive reactances; the center horizontal line represents zero reactance, or pure resistance. Two reactance arcs — $X/Z_0 = -0.9$ and $X/Z_0 = 0.5$—are darkened on the chart for easy identification. The range

$$*Z_i = Z_0 \frac{Z_L \cos \beta l + jZ_0 \sin \beta l}{Z_0 \cos \beta l + jZ_L \sin \beta l}$$

of usable values for X/Z_0 extends 0–50. Notice that each of these scaled values represents the line impedance values (R and X) divided by the characteristic impedance value Z_0. This process, *normalization*, refers all transmission line values to a common characteristic impedance of unity and thereby makes it possible to use the same chart for a line of any characteristic impedance. To obtain true impedance values, it is necessary to multiply the normalized values obtained from the chart by the actual Z_0 of the line. The normalized impedance of any point on the chart is readily evaluated by noting the R/Z_0 circle and X/Z_0 arc on which it lies.

EXAMPLE 10–7

Find the normalized impedance corresponding to points A, B, and C in Fig. 10–24.

Solution

1. Point A is at the intersection of the 0.3 resistance circle and the -0.9 reactance arc. The normalized impedance for this point is $0.3 - j0.9\ \Omega$.
2. Point B is at the intersection of the 2.0 circle and the 0.5 arc. Therefore $Z_n = 2.0 + j0.5\ \Omega$.
3. The impedance corresponding to point C is $1.0 + j0\ \Omega$.

An important feature of this chart is that any *lossless* line can be represented by a circle having its origin at the center point of the chart $(1 + j0)$, and with a radius such that the circle crosses the horizontal axis *on the right-hand side* at an R/Z_0 value equal to the standing wave ratio on the line. For example, in Fig. 10–24, the dashed circle, with point C as its origin, crosses the horizontal (right side) at $R/Z_0 = 1.6$. It therefore represents a transmission line with a standing wave ratio of 1.6. Moving around this circle corresponds to traveling along the transmission line. Now examine the outermost circle around the chart. The scale for this circle is graduated in wavelength units, starting with zero at the left-hand center of the chart (opposite the infinity point). This scale is used to measure distance between points on the transmission line—in wavelength units. Distance from the load (toward the generator) is measured in a clockwise (cw) direction, and vice versa. Notice that the wavelength scale has a maximum value of 0.5 wavelength. Distances greater than this value are readily obtained by going around the circle as many times as necessary.

Once a line has been represented by its standing wave circle, the impedance of any point on the line can be determined if the load impedance is given. This is most readily shown by a problem.

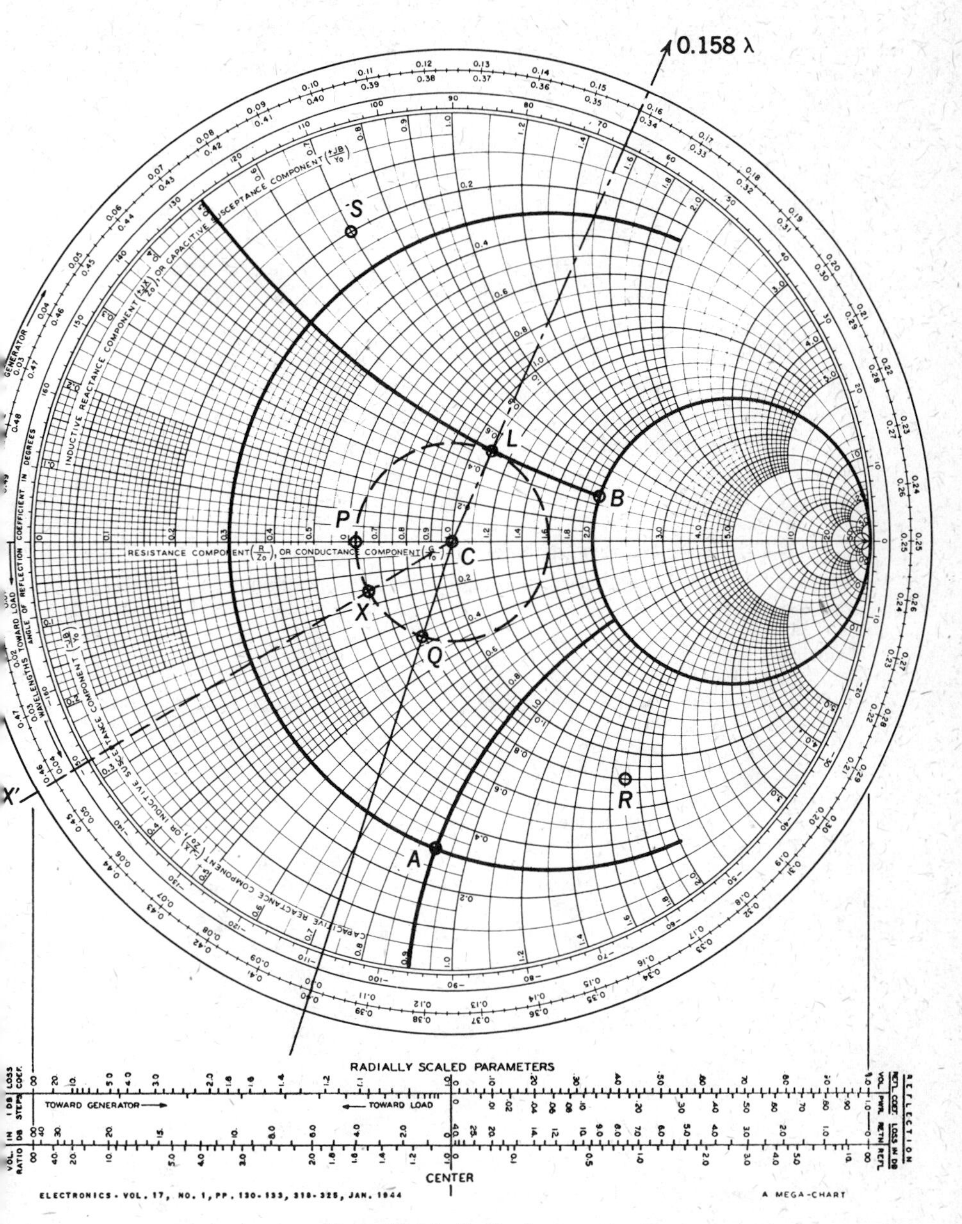

Figure 10-24 Smith chart.

EXAMPLE 10–8

A transmission line is terminated by a pure resistive load, but this load produces a standing wave ratio of 1.6. Find the load impedance value and the input impedance (normalized values) if the line is 0.40 wavelength long.

Solution

1. (See Fig. 10–24.) For a pure resistive load (assuming $R_L < Z_0$)

$$R_L = \frac{Z_0}{\text{SWR}} = \frac{1}{1.6} = 0.625 + j0$$

   Notice that this is exactly where the SWR circle of 1.6 crosses the horizontal (zero reactance) on the left side. Therefore point P is also the load end of the line.
2. Measuring clockwise around the wavelength circle to 0.40λ, connect this point to the origin (C). This line crosses the SWR circle at Q. The coordinates at Q are $0.8 - j0.36\ \Omega$. This is the input impedance of the line (normalized value).

EXAMPLE 10–9

A 72-Ω transmission line is terminated with a load $Z_L = 79.2 + j36.0\ \Omega$. This produces a standing wave ratio of 1.6. Find the impedance at a point (X) 1.30 wavelengths from the load.

Solution

1. (See Fig. 10–24) The normalized load impedance is

$$Z_{L(n)} = \frac{Z_L}{Z_0} = \frac{79.2 + j36.0}{72} = 1.1 + j0.5$$

2. On the SWR = 1.6 circle, locate point L, whose coordinates are $1.1 + j0.5$.
3. Draw a line from C through L to the wavelength circle. This (0.158λ) is the load end of the line.
4. Point X' should be 1.30λ from point L, and 1.458λ ($1.30 + 0.158$) or 0.458λ from the zero point of the scale. Locate this point; draw a line to the center of the SWR circle, crossing the circle at X.
5. The normalized line impedance at this point (1.30λ from the load) is read off as $0.66 - j0.17\ \Omega$.
6. The actual impedance is $Z_0(0.66 - j0.17) = 72(0.66 - j0.17)$

$$= 47.5 - j12.3\ \Omega.$$

In the above examples, the standing wave ratio was given to define the transmission line circle. It should be noted, however, that the impedance value at any point on the line (normalized) is sufficient to define the circle and give the standing wave ratio.

The above discussion has assumed ideal conditions, wherein the line has no losses, and the standing wave ratio is constant throughout the length of the line. However, if the line attenuation cannot be neglected, not only does the initial (incident) wave get weaker as it travels toward the load, but also the reflected wave decreases in amplitude as it travels back toward the source. Obviously the standing wave ratio decreases as we approach the source end of the line. A true standing wave representation on the Smith chart would be a spiral. Since a constant-radius circle is used, correction is necessary. Such corrections are made using the TRANSM LOSS–1DB STEPS scale along the bottom (horizontally) of the Smith chart.

EXAMPLE 10–10

A transmission line is terminated so as to produce an input impedance of $1.0 + j1.0\ \Omega$ (normalized value). Find the impedance at a point (P) 1.3λ *closer* to the load, if the line attenuation between these two points is 2.5 dB.

Solution

1. (See Fig. 10–25.) Locate point O with the given impedance ($1.0 + j1.0$).
2. Draw the SWR circle corresponding to this value, assuming no losses.
3. Draw a line from the origin through point O to intersect the wavelength circle at 0.338λ—on the ccw (counterclockwise) scale, since we are going to measure distance toward the load.
4. Locate point P' on the wavelength circle at a distance (ccw) corresponding to 1.3λ from point O.
5. Connect point P' to the origin. Point P would give the impedance value at the desired location—*if there were no line losses*. Now we must make the correction.
6. From the left-hand extremity of the SWR circle, drop a vertical down to the 1-dB transmission loss scale (point A).
7. Moving toward the load (to the left), locate a point on this scale 2.5 dB away—(point B). (Do NOT use the numbered LOSS COEF scale.)
8. Draw a second vertical back to the horizontal center-line of the chart. This specifies the corrected SWR circle radius.
9. Draw the SWR circle for the desired location (shown by dash line). The intersection P'' of this circle with line P–P' represents the impedance value at a point 1.3λ from point O, corrected for the line loss of 2.5 dB. This value is $0.27 - j1.15$ ohms (normalized).

Lines Terminated by Reactive Loads

From basic fundamentals, we know that inductors and capacitors do not dissipate energy. What happens, then, if a line is terminated by a pure reactive load? The initial energy traveling down the line is not absorbed

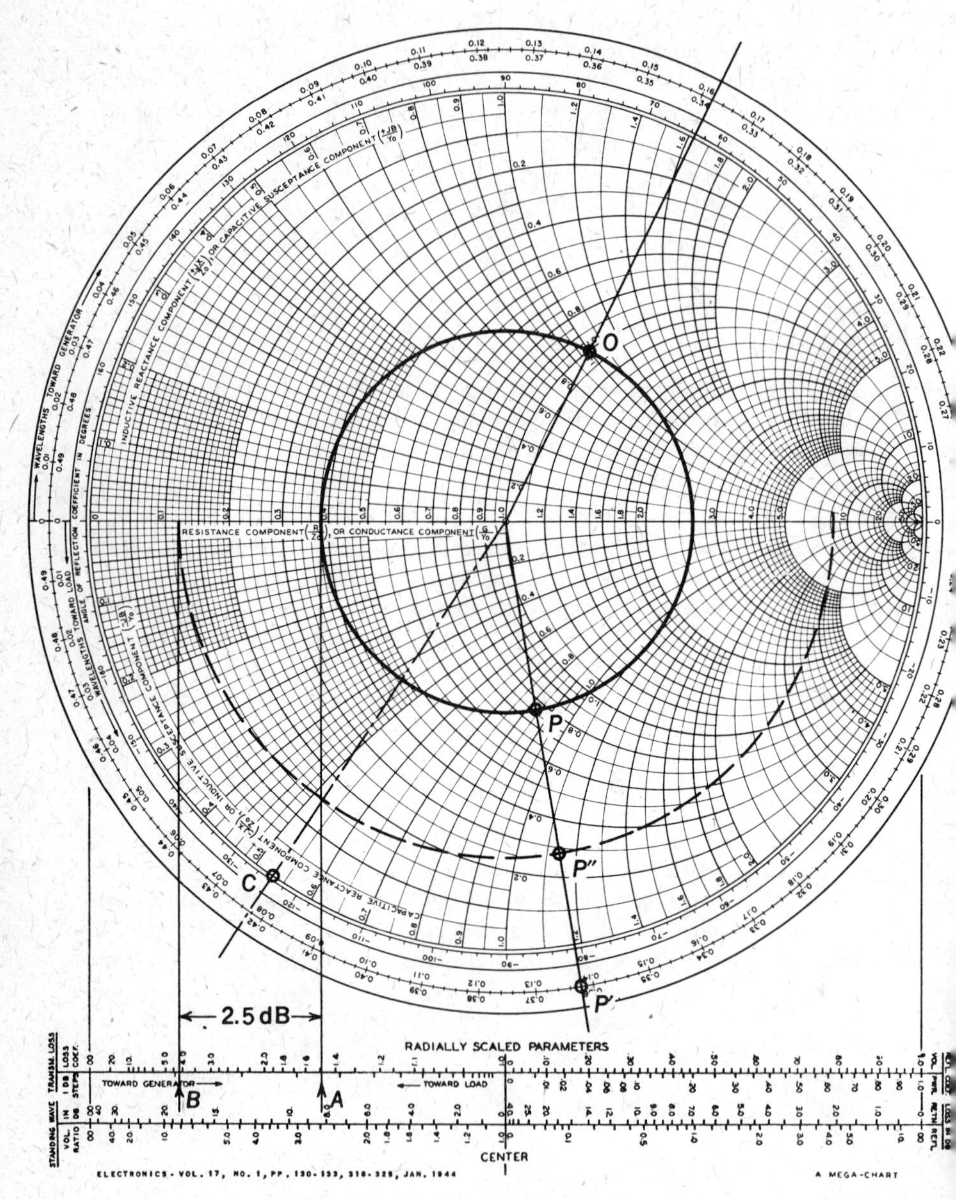

Figure 10-25 Correction for line attenuation.

by the load, but is reflected back completely, as if the line were open or shorted. However, this type of termination (*L* or *C*) will cause a shift in the standing wave pattern.

Earlier, we saw that a piece of open transmission line shorter than one quarter-wave acts as a capacitance. Therefore, terminating a line with a capacitor is similar to extending the line by some distance less than one quarter-wave and leaving the line open. The question now is, by how much is the line length seemingly increased? The added "electrical" length depends on the capacitance of the load and the operating frequency. For example, if the reactance (X_c) of the load is equal to the characteristic impedance of the line, the added electrical length is one-eighth wavelength. Referring to Fig. 10–22, the smaller the capacitance of the terminating load (and the higher its reactance), the shorter is the equivalent electrical length added to the line. This should be obvious, because if the terminal capacitance were negligible, the added length of line would also be negligible, and the line would act as a normal open line.

Now let us assume the line is terminated by an inductance. If the inductance were very low and its reactance were negligible the line would act as a shorted line. Refer back to Fig. 10–23 for a moment. From this figure you will notice that a shorted section of transmission line less than one-quarter wavelength acts as an inductance, and that the shorter the section of line, the lower the reactance, or the smaller the equivalent inductance. Therefore, if we terminate a line with an inductive load, it is equivalent to adding a section of line less than one-quarter wavelength and shorted. The increase in the "electrical" length of the line will depend on the inductance of the load and the operating frequency. The lower the reactance of the load, the shorter is the increase in electrical length.

Figure 10–26 shows the effect of reactive terminations on the standing wave pattern. In general, a line with a capacitive load acts like an open line (of increased length) with its standing wave pattern shifted towards the load. The amount of shift increases with the capacitance of the load from zero towards one-quarter wavelength. Similarly, a line with inductive load acts like a shorted line (also of increased length) with its

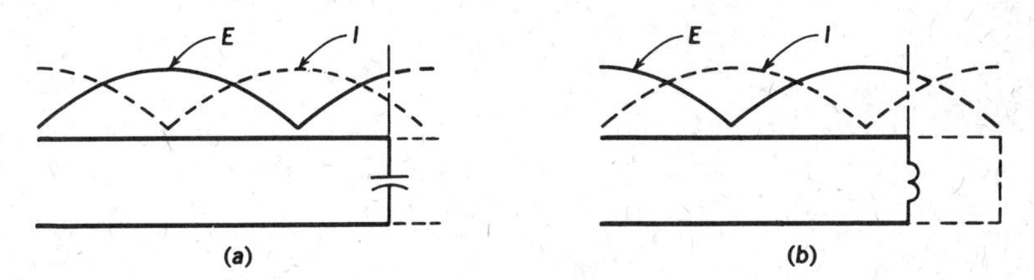

Figure 10-26 Effect of reactive loads on a standing-wave pattern.

standing wave pattern shifted toward the load. Again the amount of shift increases with the inductance of the load from zero toward one-quarter wavelength.

TRANSMISSION LINE APPLICATIONS

As mentioned at the start of this chapter, the basic function of any transmission line is to transfer energy from one unit (source) to another (load). It is now obvious that such lines should be matched to the source and to the load to prevent reflection of energy and unnecessarily high losses due to standing waves. On the other hand, there are many applications of transmission lines that depend on standing waves. Such applications will be discussed next.

Wavelength and Frequency Measurements

If an unknown frequency is fed into an open-ended section of transmission line, we can readily determine the frequency of the applied signal. Using one of the current or voltage indicators described earlier, merely measure the distance between successive minimum (or maximum) points in the standing wave pattern that results on the line. Then, using the equation for wavelength, solve for the frequency. Of course, correction must be made for velocity by using the proper velocity constant for the type of line used. More accurate results will be obtained if the minimum points of the standing wave pattern are used. Check back on Fig. 10–14 or 10–16. You will notice the maximum points are broad, while the minimum levels are abrupt changes.

Transmission line sections used specifically for this purpose are called *Lecher lines*. They are made of parallel two-wire lines and are generally one-half to five wavelengths long. For laboratory measurements, commercial units called *slotted lines* are used. Essentially, a slotted line is a section of tubular coaxial line, with a slot cut along the outer tube. Mounted in a box directly over the slot is an RF voltmeter. A probe extends down from the box, through the slot, and couples loosely to the inner conductor. The voltmeter and probe assembly can be moved along the length of the coaxial line. This assembly is shown in Fig. 10–27.

There are many uses for the slotted line. By itself, it can be used in place of a Lecher line to determine the frequency of an unknown source. By inserting a slotted line between the generator and the actual transmission line, we can determine:

1. The velocity constant of the line.
2. The impedance and phase angle of any load.
3. Correct matching between generator and line.
4. Correct matching between line and load.

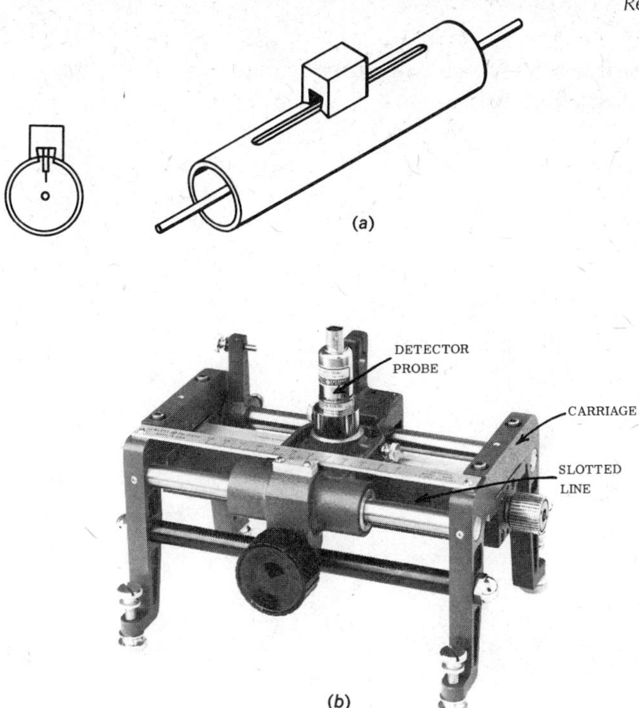

Figure 10-27 A slotted line system.

For these uses, the impedance of the slotted line must match the impedance of the actual line. The actual procedure for such measurements will be given in the instruction book for the line.

Reactance

At high frequencies it is very difficult to construct a capacitor or an inductance of low enough value. In addition, an inductor having excellent characteristics at lower frequencies may have a very poor figure of merit (Q), due to increased losses. What is even worse, it may have more capacitance than inductance. Similarly, a standard type of capacitor may also have very high losses, or end up with more inductance than capacitance. But we saw earlier that a section of transmission line shorter than one-quarter wavelength acts as a capacitor if the line is open, or as an inductance if the line is shorted. Therefore, don't be surprised if a schematic diagram of some VHF or UHF equipment shows a capacitor or inductor, and when you check the equipment itself, you find that these circuit elements may be the connecting wires, cut to accurate lengths and acting according to transmission line principles.

We already know that a one-eighth wavelength line has a reactance equal to the characteristic impedance of the line and that the reactance,

X_L, of a shorted line decreases as the line is made shorter, while X_C of an open line increases if the line is shorter. If we are to use lines for a specific reactance value, the relationship between line length and reactance must be made more definite. Mathematically, this relation is

1. $X_L = Z_0 \tan(360^\circ\, x)$ **(10–17A)**

2. $X_C = \dfrac{Z_0}{\tan(360^\circ\, x)}$ **(10–17B)**

where x is the line length in wavelengths.

EXAMPLE 10–11

What length of line is needed and how should it be terminated to act as a capacitance of 2 pF at 500 MHz? The line impedance (Z_0) is 300 Ω.

Solution

1. $X_C = \dfrac{1}{2\pi fC} = \dfrac{0.159 \times 10^{12}}{500 \times 10^6 \times 2} = 159\ \Omega$

2. $\tan(360^\circ\, x) = \dfrac{Z_0}{X_C} = \dfrac{300}{159} = 1.9$

3. $360^\circ\, x = 62.3^\circ$, and $x = \dfrac{62.3}{360}\lambda = 0.173\lambda$

4. At 500 MHz, $\lambda = \dfrac{V}{f} = \dfrac{984}{500} = 1.97$ ft

5. Length needed $= 0.173 \times 1.97 = 0.341$ ft or 4.09 in

6. For capacitive effect, the line must be open.
7. The line could also be 0.423 λ and shorted.
8. These same answers can also be obtained by using the Smith chart. The normalized value of input impedance (X_C) is 159/300 or $0 - j0.53$. Locate this point on the Smith chart (point C in Fig. 10–25). Draw a line from the center, through this point, intersecting the wavelength circle at 0.423λ. This is the length of line required for a *shorted* section. For an open-ended section, use a length one quarter-wave shorter (0.423 − 0.25) or 0.173λ.

This principle is sometimes used in FM or TV tuners. To obtain a high Q, a minimum of capacitance is used—sometimes only the transistor (or tube) capacitance and stray capacitance. To decrease losses, the inductance is a section of shorted coaxial line less than one quarter-wave at the operating frequency. Tuning is accomplished by varying the position of a shorting plunger in the line.

Tank Circuits

Let us assume that we have a section of transmission line that is exactly one-quarter wavelength, and shorted at one end. Reference to Fig. 10–23 shows that the input impedance is a maximum and resistive. This is the same condition that exists in a parallel *L-C* circuit at resonance. A little thought will also show that, at a slightly lower frequency, this section of line is now too short for a one-quarter wavelength. The input impedance decreases and becomes inductive. The same effect occurs in a tank circuit operating below its resonant frequency. On the other hand, if we raise the applied frequency above the original value, the line will now be longer than one-quarter wavelength (but shorter than one-half wavelength), and the input impedance is capacitive. The same change in impedance also applies to a parallel-resonant circuit. Since a quarter-wave shorted line acts as a parallel-resonant circuit, it is often used in place of a conventional tank circuit in VHF and UHF applications. Variation of a shorting bar, or plunger, will change the length of the section and cause it to be resonant at different frequencies. An advantage of transmission line tuners is that Q values above 1000 are readily obtainable. This is appreciably higher than can be obtained with conventional *L-C* circuits.

It should also be obvious from Fig. 10–23 that series-resonant effects can be obtained by using a shorted section of transmission line that is one-half wavelength long. It should further be obvious that either of the above resonant effects can be obtained using a section of open-ended transmission line. (See Fig. 10–22.) For series resonance, the open line section should be one-quarter wavelength, while tank circuit action (parallel resonance) requires an open-end section one-half wavelength long.

The following examples will show the advantage of a transmission line tuner over lumped *L-C* constants.

EXAMPLE 10–12

Assuming we could construct a coil of 1 μH, what value of capacitor would be needed for resonance at 500 MHz?

Solution

1. $X_L = 2\pi fL = 6.28 \times 5 \times 10^8 \times 1 \times 10^{-6} = 3140\ \Omega$

2. $C = \dfrac{1}{2\pi f X_C} = \dfrac{0.159 \times 10^{12}}{5 \times 10^8 \times 3140} = 0.1\ \text{pF}$

But such a capacitance value is impossible! Its own stray capacitance would be more. In addition the *L/C* ratio would be poor, the gain low, and the selectivity bad.

EXAMPLE 10–13

What length of transmission line should be used and how should it be terminated for use as a tank circuit at 500 MHz?

Solution

1. At 500 MHz, $\lambda = \frac{V}{f} = \frac{984}{500} = 1.97$ ft

2. $\frac{1}{4}\lambda = \frac{1.97 \times 12}{4} = 5.91$ in

3. The quarter-wave line must be shorted.

A "parallel-resonant" circuit of this type has much higher Q than a conventional L-C circuit—even if such a circuit could be made. In spite of their superiority, transmission line sections are not practical for use on the broadcast band. At 1000 kHz (1 MHz), the length of line needed would be 500 times longer, or approximately 250 ft!

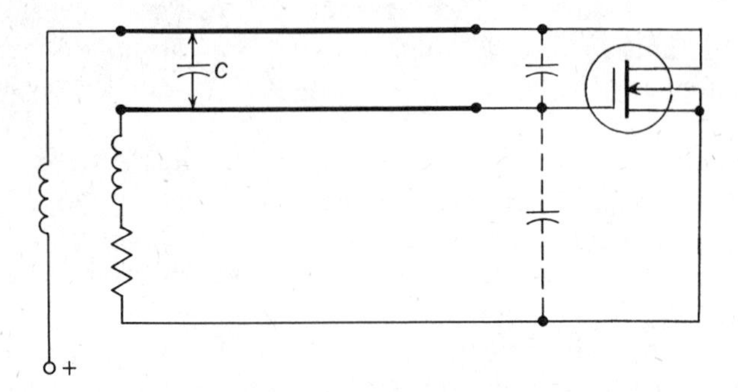

Figure 10-28 UHF oscillator using a resonant line.

Figure 10–28 shows a transmission line used in a simple high-frequency oscillator. The frequency of the oscillator is adjusted by varying the position of the shorting capacitor C. The capacitor is effectively the same as a shorting bar. An actual bar cannot be used because the dc potential is applied to the plate of the FET through the upper line.

Supporting Insulators

At ultrahigh frequencies, it is possible to use a piece of metal as an insulator. A quarter-wave section of line shorted at one end has a high impedance at the other end. If we attach the object to be insulated at the open end, it will not affect the operation of the unit. What is more, this type of insulator is superior to the best insulating material. Dirt, moisture, or other impurities cannot make it "leak." It is already a conductor. Of

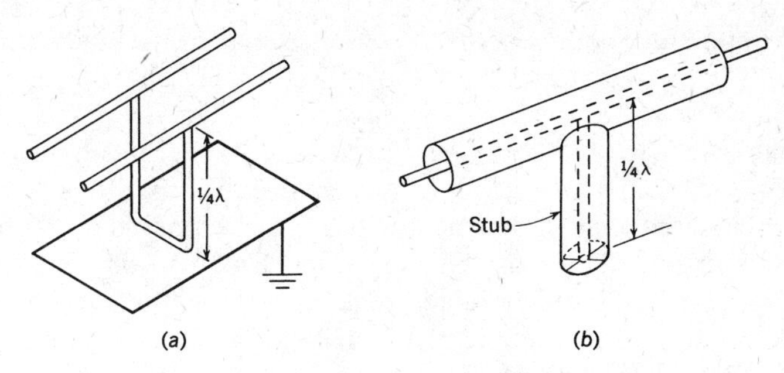

Figure 10-29 A quarter-wave stub as a metallic insulator.

course this type of insulator can only be used for a narrow band of frequencies at which it essentially remains a quarter-wave section. Figure 10–29 shows a quarter-wave "stub" used as an insulator (a) to support an open wire line and (b) to support the inner conductor of a coaxial line.

Wave Filters

For efficiency of power transfer from a tube to a loaded tank circuit, it is necessary to use relatively low Q in the resonant circuit. Since most transmitters operate in class C and have high harmonic content, harmonics may be getting out to the antenna. The worst offender would be the second harmonic. A simple means of suppressing even harmonics is by use of transmission line stubs.

Figure 10–30 shows two methods using quarter-wave stubs. At the fundamental frequency, the shorted stub in (a) presents a high impedance

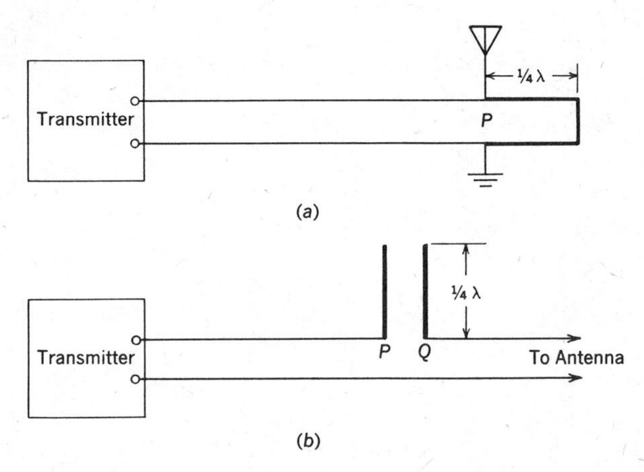

Figure 10-30 Even harmonic suppression by quarter-wave stubs.

to the antenna. Therefore, there is no loss. At the second harmonic frequency (or any even harmonic), the stub is one half-wave long. The short at the far end makes the input impedance zero. No energy is delivered to the antenna. The stub could be placed at any other location.

Now consider the action of the open stub in Fig. 10–30(b). To the fundamental frequency, an open at the far end, one quarter-wave away, means zero impedance at the main line. Energy from the transmitter travels down as if there were no break in the line. But at the second harmonic, or any even harmonic, the stub becomes an even number of quarter-waves in length and an open at the far end means high impedance at the main line. Energy at these frequencies cannot be delivered to the antenna. Combinations of (a) and (b) can be used to increase the efficiency of the suppression.

Phase-Shifting Circuits (Delay Lines)

Sometimes it is necessary to introduce a definite amount of phase shift between signals, or to feed the same signal to more than one load but with different phasing for each load. A particular example is in obtaining an antenna system to send energy out in a narrow beam, in one direction. Such an antenna system may contain several elements. The signals fed to each element must have some specific phase relation.

We know that it takes time for a signal to travel through a transmission line. At any particular time, the phase of the initial signal will depend on how far it has traveled in wavelengths. Therefore, to get different phase relations, all we need to do is feed each unit with transmission lines of different lengths! But we do not want standing waves. So we must prevent reflections by terminating each line with a load equal to its characteristic impedance. The length of line to be used for a given phase shift can be obtained from a simple ratio. One wavelength is 360°; therefore:

1. For 30°: $\frac{30}{360}$ or $^1/_{12}\lambda$

2. For 45°: $\frac{45}{360}$ or $^1/_8\lambda$

3. For 120°: $\frac{120}{360}$ or $^1/_3\lambda$

4. For 180°: $\frac{180}{360}$ or $^1/_2\lambda$

The last example brings out another point. A 180° shift in phase is inverter action. If we need phase inversion, merely feed the signal through a half-wave section of line—but the line must be a "nonresonant" line, that is, terminated by $R = Z_0$.

Television Antenna High-Low Feed

For optimum results, particularly at high frequencies, it is desirable to use an antenna that is one-half wavelength at the desired frequency. In television, a high signal-to-noise ratio is needed for clear picture reception. Therefore, proper antenna length is important. Also, to eliminate "ghosts" caused by reflected waves, proper matching of the transmission line is required. Most cities have more than one frequency channel allocated to their area. If the channels were all in one narrow frequency band, a compromise antenna cut for the center of that band would probably be satisfactory. But television channels unfortunately are scattered over the radio-frequency spectrum. For example, channels 2–6 cover frequencies of 54–88 MHz; channels 7–13 cover frequencies of 174–216 MHz; channels 14–44 operate at frequencies above 500 MHz. No one antenna would be suitable for complete coverage of these bands. Separate antenna installations and multiple transmission lines would be ideal theoretically—but would be expensive, messy as far as roof appearance is concerned, and a nuisance in changing receiver connections to each antenna.

Many manufacturers of antenna equipment have solved the problem for the two lower bands by using two antennas with a common transmission line to the receiver. The high-band antenna is cut for a compromise length in the 174–216 MHz band, while the low-band antenna length is a compromise for the lower channels 2–6 (54–88 MHz). The problem now is how to connect these antennas to a common transmission line and yet prevent interaction of one antenna on the other. The solution employing transmission line principles can be seen by reference to Fig. 10–31.

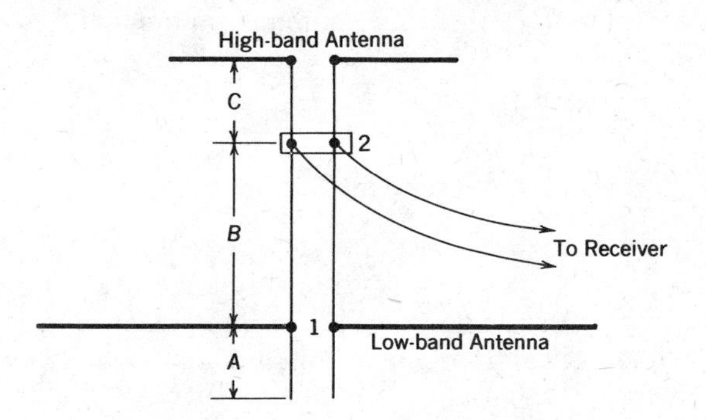

Figure 10-31 Dual-band TV antenna system.

Let us consider first the operation of this system on the high band. Section A is a quarter-wave open stub at the high frequency (approximately 195 MHz). Since it is open at the far end, the impedance at point 1 is zero. Section B is three-quarters wavelength at the high frequency. Due to the electrical short at point 1, the impedance of this line at point 2 is a maximum (approaching infinity). Therefore, the low-band antenna is effectively isolated from the transmission line to the receiver. Section C is matched to the high-band antenna and to the transmission line. It therefore acts as a nonresonant line and its length at the high-band frequencies is unimportant. Energy from the high-band antenna travels down through section C, through the transmission line, to the receiver. By making the antenna impedance, section C impedance, the transmission line impedance, and the receiver impedance as closely alike as possible (usually 300 Ω), we get a minimum of reflections and standing waves.

Now let us analyze the operation of this system on the low-frequency band. Section A is one quarter-wave at approximately 195 MHz. But at the low-band frequency (approximately 65 MHz), it is only one-twelfth wavelength. Therefore, this section now acts as a small capacitor with high reactance. This effect is practically negligible. Since the electrical short at point 1 is removed, section B is now properly terminated and acts as a nonresonant line. Energy from the low-band antenna travels through section B, through the transmission line, to the receiver.

But what is the effect of the high-band antenna at point 2? The impedance match between the high-band antenna and section C is destroyed due to change in antenna impedance. Section C now acts as a resonant line with standing waves. The impedance of section C at point 2 will depend on its length. By adjusting this length, the maximum impedance point can be made to occur at point 2. Unfortunately, since the high-band antenna impedance is not zero at the lower frequency, the impedance of section C at point 2 will not be infinite. However, this impedance can be made high enough by adjusting the length of this section to give sufficient isolation of the upper antenna on the low-frequency channels.

Impedance Matching—Stub Lines

In the above antenna illustration, it was assumed that each antenna was a perfect match ($R_L = Z_0$) to its transmission line at its operating frequency. Although an antenna of proper length acts as a resistive load, it is sometimes impractical to obtain an antenna impedance that is resistive and/or a transmission line characteristic impedance of matching value. But if proper termination is not obtained, standing waves will result, and

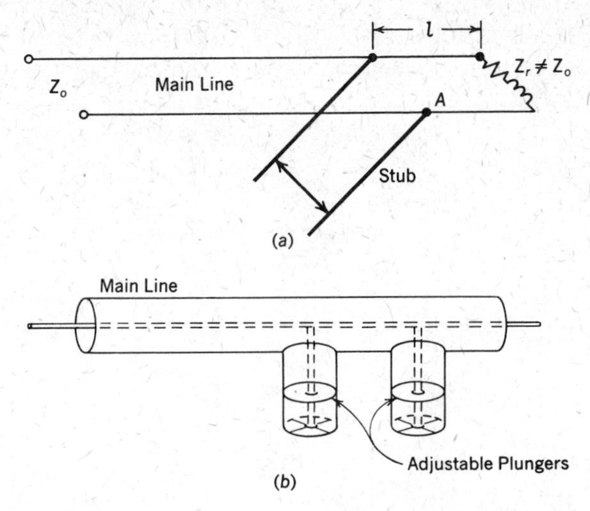

Figure 10-32 Stub-line impedance matching.

the line will have high losses. When the transmission line is long, the losses in such a line may be very serious. To remedy this situation a stub can be attached to the main line, close to the load end. By adjusting the location of the stub and the length of the stub, standing waves can be eliminated from the main line. Such a system is shown in Fig. 10–32(a). By varying the location of point A and the length of the shorted stub, the parallel impedance formed by the stub and section l of the main line can be made resistive and equal to the characteristic impedance of the line. Of course, standing waves will still exist on section l, but the rest of the line will operate as a nonresonant line.

With coaxial lines, the location of the stub cannot be changed at will. Therefore, a double stub system is used. By proper positioning of each plunger, standing waves can be eliminated from the generator side of the line.

In Fig. 10–32(a), if the load impedance and the characteristic impedance of the line are known, Smith charts can be used to calculate the proper location for the stub and the length of stub required to effect an impedance match.

EXAMPLE 10–14

An antenna has an input resistance of 46.8 Ω and an inductive reactance of 7.2 Ω at an operating frequency of 200 MHz. It is to be fed from a 72-Ω line. Find the location and length of stub needed to eliminate standing waves from the main line.

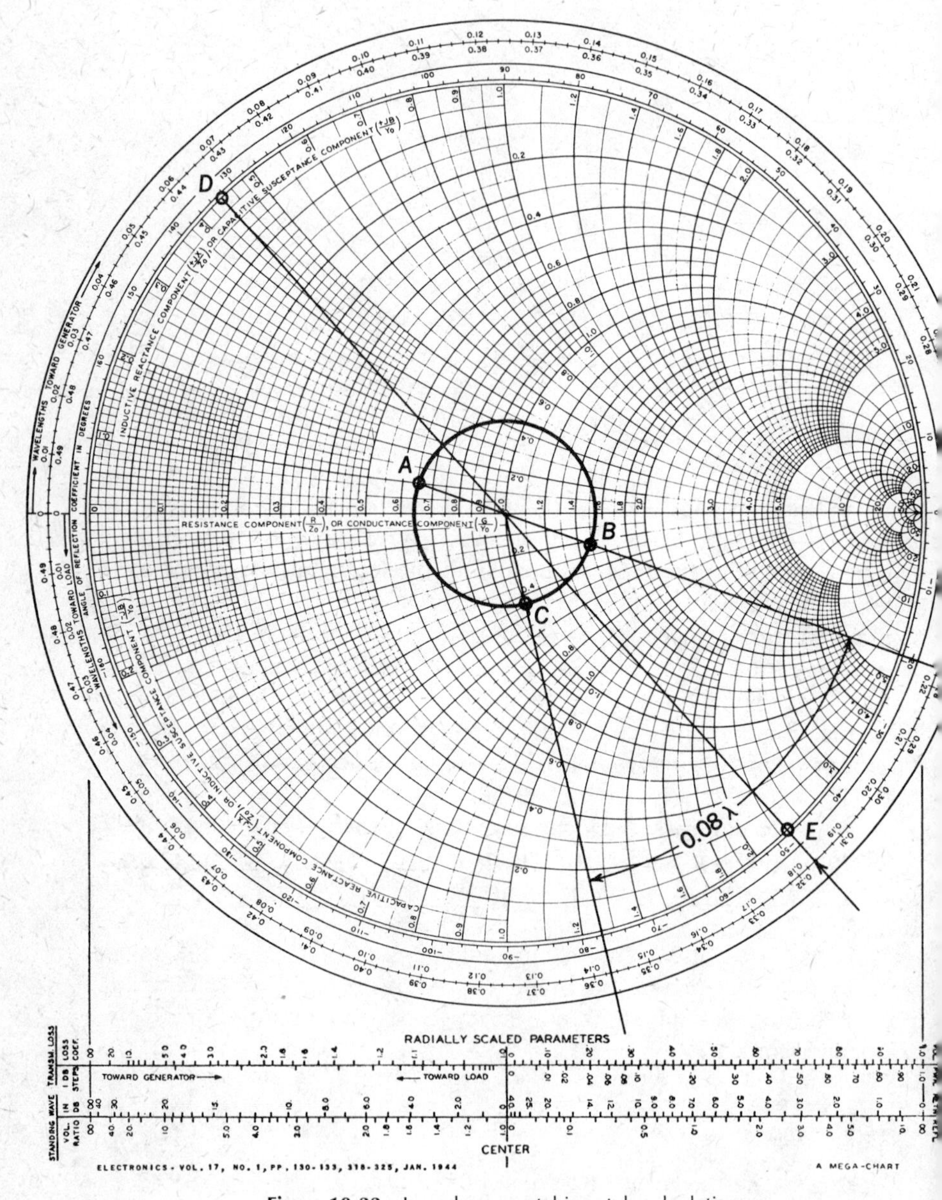

Figure 10-33 Impedance matching stub calculation.

Solution

1. The normalized impedance of the load (antenna) is

$$Z_{L(n)} = \frac{46.8 + j7.2}{72} = 0.65 + j0.10$$

2. Locate this point as A in Fig. 10–33 and draw the standing wave circle.

To match this load to the line, it is necessary to "add" another impedance, *in parallel,* so that the combined impedance becomes $1.0 + j0$. Since addition of parallel impedances requires working with reciprocals ($1/Z_T = 1/Z_1 + 1/Z_2 + \cdots$), it is at times more convenient to convert all impedances to *admittances* (Y). However, instead of using a vector-algebra solution, this conversion is readily done on the Smith chart by extending a line from point A diametrically opposite to point B. Note that the admittance value is $1.49 - j0.23$ *mho.**

3. We now have the admittance value *at the load.* By moving around the constant SWR circle from load toward generator (cw), we locate point C with a conductance value of 1.0. (This location also has a *susceptance* value of $-j0.44$.) At this point on the line, the conductive (resistive) component is matched, and all we need now is to add a *susceptance* (reciprocal of reactance) of $+j0.44$. This is a capacitive effect and can be supplied by a stub connected at this location.
4. From the wavelength circle, note that point C is 0.08λ from the load. At 200 MHz

$$l = \text{distance from load} = 0.08\left(\frac{984}{200}\right) = 0.39 \text{ ft}$$

5. Locate point D with a capacitive susceptance of $0 + j0.44$ mho. Diametrically opposite, locate point E with a capacitive *reactance* of $j2.25$ Ω. This is the required capacitance (normalized value). The stub length should be 0.316λ shorted, or 0.066λ open. At an operating frequency of 200 MHz, this corresponds to 1.56 ft or 0.325 ft.

Quarter-Wave Open Line Transformer Action

If the "load" end of a quarter-wave section is open, the input impedance is a minimum and resistive. If the load end is shorted, the input impedance is a maximum but still resistive. Obviously, then, a quarter-wave section acts as a step-up or step-down transformer. The impedance transformations are as follows:

1. If $R_L = Z_0$ 1:1 ratio, input impedance $= Z_0$
2. If $R_L > Z_0$ step-down, input impedance $< Z_0$
3. If $R_L < Z_0$ step-up, input impedance $> Z_0$

*Admittance (Y) is the reciprocal of impedance. The in-phase component (1.49) is the reciprocal of resistance, or *conductance*; the quadrature component ($j0.23$) is called *susceptance* and is the reciprocal of reactance. It should be noted that a negative susceptance represents *inductance.* (This is a reversal from reactance signs.)

The exact relationship is given by

$$Z_i = \frac{Z_0^2}{Z_L} \quad \text{or} \quad Z_0 = \sqrt{Z_i Z_L} \tag{10–18}$$

where: Z_0 is the characteristic impedance of the section,
Z_i is the input impedance, and
Z_L is the load impedance.

This gives us another means for matching a transmission line to a resistive load not equal to the characteristic impedance of the line, or of connecting two lines of different characteristic impedances. A simple illustration is shown in the following example.

EXAMPLE 10–15

Show by diagram and calculations how a quarter-wave section can be used to match a 500-Ω line to a 72-Ω dipole antenna at 200 MHz.

Solution

1.

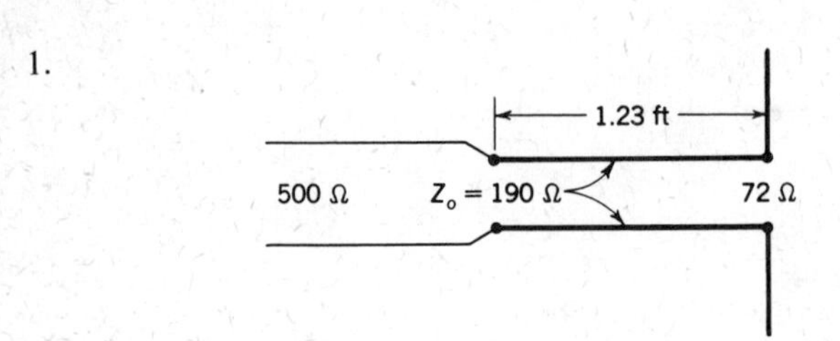

Figure 10-34 A quarter-wave matching section.

2. $Z_0 = \sqrt{Z_i Z_L} = \sqrt{500 \times 72} = 190\ \Omega$

3. $\lambda = \dfrac{V}{f} = \dfrac{984}{200} = 4.92$ ft

4. $\frac{1}{4}\lambda = \dfrac{4.92}{4} = 1.23$ ft

5. Use a line 1.23 ft long of $Z_0 = 190\ \Omega$

Impedance Matching—Quarter-Wave Shorted Sections

A quarter-wave shorted section exhibits the properties of a parallel resonant circuit. We know that the impedance of a tank circuit is high. But we also know that we can tap down on the inductance of the tank circuit to match this circuit to a lower impedance. Similarly, we can tap

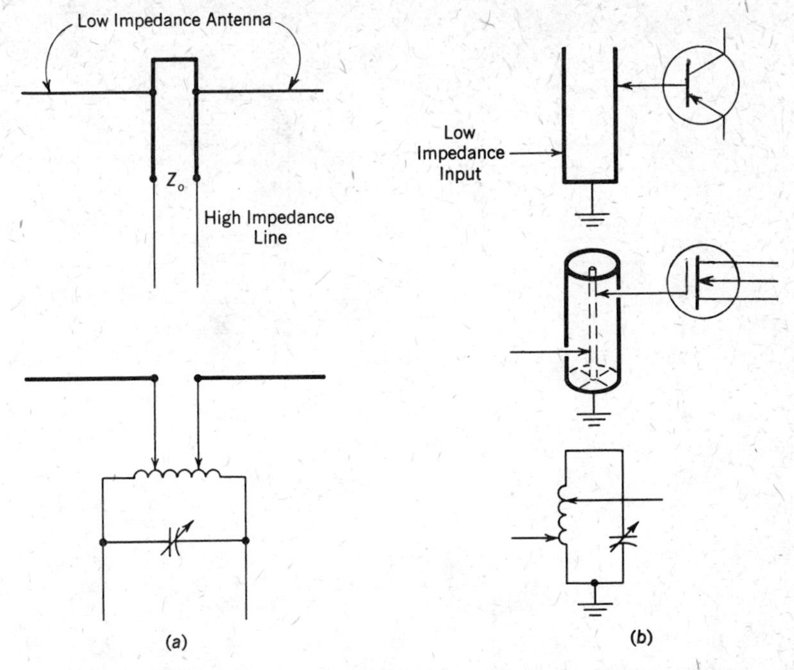

Figure 10-35 Shorted quarter-wave impedance matching.

into a shorted quarter-wave section and get the same results. Figure 10–35 shows quarter-wave sections used for impedance matching together with their equivalent *L-C* circuit.

Coaxial Line Balance Converter

The coaxial line is essentially an unbalanced feeder system, since the outer conductor is grounded. Many times it is necessary to connect from an unbalanced coaxial line to a balanced system such as an antenna array or a balanced line. A direct connection would cause discontinuity and loss of energy due to reflections and standing waves. The problem can be solved by using a section of transmission line as a balance converter.

One method places a quarter-wave section called a *bazooka* around the end of the coaxial line. This is shown in Fig. 10–36(a). The bazooka does not affect the impedance between the inner and outer conductors of the coaxial line. However, it does raise the outer conductor above ground. The outer conductor and the quarter-wave bazooka form a shorted quarter-wave section. Since they are at ground potential, shorted together at point *A*, a high impedance is developed between these two elements at point *B*. Therefore, the outer conductor now has a high impedance to ground.

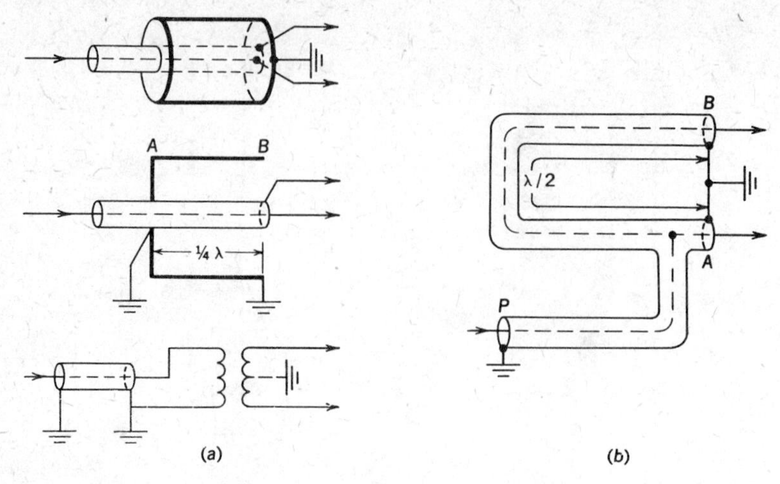

Figure 10-36 Balance-line converters.

Another method for obtaining balanced output uses a half-wave section, properly matched to prevent reflections. (See Fig. 10–36(b).) Since a nonresonant half-wave section acts as a phase inverter, the output at B will be 180° reversed from the output at A. This is exactly what we need for balanced output.

Either method can also be used in reverse to feed from any source of balanced output into an unbalanced coaxial line.

INTRODUCTION TO WAVEGUIDES

To the uninitiated, waveguides look like a plumber's heaven of pipes: round, oval, or rectangular. Actually it is another means of transmitting energy from a source to a load. Waveguides are used at ultrahigh frequencies because they are more efficient than the transmission lines previously discussed. The basic principle for a rectangular waveguide can be developed from our knowledge of transmission lines.

In Fig. 10–29 we saw how a quarter-wave section could be used to support a two-wire transmission line. Since the impedance of the quarter-wave section at its open end is infinite, it has no effect on the main line. Since it has no effect, let us connect another quarter-wave section right above the first one and short it at its far end. Now we have a half-wave section shorted at its far ends and the lines connected at the center of this half-wave rectangle. The impedance at the center of this rectangle is infinite, so there is still no effect on the main line. Now let us support the main line by many more of these sections. Such a line is shown in Fig. 10–37. The dimension a is a half wave at the operating frequency. The dimension b is not critical, but it must be sufficient to prevent voltage breakdown between the lines.

Let us add many more sections in between the ones shown—until finally the sections touch and become solid sides. We now have a rectangular pipe with the lines at the center of the side walls. Such a waveguide can be used for any frequency above the "cutoff" frequency. This cutoff is determined by dimension a and is the frequency which makes the dimension a one-half wavelength.

In general, energy is fed into (or taken out from) a waveguide by one of three means:

1. A probe inserted into the waveguide at a point of strong electrostatic field. The coupling is varied by adjusting the length of probe inserted.
2. A small loop inserted into the waveguide at a point of strong magnetic field. The coupling is varied by rotating the loop.
3. Slots or openings in the waveguide. A conductor is located close to the slot. The electrostatic (or magnetic) field around the conductor reaches into the guide (or vice versa). This method is used where a relatively small degree of coupling is desired.

Just as in our transmission lines, waveguides also have a characteristic impedance. This impedance varies with the size and shape of the guide. The minimum practical value of characteristic impedance for a circular waveguide is approximately 350 Ω. Rectangular waveguides range in impedance from zero to approximately 465 Ω, the impedance varying directly with the dimension b.

Any abrupt change in the size or shape of a waveguide will cause a change in characteristic impedance. Such discontinuities will cause reflections, standing waves, and loss of energy just as in two-wire or coaxial lines. Therefore, if it is necessary to connect from a guide of one size or

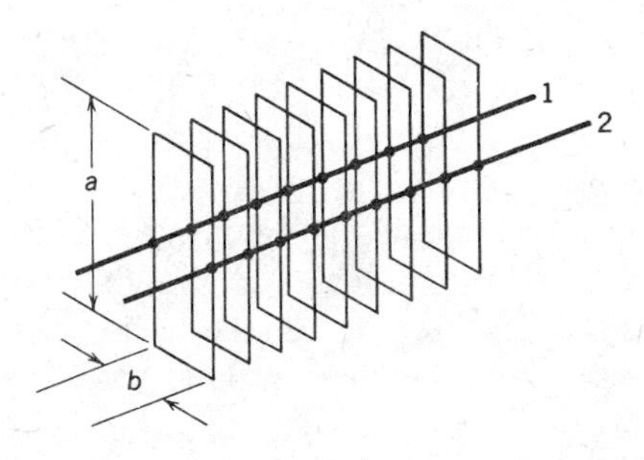

Figure 10-37 Development of a rectangular wave guide.

shape to another of different size or shape, a transition piece having a gradual expansion or contraction must be used. Even when connecting two pieces of identical size and shape, no solder must be allowed to run into the guide. The lump so formed may easily result in standing waves and loss of power.

Comparison of Waveguides and Coaxial Lines

At ultrahigh frequencies, waveguides are used in place of coaxial lines. The first application of waveguides has already been given, namely, as a means of transmitting energy from a source to the load. In addition, sections of waveguide can also be used as phase shifters, for impedance matching, or as tuned circuits.

Waveguides have several advantages over other types of transmission lines. At high frequencies, current flows in a thin layer on the outer surface of a conductor. Since the waveguide has a large surface area, this skin effect is reduced to a minimum. In other words, copper losses and attenuation are less in a waveguide than in any other type of line. Since the dielectric in the waveguide is air—and there is no inner conductor to support—dielectric losses in waveguides are less than for any other line. There are no radiation losses in waveguides. In this respect, the coaxial line is equally good.

For a given line impedance, the power that can be transmitted through a line depends on the square of the applied voltage. Higher power means higher voltage. The limiting factor is the breakdown voltage of the line. A waveguide has no inner conductor, as in a coaxial line. Therefore, for a given outside diameter, a waveguide can withstand higher voltages. For this reason, waveguides have a higher power-handling capacity than coaxial lines of equal size.

A final advantage of the waveguide is its simpler construction and greater mechanical strength as compared to the coaxial line.

On the other side of the ledger, the waveguide has two disadvantages:

1. Its use is restricted to higher frequencies. In practice waveguides are seldom seen below 3000 MHz. The physical size needed to extend the cutoff frequency below this value makes the waveguide too bulky.
2. Installation of waveguides is much more difficult than for other types of transmission lines. We already pointed out the care needed in soldering sections of guide. If bends are needed in the line, the minimum radius of bending to avoid excessive attenuation is two wavelengths. When bending, extreme care must be used to avoid dents. Such dents would produce discontinuities

and increase losses greatly. Before making a bend a waveguide section is often filled with molten dielectric. The material is allowed to cool and harden. Then the section is bent to the desired shape. After this the section is heated, allowing the filling to melt and flow out. Unless the installation is carefully made, poor joints or poor bends may easily reduce the efficiency below that of a coaxial line.*

This is merely an introduction to waveguides. A complete treatment is beyond the scope of this text. For further information, those interested should refer to texts on ultrahigh frequencies.

REVIEW QUESTIONS

1. (*a*) What is the basic purpose of a transmission line? (*b*) Name three other uses for transmission lines at high frequencies.

2. (*a*) At what speed do electrical waves travel down a line? (Use English units.) (*b*) What is this speed in metric units?

3. Refer to Fig. 10–1. The line switch is closed when the ac source voltage is at its zero value and just starting to go negative. At this instant, what is the voltage (*a*) at B–B'? Why? (*b*) at D–D'? Why?

4. In Question 3, at a time 0.5 s later, what is the voltage (*a*) at A–A'? Why? (*b*) at B–B'? Why? (*c*) at C–C'? Why? (*d*) at E–E'? Why?

5. In Question 3, at a time 1.75 s later, what is the voltage (*a*) at A–A'? Why? (*b*) at C–C'? Why? (*c*) at D–D'? Why? (*d*) at E–E'? Why?

6. What is the minimum time required for a voltage to be noticed across the load at E–E'?

7. Refer again to Fig. 10–1. (*a*) Is the voltage at C–C' constant? (*b*) Explain why this is so.

8. Refer to Fig. 10–2. (*a*) What are the axes of these two curves? (*b*) In what respect do the conditions for curve (a) differ from curve (b)? (*c*) In Fig. 10–2(a), is the voltage at location E always positive? Explain. (*d*) In Fig. 10–2(b), is the voltage at location C always zero? (*e*) Account for the change in voltage at location C for these two conditions.

9. Refer to Fig. 10–3. (*a*) What are the axes of each of these four curves? (*b*) In what respect do these four cases differ? (*c*) Account for the difference in phase shown for the curves at B–B' and at D–D'.

10. (*a*) Give a definition for the term *wavelength.* (*b*) What symbol is used to represent this quantity? (*c*) In what units could wavelength be measured? (*d*) Is the wavelength of all radio waves the same? Explain. (*e*) Give the general

* Special rigid and flexible waveguide sections are commercially available to avoid such problems. The rigid sections introduce a minimum of losses but are obviously suitable only for a fixed degree of bend.

relationship used to calculate wavelength. (*f*) Give an equation for wavelength in meters, and in feet.

11. (*a*) What determines whether a transmission line should be considered electrically long? (*b*) A line is 10 in long. Is it a long line or a short line? Explain. (*c*) What length of line may be considered short?

12. Refer to Fig. 10–4. (*a*) What do each of the following quantities represent: R_1, R_2, L, and C? (*b*) Are these discrete values found at specific intervals of length? Explain. (*c*) What is a *unit length?*

13. (*a*) What is meant by the term *characteristic impedance* of a line? (*b*) What symbol is used to represent this quantity? (*c*) If the length of a line is doubled, what happens to its characteristic impedance? Explain. (*d*) What does determine the value of characteristic impedance?

14. Refer to Fig. 10–6. (*a*) Which of these lines would be called parallel-conductor lines? (*b*) Which would be open-wire lines? (*c*) Which would be essentially air dielectric? (*d*) What is the major difference between the line shown in diagram (c) and those in diagram (b)? (*e*) What is the main advantage of this added feature? (*f*) What is the line in diagram (d) called? (*g*) What is the line in diagram (e) called?

15. Refer again to Fig. 10–6. (*a*) Which of the lines shown have the lowest radiation losses? Why? (*b*) Which line has the lowest total losses? Why? (*c*) Which lines are least affected by stray fields? Why? (*d*) Compare the lines in diagrams (a) and (e) with regard to stray-field pickup.

16. (*a*) What two physical factors determine the characteristic impedance of a two-wire parallel-conductor line? (*b*) How does this impedance vary with each factor? (*c*) Give the equation for the characteristic impedance for an air-dielectric line. (*d*) Will the type of dielectric affect the impedance value? Explain.

17. Two transmission lines of type 14-022 (Fig. 10–6(b)) are available. One is wider than the other. Is there any difference in their electrical properties? Explain.

18. (*a*) What two physical factors determine the characteristic impedance of a coaxial line? (*b*) One coaxial line has a larger cross-sectional area than another. How would their characteristic impedances compare (all other factors equal)? (*c*) Two coaxial lines have equal cross sections, but one uses a thicker inner conductor. How do their characteristic impedances compare? (*d*) Give the equation for the characteristic impedance of air-dielectric coaxial lines.

19. (*a*) Give an equation for the characteristic impedance of a line, based on its electrical parameters. (*b*) Does this equation apply to parallel-conductor or coaxial lines? Explain. (*c*) Does it apply to air-dielectric or to solid-dielectric lines? Explain.

20. (*a*) What is meant by the *velocity constant* of a line? (*b*) Does this affect any transmission line calculations? Explain. (*c*) Give the equation for wavelength (in meters), corrected for the velocity constant of the line.

21. Refer to Fig. 10–7. (*a*) What quantities are plotted in diagram (b)? (*b*) In going from A toward F in this diagram, how does time vary? Explain. (*c*) Why does the voltage vary with distance? (*d*) Why are the current variations in step with the voltage changes? (*e*) What do the arrows on the current curve indicate? (*f*) If time is "released," what happens to the current and voltage waves shown?

22. Refer to Fig. 10–8. (*a*) What do the solid lines represent? (*b*) What created this field? (*c*) What do the dashed lines represent? (*d*) What created this field? (*e*) What is the "space phasing" between these two fields? (*f*) Does this electromagnetic field exist—at this instant—at location *A* on the line in Fig. 10–7? Explain. (*g*) Give two other locations where this field would exist as shown. (*h*) Does an electromagnetic field exist at a location halfway between *A* and *B*? Why? (*i*) How does it differ from the field shown? (*j*) If time is "released," what happens to the electromagnetic field? (*k*) With time, what happens to the electromagnetic field at location *D*?

23. (*a*) Does Fig. 10–7 represent a line with losses? (*b*) If the line had appreciable loss, how would these waves be affected? (*c*) What term is used when considering the resistive losses in a line? (*d*) What is its symbol? (*e*) What unit is used in evaluating line loss? (*f*) Does this factor consider the series resistance or the shunt resistance of the line?

24. (*a*) What is meant by the *phase constant* of a line? (*b*) What symbol is used to represent this quantity? (*c*) In what units is it measured? (*d*) What is responsible for these phase variations?

25. (*a*) What term is used to describe both the resistive losses and the phase effects in a transmission line? (*b*) What symbol is used to represent this quantity?

26. Refer to Fig. 10–9. (*a*) What is the length of the line shown in diagram (a)? (*b*) What is the input impedance of this line at location *A*? (*c*) What is the impedance of the line at location *C*? Why? (*d*) What is the input impedance of the line in diagram (b)? Why?

27. A transmission line is terminated by a pure resistive load $R = Z_0$. (*a*) What happens to the energy sent down the line? (*b*) Is any reflected back to the source? (*c*) What is the impedance at the input end of this line? (*d*) What is the impedance at any point on the line? (*e*) What is the phase relation between current and voltage at any point on the line?

28. (*a*) What is the general meaning of the term *discontinuity* when it is applied to a transmission line? (*b*) Give two examples of a drastic discontinuity. (*c*) What happens to a wave traveling down a line when it meets a discontinuity?

29. What happens to the energy traveling down a line when it reaches a load that is (*a*) resistive and equal to Z_0? (*b*) resistive but not equal to Z_0? (*c*) inductive and equal to Z_0? (*d*) What term is used to evaluate the degree of reflection? (*e*) Give the equation for this term.

30. Refer to the current waves in Fig. 10–10. (*a*) What does the solid curve represent? (*b*) What does the dot-dash curve represent? (*c*) What does the x'd curve represent? (*d*) What does the dash curve represent? (*e*) What is the phasing of the dash curve with respect to the dot-dash curve?

31. In general how does a current wave reflect back—in an open line?

32. Refer to Fig. 10–10 (upper diagram). (*a*) Which curve represents the incident (initial) voltage wave? (*b*) What does the dot-dash curve signify? (*c*) Where is the reflected wave?

33. In general, how does a voltage wave reflect back – in an open line?

34. In Fig. 10–11(a) (upper diagram): (*a*) What quantity is plotted as the horizontal axis? (*b*) What does the dot-dash curve represent? (*c*) How is the reflected voltage wave drawn with respect to this dot-dash curve? (*d*) Can we compare the general phasing of initial and reflected waves from this one diagram? Why not?

35. In Fig. 10–11(b) (upper diagram): (*a*) Identify the incident wave and the reflected wave. (*b*) What does the dot-dash curve show? (*c*) How is the reflected wave drawn compared to this dot-dash curve?

36. Refer to the current diagrams of Fig. 10–11(a) and (b). (*a*) What do the dot-dash curves represent? (*b*) What do the x'd curves represent? (*c*) How are the reflected current waves drawn compared to the x'd curves? (*d*) How are they drawn compared to the dot-dash curves?

37. (*a*) What magnitude of voltage must exist at the end of a shorted line? (*b*) What phasing of the reflected voltage wave would create this condition?

38. (*a*) Would the magnitude of the current at the end of a shorted line be greater, smaller, or equal compared to a line with matched resistive load? (*b*) How is this current magnitude obtained?

39. Refer to the voltage diagram in Fig. 10–12(b). (*a*) What does the dot-dash curve section represent? (*b*) What does the x-curve section represent? (*c*) How is the reflected voltage wave drawn compared to the dot-dash curve?

40. How is the reflected current wave drawn compared to the dot-dash curve (*a*) in Fig. 10–12(a)? (*b*) in Fig. 10–12(b)?

41. Refer to Fig. 10–13. (*a*) What are the axes for each of these sets of curves? (*b*) In what respect does any one set of curves differ from the others? (*c*) What is used to indicate this time difference? (*d*) What is the magnitude of the resultant voltage at a distance of one-quarter wavelength from the open end? (*e*) If the input voltage were tripled, what would this voltage be? Why is this so? (*f*) At what other points along the line does this same condition exist? (*g*) At a distance of one-half wavelength from the open end, over what range of values does the magnitude of the resultant voltage vary? (*h*) At what other points does this condition exist?

42. Refer to Fig. 10–14. (*a*) Is the voltage at location A always constant at the magnitude given by A'? (*b*) How does this voltage vary with time? (*c*) What is the voltage shown at location B? (*d*) How does this voltage vary with time? (*e*) Is the voltage at location C constant? (*f*) Over what limits does it vary with time? (*g*) What does a voltage standing wave pattern show?

43. In an open-ended *lossless* transmission line: (*a*) What maximum voltage value will be observed on the line? (*b*) What is the maximum voltage value at the open end itself? (*c*) Where else will this value of voltage exist? (*d*) At what points are the standing wave nulls (minimum values)? (*e*) What is the maximum value of voltage at these null points? (*f*) Why are measurements made from the open end rather than from the source?

44. If a transmission line had noticeable losses, what effect would this cause (if any) to the standing wave pattern of Fig. 10–14 with regard to (*a*) the location of the null points? (*b*) the location of the standing wave maxima? (*c*) the magnitude of these maxima? (*d*) the magnitude at the null points?

45. Refer to Fig. 10–15. (*a*) What is the resultant value of current, at any time instant, at the end of this line? (*b*) Why is this so? (*c*) At a distance of one-half wavelength from the end, why is the resultant current zero at time condition (a)? (*d*) Why is it still zero at time condition (b)? (*e*) Why is it still zero at time condition (c)? (*f*) At this location, when does the resultant rise above the zero value? (*g*) Why is this so? (*h*) At what other locations is the resultant current value always zero?

46. In Fig. 10–15, at a three-quarter wavelength distance from the end of the line and (*a*) at time condition (a), what is the resultant value? Why? (*b*) At time (b), what is the resultant value? (*c*) At time (c), what is the resultant value? Why? (*d*) Over what range of values does the resultant vary (with time)? (*e*) At what other locations does the resultant vary over this full range?

47. Refer to Fig. 10–16. (*a*) Is the current value always zero at a location 1.5λ from the open end? (*b*) Is the current always at this maximum value at location *A*? Explain. (*c*) Is the current ever negative in value at any location? (*d*) What is the maximum value of current (approximately) at location *B*? (*e*) Is the current steady at this value? (*f*) Over what range does it vary?

48. Refer to Fig. 10–17. (*a*) What quantities are being plotted as the *X* and *Y* axes? (*b*) Is this the standing wave pattern for an open or shorted line? How can you tell? (*c*) Why are distances measured from the end of the line? (*d*) At what distance are the terminal conditions reversed? (*e*) At what intervals is the pattern repeated? (*f*) Are the current and voltage values at any location steady at the magnitudes indicated? Explain. (*g*) Does this pattern apply to a line with losses or to a lossless line? Explain. (*h*) Do these current and voltage curves show the phase relations between *E* and *I*? Why not?

49. How can the existence of standing waves on an open-wire line be demonstrated?

50. (*a*) How can the standing wave ratio of a line be determined? (*b*) What is the minimum SWR value a line can have? (*c*) Under what load condition would a line have this value? (*d*) What SWR value would an ideal lossless line have if it were shorted? open? (*e*) If the reflection coefficient of a line should increase, how would this affect the SWR value? (*f*) Give the relationship between SWR and Γ.

51. How can the SWR ratio be used to determine if a line is properly matched to the load?

52. A line is terminated by a resistive load, but the standing wave ratio is found to be 3. (*a*) What is a possible value of load resistance? (*b*) Is any other value also possible? (*c*) Describe how this ambiguity can be resolved without changing the load.

53. Refer to Fig. 10–18(a). (*a*) Do these curves show the phase relation between current and voltage in the line? Explain. (*b*) What do these *E* and *I* curves show? (*c*) What is the magnitude of the input impedance of this piece of line? Why?

54. Refer to Fig. 10–18(b). (*a*) What do the phasors E_0 and I_0 represent? (*b*) What do E_i and I_i represent? (*c*) Why do they lead the output phasors by 90°? (*d*) Which are greater in amplitude, E_0, I_0 or E_i, I_i? Why? (*e*) What does E_r represent? (*f*) At the instant shown, where is this E_r voltage? (*g*) Why does it lag E_0 by 90°? (*h*) What symbol represents the net voltage at the source? (*i*) How is this value obtained? (*j*) Why is it in phase with E_i? (*k*) Could this value approach zero? Explain. (*l*) What does I_r represent? (*m*) How is this phasing obtained? (*n*) What does I_s represent? (*o*) What would this value be in a lossless line? (*p*) Is the input impedance resistive, inductive, or capacitive? Explain.

55. From Fig. 10–19(a): (*a*) Can we tell if the input impedance is high or low? Explain. (*b*) Can we tell if this impedance is resistive, inductive, or capacitive? Explain.

56. Refer to Fig. 10–19(b). (*a*) Which phasors represent the initial current and voltage waves at the open end and at the source end? (*b*) How are these waves related in *time*? (*c*) How do the initial waves at the input compare to the initial waves at the output as to phase? Why is this so? (*d*) What does I_s represent? (*e*) How is this value obtained? (*f*) What does E_s represent? (*g*) How is this value obtained? (*h*) What is the character (R,L,C) of the input impedance? Explain.

57. Refer to Fig. 10–20. (*a*) Why does E_i lead E_0 by 45°? (*b*) Why does E_r lag E_0, and why by the same angle? (*c*) What does E_s represent? (*d*) What does $-I_0$ represent? (*e*) Why is it in phase opposition to I_0? (*f*) What is the character (R,L,C) of the input impedance? Explain. (*g*) How does the magnitude of this impedance compare with input impedance of a quarter-wave line and a half-wave line?

58. Refer to the phasor diagram of Fig. 10–20. As the length of the line is increased from zero up to, but not quite, one-quarter wavelength, how would this affect (*a*) the phase of E_s? Why? (*b*) the magnitude of E_s? Explain. (*c*) the phase and magnitude of I_s? Explain. (*d*) the character of the input impedance? Explain. (*e*) the magnitude of the input impedance? Explain.

59. Refer to Fig. 10–21. (*a*) By what angle do the initial waves at the sending end lead (or lag) the initial waves at the open end? Why? (*b*) By what angle does the reflected voltage *at the open end* lead (or lag) the initial voltage at the open end? (*c*) By what angle does the reflected voltage at the source lead (or lag) the reflected voltage at the open end? Why? (*d*) What phasor represents the reflected current at the open end? (*e*) What is its phase relation to the initial current at the open end? Why? (*f*) Explain the phase of I_r. (*g*) What is the character of the input impedance?

60. Refer to the phasors in Fig. 10–21. If the line length is varied from just over $\frac{1}{4}\lambda$ to not quite $\frac{1}{2}\lambda$, explain what happens (*a*) to the phase of E_s, (*b*) to the magnitude of E_s, (*c*) to the phase of I_s, (*d*) to the magnitude of I_s, (*e*) to the character of Z_i, (*f*) to the magnitude of Z_i.

61. Refer to Fig. 10–22. (*a*) What quantities are plotted as the X and Y axes? (*b*) From what point are the distance measurements made? (*c*) What lengths of line would have resistive input impedances? (*d*) What is the impedance of any odd quarter-wavelength line? (*e*) What is the impedance of any even quarter-wavelength line? (*f*) If the line were ideally lossless, what would the impedance of a $\frac{3}{4}\lambda$ line be? (*g*) What lengths of line would have a capacitive input impedance? (*h*) What length of line would have a capacitive input impedance equal to Z_0? (*i*) For what length of line would $Z_i = Z_0$ and be inductive?

62. Refer to Figs. 10–22 and 10–23. (*a*) How does the input impedance for any length of shorted line compare with the input impedance of the same length of open line? (*b*) Why is this so? (*c*) Give a qualitative value (high, low, or equal to Z_0, and L, C, or R) for the input impedance of the following lines—$\frac{3}{4}\lambda$ shorted; $\frac{3}{4}\lambda$ open; $\frac{5}{16}\lambda$ shorted; $\frac{5}{16}\lambda$ open; $\frac{5}{8}\lambda$ shorted; $\frac{5}{8}\lambda$ open.

63. Refer to Fig. 10–24. (*a*) Where is the "infinity point" of this chart? (*b*) What quantity is represented by the circles tangent at this point? (*c*) What does the numerator in this ratio stand for? (*d*) What is the value of R/Z_0 for the small darkened circle? (*e*) What is the R/Z_0 value for the larger circle, two-thirds of which is darkened? (*f*) What quantity is represented by the arcs emanating from the infinity point and curving upward or downward? (*g*) What does the numerator in this ratio stand for? (*h*) What is the X/Z_0 value for the upper darkened arc? Is it inductive or capacitive? (*i*) What is the X/Z_0 value for the lower darkened arc? Is it inductive or capacitive?

64. (*a*) With reference to Smith chart applications, what is meant by the *normalized impedance* of a line? (*b*) How are normalized values obtained?

65. In Fig. 10–24, what is the normalized impedance corresponding to (*a*) point *R*? (*b*) point *S*? (*c*) point *P*? (*d*) point *Q*?

66. In a Smith chart, what is the significance of any circle drawn with the center of the chart as the origin?

67. Refer to Fig. 10–24. (*a*) What specific transmission line condition determines the radius of the dashed circle? (*b*) What is the value of this quantity for this particular circle? (*c*) What is the purpose of the outermost circle on the Smith chart? (*d*) In what units are these measurements made? (*e*) Why does this circle have two scales—in opposite directions? (*f*) What is the maximum line length directly measurable on these scales? (*g*) How can this scale be used for a line length of 1.2λ? (*h*) Why is this technique possible?

68. Refer to Example 10–8 and Fig. 10–24. (*a*) Can the dashed circle be used to represent the given line? Explain. (*b*) What is the impedance of point *P* on this circle? (*c*) Can it represent the load end of the line? Explain. (*d*) What length of line is represented in going clockwise around the dashed circle from *P* to *Q*? (*e*) Why did we move in a clockwise direction? (*f*) Why does point *Q* represent the input impedance of the given line? (*g*) What is the normalized impedance at point *Q*?

69. Refer to Example 10–9 and Fig. 10–24. (*a*) Why can the same dashed circle be used to represent this new line? (*b*) Does point *P* represent the load end of this line? Explain. (*c*) What impedance does point *L* represent? (*d*) Why does point *L* represent the load end of the line? (*e*) How far should point *X* on the line be from point *L*? (*f*) Is the *X′*-*C* line the correct distance away? Explain. (*g*) What is the impedance represented by point *X*? (*h*) Is this the true desired impedance? (*i*) How is the true value obtained?

70. Refer to Fig. 10–25. (*a*) What is the standing wave ratio for the line represented by this circle? (*b*) Does this represent a line with or without losses? Explain. (*c*) When a line has losses, how does the standing wave ratio vary from source to load? Explain.

71. Refer to Example 10–10 and Fig. 10–25. (*a*) What is the normalized impedance at point *O*? (*b*) Neglecting line losses, does the circle shown otherwise represent the given line? Explain. (*c*) In locating point *P*, should the measurement be scaled in a cw or ccw direction? Explain. (*d*) What is the ccw distance from point *O* to *P*? (*e*) Is this the correct location? Explain. (*f*) Is the impedance at point *P* the correct value? Explain. (*g*) How far apart are points *A* and *B* as measured on the *transmission-loss 1-dB* scale? (*h*) What is the standing wave ratio corresponding to the dashed circle? (*i*) Should the standing wave ratio be higher or lower at point *P* as compared to the SWR at the input? (*j*) Why is the impedance at P'' the corrected value?

72. A transmission line is terminated with a pure capacitive load such that $X_C = Z_0$. Will there be any standing waves on this line? Explain.

73. Refer to Fig. 10–26. (*a*) In diagram (a), what do the dashed lines extending beyond the capacitor signify? (*b*) What is the effect of this capacitor on the standing wave pattern? (*c*) Which causes a greater increase in the electrical length, a smaller or larger capacitance value? Explain. (*d*) What is the effect of the load in diagram (b) when $X_L = Z_0$? Explain. (*e*) A line is terminated with a reactive load. The voltage at the load is high, but rises to a maximum a short distance away from the load. Is the load capacitive or inductive? Explain.

74. (*a*) What is a Lecher line used for? (*b*) Describe its construction.

75. (*a*) When using a Lecher line to measure frequency, how should the line be terminated? (*b*) Why? (*c*) What is the important measurement? (*d*) Why is it preferable to use null points rather than maximum points? (*e*) How is frequency obtained from the above measurement?

76. (*a*) What other device is more accurate for such measurements? (*b*) Give three other uses for this commercial unit.

77. (*a*) Can a transmission line section be used in place of an inductor? Explain. (*b*) What length should it be, and how must it be terminated? (*c*) State two advantages of a transmission line section over a conventional inductor for UHF applications.

78. (*a*) Can a transmission line section be used as a capacitor? (*b*) Should it be shorted or open-ended? Explain.

79. (*a*) What length of shorted line acts as a tank circuit? (*b*) Why is this so? (*c*) Can an open line be used as a tank circuit? Explain. (*d*) For variable-frequency operation, is there any preference? Explain. (*e*) What advantage do transmission line tuners have over lumped-constants tank circuits?

80. Can series-resonant action be obtained from (*a*) a short-circuited transmission line? Explain. (*b*) an open-ended line? Explain. (*c*) What characteristic impedance should be used to maintain proper impedance match?

81. (*a*) Where is the tank circuit for the oscillator in Fig. 10–28? (*b*) What is the function of capacitor C? (*c*) Why isn't a shorting bar used? (*d*) What basic oscillator circuit is this?

82. Refer to Fig. 10–29. (*a*) What is the U-shaped bar in diagram (a) made of? (*b*) Does it short-circuit the line? Explain. (*c*) Does the center rod of the stub in diagram (b) make electrical connection to the outside wall at the bottom? (*d*) Doesn't this short-circuit the inner and outer conductors? Explain.

83. Refer to Fig. 10–30. (*a*) What is the purpose of the stubs shown? (*b*) At what frequency are they $\frac{1}{4}\lambda$? (*c*) What is the line impedance at location P at the fundamental frequency? at the second harmonic? at any odd harmonic? at any even harmonic? (*d*) In diagram (b), how does energy get through to the antenna at the fundamental frequency? (*e*) Would energy get through at any other frequency? (*f*) Is flow of energy blocked at any frequencies? Explain.

84. When a transmission line section is used to produce a phase shift: (*a*) How should it be terminated? (*b*) How is the amount of phase shift changed? (*c*) How can phase inversion be obtained?

85. Refer to Fig. 10–31. (*a*) From the physical appearance, how can you tell which antenna is for which frequency band? Explain. (*b*) If section A is $\frac{1}{4}\lambda$ at the high-frequency band, what is the impedance at location 1? (*c*) Why doesn't this short-circuit the transmission line at location 2? (*d*) How does energy from the high-band antenna reach the receiver? (*e*) How is the short-circuiting effect of stub A "removed" for the low-frequency signals? (*f*) How is the high-band antenna isolated form the transmission line at this lower frequency?

86. Refer to Fig. 10–32(a). (*a*) What is the purpose of putting a stub on the transmission line? (*b*) What two variables must be adjusted to obtain proper matching? (*c*) When proper match is established, are there any standing waves in the stub section? in the l section? in the main line? (*d*) Since section l still has standing waves, of what value is this technique?

87. Why use a double stub in Fig. 10–32(b), instead of using a single stub and varying its location as in Fig. 10–32(a)?

88. (*a*) What length of transmission line should be used to get "transformer action"? (*b*) Should the line be open or shorted? Explain. (*c*) How is proper matching between source and load effected? (*d*) Give the equation for the required Z_0 of the matching section.

89. Refer to Fig. 10–34. (*a*) What is the problem involved? (*b*) What does the section marked "$Z_0 = 190\ \Omega$" represent? (*c*) Why is this section 1.23 ft long? (*d*) Can this technique be used with reactive loads? Explain.

90. Refer to Fig. 10–35(a). (*a*) What is the dimension of the inverted U section shown in dark outline? (*b*) What is the purpose of this section? (*c*) If it were used with another antenna of higher resistance, what change is necessary from the condition shown? (*d*) If the impedance of the feed line were lower, what change would be necessary?

91. In Fig. 10–35(b), if the input were connected to a higher point on the coaxial line, what effect would it have on circuit operation?

92. (*a*) What is a bazooka used for? (*b*) Essentially, what is it?

93. Refer to Fig. 10–36(a). (*a*) What portion of this diagram is the bazooka? (*b*) What is the impedance between the outer coaxial conductor and ground at point *A*? Why? (*c*) What is this impedance at point *B*? (*d*) What is the impedance between the inner conductor and ground at point *A*? at point *B*?

94. Refer to Fig. 10–36(b). (*a*) What is the purpose of this device? (*b*) Is the length from *P* to *A* critical? Explain. (*c*) Is the length from *A* to *B* critical? Explain. (*d*) Can this technique be used with an unmatched load? Explain.

95. (*a*) What does a waveguide look like? (*b*) Basically, what is it used for? (*c*) Can a given waveguide be used at any frequency? Explain. (*d*) Why aren't they used for AM broadcast-band applications?

96. (*a*) Does a waveguide have a characteristic impedance? (*b*) What determines the value? (*c*) Can standing waves be created in waveguides? (*d*) Give two conditions that could cause them.

97. Compare waveguides and coaxial lines as to (*a*) resistance losses, (*b*) dielectric losses, (*c*) radiation losses, and (*d*) power-handling ability, for the same outer dimension.

PROBLEMS AND DIAGRAMS

1. Using the transmission line of Fig. 10–1 and the text data pertaining to it, plot a curve showing (*a*) voltage conditions on the line 1.50 s after the line is energized and (*b*) the voltage variations at a location halfway between *B* and *C* during the time interval 0.5–2.0 s after the line is energized.

2. (*a*) An RF signal has a frequency of 54 MHz. Calculate its wavelength in meters. (*b*) Calculate the wavelength, in feet, corresponding to frequencies of 1200 kHz and 2000 MHz. (*c*) A dipole antenna has sections 0.6 ft in length. For what frequency is this antenna designed? (Each section should be $\frac{1}{4}\lambda$.)

3. In the transmission line of Fig. 10–5, calculate the input impedance (*a*) when section *k* is added. (*b*) when sections *k* and *l* are added.

4. Calculate the characteristic impedance of an open-wire, parallel-conductor line spaced 0.736 in apart and using No. 12 wire.

5. Calculate the characteristic impedance of a coaxial line (air dielectric) whose inner conductor is a No. 4 wire and whose outer conductor has an inside diameter of 2.5 in.

6. A piece of coaxial line (air dielectric) is required to have an electrical length of 0.5λ at a frequency of 2500 MHz. What physical length (in inches) is needed? (Correct for the velocity constant.)

7. Show the initial, reflected, and resultant voltage waves on an open-ended transmission line $1\frac{1}{4}\lambda$ long, when the initial wave at the source starts at zero and is going positive.

8. Repeat Problem 7 with the initial wave starting at maximum positive.

9. Repeat Problem 7 for a line $1\frac{5}{8}\lambda$ long.

10. Repeat Problem 7 for current waves.

11. Repeat Problem 7 for current waves, but with the initial wave starting with a 45° lead.

12. Draw the standing wave patterns (voltage and current) for an open line $1\frac{7}{8}\lambda$ long.

13. Repeat Problem 12 for a shorted line.

14. The amplitudes of the incident and reflected waves on a transmission line are found to be 25 V and 5 V, respectively. Find (*a*) the reflection coefficient and (*b*) the SWR value.

15. A 500-Ω line is terminated with a 200-Ω resistive load. Find (*a*) the SWR value and (*b*) the reflection coefficient.

16. Develop suitable phasor diagrams to show the nature (*L*, *C*, or *R*) and the qualitative value of the input impedance for (*a*) an open line $\frac{1}{6}\lambda$ long; (*b*) a shorted line $\frac{1}{2}\lambda$ long; (*c*) a shorted line $\frac{1}{8}\lambda$ long.

17. A 150-Ω transmission line is terminated in a mismatched resistive load. The standing wave ratio is 2.0. Assuming the load resistance is too low, find (*a*) the load resistance and (*b*) the input resistance of the line if it is 10.3λ long. (Assume no losses.)

18. Repeat Problem 17, assuming the load resistance is too high.

19. A 150-Ω transmission line is terminated with a load impedance having a resistance of 150 Ω and an inductive reactance of 60 Ω. Find (*a*) the standing wave ratio and (*b*) the input impedance for a line length of 4.6λ. (Assume no attenuation.)

20. The input impedance of a 72-Ω line has a resistive component of 151 Ω and a capacitive reactance of 115 Ω. Find (*a*) the standing wave ratio and (*b*) the load impedance, if the line is 1.8λ long. (Assume no attenuation.)

21. Repeat Problem 19(*b*) for a line loss of 4.0 dB.

22. Repeat Problem 20(*b*) for a line loss of 1.5 dB.

23. A shorted section of coaxial line is to act as an inductance of 0.3 μH at a frequency of 300 MHz. The coaxial line has a characteristic impedance of 72 Ω. What length of line will be needed?

24. Repeat Problem 23 using a Smith chart.

25. A piece of coaxial line having a capacitive reactance of 162 Ω is needed for impedance matching at 200 MHz. Using 72-Ω line, what length of line would be needed, and how should it be terminated?

26. The above 72-Ω line is to be used as a series-resonant trap circuit. What length of line should be used to cause a short circuit at 600 MHz? How should it be terminated?

27. A coaxial-line tank circuit is to resonate at 1200 MHz. What length, what termination, and what characteristic impedance should be used?

28. A 50-Ω coaxial transmission line is connected to an antenna of 20 Ω resistance and 30 Ω inductive reactance. The operating frequency is 500 MHz. Find the length and location for a stub to correct the mismatch.

29. An antenna has an input impedance of $50 - j80$ at a frequency of 1000 MHz. Find the stub requirement to effect an impedance match with a 200-Ω line.

30. It is necessary to connect a 75-Ω line to a 50-Ω line. Give the constants for a suitable "transformer" matching section for use at 800 MHz.

31. A yagi antenna for use at 65 MHz has an input resistance of 30 Ω. It is to be fed from a 300-Ω line. Specify length and impedance for a suitable matching section.

32. Show by diagram how a shorted quarter-wave line can be used for impedance matching.

TYPICAL FCC QUESTIONS

3.220 What is the velocity of propagation of radio-frequency waves in space?

3.262 If the period of one complete cycle of a radio wave is 0.000001 second, what is the wavelength?

3.306 Indicate by a drawing two cycles of a radio-frequency wave and indicate one wavelength thereof.

3.500 Discuss Lecher wires, their properties, and use.

3.502 What are waveguides? cavity resonators?

3.530 Describe briefly the construction and purpose of a waveguide. What precautions should be taken in the installation and maintenance of a waveguide to ensure proper operation?

3.533 Describe three methods for reducing the radio-frequency harmonic emission of a radiotelephone transmitter.

4.062 Draw a simple schematic diagram showing a method of coupling the radio-frequency output of the final power amplifier stage of a transmitter to a two-wire transmission line, with a method of suppression of second- and third-harmonic energy.

4.063 An antenna is being fed by a properly terminated two-wire transmission line. The current in the line at the input end is 3 amperes. The surge impedance of the line is 500 ohms. How much power is being supplied to the line?

4.073 What is the ratio between the currents at the opposite ends of a transmission line one-quarter wavelength long and terminated in an impedance equal to its surge impedance?

4.074 The power input to a 72-ohm concentric transmission line is 5000 watts. What is the rms voltage between the inner conductor and sheath?

4.075 A long transmission line delivers 10 kilowatts into an antenna; at the transmitter end, the line current is 5 amperes and at the coupling house it is 4.8 amperes. Assuming the line to be properly terminated and the losses in the coupling system negligible, what is the power lost in the line?

4.076 The power input to a 72-ohm concentric line is 5000 watts. What is the current flowing in it?

4.077 What is the primary reason for terminating a transmission line in an impedance equal to the characteristic impedance of the line?

4.143 What is the power that is actually transmitted by a standard broadcast station termed?

4.198 If the spacing of the conductors in a two-wire radio-frequency transmission line is doubled, what change takes place in the surge impedance of the line?

4.199 If the conductors in a two-wire radio-frequency transmission line are replaced by larger conductors, how is the surge impedance affected, assuming no change in the center-to-center spacing of the conductor?

4.200 Why is an inert gas sometimes placed within concentric radio-frequency transmission cables?

4.202 Explain the properties of a quarter-wave section of a radio-frequency transmission line.

4.211 If the two towers of a 950-kilohertz directional antenna are separated by 120 electrical degrees, what is the tower separation in feet?

6.084 What is the formula for determining the wavelength when the frequency, in kilohertz, is known?

6.085 If the period of one complete cycle of a radio wave is 0.000001 second, what is the wavelength?

6.093 Indicate by a drawing two cycles of a radio-frequency wave and indicate one wavelength thereof.

6.102 What is the velocity of propagation of radio-frequency waves in space?

6.207 What are waveguides and in what type of radio circuits do they find application?

6.501 What should be the approximate surge impedance of a quarter-wavelength matching line used to match a 600-ohm feeder to a 70-ohm antenna?

6.502 What determines the surge impedance of a two-wire nonresonant radio-frequency transmission line?

6.600 How may harmonic radiation of a transmitter be prevented?

S-3.213 What is meant by the "characteristic" (surge) impedance of a transmission line; to what physical characteristics is it proportional?

S-3.214 Why is the impedance of a transmission line an important factor with respect to matching "out of a transmitter" into an antenna?

S-3.215 What is meant by "standing waves," "standing wave ratio (SWR)," and "characteristic impedance," as referred to transmission lines? How can standing waves be minimized?

S-3.216 If standing waves are desirable on a transmitting antenna, why are they undesirable on a transmission line?

S-3.217 What is meant by "stub tuning"?

S-3.218 What would be the considerations in choosing a solid-dielectric cable over a hollow pressurized cable for use as a transmission line?

11

Antennas and Propagation

The function of any antenna is to radiate into space the power delivered to it from the transmitter. In some cases, as in broadcasting, it may be desirable to radiate this power equally well in all directions. On the other hand, point-to-point communication requires an antenna system that will concentrate the transmitted energy into the narrowest possible beam. How well these effects are accomplished depends on the characteristics of the specific antenna system.

BASIC ANTENNA PRINCIPLES

In Chapter 10, we saw that if a transmission line is open at the load end, standing waves are produced, with the voltage standing wave a maximum at the open end, and the current standing wave a minimum. We also saw (in Fig. 10–8) that an electromagnetic field is created around the wires, that the electric and magnetic fields are mutually perpendicular, and that the field is concentrated mainly in the region between the wires. Since the electromagnetic field is concentrated between the lines, relatively little energy is radiated out into space.

Now consider a quarter-wave transmission line connected to an RF source—but this time the transmission line wires will be "fanned out" until they are diametrically opposite each other, or in line. Such a condition is shown in Fig. 11–1(a). The standing wave patterns will be as before. The maximum voltage will be at the ends, and since each person is

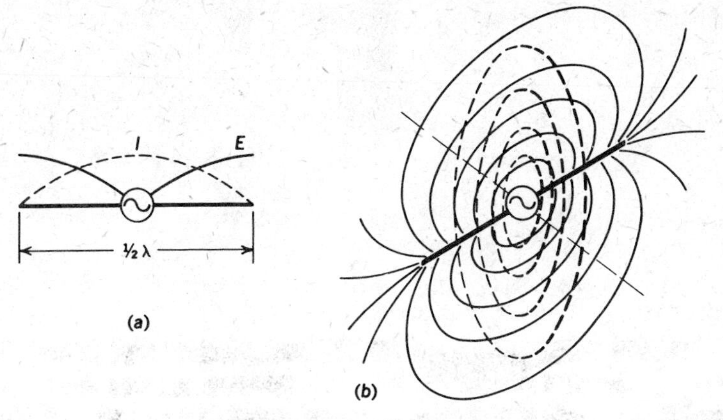

Figure 11-1 A half-wave antenna.

one quarter-wave long, maximum current is at the center. Again, electric and magnetic fields are created. The electric field (solid lines) is still from one point on one line to the identical point on the other line, but now notice how this field stretches out into space. The field strength is a maximum from end to end and becomes weaker toward the generator (center). On the other hand, the current standing wave is a maximum at the center, and the magnetic field (broken lines) is strongest in this region. These fields are shown in Fig. 11-1(b).

Since the energy is supplied from an RF source, the amplitude of the standing waves (voltage and current) will vary *with time* from the maximum values shown, to zero, to the same maximum with reversed polarity, and again to zero, at the RF rate. Obviously, the strength of the electric and magnetic fields will vary accordingly. At very low frequencies, it is assumed that all the energy in these fields is returned to the circuit, as the line voltage and current drop to zero and the fields collapse. This is generally true when the length of the conductors involved and the spacing between conductors are a very small fraction of a wavelength. However, at high frequencies, it is necessary to consider these fields from two aspects—as *induction fields* and as *radiation fields*.

The induction field is the phenomenon observed at low frequencies. Energy stored in the induction field is returned to the circuit as the fields collapse. This field strength is inversely proportional to the *square* of the distance from the conductor. Consequently, the induction field is limited to the zone immediately surrounding the antenna, and it drops to a negligible value at distances beyond (approximately) one-half wavelength from the antenna.

Field Intensity

The radiation field is a "traveling-wave" field that reaches much farther out into space. Even when the antenna current, and the voltage that caused it, die out, the wave still continues to travel out into space. This effect is similar to the water waves produced by a speed boat in the middle of a large lake. The water waves spread out and travel towards the shores, even after the boat has passed. As additional boats pass, a series of waves are created, all traveling out in sequence toward the shores, but getting weaker the farther they travel from the originating source.

The strength of the radiation field, or *field intensity*, at any remote location is evaluated in terms of the voltage this changing field will induce across a wire that is parallel to the electric field lines. Since the magnitude of the induced voltage is affected by the length of wire used, a standard length of 1 meter is specified. The field strength is then measured in *volts per meter*. Usually, since the field strengths involved are quite weak,* the specific unit will be microvolts or millivolts per meter. Special field-strength meters are available for such measurements.

The field intensity at any location depends on the distance from the transmitter and on the radiated power. Neglecting ground losses and considering only ground-wave coverage,† the field intensity varies inversely with distance. With good ground conductivity (such as over seawater), this inverse relationship applies up to about 100 miles. Beyond this range —or with poor ground conductivity—the ground losses are very high, and the actual field strength can be much lower than as calculated by the inverse relation.

If the power radiated by the transmitter is increased, it is obvious that the field strength at any location will also increase. Since power is proportional to the square of the voltage, it follows that the increase in field strength is proportional to the square root of the power change, i.e., doubling the radiated power output will increase the field intensity by $\sqrt{2}$ or 1.41. Conversely, to increase the field intensity by three, it is necessary to increase the radiated power by nine.

Polarization

As mentioned earlier, the electromagnetic waves radiated from any antenna have two component fields (the electric field and the magnetic field) that are mutually perpendicular. These fields were shown in Figs.

* Field strengths as low as 10 μV per meter will provide good quality broadcast-program reception in a low-noise location. For commercial communications, even weaker fields can produce satisfactory results.

† Propagation of waves is discussed later in the chapter.

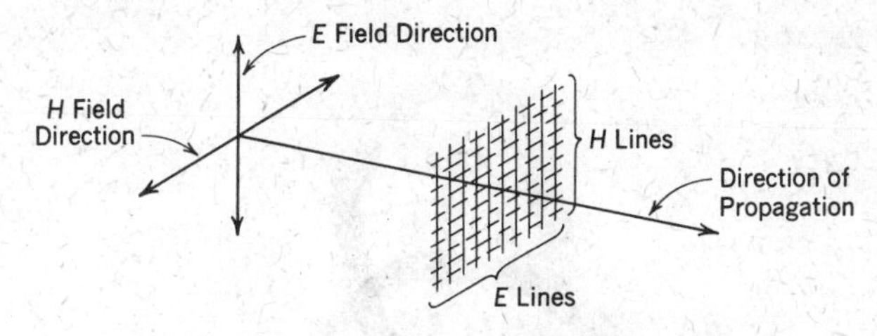

Figure 11-2 Polarization of an electromagnetic wave.

10–8 and 11–1, with curved (or circular) lines. However, as the fields expand into space, only a small portion of their *wavefront* is effective in inducing a voltage at the receiving antenna. With regard to such small sections, the field lines are straight and still mutually perpendicular; in turn, both fields are perpendicular to the direction of propagation of the electromagnetic wave. These relations are shown in Fig. 11–2.

In Fig. 11–2, the electric field lines are in a vertical direction. Such a field pattern is called a *vertically polarized field.* If the pattern were rotated 90°, the magnetic field lines would be vertical, and the electric field pattern would be horizontal. This field would be horizontally polarized. Obviously the direction of the electric field is used to specify polarization. As we will see later, a vertical antenna produces a vertically polarized field, while a horizontal antenna produces horizontal polarization.* The importance of this is that a field will induce a maximum voltage in a receiving antenna that is parallel to the direction of polarization. Theoretically, the voltage induced in a horizontal antenna would be zero if the field is vertically polarized.

Radiation Patterns

If an antenna could be considered as a point source that radiated energy equally well in *all* directions, then the field strength, as measured *at some fixed distance* from the antenna, would be constant regardless of the angle—horizontal or vertical—that the receiving location makes with the antenna. Such an antenna is called an *isotropic* antenna and would be *nondirectional* or *omnidirectional* in both the horizontal and vertical planes. However, practical antennas do not have such a radiation characteristic. In fact, isotropic distribution would result in a waste of energy. It is generally more desirable to restrict a transmitter's energy either in the horizontal plane, or in the vertical plane, or both. It is therefore necessary to use some means to describe (or compare) the variation

*It should be noted that a change (rotation) in the direction of polarization may occur before the wave reaches a receiving location.

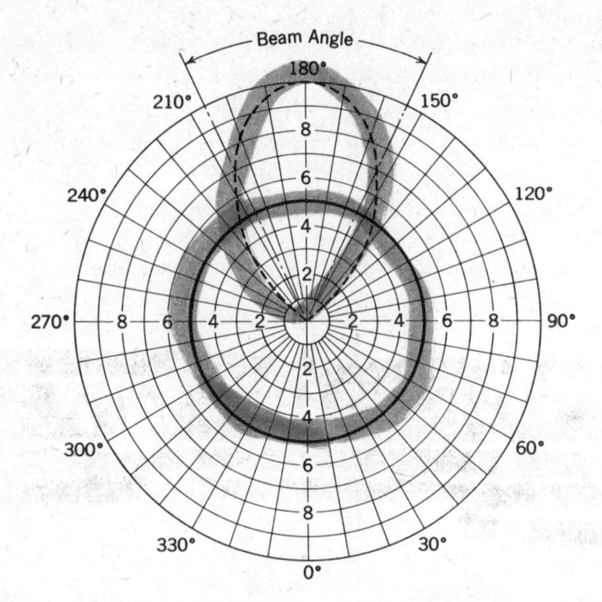

Figure 11-3 Antenna radiation patterns.

in field strength at various directions with respect to the antenna. Antenna *radiation patterns* are used for this purpose. Because of the three-dimensional aspect of the energy fields, radiation patterns are taken in the horizontal plane and in the vertical plane, if the full field distribution pattern is desired.

Two possible radiation patterns are shown in Fig. 11-3. They are shown not to study a specific antenna characteristic, but rather to explain how to interpret this type of representation. Let us assume that these are the *horizontal* radiation patterns for two different types of antennas. Notice that the curves are plotted on *polar-coordinate* paper. The coordinates for these curves are "angle of radiation—in degrees" (with respect to some reference direction—zero degrees), and "relative field strength." The numbers shown on the 0°, 90°, 180°, and 270° radials are for the field-strength scale. It must be clearly understood that field-strength patterns are taken at some *fixed* distance from the antenna, and these curves do *not* in any way represent the effect of change in distance.

Now examine the solid curve. It shows a relative field strength of 5* that is constant regardless of the direction of radiation. This antenna is obviously nondirectional in a horizontal plane. The broken-line radia-

*This does not mean an absolute field strength of 5 mV or 5 μV per meter. It is a reading that could be obtained on any uncalibrated instrument. Its value is significant only in relation to comparative readings in other directions.

tion pattern, on the other hand, is for a highly directional antenna. Maximum radiation is obtained in a direction 180° from the reference. Notice that all the radiation is confined to a region of 100° ranging (counterclockwise) 130–230° from the reference direction. Considering the field strength in the 180° direction as 10 units, the field strength in a direction 200° from the reference is 8.2, or (approximately) 80 percent of maximum. Similarly, note that the field strength in a 145° direction is 5, or only 50 percent of maximum.

EXAMPLE 11–1

With reference to the broken-line pattern of Fig. 11–3, in what directions does the field strength drop to 70 percent of maximum?

Solution

The circle for relative field strength 7 crosses the radiation pattern at 154° and at 206°. These are the directions where the field strength falls to 70 percent of maximum.

To obtain the radiation pattern of an antenna, the antenna should be located in an open area so as to avoid reflections (or absorption) from nearby objects. For a true "free-space" pattern, the antenna should also be several wavelengths above ground. A receiver or other suitable field-strength indicator is then located at some distance from the transmitting antenna. This distance should be at least 0.6 wavelength and preferably several wavelengths from the antenna so as to minimize any effect from the induction field. Keeping this distance fixed, the antenna is rotated and field-strength measurements are taken at intervals of 10–30°.

BASIC ANTENNAS

Many variations of antenna designs have been used to obtain extra gain, or a particular directive radiation pattern, or to achieve a certain vertical angle of radiation. However, all versions are derived from one of two basic types—the half-wave isolated antenna (also known as the Hertz antenna or dipole antenna) and the grounded quarter-wave antenna (also known as the Marconi antenna).

Half-Wave Antennas

The half-wave or dipole antenna is the most widely used type of antenna—particularly for frequencies above 2 MHz. As the name implies, it consists of two quarter-wave sections, for an overall length of one-half wavelength. This length refers to the *electrical* length. Because of end effects (capacitive), and because of the velocity constant of the line, the actual physical length is 3 to 8 percent shorter, depending on frequency. Such an antenna was shown in Fig. 11–1(a).

Input Impedance

As explained at the start of this chapter, a dipole antenna can be considered similar to an open-ended transmission line. Energy traveling to the end of the line is reflected back to the source, causing standing waves. Therefore, in accordance with transmission line theory, if the antenna is of resonant length (quarter-wave elements or multiples thereof), the input impedance is resistive. For any other lengths, the input impedance is inductive or capacitive, depending on length. There is, however, one difference between antennas and transmission lines. An efficient transmission line has negligible losses; its standing wave nulls approach zero; and its input impedance approaches zero (for a quarter-wave section) and infinity (for a half-wave section). On the other hand, an antenna is designed to radiate power into space. This power comes from the transmitter, and so far as the transmitter is concerned, power is dissipated or "lost" much as if it fed a line with appreciable losses. Consequently, because of radiation, the standing wave nulls on an antenna do not drop to zero, nor does the input impedance drop to zero or approach infinity. The input impedance of a half-wave dipole antenna in free space, measured at the center (current maximum, voltage minimum), is approximately 73 Ω.* If the antenna is end-fed (voltage maximum, current minimum), the input impedance is approximately 2400 Ω. In either case, if the antenna is resonant, this input impedance is resistive.

The total antenna resistance can be considered to consist of a resistance due to the useful power radiated into space, and a resistance due to losses. These losses will include the I^2R loss in the antenna wires, leakage and dielectric losses in supports and insulators, eddy current and hysteresis losses in nearby metallic objects, and corona losses due to high-voltage electrical discharges. The total *antenna resistance* (radiation + losses) can be computed from the power relation $P = I^2R$, where P is the total power delivered to the antenna, and I is the antenna current (rms) measured at the point where the antenna if fed.†

The *radiation resistance* of an antenna represents that portion of the antenna resistance due to the actual power radiated. Mathematically

* If the antenna is close to ground (less than 2λ), reflections from the ground surface can cause this resistance to vary from a high of 95 Ω at 0.3λ above ground to a low of 58 Ω at 0.6λ above ground.

† If nonresonant lines are used to feed the antenna (no standing waves on the transmission line), this current reading can be made at the transmitter end of the line.

it is the ratio of the power radiated divided by the square of the antenna current (rms) at the feed point. The radiation efficiency of an antenna (power radiated/total power supplied) will obviously depend on the ratio of radiation resistance to total antenna resistance. For good efficiency, the radiation resistance should be high compared to other loss effects, that is, the losses should be kept low. Efficiencies ranging 75–95% are common for antenna lengths of one-quarter wavelength or longer.

Horizontal Antenna-Radiation Patterns

The free-space pattern for a half-wave dipole antenna will vary depending on whether the antenna is horizontal or vertical or on—what amounts to the same thing—the plane of the radiation field with respect to the antenna. Also, if the antenna is close to earth, ground effects (reflections) can cause a change in the field distribution.

Figure 11–4(a) shows the *horizontal* radiation pattern for a half-wave dipole that is mounted horizontally in free space.* Notice that the direction for maximum radiation is *broadside* (at right angles) to the line of the antenna, as shown by the lengths of vectors P and P'. The field strength decreases as the angle θ between the direction of radiation and

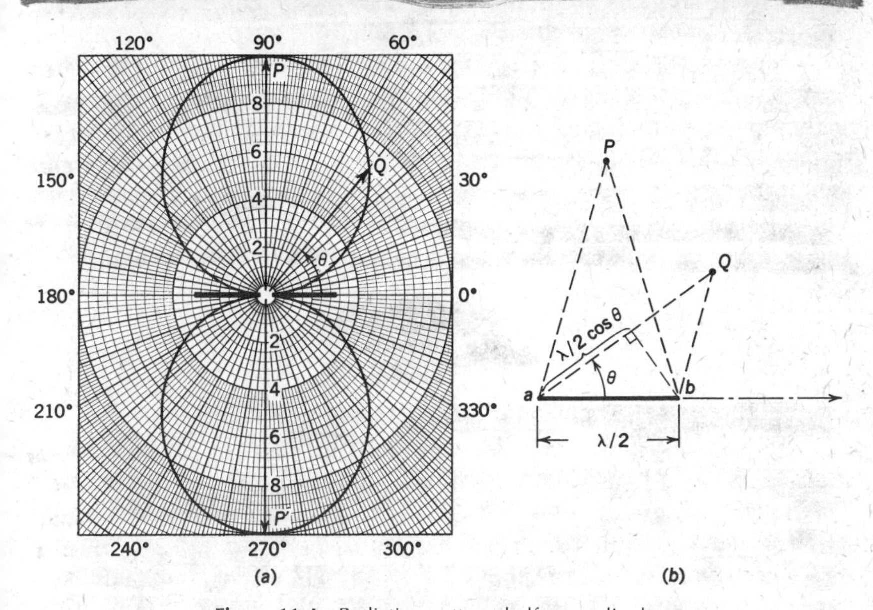

Figure 11-4 Radiation pattern, half-wave dipole.

*The patterns shown (*lobes*) are not circular. Circular lobes would be obtained only if the current were constant throughout the length of the antenna.

the axis of the antenna decreases. For example, if the antenna itself is pointed in an east-west direction, maximum radiation is in the north-south direction, falling off to a theoretical value of zero in the east-west direction. The reason for this pattern can be seen from Fig. 11–1 and Fig. 11–4(b). Figure 11–1 brings out two points: first, that the magnetic field extends out at right angles to the antenna wire; second, that the current in the antenna is in phase throughout its length (with the amplitude at any location varying as shown by the standing wave). Now consider point P in Fig. 11–4(b), at a distance many wavelengths from the antenna. The distance from this point to any point on the antenna is essentially identical. Therefore, electromagnetic radiations arriving at this point from any location on the antenna are in-phase and additive. The energy radiated in this direction is a maximum. With regard to any other direction, such as toward point Q, two factors combine to cause a reduction in field strength. The magnetic field is at right angles to the antenna wire, and the distances from the ends of the antenna to point Q are not the same. An electromagnetic wave from point a on the antenna must travel an additional distance equal to $(\lambda/2) \cos \theta$ and will be out of phase with the radiation from end b. If this distance $(\lambda/2) \cos \theta$ is greater than one-third wavelength, there will be a decrease in the resultant signal strength. As angle θ decreases, the interference increases, so that for directions in line with the antenna, the resultant signal strength is zero.

In general, the field strength produced at some distant point P by a given antenna is a function of the length of the antenna (in wavelengths at the operating frequency), the variation in the maximum current value throughout this length, the direction angle, θ to point P measured from the line of the antenna wire, and the distance from the antenna to the point. This relation can be analyzed quantitatively.* For a half-wave dipole, and at a fixed distance to the antenna, the pattern factor reduces to

$$k = \frac{\cos\left(\frac{\pi}{2} \cos \theta\right)}{\sin \theta} \tag{11–1}$$

where $(\pi/2) \cos \theta$ is an angular measurement in radians.

Now let us study the field-strength distribution in a *vertical* plane. First, consider a plane that is perpendicular to the axis of the antenna. This is the field distribution that would be seen from an end view in Fig. 11–4(b), looking in the direction a towards b (or vice versa). Since all directions of propagation in this view are at right angles to the line of the antenna, maximum radiation will be obtained at any angle to the horizon.

* R. G. Brown, R. A. Sharpe, and W. L. Hughes, *Lines, Waves and Antennas* (New York, Ronald Press, Inc., 1961).

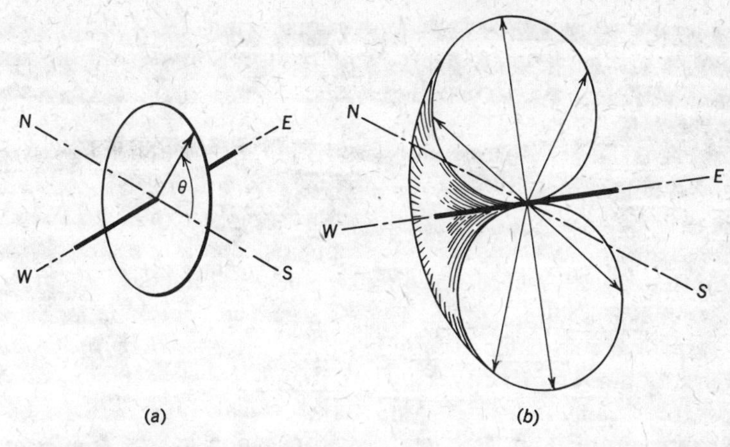

Figure 11-5 End view and cross-sectional view of the full radiation pattern for a horizontal dipole.

The radiation pattern will be a circle, with the antenna at the center, as shown in Fig. 11–5(a).

However, this one diagram does not show the full picture of the field distribution in vertical directions. For example, if we study the pattern in a vertical plane *parallel* to (or through) the line of the antenna, we will get another "figure-eight" type of pattern, as shown in Fig. 11–4(a). This time, of course, direction P–P' would be up and down, and direction Q an angle of *elevation* of $\theta°$. Obviously, the full radiation pattern for this dipole antenna is three-dimensional and can be visualized by rotating the figure-eight pattern of Fig. 11–4(a) around the antenna line as an axis. The result is a doughnut-shaped field with the antenna running through the hole. Half of this pattern is shown in Fig. 11–5(b).

Vertical Antenna-Radiation Patterns

If a half-wave dipole antenna is mounted vertically (rotated 90° from the horizontal), the *free-space* radiation patterns will be the same as for the horizontal antenna but rotated through 90°. In other words, the

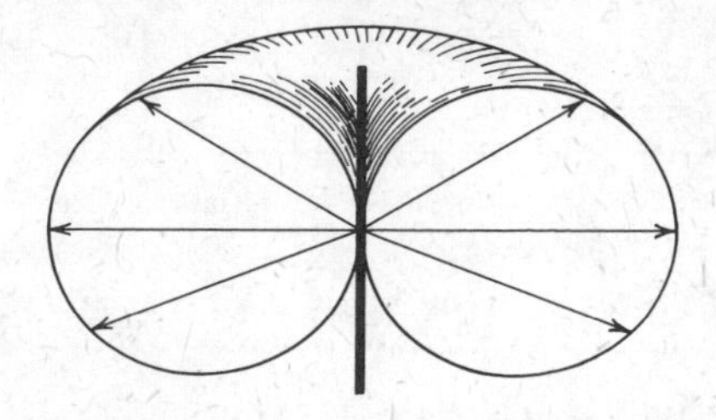

Figure 11-6 Free space radiation pattern for a vertical half-wave dipole.

horizontal distribution is now omnidirectional (circular), while the vertical distribution is the familiar figure-eight pattern. The three-dimensional doughnut pattern for a vertical half-wave dipole, in free space, is shown in Fig. 11–6.

There are several reasons for using a vertical antenna, in preference to horizontal mounting. One reason should be fairly obvious. If a nondirectional radiation field is desired, as for example in radio broadcasting, a vertical antenna is necessary. Another consideration is that at low to medium frequencies, the required antenna length can become a problem. For example, at 1000 kHz, a half-wave antenna is approximately 500 ft long. It may be quite difficult to support such an antenna in free space. A vertical structure can be easier to install and—as we will see shortly—will have excellent radiation characteristics with only half the height. Also, at low to medium frequencies, propagation is mainly via *ground waves*,* and this type of propagation requires vertically polarized waves, as obtained from vertical antennas.

Ground Effects

In the above discussions, the radiation patterns were characterized as free-space patterns. Yet it is not always possible to mount an antenna far enough above ground to obtain a free-space condition, and reflections from ground can alter the vertical plane patterns considerably. The effect of ground reflection can be seen from Fig. 11–7. The antenna is a half-wave horizontal dipole mounted some short distance h above ground. The field strength at any point in space, such as P, is the resultant of the direct radiation and the reflected radiation. This reflected wave can be considered as emanating from an *image antenna* at a distance h below the ground surface (mirror image). In arriving at a point remote from the

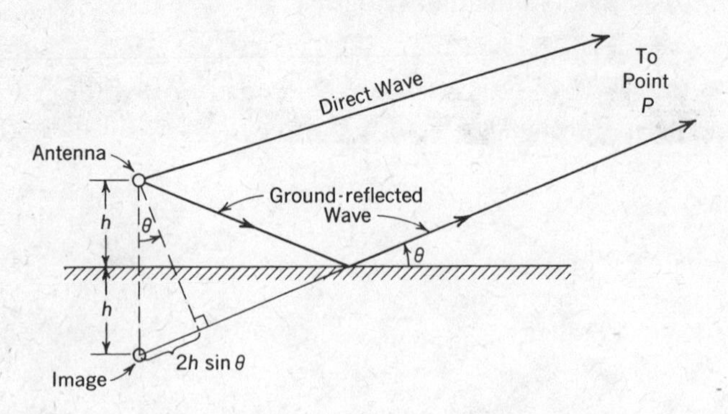

Figure 11-7 The effect of ground reflection.

*These are discussed later in the chapter.

antenna, the reflected wave travels a distance equal to $2h \sin\theta$ farther than the direct wave. Also, this wave is inverted 180° upon reflection. From this, it is a simple matter to calculate the *ground-reflection factor* for any angle of radiation and for any height above ground.*

EXAMPLE 11–2

Calculate the ground-reflection factor and relative field strength for a half-wave horizontal dipole mounted $\frac{1}{4}\lambda$ above ground, for vertical radiation angles of 0, 15, 30, 45, and 90 degrees.

Solution

1. For a vertical angle of 0°, $\sin\theta = 0$, $2h \sin\theta = 0$, the reflected wave is 180° out of phase with the direct wave, and the reflection factor and resultant are zero.
2. (*a*) For $\theta = 15°$, $\sin\theta = 0.259$, $2h = \frac{1}{2}\lambda$, and $2h \sin\theta$ is equivalent to a phase shift of 46.6°, (180° × 0.259). The reflected wave lags by 180° + 46.6° or 227°. The resultant of two waves, amplitude 10 (assumed) and 227° apart, is 8.0.
   (*b*) The reflection factor is 0.8.
3. (*a*) For $\theta = 30°$, $\sin\theta = 0.5$ and $2h \sin\theta$ is equivalent to 90°. The direct and reflected waves are 90° apart. Their resultant (for assumed amplitudes of 10) is 14.1.
   (*b*) The reflection factor is 1.41.

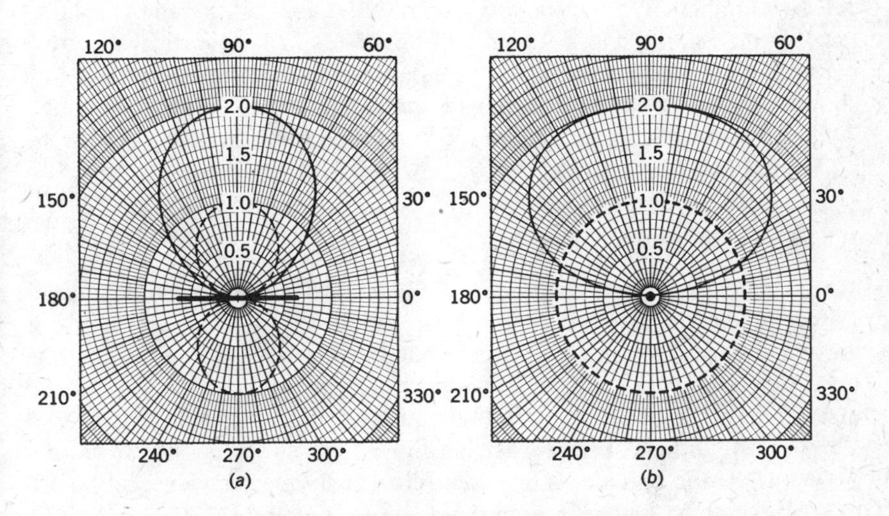

Figure 11-8 The effect of ground on vertical-plane radiation patterns.

*The value of the ground-reflection factor for various heights of antennas can also be obtained from specially prepared graphs and charts.

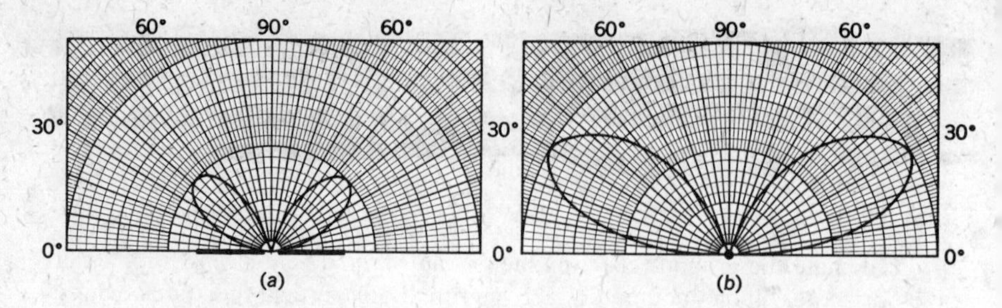

Figure 11-9 The effect of ground, for antenna height $\frac{1}{2}\lambda$ (vertical plane).

4. (*a*) For $\theta = 45°$, $\sin\theta = 0.707$ and $2h \sin\theta = 127°$. The reflected wave lags by $180° + 127°$ or $307°$ (or leads by $53°$). Their resultant is 17.9.
   (*b*) The reflection factor is 1.79.
5. (*a*) For $\theta = 90°$, $\sin\theta = 1$ and $2h \sin\theta = 180°$. The reflected wave will be in phase with the direct wave. The field strength is doubled (20).
   (*b*) The reflection factor is 2.0.

Figure 11–8 shows the vertical radiation patterns for a horizontal half-wave dipole, one quarter-wave above ground. Notice how the lower lobe is completely gone, while the field strength directly upward is doubled. Diagram (a) shows the vertical distribution in a plane parallel (and through) the antenna, while diagram (b) shows the vertical distribution in a plane at right angles to the antenna.

In similar fashion, radiation patterns for any other antenna height can be obtained. Obviously, the ground reflection factors and the direction for maximum field strength will be affected. The vertical radiation patterns for a horizontal dipole one-half wavelength above ground are shown in Fig. 11–9. Notice that the pattern has broken into two lobes, while the direction for maximum radiation (end view) is now at 30° to the horizontal instead of straight up.

Grounded Antennas

The Marconi antenna is essentially a quarter-wave monopole vertical antenna with the lower end either connected directly to ground, or grounded through the transmitter coupling network. (Quite often, the tower itself is the antenna.) However, its characteristics are similar to the half-wave dipole because ground reflections create an image antenna that supplies the other quarter-wave section. This effect is shown in Fig. 11–10(a). If the antenna is an exact (electrical) quarter wave, the standing wave pattern for the antenna *and image* is exactly the same as for a half-wave antenna. Notice that a current maximum occurs at the grounded end. This means that high currents will flow through ground.

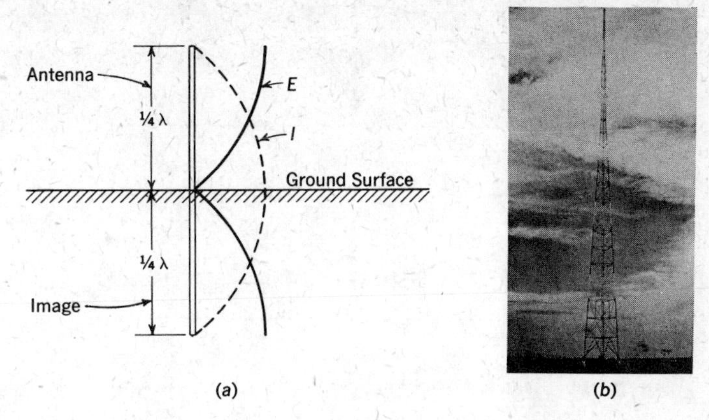

Figure 11-10 A quarter-wave grounded antenna.

To keep losses (I^2R) at a minimum, the ground should have good conductivity.* To reduce such losses, high-powered stations will generally use a *ground system* consisting of heavy copper wires spread out radially directly below the antenna tower. These rods are spaced as close as every 3° and are approximately one-half wavelength long. (Alternate radials are sometimes made shorter.) The radials are welded together by circular straps spaced at approximately one-third and two-thirds of the radial lengths, and the entire system is then buried 6–12 in deep. If the ground is too rocky (or too sandy), a *counterpoise* or capacitive ground system is used. This is similar to the radial system, but it is suspended approximately 5–10 ft above ground. The area of a counterpoise should be as large as local conditions will permit.

If we examine the standing wave pattern of the grounded antenna (Fig. 11–10), we can see that at the input (the ground end) current is a maximum, and voltage a minimum. Therefore the input impedance is low. For an exact quarter wave, this value is 36–38 Ω and resistive. This in turn means that, for a given antenna current, the quarter-wave antenna will radiate only half as much power as a free-space half-wave antenna.

Grounded-Antenna Radiation Patterns

In the *horizontal* plane, the radiation pattern for a vertical grounded antenna will be omnidirectional (Fig. 11–3, solid curve), the same as for a vertical dipole. However, because of ground reflections, the vertical radiation will not be a figure-eight pattern. The radiation pattern for a

* In this respect flat, rich (loamy) soil has fair conductivity, whereas sandy or rocky terrain is poor. The conductivity of the ground can be improved by chemical treatment.

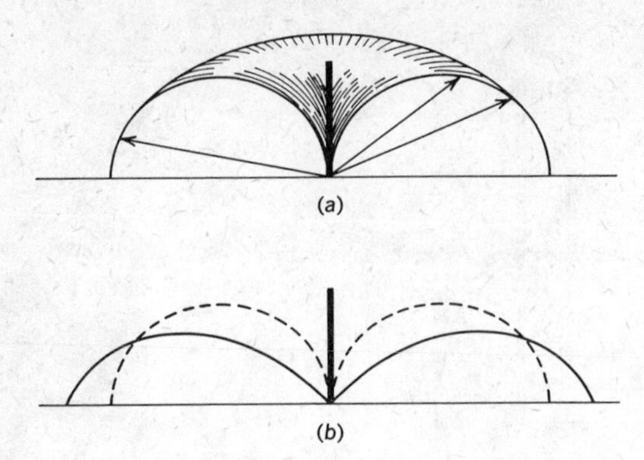

Figure 11-11 Radiation patterns (vertical plane) for grounded antennas.

quarter-wave Marconi antenna is shown in Fig. 11–11(a). Notice that the lower half of each lobe has been canceled by the ground-reflected waves. This loss of radiation in a downward direction is of no consequence, since actually the signal strength in the horizontal and upward direction is increased.

Although the nondirectional *horizontal* radiation pattern for this antenna is quite satisfactory for many radio broadcasting applications, Fig. 11–11(a) shows one weakness. An appreciable amount of energy is directed upward (sky waves*). Yet for area coverage (and for reliable daytime long-distance coverage), radiation along the earth's surface (ground waves) is desirable. An increase in antenna length will improve the horizontal field strength at the expense of sky-wave radiation. The solid curve in Fig. 11–11(b) shows the improvement obtained with an antenna length of one half-wave. (The antenna is still grounded.) Optimum ground-wave strength is obtained with a length just under $\frac{5}{8}\lambda$. Any further increase in length produces high-angle lobes of increasing strength, while the horizontal radiation is weakened. With an antenna height of 1λ, there is no ground wave.

Loaded Antennas

At low frequencies (for example, below 500 kHz), even a quarter-wave Marconi may be an impractical length. (For that matter, in mobile communications, a one-quarter wavelength may be an impractical dimension even at higher frequencies.) This can seriously impair the radiation. The antenna is no longer resonant; its input impedance has a high capacitive reactance; the antenna current decreases; radiation decreases. Another way of looking at this is that a reactive load will not accept power. Simultaneously, the radiation resistance decreases from 36 Ω for

* Sky waves, ground waves, and propagation are discussed later in the chapter.

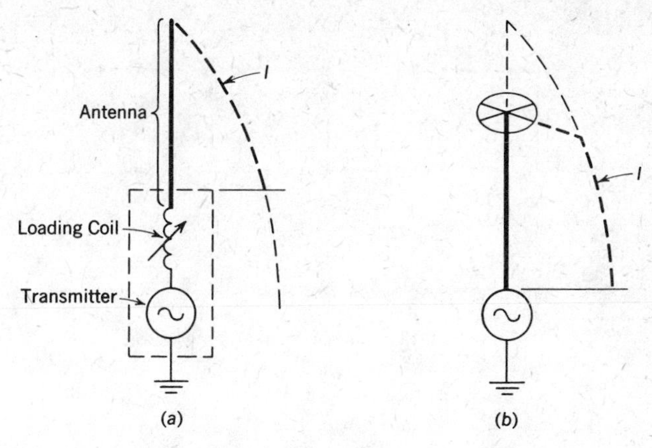

Figure 11-12 Standing-wave patterns on loaded antennas.

the quarter-wave antenna, towards zero as the length is reduced. This also indicates a loss of radiated power. Finally, because of the short length, the standing-wave current maximum will not occur on the antenna wire, but on the transmission line. The loss of this high-current peak on the antenna drastically reduces the radiated power.

Several techniques are used to minimize these defects. Figure 11–12(a) shows the use of a *loading coil* (inductance) to counteract the excessive antenna capacitive reactance. For convenience, the coil is connected at the base of the vertical antenna. This makes it possible to "tune" the antenna to the correct electrical length (resonance). The increase in antenna current because of resonance will increase the radiated power. (The antenna radiation resistance will increase by approximately 5 Ω.) However, this "cure" is not too effective for several reasons. Notice from Fig. 11–12(a) that the *loop* or high-current portion of the standing wave is in the loading coil. This does not add to radiated power. On the other hand, this high current will cause heavy I^2R losses in the coil, reducing the radiation efficiency, while the high voltage (standing wave) at the antenna end of the loading coil can create insulation and safety problems.

These difficulties are avoided by using *top loading*. A cylinder or spoked wheel placed at the top of the antenna acts to increase the shunt capacitance to ground. The increased capacitance reduces the capacitive reactance, bringing the antenna into resonance.* This technique is shown in Fig. 11–12(b). Notice that the standing wave pattern has been "pulled up" along the antenna, and the current maximum is now at the base. Top

* This is the same effect as shunting a capacitor across the open end of a transmission line to increase its electrical length.

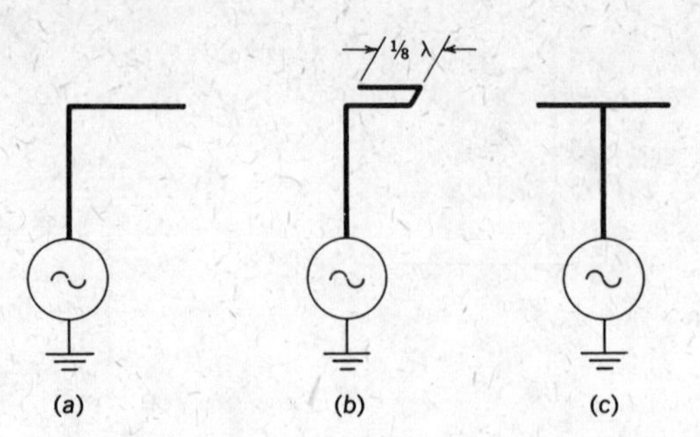

Figure 11-13 Flat-top vertical antennas.

loading results in appreciably higher radiation resistance and radiation efficiency. It also reduces the voltage at the base. Unfortunately, top loading is not readily adaptable for mobile use.

Still better radiation efficiency (and better radiation pattern) can be obtained if the current loop of the standing wave can be raised toward the top of the vertical radiator. This can be achieved by adding a "flat top" to a vertical antenna. When the vertical section can't be made longer, the top is folded over to form an inverted L or a T, as shown in Fig. 11–13. If the flat-top and vertical portions are each a quarter wave, then the current loop will occur at the top of the vertical radiator. In Fig. 11–13(b) and (c), radiation from the flat-top portions is minimized due to the cancellation from the out-of-phase currents. Obviously, if the total length (flat top plus vertical) is only one quarter-wave, the current loop is at the base of the vertical section, and the effect is similar to the top-loaded antenna.

METHODS OF FEEDING ANTENNAS

From the foregoing discussions, it is obviously both impractical and undesirable to connect the antenna directly to the transmitter (or vice versa). It is therefore necessary to use a transmission line for interconnection. Many variations are possible in the method used to feed the antenna. For example, the transmission line could be resonant or nonresonant, or the antenna could be fed at the center or at one end. Some of the more commonly used techniques will be discussed below.

Resonant-Line Feed

If a given antenna is to be used at more than one transmitting frequency (multiband operation), the input impedance of the antenna will vary, depending on its length and operating frequency. Obviously, no one transmission line can maintain an impedance match at the antenna, and standing waves will appear on the line. The standing wave ratio will depend on the degree of mismatch. Such a line is called a *resonant line*. Because of the standing waves, the losses in the line will be higher than for a matched (nonresonant) line. Consequently, the use of resonant feeders should be limited to line lengths not in excess of one wavelength.

Resonant lines can be connected to an antenna either at the center (center-fed) or at one end (end-fed). Two possible lines and coupling networks are shown in Fig. 11–14. Diagram (a) shows a half-wave antenna used with a resonant line also one half-wave long. Notice that a current maximum occurs at the transmitter coupling network. The transmission-line impedance at this point is low and requires a series-tuned network for impedance matching. The coupling is adjustable so as to reflect the proper load value to the power amplifier stage. On the other hand, if the transmission line were one (or three) quarters-waves long, the input impedance to the line would be high, and a parallel-resonant tank circuit would be necessary to effect an impedance match.

In the above illustrations, because the antenna and line are exact one-quarter wavelengths (or multiples thereof), the input impedance at the transmitter end of the line is resistive. Yet, for multifrequency operation—unless all frequencies are harmonically related—such a situation would merely be a coincidence. Therefore, depending on the operating frequency, the input impedance at the transmitter end of the line may contain inductive or capacitive components. It is therefore necessary to insert multitap loading coils into each leg of the coupling network in series with C_1 and C_2 of Fig. 11–14(a). A proper combination of loading coil and/or capacitor settings will tune the antenna and line into resonance.

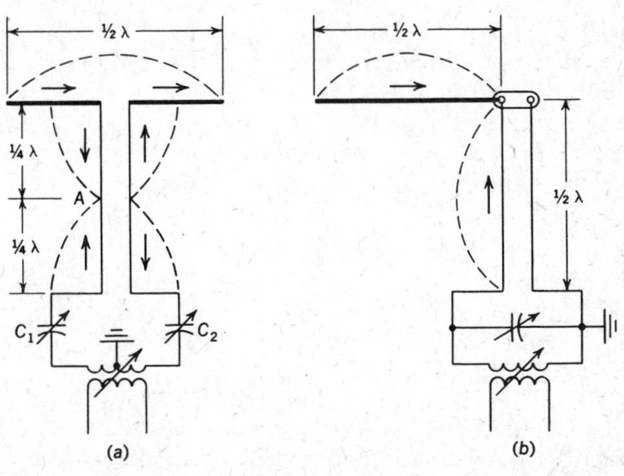

Figure 11-14 Resonant-line feeders.

Figure 11–14(b) shows an end-fed antenna. Again, the antenna and feeders are each one-half wavelength. However, because the line is connected to the high-impedance end of the antenna, the coupling network end of the line is also a high-impedance point. Notice that a parallel-tuned network is used. Of course, if the line length were changed to one-quarter or three-quarter wavelengths, a series-tuned coupling network would probably be needed to match the antenna impedance to the power amplifier stage. Also, for any length other than exact quarter-wave multiples, loading coils and/or additional capacitance may be necessary to balance out the reactive component of the line impedance.

Nonresonant Feeders

When an antenna is used for only one operating frequency, it is a relatively simple matter to match the antenna impedance to the characteristic impedance of the transmission line—particularly so, if the antenna is the basic half-wave dipole or quarter-wave Marconi. If the standing wave ratio is less than 1.5, the line is considered a nonresonant line. Any of the methods shown in the previous chapter can be used for matching. (See Figs. 10–32, 10–34, and 10–35.) Either center feed or end feed* can be used. Another technique, the *delta match,* is shown in Fig. 11–15. It can be used when the characteristic impedance of the line is higher than the radiation resistance of the antenna. Since the current standing-wave amplitude decreases with movement away from the center (while the voltage standing-wave amplitude increases), the antenna resistance increases. By spreading the feeder lines apart, a good impedance match can be obtained.

Open-wire lines are readily used with any of the above feed systems. Coaxial lines, with the outer conductor grounded, are obviously well suited for end-fed systems. However, the coaxial line is an unbalanced line, and balance-line converters (Fig. 10–36) must be used at the antenna

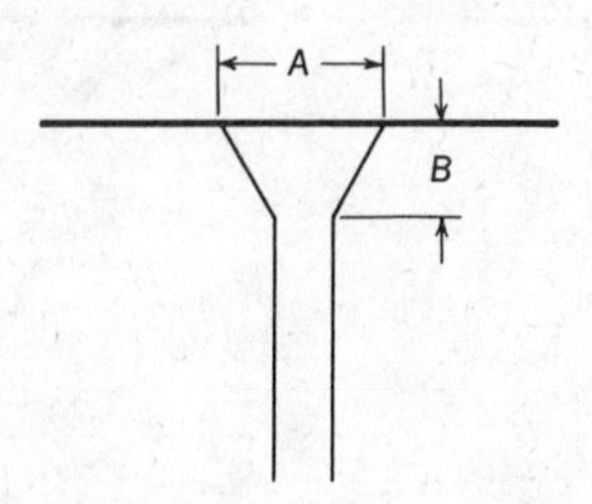

Figure 11-15 A delta-matched nonresonant line.

* Because of the current maximum, a center-fed system is also called *current-fed,* while the end-fed (voltage maximum) is known as a *voltage-fed* system.

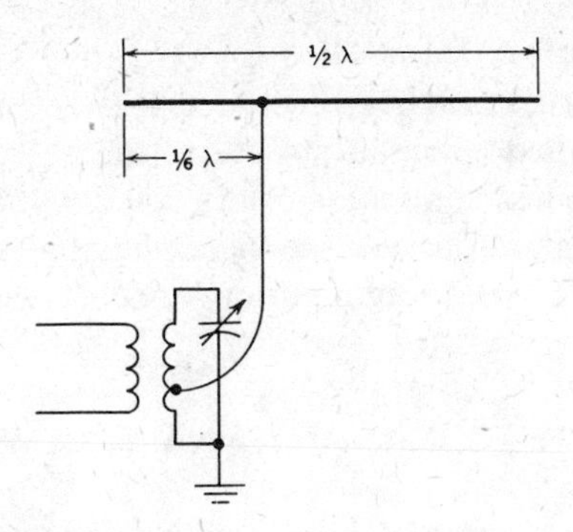

Figure 11-16 A single-wire feed system.

end of a coaxial line for center-fed systems. Another type of line that is used because of its simplicity and ease of construction is a *single-wire* feed. This is shown in Fig. 11–16. Since the characteristic impedance of a single-wire line is approximately 500 to 600 ohms,* an impedance match will be obtained by connecting the line at a point approximately $\frac{1}{6}\lambda$ from the end. The exact matching point is found experimentally. To minimize radiation interference, the feeder should drop straight down from the antenna for a distance of at least $\frac{1}{6}\lambda$. This system has two disadvantages. The radiation losses in the feeder wire are high, and the ground losses are high. When used, it should be used only with high-conductivity grounds.

Vertical antennas are readily fed with coaxial lines, since they are essentially unbalanced loads. The outer conductor of the line is grounded. Then if the antenna structure is grounded, the inner conductor of the coaxial line is brought up along the antenna to obtain an impedance match.

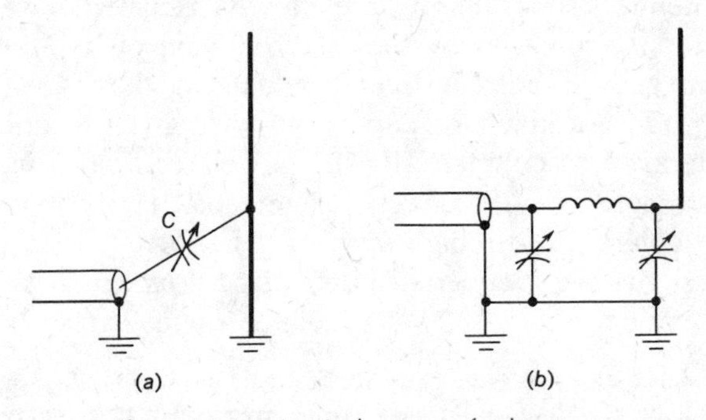

Figure 11-17 Vertical antenna feed systems.

*Z_0 varies with height of antenna and with the diameter of the wire used for the line.

This shunt-fed system is shown in Fig. 11–17(a). The capacitor is adjusted to neutralize the inductive reactance created by the shunt-fed circuit. In an ungrounded antenna structure, the tower is supported on insulators, and the inner conductor of the coaxial line is fed to the base of the antenna through an impedance-matching network. This technique is shown in Fig. 11–17(b), using a pi network.

DIRECTIONAL ANTENNA SYSTEMS

In many transmitting situations, it is desirable to restrict the direction of propagation of the radiated wave within some specific limits. This need is quite obvious for commercial point-to-point communication. Any power radiated except in the direction of the receiving station would be wasted. In radar units, the accuracy of tracking depends on the narrowness or *directivity* of the transmitted beam. Even in broadcasting, some degree of directivity may be desirable, depending on population distribution in relation to the location of the transmitting antenna. When the radiation pattern of an antenna system is essentially unidirectional (or bidirectional), the term *beam width* or *beam angle* is used as a measure of directivity. This quantity is evaluated from the radiation pattern as the angular distance between the points (to either side of the direction of maximum radiation) where the signal strength drops to 0.7 (70.7%) of the maximum value. This is shown, for the broken-line curve of Fig. 11–3, as 52°.

An antenna system with a narrow beam width has the further advantage of providing a *directive gain.* Since the radiated power is confined within a relatively small angle, the signal strength in the favored direction can be much greater than would be obtained from an omnidirectional antenna. Consequently, the effective radiated power (ERP) of any transmitter is increased by the directive gain of its antenna. This directional gain of an antenna system is evaluated as the ratio of the power required to produce a given signal strength at a given location, using a reference antenna,* as compared to the power required to produce the same signal strength from the directional antenna. Tremendous gains are possible, particularly at ultrahigh frequencies. For example, a 100-mW commercial microwave relay transmitter using a *parabolic reflector* antenna system with a gain of 8000 can produce a signal strength equivalent

*Two reference antennas have been used for gain evaluation. The theoretical isotropic antenna may be considered as an "absolute" reference. However, many prefer to use the practical dipole (half-wave) antenna as the reference. Although the dipole has a gain of 1.64 (or 2.15 dB) over the isotropic antenna, conversion from one standard to the other is generally made by using the rounded-off figure of 2.0 dB.

to an 800-W transmitter with an isotropic radiator. The ERP is now ($8000 \times 100 \times 10^{-3}$) or 800 W.

Directivity is obtained, at below microwave frequencies, by using combinations of two or more antenna elements to form an *antenna array.* Depending on the method used to excite the additional elements, antenna systems are classified as *parasitic* or *driven arrays.*

Parasitic Arrays

An antenna element that is not connected to the transmission line will develop a voltage by induction. Such an element is called a *parasitic element.* Figure 11–18 shows a parasitic element located 0.25λ from the dipole antenna or *driven* element. Both elements are 0.5λ long (resonant). The radiation pattern for the dipole antenna (by itself) is essentially bidirectional, with maximum radiation at right angles to the line of the antenna. Consider the energy traveling toward the parasitic element. The electromagnetic wave travels 0.25λ in space; therefore, it reaches the parasitic element when the wave at the antenna has gone through a 90° phase change (with time). As the wave cuts the parasitic element, a voltage is induced, 180° reversed with respect to the wave that induced it. Current flows through the element (in phase with the induced voltage), and the element radiates. In the direction beyond the parasitic element, the field radiated by this element is opposite to the field from the antenna (and almost equal in magnitude). The radiated power in that direction is negligible. But the parasitic element also radiates in the direction toward the antenna. By the time this field reaches the antenna, another 90° time delay has occurred, so that the field being radiated by the antenna *at this instant* is 180° reversed from the original radiation that induced a voltage in the parasitic element. The present field from the antenna and the field

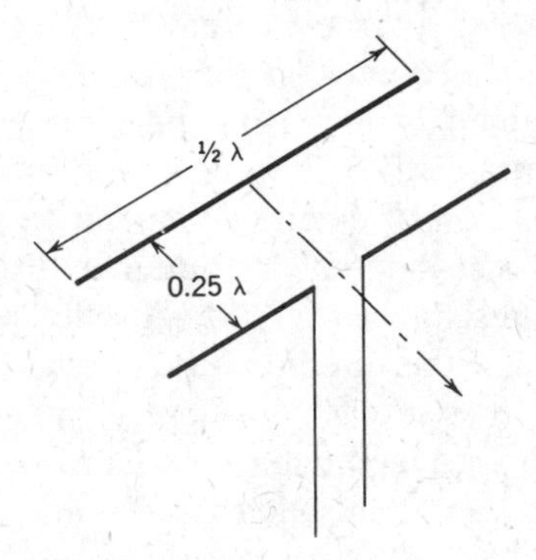

Figure 11-18 A dipole antenna with reflector.

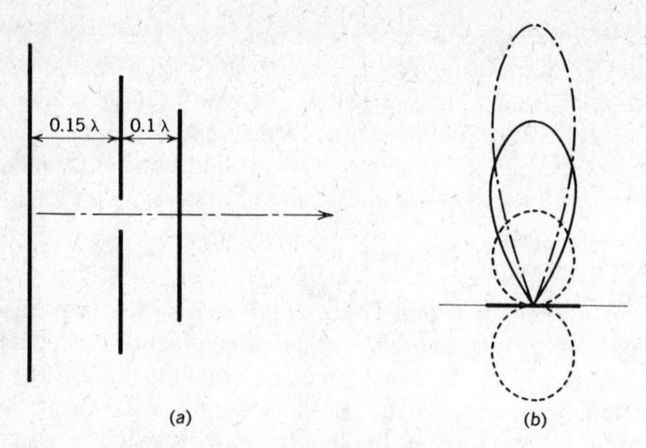

Figure 11-19 A parasitic antenna system.

arriving from the parasitic element are in phase. The total field strength—in the direction of the arrow (Fig. 11–19)—is increased. This direction is called the *forward* direction, and the parasitic element is called a *reflector*.

In general, the radiation pattern obtained with parasitic elements depends on the magnitude and phase of the current in the parasitic element. This, in turn, depends on the spacing between the parasitic and driven elements, and on the length (and diameter) of the parasitic element. It has been found that optimum reflector action (maximum gain) is obtained with a spacing of approximately 0.18–0.2λ and an element length approximately 5% longer than 0.5λ. Furthermore, if the parasitic element is made shorter than the dipole antenna (approximately 5% less than 0.5λ), and the spacing is reduced to approximately 0.1λ, reinforcement will be in the direction of the parasitic element. In this case the element is called a *director*.

Figure 11–19 shows a three-element Yagi* array together with the relative effect of these elements on the radiation pattern. Still sharper directivity can be obtained by increasing the number of parasitic elements. However, there is a practical limitation to the number of elements that can be used. The addition of elements (similar to adding parallel loads) reduces the radiation resistance of the antenna. For example, the dipole alone has a radiation resistance of 72 Ω. A reflector lowers this value to approximately 60 Ω and the director further reduces it to 30 Ω. Meanwhile, the added I^2R loss in each of these elements increases the total losses. A point is soon reached where the increase in directional gain is offset by the reduced radiation efficiency. Multielement Yagi antennas

* Named after the inventor.

have another limitation. Since their effect is dependent on element length and spacing, their use is restricted to a narrow band of frequencies around the design value.

Driven Arrays

When all the elements of an antenna system are fed from the transmission line, the structure is called a *driven array.* A variety of radiation patterns can be obtained, depending on the orientation of the elements, the spacing between elements, and the phasing of the currents in adjacent elements. Three specific arrays are of special interest. These are known as the collinear, the broadside, and the end-fire array.

Collinear Arrays

A collinear array contains two or more elements placed end-to-end as shown in Fig. 11–20. Each element is $\frac{1}{2}\lambda$ long, and they are fed so that the current in each element is in phase. Consider the two-element array in Fig. 11–20(a). In a direction in line with the antenna (left or right), the

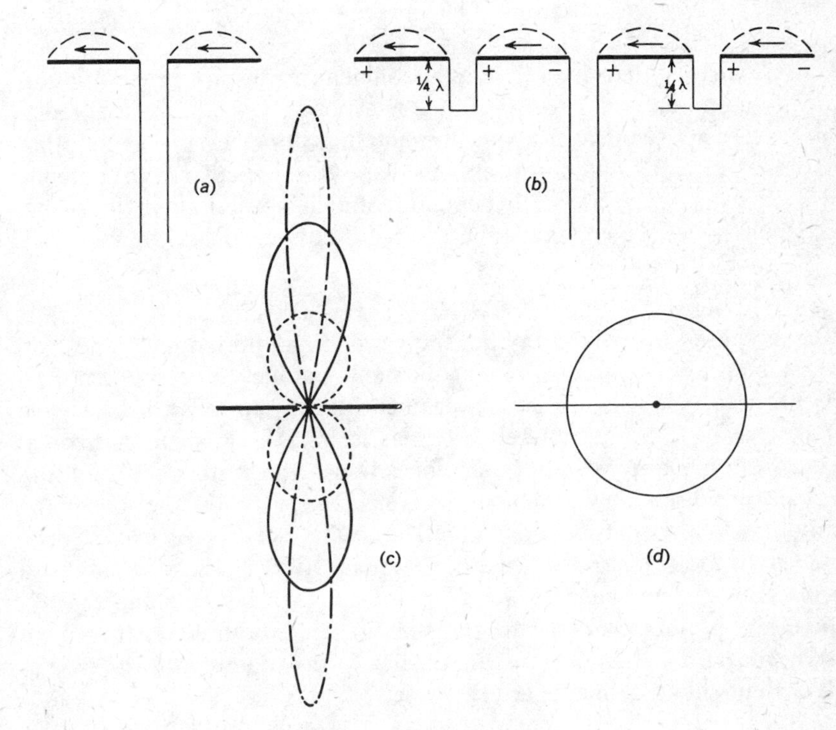

Figure 11-20 Collinear arrays.

field that would be produced by any small section of one element is canceled by the 180° lagging field produced by the identical small section of the other element. This is so because the currents are of equal magnitude and in phase, but the field from the "rear" section must travel an additional $\frac{1}{2}\lambda$. On the other hand, in a direction perpendicular to the line of the antenna, the distance traveled by the fields from each section is identical, and the total field strength is doubled. Obviously, the full (three-dimensional) radiation pattern would be doughnut-shaped (as shown in Fig. 11–5 for the simple dipole), but the doughnut would be flatter. A comparison between the horizontal-plane pattern for the basic dipole (dotted curve) and the two-element collinear (solid curve) is shown in Fig. 11–20(c). (For simplicity, minor lobes are not shown.) The pattern is essentially bidirectional.

Sharper beam angles can be obtained by using additional elements. Figure 11–20(b) shows a four-element collinear array. Notice the use of the folded half-wave sections to interconnect the elements while maintaining the necessary in-phase current relationship. Because of the in-phase currents and half-wave spacing, the cancellation at small angles from the line of the array is increased, while the reinforcement for angles approaching 90° is increased. This increased directivity (and gain) is shown by the dot-dash curve of Fig. 11–20(c).

Because of their overall length, collinear arrays are generally horizontally mounted, and Fig. 11–20(c) corresponds to the horizontal radiation pattern as seen from above the antenna (top view). On the other hand, the vertical radiation pattern in a plane perpendicular to the antenna (end view) is circular, indicating omnidirectional distribution in this plane. For point-to-point communication this is an obvious disadvantage.

Broadside Arrays

Increased directivity in both the vertical and horizontal planes can be obtained by placing the antenna elements parallel to each other. Figure 11–21(a) shows such an arrangement using four vertical half-wave dipoles.* The spacing between antennas is $\frac{1}{2}\lambda$. Although the horizontal pattern for any one element is omnidirectional, the pattern for the array is sharply bidirectional, and the more elements used (or the longer the array length), the sharper is the directivity. This can be shown as follows. First, note the transmission line connection. By the time a signal is fed to element 2, the phase of the signal at element 1 has changed 180°. Therefore, by crisscrossing the transmission line, both elements will be fed in phase. The same technique applies to the other elements, so that the currents in all elements are in phase.

* Grounded vertical antennas are used in similar fashion.

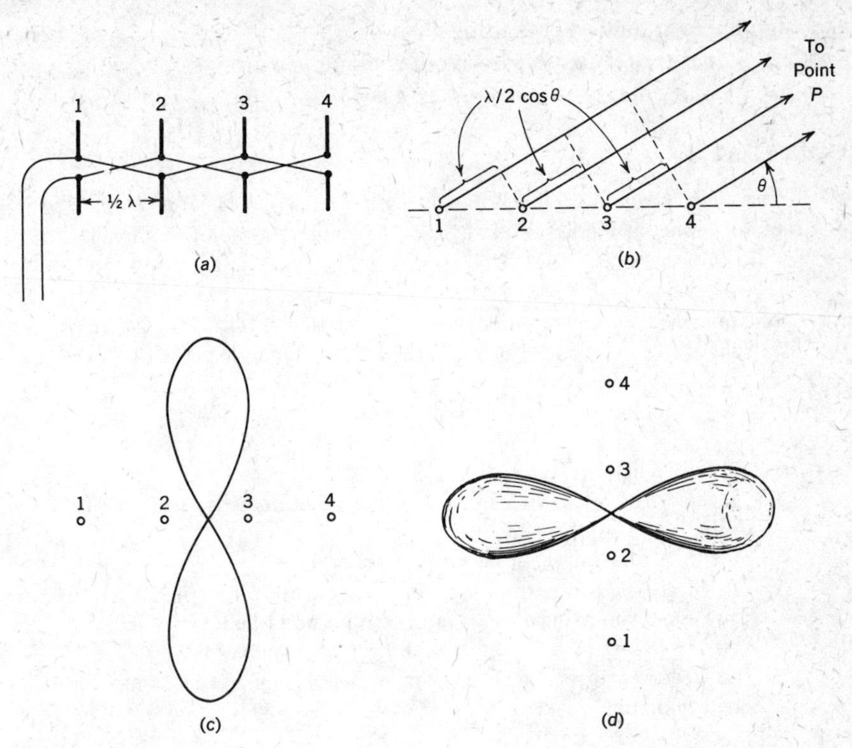

Figure 11-21 A vertical broadside array. (a) Front view. (b) and (c) Top view. (d) End view.

Now, assume that the array is aligned in an east-west direction. Toward the east or west, the radiated fields from adjacent elements will cancel because of the half-wave spacing. On the other hand, in a north-south direction, or *broadside* to the plane of the array, the signal from each element travels an equal distance to reach a point in space. The fields will be in phase, increasing the total field strength directly. For any other direction at some angle θ to the line of array, the signal from each next adjacent element must travel an additional distance of $(\lambda/2)\cos\theta$, and the resultant in that direction is the vector sum of the individual fields. For small angles (less than 45°), the resultant field will be reduced considerably. The horizontal plane radiation pattern is shown in Fig. 11–21(c).

So far the effect is similar to the collinear array. However, the broadside array also provides sharp directivity in the vertical plane. Each element, by itself, has reduced radiation at high vertical angles. (See Fig. 11–6.) The combined effect of the several elements will further weaken the high-angle radiation, resulting in a sharp figure-eight pattern in the vertical plane as well as the horizontal plane. A three-dimensional pattern for this antenna is shown in Fig. 11–21(d).

EXAMPLE 11–3

Calculate the gain of the broadside array of Fig. 11–21(a), compared to one radiator alone, at a horizontal angle of 70° from the line of the array.

Solution

1. Compared to the radiation field ϵ_4 from element 4 (Fig. 11–21(b)), the field ϵ_3 from element 3 must travel an additional distance of

$$\Delta d = \frac{\lambda}{2} \cos \theta = \frac{\lambda}{2} \cos 70° = 0.171\lambda$$

2. Therefore field ϵ_3 lags ϵ_4 by 61.5°.
3. The field from element 2 must travel an additional distance of 2(0.171λ), so that ϵ_2 lags ϵ_3 by 123°.
4. By similar reasoning, ϵ_1 lags ϵ_4 by 184.5°.
5. The total field strength is the vector sum of the four fields. By vector representation, using ϵ_4 as reference, and unity field strength: $\epsilon_1 = 1 \angle 184.5°$, $\epsilon_2 = 1 \angle 123°$, $\epsilon_3 = 1 \angle 61.5°$, and $\epsilon_4 = 1 \angle 0°$
6. The vector sum is 1.65. The array has a gain of 1.65 compared to a single radiator.

End-Fire Arrays

If the transmission line connections in Fig. 11–21(a) are not criss-crossed, the current in each adjacent element will be reversed. Therefore, the radiated field of the elements will be additive in the line of the array. Such a system is called an *end-fire* array. Conversely, broadside to the plane of the array – since adjacent elements are oppositely phased – the field strength is zero. The radiation pattern for this array is identical to that of the broadside array, except that the lobes are now in line with the axis of the array, as shown in Fig. 11–22. Again, the directivity increases with the number of elements used. As with the broadside, this array can be mounted in a vertical or horizontal plane.

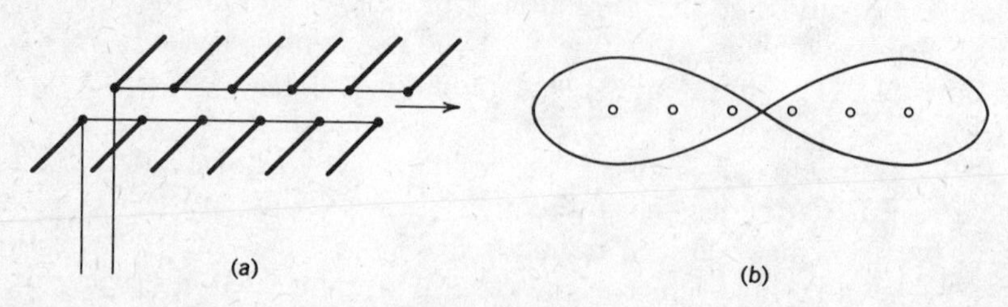

Figure 11-22 (a) Horizontal end-fire array. (b) Side view.

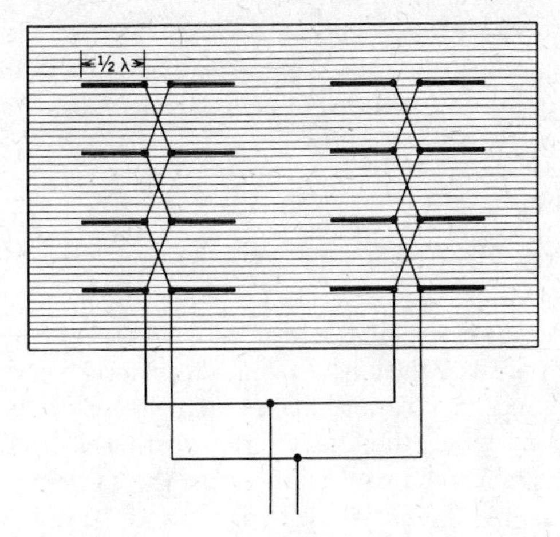

Figure 11-23 The "mattress" antenna.

Combination Arrays

Each of the driven arrays previously discussed has a bidirectional pattern. For point-to-point communication, one half of the radiated energy would be wasted. At high frequencies, with horizontal arrays, this is readily corrected by adding parasitic reflectors. One such antenna system, Fig. 11–23, combines collinear and broadside arrays with a wire screen for the parasitic reflector. Quite appropriately, it earned the name *mattress antenna*.

Unidirectional patterns can also be achieved using driven elements. All that is necessary is to adjust the spacing between elements and the phasing of the respective currents so as to cancel the radiated field in one of the previous two directions. This technique is especially useful when using grounded (vertical) antennas. For example, consider four vertical towers aligned in an east-west direction, with proper spacing and phasing to form a broadside array (row 1 of Fig. 11–24). By themselves these would have a bidirectional north-south radiation pattern. Now, let us add a second row, also $\frac{1}{2}\lambda$ apart and fed in phase. These by themselves will have a similar pattern. However, the currents in this second row *lead* the antenna currents in row 1 by 90°. In a southerly direction, because of the

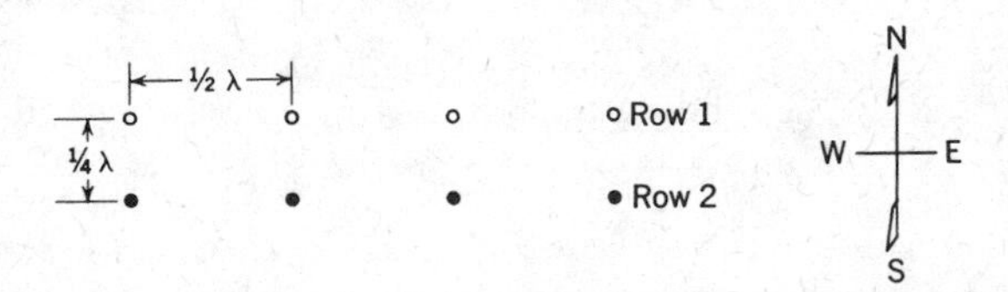

Figure 11-24 A unidirectional vertical array.

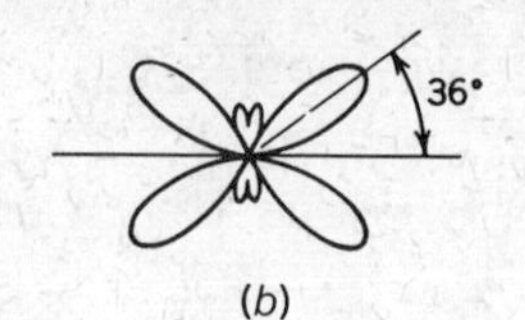

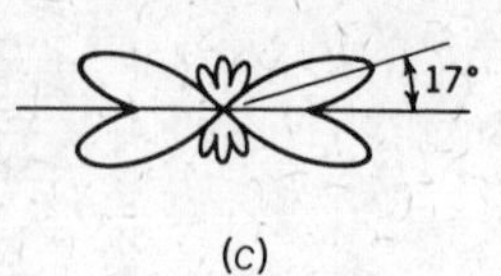

Figure 11-25 The effect of length on a radiation pattern.

spacing, there is a time lag of $\frac{1}{4}\lambda$ before the fields from row 1 reach row 2. In this time, the phase of the currents in these antennas has advanced 90°. But they already led by 90°, so that the fields are now 180° out of phase and will cancel. On the other hand, in a northerly direction, when the fields from row 2 reach the row 1 antennas, the $\frac{1}{4}\lambda$ delay will counteract their 90° lead. The two fields will be in phase and the radiation is strengthened. The result is a unidirectional pattern.

Rhombic Antennas

In the above discussions, the antenna has been a resonant structure ($\frac{1}{2}\lambda$ or equivalent). The radiation pattern for the basic $\frac{1}{2}\lambda$ was the figure eight, perpendicular to the axis of the antenna. If the length of the antenna is doubled, since each half now carries currents in opposite directions, the radiated field at right angles to the wire drops to zero (the field from one half of the antenna cancels the other); the field strength along the line of the antenna is still zero; but now four major lobes appear with the direction of maximum radiation at an angle of 54° to the line of the antenna axis.* This is shown in Fig. 11–25(a). As the length is increased further, the angle of these major lobes approaches the axis of the antenna wire. Diagrams (b) and (c) show this effect for lengths of two and eight wavelengths, respectively. Although minor lobes also appear, it has been found that a long-wire antenna will radiate more power along the axis of the wire than the dipole radiates in its optimum direction. Furthermore, if a long-wire antenna is terminated with a resistive load so that there are no standing waves on the antenna wire, the radiation pattern becomes essentially unidirectional in the direction of the load end.

Several applications have been made of these principles. The most widely used is the rhombic antenna shown in Fig. 11–26. The antenna consists of four "legs," each several wavelengths long. The antenna is mounted horizontally at a height of at least $\frac{1}{2}\lambda$ above ground. The optimum height varies, depending on the length of each leg and the desired vertical angle of radiation.

* Considering the three-dimensional aspects of the pattern, this angle applies to both the horizontal and vertical planes.

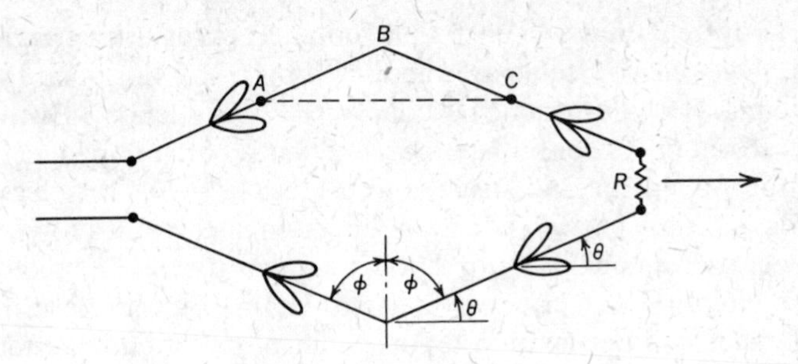

Figure 11-26 Rhombic antenna.

To get maximum radiation in a direction through the centerline of the rhombus (see Fig. 11–26), it is necessary to fix the *tilt angle* ϕ, in relation to the pattern angle θ. (This latter angle, as shown in Fig. 11–25, depends on the length of wire used in each leg.) To align one lobe of the radiation pattern for each leg in the desired direction, the tilt angle must be fixed at $(90 - \theta)$ degrees. Simultaneously, the fields radiated in the direction of the second lobe of each leg will cancel each other, resulting in a unidirectional overall pattern. However, there is still another requirement to be satisfied. The distance along the wires from the midpoint of one leg to the midpoint of the other (A-B-C) must be $\frac{1}{2}\lambda$ longer than the direct distance A-C, in order for the fields from the back legs to reinforce the fields from the front legs.

Rhombic antennas are widely used for point-to-point communications at operating frequencies 2–30 MHz. Since they are not resonant, they will give high efficiency over a broad frequency range and are therefore useful in multifrequency transmission. In addition, they are relatively simple to construct and can handle high power levels. Gains of over 40 (16 dB) have been obtained with rhombic antennas as compared to a half-wave dipole.

UHF Antennas

The antennas and arrays described thus far are theoretically usable at any frequency. However, as the frequency of operation gets higher and higher, the small size of the elements and the closer tolerances required in construction create some problems. Wire (or rod) antennas still find some application in the VHF band and at the low end of the UHF band. For example, the mattress antenna has been used in long-range, early-warning radar units operating up to about 500 MHz. Also, the *turnstile* antenna used in television broadcasting is nothing more than two half-wave dipoles crossed at right angles and fed 90° out of phase, to produce an (almost) omnidirectional pattern. On the other hand, other types of antenna systems have been found more practical at microwave frequencies. Two such systems are the electromagnetic horn and the parabolic antenna. The horn may be considered as a section of waveguide that is flared out (in one or both planes) to effect a better impedance match between the waveguide and space, and it literally spews the electromagnetic fields out into space. The parabolic antenna is really a reflector that gets its energy from one or more dipole antennas (or a horn) and concentrates this energy into a narrow (focused) beam. Further description of UHF antennas will be found in texts specifically written on antennas or UHF fields.

Receiving Antennas

In general, an antenna that is efficient as a radiating antenna will be equally efficient in receiving RF energy. In many cases—particularly at the higher frequencies—the receiving antenna is identical to the transmitting antenna. For point-to-point communications, rhombics and arrays are often used. In fact, for optimum reliability, several antennas with different characteristics, or located short distances apart, may be used for one receiver, to overcome possible signal fade. However, in many applications, the receiving antenna may be a very simple structure such as a relatively short vertical (or horizontal) wire of random length. A typical example is the "whip" used with automobile radios.

WAVE PROPAGATION

Except for highly directional antennas, the energy radiated from a transmitting antenna travels into space in many directions. As the distance from the antenna increases, the energy field spreads out and the field intensity decreases. However, there are other factors that can greatly affect the field strength at the receiving location. These factors are related to the path (or paths) by which the signal reaches the receiving location. Depending on the vertical angle of radiation from the antenna, there are three broad classifications—the *ground wave,* the *space wave,* and the *sky wave*.

Ground Waves

The ground wave is a vertically polarized wave that travels along the surface of the earth. It is, therefore, also known as a *surface wave*. Vertical antennas must be used if propagation by surface waves is desired.*

*The electric field in a horizontally polarized wave would be parallel to the earth, and such waves would be short-circuited by the conductivity of the earth.

The changing field (with time and with travel of the wave) induces varying voltages in the earth, and currents flow in much the same manner as in a transmission line or waveguide. This, in turn, gives rise to resistance and dielectric losses in the ground. The energy to supply these losses must come from the signal. In other words, the ground wave is attenuated as it travels away from the antenna. To keep the losses to a minimum, the transmission path should be over a "ground" with high conductivity. In this respect, seawater has the best conductivity; flat rich soil or large bodies of fresh water would be a poor second; while sandy or rocky soil is even worse. For this reason, high-power low-frequency (ground-wave) transmitters are preferably located on ocean fronts.

Ground-wave transmission is generally limited to the lower frequencies because the ground losses increase rapidly with frequency. At frequencies above 2 MHz these losses are excessive. However, ground-wave transmissions have the advantage of excellent reliability. Reception is not affected by daily or seasonal changes. Given sufficient power and low enough frequency, communications can be maintained with any place in the world.*

Space Waves

Those portions of the radiated energy that travel in the region up to a few miles above the earth (the troposphere) fall into the category of *space waves.* Two such waves are the *direct wave* and the *ground-reflected wave*,† shown in Fig. 11–27. The field strength at the receiving location is obviously dependent on the phase relation between the two signals. This, in turn, depends on the difference in the distance traveled by each wave. For maximum reinforcement, the distance traveled by

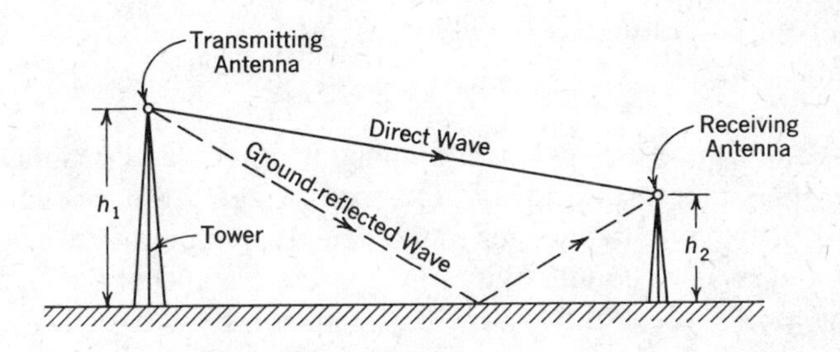

Figure 11-27 Space-wave propagation.

*The U.S. Navy *Fox* stations use frequencies of around 18 kHz and power in excess of 1 MW for this purpose.

†These waves are sometimes classified together with the surface wave as ground waves.

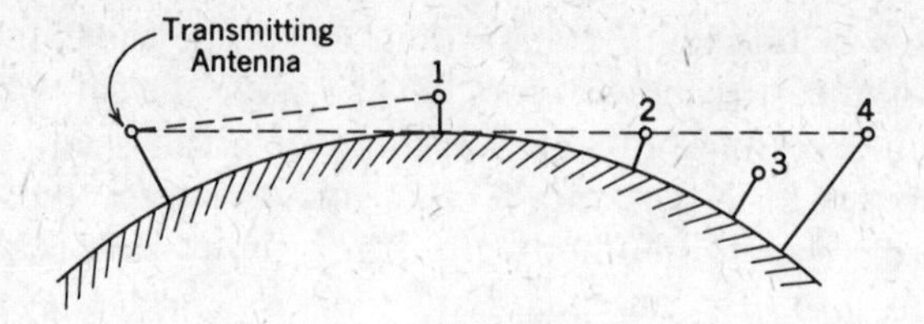

Figure 11-28 The effect of antenna height on communication range.

the ground-reflected wave should be $\frac{1}{2}\lambda$ (or any odd multiple thereof) longer than the direct path. At low frequencies it may not be possible to obtain sufficient antenna height to create a favorable phase relation. However, space-wave transmission becomes of major importance at frequencies above 30 MHz, where the other methods (ground and sky wave) are of negligible value. At these high frequencies—and particularly at UHF—it is possible that a small change in antenna height can move the antenna from a point of weak signal level to a point of maximum signal strength (or vice versa).

Antenna heights are also important from another aspect. High-frequency waves travel essentially in a line-of-sight path. Therefore, because of the curvature of the earth's surface, the maximum communicable distance depends on antenna height. This effect can be seen from Fig. 11–28. At receiving location 1, the antenna height is of no consequence to the line-of-sight distance. For the same antenna height, location 2 would be the maximum distance that can be spanned. Yet, notice that with a higher receiving antenna, the communication range can be increased to reach location 4.

In practice, the actual communicable distance is 20–35% greater than the straight-line (or optical) distances shown in Fig. 11–28. The corrected line-of-sight range is readily calculated as

$$d = \sqrt{2h_1} + \sqrt{2h_2} \tag{11-2}$$

where h_1 and h_2 are the two antenna heights in feet, and d is the distance in miles. This increase in range is caused by refraction of the waves in the troposphere, due to changes in the density, temperature, water-vapor content, and relative conductivity (dielectric constant) of the layers of the atmosphere. As a result, the wave path tends to bend back toward earth. Since the atmospheric conditions are not fixed, the degree of refraction will vary, at times permitting an abnormally long-distance communication range. One phenomenon known as *duct propagation* occurs when electromagnetic waves are trapped between two layers of air and travel with little attenuation, following the earth's curvature as though in a giant waveguide. A receiving antenna within this region would intercept a strong signal far beyond the refraction range. Duct propagation is more likely to occur over ocean areas or over large flat masses when the moisture content of the air layers close to ground is high.

Sky Waves and the Ionosphere

That portion of the radiated energy that is directed above the horizon level constitutes *sky waves.* Except for communications from land bases to aircraft, this energy would be lost were it not for the action of ionized layers in the upper atmosphere. Ultraviolet radiations from the sun can cause ionization of the air particles into free electrons, positive ions, and negative ions. The region so affected is called the *ionosphere* and ranges 30–250 miles above the earth's surface. Because of the variation of the composition of the air, there are several regions of high density within this overall span, dividing the ionosphere into layers. The formation of layers and the height of these layers are affected by the time of day, season of the year, and the year-to-year variations in the sun's activity (an 11-year cycle).

The lowest of these layers, called the D layer, exists only in the daytime at an altitude of approximately 30–50 miles above the earth's surface. Ionization in this region is relatively weak and does not affect the direction of travel of radio waves. However, the ionized particles absorb appreciable energy from electromagnetic waves. The next layer, the E layer, is in a region of about 55–85 miles, with the maximum density at about 70 miles. It has a maximum density at noon, but is only weakly ionized at night. The last layer, the F layer, is quite variable. At night it exists as a single layer in the region of approximately 110–250 miles above the earth, while in the daytime, it splits into two layers—F_1 and F_2. The F_1 layer ranges 85–155 miles. The F_2 layer is even more variable, extending as low as 90–185 miles on a winter day, or as high as 155–220 on a summer day.

The action of these ionized layers on a radio wave is to refract or bend the wave back toward the region of lower density in much the same fashion as light waves are bent when traveling through different media. The amount of refraction, and whether or not a radio wave is bent back toward earth, will vary depending on the frequency of the radiated wave, the density of the ionized layer, and the angle of incidence at which the wave enters the ionosphere.

Critical Frequency and Critical Angle

The virtual height (see Fig. 11–29) of the various layers just described can be calculated from the time it takes a pulse of RF energy—directed vertically upward—to be returned to earth. However, it is found

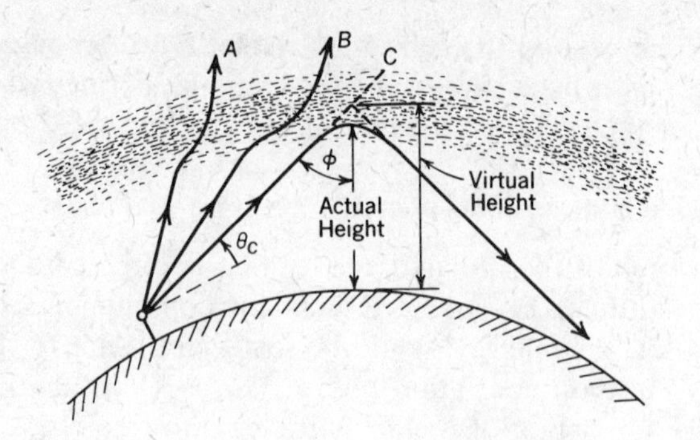

Figure 11-29 The effect of radiation angle on refraction from the ionosphere.

that if the signal frequency is raised above a certain value, the wave is no longer returned. This *critical frequency* is defined as the highest frequency that will be returned to earth when transmitted in a vertical direction. Since the specific frequency value depends on the ionization density,* it will obviously vary with the time of day and season of year. Furthermore, it is possible for a particular frequency to pierce the E layer but still be returned from the F layer because it has a higher density. Of course, a still higher frequency will pierce both layers and be lost.

On the other hand, if the angle of radiation (vertical) is lowered, the wave will travel longer through a given layer and will be refracted to a greater degree. Therefore, even signals above the critical frequency can be returned to earth. Again, there is a limitation. For any given frequency there is a *critical angle* beyond which the signal will not be refracted sufficiently to return to earth. This is shown in Fig. 11–29. The critical angle is variously taken as θ, the vertical radiation angle, or as ϕ $(90° - \theta)$, the angle of incidence between the wave and the ionosphere. Notice that, for radiation angles greater than θ_c, the wave is not returned. In general, the lower the operating frequency, the higher the angle of radiation can be. Unfortunately, however, at frequencies above (approximately) 40 MHz, the layer density is not sufficient—at any radiation angle—and these waves are not refracted. Communications above this frequency are entirely by surface wave.†

Skip Distance—Maximum Usable Frequency

The solid wave (path X) in Fig. 11–30 represents the critical-angle path for a particular operating frequency. So far as a sky wave is concerned, there can be no reception of this signal closer than receiving

*$f_c = \sqrt{81N}$ where N is the electron density. **(11–3)**

†An exception to this is *forward-scatter propagation*, explained later in this chapter.

location *B*. The ground distance from the transmitter to location *B* is called the *skip distance*. If the ground-wave range at this frequency is location *C*, no reception is possible in the region from *C* to *B—at this frequency.* This region is called the *skip zone*.

Suppose we wish to communicate with a receiving station at location *D*. Obviously, this cannot be done at the frequency used in the above illustration. Of course, if the operating frequency is reduced, a higher radiation angle can be used, and the wave will return to earth at a location closer to the transmitter. If the broken-line path in Fig. 11–30 represents the critical angle for some lower transmitting frequency, this is the *maximum usable frequency* that can be used to effect transmission with location *D*. It should be obvious that by using a still lower frequency the skip zone will be eliminated completely. Yet this is not done in commercial communication. As the operating frequency is reduced, the energy absorbed in the ionosphere increases rapidly, and the signal is drastically attenuated.* The maximum signal level, for a given radiated power, is obtained when operating at the maximum usable frequency.

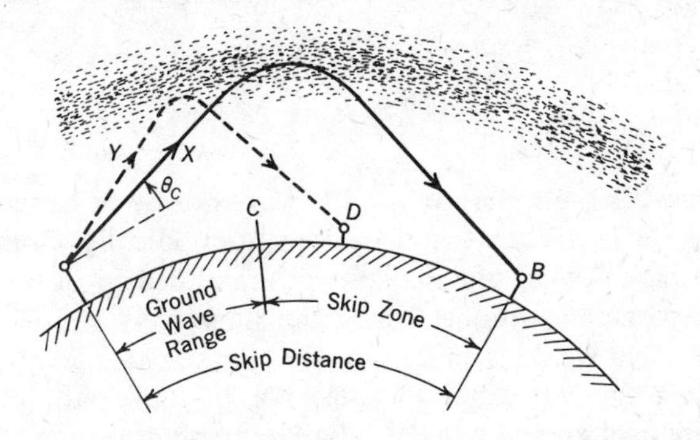

Figure 11-30 Skip distance.

Since the maximum usable frequency depends on the density and height of the ionosphere layers, it will vary from hour to hour, from day to day, and from location to location. Transmission would also vary, depending on direction (north, east, south, or west). This could reduce sky-wave communication to almost a hit-and-miss proposition. To remedy this situation, charts are available that predict the maximum usable frequency for every hour of the day, over any path on earth, during a given month.†

* It has been found that maximum attenuation in the ionosphere occurs at a frequency of approximately 1400 kHz.

† *Basic Radio Propagation Prediction* charts are available three months in advance from the Superintendent of Documents, U.S. Government Printing Office, Washington, D.C., 20025.

Multiple-Hop Transmission

In Fig. 11–30, if the signal returned to earth at location D (or at location B) has sufficient strength, it can be reflected by the ground back up to the ionosphere, where it is refracted again, and it returns to earth at some more distant location. This is known as *multiple-hop* transmission. Such transmission is not limited to the above two-hop path. The number of hops will depend on the radiated power, ground losses, and absorption in the ionosphere.

Multiple-hop propagation makes possible long-distance communication via sky waves. Under average conditions the maximum distance that can be covered by single-hop transmission is approximately 2500 miles if refraction is from the F layer, and only 1250 miles from the E layer. (This assumes a vertical radiation angle of zero degrees.) The limitation is caused by the curvature of the earth. For greater distance coverage, as needed for transoceanic communications, multiple-hop propagation is necessary. However, since each hop increases the attenuation, the radiation angle should be adjusted for a minimum number of hops.

Fading

When signals are received via sky waves, it often happens that the signal strength will increase and decrease periodically. Sometimes the amount of change is small and may escape unnoticed because of the AGC action of the receiver. In other cases, the signal may be lost completely or may become so weak as to be drowned in the noise level. This effect is known as *fading*. It is generally caused by multiple-path reception and changes in ionospheric conditions. For example, consider a receiving location that is within ground-wave range and also receives sky waves. The signal strength will be the resultant of these two waves. Because the lengths of these paths differ, the signals may be in phase or out of phase by some angle. As conditions in the ionosphere change, the phasing of the sky wave will shift, and the resultant field strength will vary.

Fading will also occur at locations beyond the ground-wave range. The received signal can be the resultant of waves reflected from the E and F layers, or of single-hop and multiple-hop paths. In either case, the lengths of paths and the phasing of the received signals will vary with ionospheric conditions. Another condition that causes fading occurs when a receiving location is on the edge of the skip distance. A slight rise and fall in the refracting region will alternately place the receiver inside or beyond the skip distance. In any case, fading may be rapid or slow, depending on the stability of the ionospheric conditions.

Ionospheric Variations

The height and density of the ionized layers are affected by normal and by abnormal conditions. Normal variations include the previously mentioned diurnal (time of day) and seasonal changes. For example, the ionosphere layers are lower, and the ionization density is greater during the daytime. This makes possible the use of higher frequencies and higher radiation angles to reduce attenuation and minimize skip distance. Another normal variation is the *11-year sunspot cycle*. During periods of maximum sunspot activity, the F_1 and F_2 layers attain maximum height and density. At such times, higher frequencies can be used without exceeding the critical value. This reduces attenuation and skip distances, providing improved reliability and long-distance communication. This effect recurs at 11-year intervals.

The ionosphere is also subject to abnormal variations that can cause unpredictable changes in wave propagation. One such condition is known as *sporadic-E ionizations*. These are erratic areas of high-density ionizations that sometimes occur at E-layer heights. Refractions from such areas can result in excellent signals in what would normally be a skip zone. It also makes possible transmissions at higher-than-normal frequencies and beyond normal distance ranges. Another abnormality is *ionospheric storms* (also caused by some type of sunspot activity). At the height of these disturbances—which can last from a few minutes to several hours—the absorption is high enough to cause radio blackout, particularly at the higher frequencies. The effect disappears gradually but may even be noticed for days.

Propagation Characteristics at Various Frequencies

By proper selection of frequencies, it is possible to maintain high-reliability communications over any distance. The utmost reliability is obtained at low and very low frequencies (below 30–300 kHz). This is ground-wave coverage, and it will provide continuous communication regardless of time or season. However, for long distances, very high power is necessary.

In the medium-frequency range (300-3000 kHz) the ground wave is getting weaker because of increased ground losses. It is very dependable for distances of 50–200 miles (depending on transmitting power and frequency). Sky-wave propagation is poor in the daytime because of high absorption, but it improves appreciably at night—particularly in winter. Reception to 3000 miles is then possible. Heavy fading can result at these frequencies at night, due to interference between ground and sky waves.

Ground-wave communication is of little value in the high-frequency (3–30 MHz) range because of excessive attenuation. On the other hand, excellent sky-wave propagation is possible. The higher frequencies can be used in the daytime to minimize losses. Transmission ranges beyond 10,000 miles are readily possible. Above 30 MHz, ground waves cannot propagate, whereas sky waves tend to pierce through the ionosphere and are therefore unreliable. Communication is restricted to surface waves and relatively short distances because of their line-of-sight characteristic.

Using high power and highly directive antenna systems, reliable communications can be achieved at frequencies in the order of 100 MHz, and at distances up to 1200 miles, by a phenomenon known as *forward-scatter propagation.* This is believed to occur in the *E* layer because of heavy ionization trails produced by meteors entering the earth's atmosphere. Turbulence in the ionosphere and auroral effects contribute to this action.

REVIEW QUESTIONS

1. (*a*) Is there appreciable radiation from a parallel-wire transmission line? Explain. (*b*) How can the amount of radiation be increased?

2. (*a*) What does Fig. 11–1(a) represent? (*b*) Why does it have standing waves of current and voltage?

3. (*a*) What does the solid-line pattern in Fig. 11–1(b) represent? (*b*) What does the broken-line pattern represent? (*c*) How are these fields related to each other? (*d*) Is this field capable of radiating into space? Why?

4. (*a*) What is meant by the term *induction field*? (*b*) Does it radiate into space? (*c*) How does its field strength vary with distance from the source?

5. (*a*) What is the essential difference between a radiation field and an induction field? (*b*) How does the field strength of a radiation field vary with distance from the source? (*c*) with radiated power?

6. (*a*) What term is used as a measure of the strength of a radiation field? (*b*) How is the value obtained?

7. Refer to Fig. 11–2. (*a*) What do the horizontal broken lines represent? (*b*) What do the vertical lines represent? (*c*) What is the space relationship between these fields and the direction of wave propagation? (*d*) In what direction is this wave polarized? (*e*) What determines whether the wave is considered horizontally or vertically polarized? (*f*) How can a horizontally polarized field be produced?

8. (*a*) In which direction does an isotropic antenna radiate best? Explain. (*b*) Is isotropic radiation desirable? Explain. (*c*) An antenna is described as nondirectional in the horizontal plane. What does this mean? (*d*) Give another term used to signify nondirectional distribution.

9. (*a*) What means are used to describe the radiation characteristics of an antenna? (*b*) What two views are required for a full representation? Why?

10. Refer to Fig. 11–3. (*a*) What kind of graph paper is this? (*b*) What are the coordinates used in these plots? (*c*) In what plane are these patterns taken? (*d*) Which (if either) of these curves would represent an omnidirectional antenna? Explain.

11. Refer to the broken-line radiation pattern of Fig. 11–3. (*a*) How does the distance from the antenna vary with angle in this pattern? Explain. (*b*) What is the direction of maximum radiation? (*c*) What is the change in field strength at a direction of 154°? 190°? 220°? 120°?

12. (*a*) Explain how the radiation pattern for an antenna can be obtained. (*b*) What is the minimum distance from the antenna at which field-strength patterns should be taken? Why?

13. (*a*) Name the two basic types of antennas. (*b*) Describe each briefly.

14. (*a*) What is the electrical length of the basic dipole antenna? (*b*) What is the physical length? Why? (*c*) Why is it called a dipole? (*d*) Describe the current standing-wave pattern on this antenna. (*e*) Describe the voltage pattern. (*f*) Do these patterns differ in any way from the standing wave patterns for an open-ended half-wave lossless transmission line? Explain.

15. (*a*) What is the nominal input impedance at the center of a half-wave dipole? (*b*) Is this value affected by antenna height? Explain. (*c*) Is this value resistive or reactive? Explain. (*d*) Why is the input impedance higher if the antenna is end-fed? (*e*) What is the value?

16. (*a*) What does the term *radiation efficiency* mean? (*b*) What two factors are involved in this efficiency? (*c*) Name four types of losses that would impair the efficiency of an antenna. (*d*) All other factors being equal, will a high or low radiation resistance improve efficiency? Explain.

17. Refer to Fig. 11–4. (*a*) Is this a horizontal or vertical radiation pattern? (*b*) In what direction is the radiated power a maximum? (*c*) Explain why this is so. (*d*) What is the field strength in the direction b to a? (*e*) Why is this so? (*f*) What is the relative field strength in a direction 230° from the zero reference?

18. Refer to Fig. 11–5(a). (*a*) Is this a horizontal or vertical radiation pattern? (*b*) What is the plane of this pattern with respect to the antenna? (*c*) How does the radiation strength at an angle θ skyward compare to the horizontal radiation in a northerly direction? (*d*) What would this shape pattern be called?

19. Refer to Fig. 11–5(b). (*a*) Does the front face of this doughnut-shaped pattern represent a horizontal or vertical plane? (*b*) What is the direction of maximum radiation? (*c*) What is the direction of minimum radiation? (*d*) Does the signal strength vary with elevation angle in this view? Explain. (*e*) How is the doughnut-shaped pattern obtained?

20. Refer to Fig. 11–6. (*a*) How does this pattern differ from that of Fig. 11–5(b)? (*b*) What causes this change? (*c*) How does the field strength produced by this antenna vary in a horizontal plane? (*d*) What is this type of pattern called? (*e*) For what type of service could this pattern be desirable? Explain. (*f*) Does this antenna have nondirectional characteristics in any vertical plane? (*g*) Would this antenna be suitable for communication with high-flying planes? Explain.

21. (*a*) Why is the radiation pattern of Fig. 11–6 called a "free-space" pattern? (*b*) Is such a condition practical at low operating frequencies? Explain.

22. Refer to Fig. 11–7. (*a*) Does this diagram apply to a horizontal or vertical antenna? (*b*) What view of the antenna is this? (*c*) What does the circle marked *Image* represent? (*d*) Which wave, direct or reflected, travels a longer distance? (*e*) How much longer? (*f*) State two possible effects of these dual signals. (*g*) For what value of $2h \sin \theta$ will the signals reaching a point in space cancel?

23. Refer to Example 11–2. (*a*) Why is the resultant zero for vertical angles of 0°? (*b*) For $\theta = 15°$, why does the ground-reflected wave lag by 227°? (*c*) How is a resultant of 8 obtained for this case? (*d*) If the ground were not a good conductor, would the resultant be greater or smaller than 8? Explain. (*e*) What is the direction for maximum field strength?

24. Refer to Fig. 11–8(a). (*a*) Does this pattern apply to a horizontal or vertical antenna? How can you tell from the diagram? (*b*) What is the plane of the pattern with respect to the antenna? How can you tell from the diagram? (*c*) What does the broken-line pattern represent? (*d*) What does the solid-line pattern represent? (*e*) How are the values for this solid curve obtained? (*f*) Would this antenna be suitable for communication with land-based stations? Explain.

25. (*a*) How does the antenna structure for Fig. 11–8(b) differ from that of diagram (a)? (*b*) Why are the patterns different? (*c*) What does the broken-line curve represent? (*d*) What does the solid-line curve represent?

26. Refer to Fig. 11–9. (*a*) Do these patterns apply to a vertical or horizontal antenna? (*b*) Are the patterns taken in the vertical or horizontal plane? (*c*) Why does diagram (a) differ from diagram (b)? (*d*) Why are these patterns different from the corresponding diagrams of Fig. 11–8? (*e*) Would this antenna be suitable for communication with land-based stations? (*f*) for long-range aircraft communications? (*g*) for aircraft in the immediate vicinity?

27. What type of antenna is referred to as a Marconi antenna?

28. What type of horizontal-plane radiation pattern does a Marconi antenna have? Why?

29. (*a*) Why is it necessary for $\frac{1}{4}\lambda$ vertical antennas to have excellent ground conductivity? (*b*) How is ground conductivity improved? (*c*) What is a counterpoise and when is it used?

30. Compare the Marconi and Hertz antennas as to (*a*) length, (*b*) radiation resistance, and (*c*) power radiation for the same value of antenna current.

31. Refer to Fig. 11–11(a). (*a*) To what kind of antenna does this pattern apply? (*b*) Is this a horizontal-plane or vertical-plane pattern? (*c*) Why isn't it a figure-eight pattern?

32. Refer to Fig. 11–11(b). (*a*) Are these horizontal-plane or vertical-plane radiation patterns? (*b*) Are they obtained with horizontal or vertical antennas? (*c*) Why are the patterns different? (*d*) Which is preferable for ground-to-ground communications? (*e*) What is the optimum antenna height for ground-to-ground communications? (*f*) Describe the horizontal-plane radiation pattern for these antennas.

33. (*a*) If an antenna is shorter than $\frac{1}{4}\lambda$, what effect will this have on the power radiation capabilities? (*b*) Give two reasons for this.

34. (*a*) Give a situation where the antenna length may have to be shorter than $\frac{1}{4}\lambda$. (*b*) Give a general term used to describe a method for increasing the radiation from such antennas.

35. Refer to Fig. 11–12. (*a*) What is the purpose of the variable inductance shown in diagram (a)? (*b*) How does this improve radiation? (*c*) What are three disadvantages of this loading method? (*d*) Which current standing wave pattern is preferable, the one in diagram (a) or (b)? Why? (*e*) How is the effect in diagram (b) obtained?

36. Refer to Fig. 11–13. (*a*) In diagram (a), if the vertical and flat-top sections are each $\frac{1}{4}\lambda$, where will the minimum and maximum standing wave current values be? (*b*) Of what advantage is this? (*c*) What added feature is obtained by the antennas in diagrams (b) and (c)?

37. (*a*) Under what condition is the transmission line feeding an antenna called a *resonant line*? (*b*) How do the losses on a resonant line compare to the losses on a properly matched line (for the same power radiated)? Why? (*c*) Why isn't the line matched to reduce losses?

38. What is the difference between a center-fed and an end-fed antenna?

39. Refer to Fig. 11–14(a). (*a*) Is this a resonant line? (*b*) What makes it so? (*c*) Are the standing waves shown current or voltage waves? (*d*) What do the arrows signify? (*e*) Why is the current standing wave a maximum at the top of the transmission line? (*f*) Why is a series-tuned circuit used in the antenna coupling unit? (*g*) Why is the secondary coil center-tapped and grounded? (*h*) Why is the coupling between coils adjustable? (*i*) If the antenna coupling unit is connected to the transmission line at location *A*, would any change be necessary in the coupling network? Explain. (*j*) In either location, is the load impedance "seen" by the coupling network resistive, inductive, or capacitive? Explain.

40. (*a*) If the antenna system of Fig. 11–14(*a*) is used for multifrequency operation, is the load seen by the coupling unit always resistive? Explain. (*b*) How is this condition "satisfied"?

41. In Fig. 11–14(b). (*a*) Why does the coupling unit use a parallel-resonant tank circuit? (*b*) If the transmission line were $\frac{1}{4}\lambda$ longer, would any change in the coupling network be necessary? Explain.

42. (*a*) How can a transmission line be made to act as a nonresonant feeder? (*b*) Must a perfect impedance match be obtained? Explain.

43. (*a*) What is a "current-fed" antenna? (*b*) How does it get this name? (*c*) What is another name for a voltage-fed antenna?

44. (*a*) Is the delta-match technique shown in Fig. 11–15 used with transmission line impedances of 72 Ω, less than 72 Ω, or more than 72 Ω? (*b*) Explain why a match is obtained. (*c*) Will this technique produce a match if the antenna is not $\frac{1}{2}\lambda$? Explain.

45. Can a coaxial transmission line be used for either of Fig. 11–14(a) or (b)? Explain.

46. Refer to Fig. 11–16. (*a*) What is the approximate characteristic impedance of the single-wire line? (*b*) How is an impedance match obtained? (*c*) Why is the line connected to a tap on the antenna coupling unit? (*d*) Can this type of feeder be used in any terrain? Explain. (*e*) What is another disadvantage of this type of feed stystem? (*f*) Why is it used?

47. Refer to Fig. 11–17. (*a*) Which of these systems would be classified as a shunt-fed system? (*b*) What makes shunt feed necessary? (*c*) How is an impedance match obtained? (*d*) Why is a variable capacitor also used? (*e*) How is impedance matching obtained in diagram (b)?

48. (*a*) Give two examples where directional antenna systems are desirable. (*b*) What term is used to designate the directivity of an antenna? (*c*) How is this quantity evaluated?

49. (*a*) With respect to antennas, what does the term "directional gain" mean? (*b*) How does an antenna achieve a gain? (*c*) Give two references used in evaluating this gain. (*d*) How do these references compare? (*e*) What does "ERP" stand for? (*f*) Can this value be greater than the actual radiated power? Explain.

50. (*a*) What is the purpose of using antenna arrays? (*b*) Name two broad categories of arrays.

51. (*a*) How does a parasitic element differ from a driven element? (*b*) Name two classifications of parasitic elements.

52. Refer to Fig. 11–18. (*a*) If the two elements are of identical length, what is the proper separation between them? (*b*) What is the direction of maximum radiation? (*c*) Why is this so? (*d*) Account for the cancellation in the opposite direction. (*e*) In practice, what changes are made in the reflector length and spacing? (*f*) What is an optimum spacing?

53. (*a*) Where is a director placed with respect to the driven element? (*b*) What is an optimum spacing? (*c*) How does the director length compare with the length of the driven element?

54. Refer to Fig. 11–19. (*a*) What do the three patterns represent? (*b*) How can sharper directivity be obtained? (*c*) Are there any limitations to the number of parasitic elements that can be added? Explain.

55. (*a*) When driven arrays are used, what determines the directivity of the radiation pattern that can be obtained? (*b*) Name three broad classifications of driven arrays.

56. Refer to Fig. 11–20. (*a*) What is the overall length of the two-element antenna? Of the four-element antenna? Explain. (*b*) What do the arrows over each element signify? (*c*) How is this uniform current direction obtained in diagram (b)? (*d*) Why doesn't radiation from the $\frac{1}{4}\lambda$ stubs cancel the radiation from the antenna elements? (*e*) What do the three patterns in diagram (c) represent? (*f*) In what plane are these patterns taken? (*g*) Explain why the patterns become sharper and stronger at right angles to the antenna axis. (*h*) What does diagram (d) represent? (*i*) Why isn't this also directional?

57. Refer to Fig. 11–21. (*a*) What is the relative phase of the current in each element? (*b*) How is this phasing obtained? Explain. (*c*) Why is the field strength zero along the line of the antenna array? (*d*) Why is this array called a broadside array? (*e*) Why is the field strength a maximum in this direction? (*f*) In what plane is the pattern of diagram (c) taken? (*g*) Is this antenna system omnidirectional in the vertical plane? Explain.

58. In Fig. 11–21(b): (*a*) Is angle θ a vertical or horizontal angle? (*b*) What direction is used as reference? (*c*) If all the rays are going to a common point, *P*, why are they shown as parallel? (*d*) How much extra distance does the field from element 1 travel, as compared to the field from element 4?

59. In the solution to Example 11–3, explain: (*a*) Why does the field from element 2 lag by 90°? (*b*) Why is the resultant field zero?

60. Refer to Fig. 11–22. (*a*) What spacing is used between these elements? (*b*) What is the relative phase of the currents in each element? (*c*) How is this phasing obtained? (*d*) If the array in diagram (a) were aligned in an east-west direction, in what direction do the elements point? (*e*) In what direction is maximum radiation obtained? Explain. (*f*) In what plane is the radiation pattern of diagram (b) taken? (*g*) Does this antenna have strong radiation at high elevation angles? Explain. (*h*) What would the three-dimensional pattern look like?

61. Refer to Fig. 11–23. (*a*) What two basic arrays are combined in this antenna system? Explain. (*b*) Can the feed lines at the bottom of either stack be reversed? Explain. (*c*) Does this system produce a bidirectional pattern? Explain.

62. Refer to Fig. 11–24. (*a*) What type of antenna system does row 1 by itself produce? Why? (*b*) Is row 2 driven or parasitic? (*c*) What is the phase of the currents in row 2 as compared to row 1? (*d*) What is the effect of this second row on the overall pattern? (*e*) Explain why this is so.

63. In Fig. 11–25: (*a*) Are these horizontal- or vertical-plane radiation patterns? (*b*) What causes the change in the lobe angles? (*c*) What antenna length applies to each of the diagrams shown? (*d*) How would these patterns be altered if the current standing wave ratio were reduced to one, by a suitable resistor at the right-hand end of each antenna?

64. Refer to Fig. 11–26. (*a*) Why is this called a rhombic antenna? (*b*) What is the purpose of resistor R? (*c*) What does angle θ represent? (*d*) What determines this angle? (*e*) What is the tilt angle? (*f*) What should this value be? Why? (*g*) What relation must apply to distance $AB + BC$ as compared to AC? Why?

65. Give two advantages of a rhombic antenna over antenna arrays.

66. Name two antennas used specifically in UHF applications.

67. (*a*) What changes should be made to transmitting antennas to make them suitable for reception? (*b*) Are elaborate receiving antennas always used? Explain. (*c*) Why may more than one antenna be used with one receiver? Explain.

68. Name three propagation paths by which signals can reach a receiving location.

69. (*a*) What is a ground wave? (*b*) Give another name also used to describe this propagation path. (*c*) What type of polarization is necessary for ground-wave propagation? Why? (*d*) What causes attenuation of the ground wave? (*e*) What type of terrain will minimize this loss? Why? (*f*) What is the effect of operating frequency on ground-wave propagation? (*g*) Up to what frequency are ground waves generally useful?

70. Refer to Fig. 11–27. (*a*) Is either of these a space wave? (*b*) How do these two waves affect the signal strength received? (*c*) For maximum signal strength, how should the two path lengths compare? Why? (*d*) How would a very low height affect the signal strength received? Explain. (*f*) At VHF, is antenna height important for optimum signal strength? Explain.

71. Refer to Fig. 11–28. (*a*) Can a signal be received at location 3? Explain. (*b*) What change would be necessary to make reception possible?

72. (*a*) Is communication via surface waves strictly limited to the line-of-sight range? (*b*) How much increased range is generally possible? (*c*) What causes this increase?

73. (*a*) What is *duct propagation*? (*b*) What is the effect?

74. (*a*) What constitutes a sky wave? (*b*) What causes its return to earth?

75. (*a*) What is the ionosphere? (*b*) Is this a uniform region? Explain.

76. (*a*) What region of the ionosphere does the *D* layer occupy? (*b*) What is the effect of this layer on radio waves? (*c*) Is its effect constant or does it vary throughout the day? Explain.

77. (*a*) What region of the ionosphere does the *E* layer occupy? (*b*) How is it affected over a 24-hour period?

78. (*a*) What region of the ionosphere does the *F* layer occupy? (*b*) How is it affected daily? (*c*) How is it affected seasonally?

79. (*a*) What is the effect of the *E* and/or *F* layer on a radio wave? (*b*) What two factors determine the amount of this action?

80. (*a*) What does the term "critical frequency" mean? (*b*) In general, what determines this value? (*c*) How is the critical frequency affected between day and night? Explain. (*d*) How is it affected between winter and summer? Explain (*e*) Which layer (*E* or *F*) will have the higher critical frequency? Why?

81. At a certain time of day, the critical frequency for the *F* layer is 21 MHz. Can a wave at this frequency be made to return to earth? Explain.

82. Refer to Fig. 11–29. (*a*) How do the frequencies of waves *A*, *B*, and *C* compare? (*b*) Why are waves *A* and *B* not returned to earth? (*c*) Under what condition would the radiation angle θ_c represent the critical angle? (*d*) What does angle ϕ represent? (*e*) How are these two angles related?

83. What effect (if any) does the frequency of operation have on the critical angle value?

84. Refer to Fig. 11–30. (*a*) Can signal *X* be made to return to earth at a point nearer to the transmitter than location *B*? Explain. (*b*) What does the term "skip distance" mean? (*c*) What does the term "skip zone" mean? (*d*) Why does signal *Y* return to earth closer to the transmitter? (*e*) What is the skip distance for signal *Y*? (*f*) Assuming no change in ground wave, what is the skip zone for signal *Y*? (*g*) How can the skip zone be eliminated completely?

85. (*a*) What is meant by the term "maximum usable frequency"? (*b*) What happens if a higher frequency is used? (*c*) Can communication be established at a lower-than-maximum frequency? (*d*) Why is it desirable to operate at close to the maximum usable frequency? (*e*) Is the maximum usable frequency between two locations always the same? Explain.

86. (*a*) What is meant by double-hop transmission? (*b*) Is transmission limited to double-hop? (*c*) How is the signal strength affected by multiple-hop propagation? (*d*) Is it ever necessary to depend on multiple-hop transmission? Explain.

87. (*a*) What is meant by the term "fading"? (*b*) In general, what causes this effect? (*c*) Give three specific conditions that can cause fading.

88. (*a*) What is meant by the 11-year sunspot cycle? (*b*) What is its effect? (*c*) Give two other normal causes for variations in the ionosphere. (*d*) Describe briefly one abnormal variation. (*e*) What is its effect? (*f*) Describe another abnormal occurrence and give its effect.

89. In the frequency range below 300 kHz (*a*) how good is sky-wave propagation? Why? (*b*) How good is ground-wave propagation? (*c*) Can long-distance communication be obtained? Explain.

90. (*a*) What frequency band does the medium-frequency range include? (*b*) Can long-distance ground-wave communication be obtained? Why? (*c*) Can long-distance sky-wave communication be obtained? Why? (*d*) Is sky-wave propagation better during the day or night? Why?

91. (*a*) What frequency spectrum is called the high-frequency band? (*b*) Is it suitable for ground-wave communication? Explain. (*c*) Is it suitable for sky-wave communication? Explain.

92. (*a*) Is sky-wave communication efficient at above 30 MHz? Explain. (*b*) What is the principal means of propagation? (*c*) How can reliable communication appreciably beyond the line-of-sight range be maintained?

93. (*a*) What is forward-scatter propagation? (*b*) What transmitter requirements are necessary? (*c*) Is this an efficient means of communication? (*d*) Up to what frequency can reliable contact be maintained? (*e*) Up to what distances?

PROBLEMS AND DIAGRAMS

1. Draw the current and voltage standing waves for a half-wave antenna.

2. Show (by diagram) the electric and magnetic fields around a half-wave antenna.

3. Calculate the length of antenna required for a basic half-wave antenna with operating frequencies of (*a*) 600 kHz, (*b*) 10 MHz, and (*c*) 2000 MHz. (Allow 5% reduction up to a frequency of 30 MHz and 6% at higher frequencies, for velocity constant and end effects.)

4. The antenna current for a center-fed half-wave dipole is 7.5 A. Neglecting losses, how much power does this antenna radiate?

5. The final power amplifier has a plate efficiency of 70%, a plate voltage of 3000 V, and plate current of 1.0 A. The current fed to a half-wave dipole antenna is

5.30 A. Neglecting antenna circuit losses, find the radiation resistance of the antenna.

6. In Problem 5, if the radiation efficiency of the antenna is 95%, how much useful power is actually radiated?

7. A dipole antenna has a radiation resistance of 73 Ω and a "loss" resistance of 3 Ω. Find the radiation efficiency.

8. In Problem 5, if the radiation efficiency is 93%, find the radiation resistance.

9. An infinitesimal (short-dipole) horizontal antenna has a horizontal radiation pattern that varies as the sine of the angle between the direction of radiation and the line of the antenna. Plot this pattern on polar-coordinate paper for the full 360°. (*Hint:* sine values repeat every 90°.)

10. The radiated field for a horizontal half-wave dipole antenna varies in accordance with Eq. (11-1). Plot this pattern on polar-coordinate paper.

11. Draw the free-space radiation pattern for a half-wave horizontal dipole taken in a vertical plane and parallel to the line of the antenna.

12. Draw the free-space radiation pattern for a half-wave vertical antenna in a horizontal plane (perpendicular to the line of the antenna).

13. Draw the free-space radiation pattern for the above antenna in a vertical plane.

14. Using the technique of Example 11–2, calculate the ground-reflection factor and the vertical-plane relative field strength for a half-wave horizontal dipole antenna mounted $\frac{1}{2}\lambda$ above the ground. Plot the radiation pattern.

15. Repeat Problem 14 for an antenna height of 1λ. Use 10° intervals.

16. Draw the current and voltage standing wave patterns for a Marconi antenna.

17. Draw the horizontal-plane radiation pattern for a Marconi antenna.

18. (*a*) Using the pattern factor [Eq. (11–1)], plot the free-space radiation pattern for a Marconi antenna in the vertical plane. (*b*) Plot a second curve, correcting for ground effects.

19. Draw the current and voltage standing wave patterns for a vertical antenna $\frac{3}{16}\lambda$ in length which is (*a*) base-loaded with a loading coil, or (*b*) top-loaded, capacitively.

20. Draw the current standing wave patterns for each of the antennas in Fig. 11–13 if the vertical and flat portions are each $\frac{1}{4}\lambda$ long.

21. Draw the current and voltage standing wave patterns for a center-fed $\frac{1}{2}\lambda$ dipole with resonant feeders $\frac{3}{4}\lambda$ long.

22. Repeat Problem 21 for an end-fed antenna with resonant feeders 1.25λ long.

23. Draw the current standing wave pattern for a center-fed dipole that is $\frac{3}{8}\lambda$, with resonant feeders $\frac{3}{4}\lambda$ long.

24. When using a delta match, the recommended spread for a 600-Ω line (see Fig. 11–15) is A (in feet) = $123/f$ (MHz) and B (in feet) = $148/f$ (MHz). Calculate the proper dimensions for an operating frequency of 230 MHz.

25. A single-wire feed system (as in Fig. 11–16) is used at an operating frequency of 14 MHz. Calculate the length of the antenna and the location of the tap for the feed point.

26. Calculate the beam width for the radiation pattern in (*a*) Fig. 11–4, (*b*) Fig. 11–8, and (*c*) Fig. 11–9.

27. Calculate the directional gain of a parabolic antenna system if a radiated power of 100 mW can produce a signal level equal to that of an 800-W transmitter with isotropic radiator. Express the gain in decibels.

28. A rhombic antenna has a gain of 16 dB (compared to a half-wave dipole). How much saving in radiated power can be realized (*a*) compared to a dipole antenna? (*b*) compared to an isotropic antenna?

29. Draw the radiation pattern for a dipole antenna and show the effect on this pattern when a reflector and director are added.

30. A dipole antenna has a resistance of 73Ω, of which 4 Ω is due to losses. When parasitic elements are added, the input resistance drops to 30 Ω, and 6 Ω of this is due to losses. Calculate the radiation efficiencies.

31. Draw a horizontal broadside array and show the direction(s) of maximum radiation.

32. Calculate the complete horizontal radiation pattern for the broadside array of Fig. 11–21 using the method shown in Example 11–3. Plot on polar-coordinate paper.

33. Calculate the complete radiation pattern for a two-element end-fire array. Plot on polar-coordinate paper.

34. Calculate and plot the radiation pattern for two vertical antennas spaced $\frac{1}{4}\lambda$ apart and fed with currents 90° out of phase.

35. What is the maximum communicable range by space wave if the transmitting and receiving antennas are at heights of 100 ft and 50 ft, respectively?

36. For so-called line-of-sight transmission, what transmitting antenna height will be needed for reliable communication at a location 100 miles distant, and a receiving antenna height of 200 ft?

37. Show by diagram how multiple-hop transmission can take place.

38. Show by diagram how simultaneous refraction from the *E* and *F* layers can cause fading.

TYPICAL FCC QUESTIONS

3.218 What is the effect on the resonant frequency of connecting an inductor in series with an antenna?

3.219 What is the effect on the resonant frequency of adding a capacitor in series with an antenna?

3.221 What is the relationship between the electrical and physical length of a Hertzian antenna?

3.222 If you desire to operate on a frequency lower than the resonant frequency of an available Marconi antenna, how may this be accomplished?

3.223 What will be the effect upon the resonant frequency if the physical length of a Hertzian antenna is reduced?

3.224 Which type of antenna has a minimum of directional characteristics in the horizontal plane?

3.225 What factors determine the resonant frequency of any particular antenna?

3.226 If the resistance and the current at the base of a Marconi antenna are known, what formula could be used to determine the power in the antenna?

3.243 What radio frequencies are useful for long-distance communications requiring continuous operation?

3.244 What frequencies have substantially straight-line propagation characteristics analogous to those of light waves and unaffected by the ionosphere?

3.245 What effects do sunspots and aurora borealis have on radio communications?

3.252 Describe the directional characteristics of the following types of antennas: (*a*) horizontal Hertz antenna, (*b*) vertical Hertz antenna, (*c*) vertical loop antenna, (*d*) horizontal loop antenna, and (*e*) vertical Marconi antenna.

3.359 Draw a simple schematic diagram of a system of coupling a single electron tube employed as a radio-frequency amplifier to a Hertz-type antenna.

3.361 Draw a simple schematic diagram of a push-pull, neutralized radio-frequency amplifier stage, coupled to a Marconi-type antenna system.

3.490 What is meant by horizontal and vertical polarization of a radio wave?

3.491 How should a transmitting antenna be designed if a vertically polarized wave is to be radiated, and how should the receiving antenna be designed for best performance in receiving the ground wave from this transmitting antenna?

3.503 What is the purpose of a diversity-antenna receiving system?

3.504 Why are insulators sometimes placed in antenna guy wires?

3.517 Show by a diagram how a two-wire radio-frequency transmission line may be connected to feed a Hertz antenna.

3.534 A ship radiotelephone transmitter operates on 2738 kHz. At a certain point distant from the transmitter, the 2738-kHz signal has a measured field of 147 millivolts per meter. The second-harmonic field at the same point is measured as 405 microvolts per meter. To the nearest whole unit in decibels, how much has the harmonic emission been attenuated below the 2738-kHz fundamental?

4.059 If the field intensity of 25 millivolts per meter develops 2.7 volts in a certain antenna, what is its effective height?

4.061 In what units is the field intensity of a broadcast station normally measured?

4.070 The ammeter connected at the base of a Marconi antenna has a certain reading. If this reading is increased 2.77 times, what is the increase in output power?

4.078 If a vertical antenna is 405 feet high and is operated at 1250 kilohertz, what is its physical height expressed in wave lengths? (One meter equals 3.28 feet.)

4.079 What must be the height of a vertical radiator one-half wavelength high if the operating frequency is 1100 kilohertz?

4.201 What is the direction of maximum radiation from two vertical antennas spaced 180 degrees and having equal currents in phase?

4.203 How does the field strength of a standard broadcast station vary with distance from the antenna?

4.206 Why do some standard broadcast stations use top-loaded antennas?

4.209 What effect do broken ground-conductors have on a standard broadcast antenna?

4.258 What is the effective radiated power of a television broadcast station if the output of the transmitter is 1000 watts, antenna transmission line loss is 50 watts, and the antenna power gain is 3?

4.273 What is meant by antenna field gain of a television broadcast antenna?

6.476 If the power of a 500-kilohertz transmitter is increased from 150 watts to 300 watts, what would be the percentage change in field intensity at a given distance from the transmitter? What would be the dB change in field intensity?

6.488 If a 500-kilohertz transmitter of constant power produces a field strength of 100 microvolts per meter at a distance of 100 miles from the transmitter, what would be the theoretical field strength at a distance of 200 miles from the transmitter?

6.489 If the antenna current at a 500-kilohertz transmitter is reduced 50%, what would be the percentage change in the field intensity at the receiving point?

6.514 What is the difference between a Hertz and a Marconi antenna?

6.515 Draw a diagram showing how current varies along a half-wave length Hertz antenna.

6.516 What is meant by polarization of a radio wave? How does polarization affect the transmission and reception of a radio wave?

6.517 What are the lowest radio frequencies useful in radio communication?

6.519 What frequencies have substantially straight-line propagation characteristics analogous to those of light waves and unaffected by the ionosphere?

6.522 Why does harmonic radiation from a transmitter sometimes cause interference at distances from a transmitter where the fundamental signal cannot be heard?

6.529 In general, what advantages may be expected from the use of high frequencies in radio communication?

6.648 What is the directional reception pattern of a loop antenna?

6.649 What is the reception pattern of a vertical antenna?

S-3.193 Explain the voltage and current relationships in a one wavelength, one-half wavelength (dipole) antenna, and one-quarter wavelength "grounded" antenna.

S-3.194 What effect does the magnitude of the voltage and current, at a point on a half-wave antenna in "free space" (a dipole) have on the impedance at that point?

S-3.195 How is the operating power of an AM transmitter determined using antenna resistance and antenna current?

S-3.196 What kinds of fields emanate from a transmitting antenna, and what relationships do they have to each other?

S-3.197 Can either of the two fields that emanate from an antenna produce an EMF in a receiving antenna? If so, how?

S-3.198 Draw a sketch and discuss the horizontal and vertical radiation patterns of a quarter-wave vertical antenna. Would this also apply to a similar type of receiving antenna?

S-3.200 In speaking of radio transmissions, what bearings do the angle of radiation, density of the ionosphere, and frequency of emission have on the length of the skip zone?

S-3.201 Why is it possible for a sky-wave to "meet" a ground wave 180 degrees out of phase?

S-3.202 What is the relationship between operating frequency and ground-wave coverage?

S-3.203 Explain the following terms with respect to antennas (transmission or reception): (*a*) field strength, (*b*) power gain, (*c*) physical length, (*d*) electrical length, (*e*) polarization, (*f*) diversity reception, (*g*) Corona discharge.

S-3.204 What would constitute the ground plane if a quarter-wave grounded (whip) antenna, 1 meter in length, were mounted on the metal roof of an automobile? mounted near the rear bumper of an automobile?

S-3.205 Explain why a "loading coil" is sometimes associated with an antenna. Under this condition, would absence of the coil mean a capacitive antenna impedance?

S-3.207 What type of modulation is largely contained in "static" and "lightning" radio waves?

S-3.208 Will the velocity of signal propagation differ in different materials? What effect, if any, would this have on wavelength or frequency?

S-3.209 Discuss series and shunt feeding of quarter-wave antennas with respect to impedance matching.

S-3.210 Discuss the directivity and physical characteristics of the following types of antennas: (*a*) single loop, (*b*) V beam, (*c*) corner-reflector, (*d*) parasitic array, (*e*) stacked array.

S-3.211 Draw a sketch of a coaxial (whip) antenna; identify the positions and discuss the purposes of the following components: (*a*) whip, (*b*) insulator, (*c*) skirt, (*d*) trap, (*e*) support mast, (*f*) coaxial line, (*g*) input connector.

S-4.057 Define field intensity. Explain how it is measured.

S-4.064 What is effective radiated power? Given transmitter power output, antenna resistance, antenna transmission line losses, transmitter efficiency, and power gain, show how ERP is calculated.

S-4.066 How does a directional antenna array at an AM broadcast station reduce radiation in some directions and increase it in other directions?

S-4.067 What factors can cause the directional pattern of an AM station to change?

S-4.069 Define polarization as it refers to broadcast antennas.

S-4.070 What is the importance of the ground radials associated with standard broadcast antennas? What is likely to be the result of a large number of such radials becoming broken or seriously corroded?

12

Additional Transmitter-Receiver Circuits

In previous chapters, basic AM and FM transmitter circuits were discussed. However, there are a variety of other modulation techniques that are used either to improve efficiency of operation, reduce bandwidth requirements, or improve transmission reliability (from a signal-to-noise aspect). The more important of these techniques will be discussed in this chapter.

THE AMPLIPHASE AM TRANSMITTER

In high-level AM transmitters, class-C amplifiers can be used in all stages, resulting in high-efficiency operation. The drawback is the need for high modulating-signal power. For high-power transmitters, the technical problems and physical size of the modulation transformer make this a serious drawback. Conversely, low-level modulation conserves on modulation power, but sacrifices RF efficiency. Yet, in FM transmitters, because the amplitude of the modulated wave is constant, it is possible to use low-level modulation and still retain class-C operation for maximum efficiency. The *ampliphase* AM system capitalizes on this principle. The carrier is first *phase*-modulated at low level. The modulated wave is amplified to the full desired power-output level, using class-C amplifiers, and it then is converted to amplitude modulation. The name *ampliphase* is coined from the *phase-to-amplitude* nature of the circuit action.

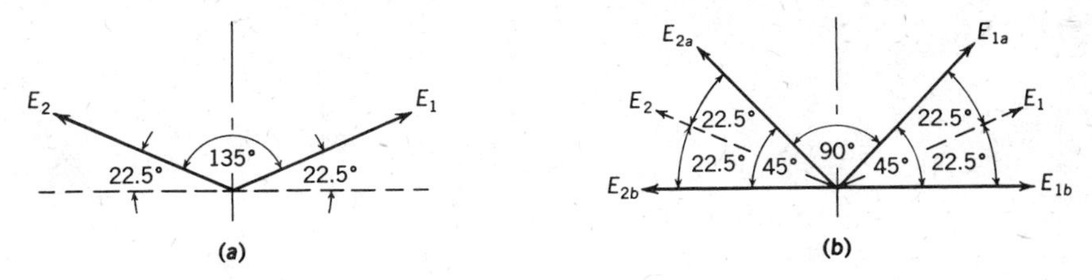

Figure 12-1 Channel phasings in an ampliphase transmitter.

Among the advantages claimed for this system are: as much as 50% reduction in floor space requirements; reduced original cost, installation cost, and operational costs; lower distortion; and flat frequency response. Most of this is due to the elimination of the costly and bulky iron-core modulation transformers.

The basic principle of the ampliphase system can be seen from the phasor diagrams in Fig. 12–1. The transmitter has two RF channels—with carriers E_1 and E_2 at the same frequency, but differing in phase by 135° before modulation. When modulation is applied, it phase-modulates each carrier, and full modulation (100%) produces a ±22.5° shift. However each carrier shifts in opposite directions, so that when carrier E_1 is shifted to E_{1b} (diagram (b)), E_2 is at E_{2b}. The phasor sum at this instant is zero. Conversely, when carrier E_1 is phase-modulated to position E_{1a}, carrier E_2 is in position E_{2a}, and the phasor sum is now at its crest value. Therefore, by proper phasing and adding of the outputs from the two channels, an amplitude-modulated wave results. Yet, all amplifier stages operate at constant amplitude and can be class-C, while the phase modulation occurs at low level and needs very little power.

Block Diagram Analysis

An overall understanding of the ampliphase system is best achieved from a block-by-block analysis. (A complete block diagram will be shown later. See Fig. 12–3.) The first block, the *crystal oscillator* block, is fully solid-state. Two crystals are used (one as a standby) in a FET oscillator circuit. The crystal units are temperature-controlled and will maintain frequency within ±5 Hz. The output from the oscillator stage is fed to an IC circuit (similar to the ECG 371 of Chapter 3), which acts as a saturated (limiter-clipper) amplifier. Its output, a square wave, is fed to a bipolar transistor amplifier. For additional stability, the power supply voltage is regulated using another IC. The entire unit is built on a printed circuit plug-in panel for easy servicing. The frequency of the oscillator-block output voltage is the same as the desired final transmitter frequency.

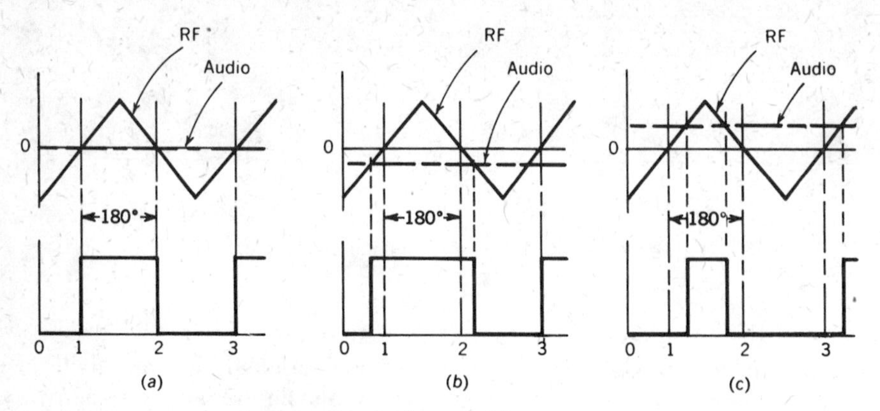

Figure 12-2 Phase modulator action.

The square-wave output from the oscillator section is fed to the *phase modulator* block (which is also known as a *linear-ramp comparator circuit*). A bipolar transistor, acting as an integrator-amplifier, converts this square wave into a triangular wave. (This provides the linear ramp voltages.) This triangular wave is then fed to the comparator circuit together with the audio modulating signal. In the comparator circuit, the instantaneous audio voltage is used as a reference level, and an output pulse (or ramp) is started when the triangular wave rises to equal the audio amplitude at that instant. The ramp is reset to zero—and the output pulse ends—when the triangular-wave amplitude falls below the audio amplitude level. This effect is shown in Fig. 12–2 for three different instants of time. Diagram (a) is for an audio amplitude of zero. Notice, at time instants of 1 and 3, that the pulse is started when the triangular wave rises to the audio level (of zero). Also, at time instant 2, the pulse is ended when the RF (triangular wave) value falls to the audio level. Now study diagram (b). The audio signal is at (approximately) its negative maximum value. Yet it is shown as a straight line. It must be realized that, in a time span of 1.5 cycles of the RF wave, the audio wave is only a minute fraction of a cycle and appears as a straight line.* Notice that the output pulse in diagram (b) has become wider. Now examine diagram (c). The audio, this time, is at (approximately) its positive maximum value, and the output pulse is appreciably narrower. Therefore, the audio amplitude determines the pulse width—or the pulse has been width-modulated, and phase modulation information is contained in the leading and trailing edges of these pulses.

*As an example, consider an RF wave of 1500 kHz. The time for 1.5 cycles of RF is only 1.0 μs. During this time interval, an audio of 1000 Hz would have completed only one-thousandth of a cycle. Obviously, this small portion of a wave appears as a straight line.

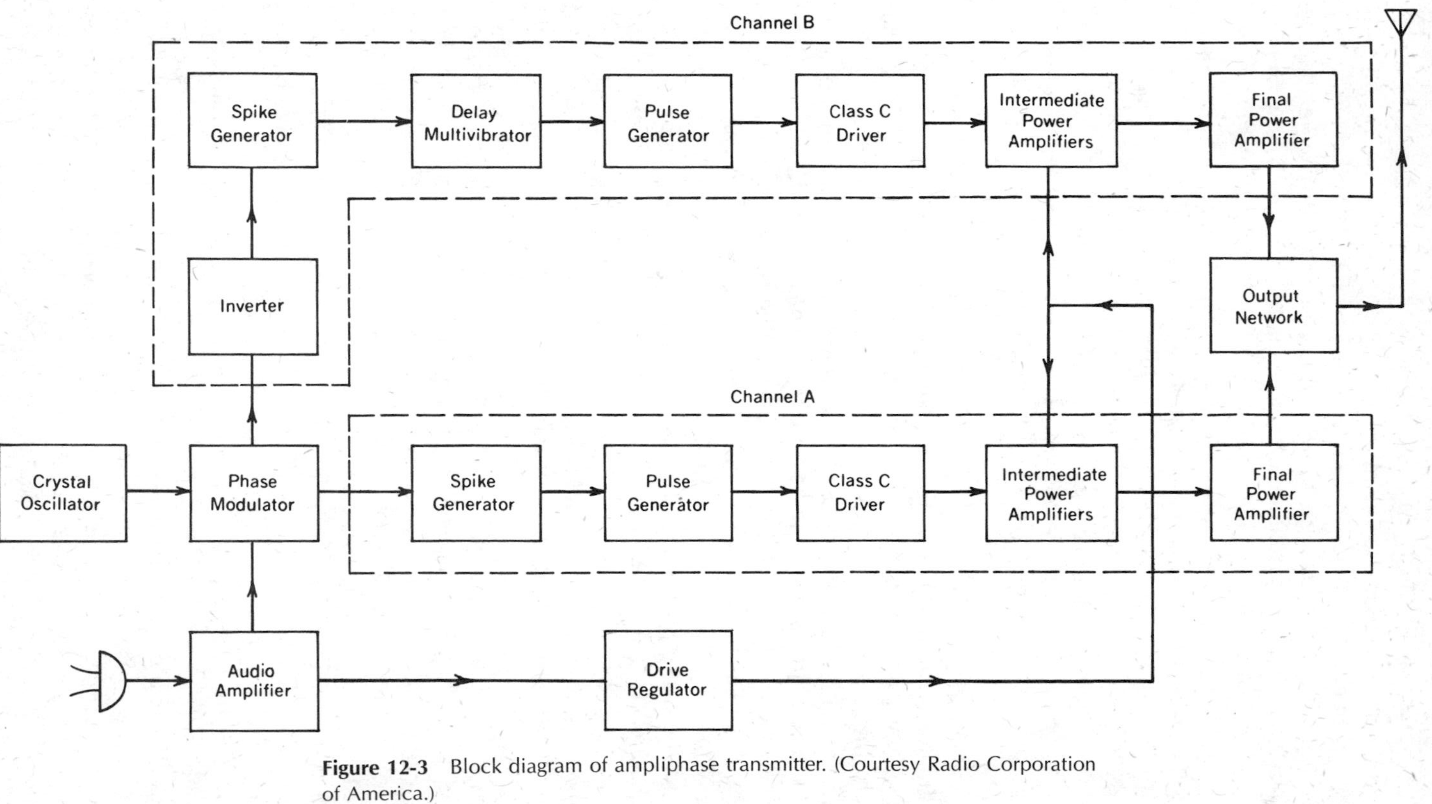

Figure 12-3 Block diagram of ampliphase transmitter. (Courtesy Radio Corporation of America.)

We are now ready to examine the complete block diagram. See Fig. 12-3. Remember that we wish to develop two RF outputs that are 135° out of phase when the modulation level is zero. We start by feeding the output of the phase modulator to two channels. Let us follow channel A. The pulses from the phase modulator are fed to a *spike generator,* which is triggered by the *positive*-going *leading* edge of the input signal. These spikes are then fed to a *pulse generator.* One pulse is produced for each input spike. The width of these pulses is adjustable and is set for best operation of the class-C driver that follows. (The usual pulse width is approximately 120°.) It is in this *class-C driver* that the RF sine-wave transmitter carrier is first produced. The phasing of its output voltage is timed to match the positive-going leading edge of the phase-modulator output.

Channel B is to be energized from the trailing edge of the phase-modulator output. But this trailing edge is negative-going and will not trigger the spike generator. It is therefore first fed to an *inverter* circuit where the negative-going trailing edge becomes a positive-going trailing edge; then it is fed to the spike generator. For a no-modulation condition, the spike produced here will be 180° from the spike produced in channel A. Yet, we want only a 135° shift between the two channel outputs. To effect this phasing, the channel B spike is fed to a monostable one-shot multivibrator with an adjustable time delay. Here the pulse duration is adjusted so as to produce a new spike at the required 135° from the spike in channel A. This new spike triggers the pulse generator of channel B, and a pulse is produced to drive the class-C amplifier of channel B. This portion of the transmitter is all solid-state.

Each channel's output is further amplified in successive stages until the desired final power output is obtained. By adjustment, the outputs from the two channels are 135° apart under a no-modulation condition. Then, as modulation is applied, the timing of the leading and trailing edges of the phase modulator changes, producing the phase relations previously shown in Fig. 12–1. The two phase-modulated outputs are then combined in the *output network* and the plate currents of the two power amplifiers, flowing through a common load (the output network) produce an AM output. This is shown in Fig. 12–4. When there is no modulation, the two plate currents, I_1 and I_2, are 135° out of phase, and their resultant I_c, flowing through the output network, produces the unmodulated carrier amplitude E_c. At one peak of the modulating signal, the phase angle between the two plate currents decreases to 90° (I_{1a} and I_{2a}). The resultant current rises to a maximum value, and the modulated output rises to its crest value E_{max}. Conversely, at the opposite polarity peak of the modulating signal, the two plate currents are 180° out of phase; the resulting current is zero (for 100% modulation); and the modulated output drops

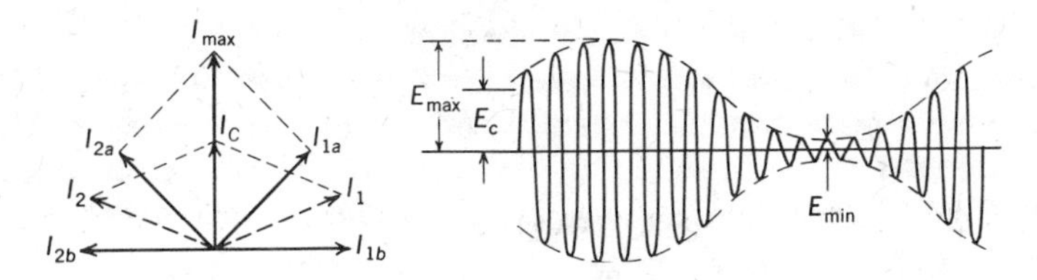

Figure 12-4 AM from PM.

to its minimum (trough) value. The output is amplitude-modulated.

In the phasor diagram of Fig. 12–4, the RF plate current for each tube is constant regardless of the degree of modulation. Yet, considering that, as the modulation trough is approached, the instantaneous power decreases, these plate currents are unnecessarily high, and the losses are high. Furthermore, to produce an AM modulation trough of zero at 100% modulation, perfect balance is required. Both of these weaknesses are corrected by use of *drive regulation.* This applies the modulating signal to the last intermediate power amplifier stage, in series with the dc bias (grid modulation). The modulating signal is adjusted so that the *output* tubes are approximately at cutoff at the modulation trough for 100% modulation. At the positive peak of the audio signal, the drive regulation provides the high value of drive needed by the power amplifiers to deliver full power output for the modulation crest.

SINGLE-SIDEBAND TRANSMITTERS

Although the ampliphase system improves operating efficiency, it does not correct the low *transmission* efficiency inherent in any AM system. For example, if a transmitter has a fully modulated (100%) output of 15 kW, the carrier power is 10 kW, and only 5 kW, in the sidebands, represents intelligence. Furthermore, the peak envelope power (PEP) at the modulation crest is four times the carrier power or 40 kW. Tubes, transistors, and components must be designed to handle these peak power and voltage levels. On the other hand, consider that the same intelligence is present in both (upper and lower) sidebands. Therefore, if we transmit one sideband (without a carrier), only 2.5 kW would be required to transmit the same effective intelligence power. Compared to the average power at the modulation crest, this is a saving of 8/1 or 9 dB. Such a system is called *single-sideband transmission (SSB).** Since the carrier is

*Also known as single-sideband suppressed-carrier (SSSC).

suppressed at low level, there is a considerable saving in operating efficiency. SSB transmitters—for equivalent transmission capability—require appreciably less ac power input and are appreciably smaller in size and weight. This is of particular advantage in aircraft and missile applications. Conversely, for the same size, weight, and total power rating, an SSB transmitter will have greater communication range.

There are other advantages obtained from SSB operation. With only one sideband, the bandwidth requirement for any transmission is cut in half. Considering the overcrowding in the 2–30 MHz spectrum range and the increasing demands for channel allocations, SSB operation proved so advantageous that in 1955 the Federal Communications Commission recommended that use of SSB transmission be required for fixed radiotelephone service and for aeronautical mobile service below 25 MHz.* (Later in 1970,† the FCC set January 1, 1970 as the date for no *new* installations of double-sideband AM, *A*3, transmission, and 1977 for complete phase-out of such equipment.) Simultaneously, the reduced bandwidth, with SSB transmission, also means less noise at the receiver and better signal-to-noise ratios. Elimination of the carrier further improves reception in that the annoying whistles (beat frequencies) produced by adjacent carriers are gone. In long-range communication, use of SSB transmissions will reduce distortion and fading due to multipath reception. With conventional AM transmission, it is possible for each sideband and the carrier components to incur different amounts of phase shift. This will cause an imbalance in the relative component amplitudes, or it may even cause complete cancellation of the carrier and/or one sideband, resulting in severe distortion from a conventional detector. With SSB operation, since there is only the one component, it can only be weakened by multipath reception, but there will be no distortion.

Unfortunately SSB operation brings on problems of its own. The main disadvantage is the added cost and complexity for better frequency stability. AM transmitters have stabilities in the order of 100 parts per million (ppm) or 0.01%. (At 30 MHz, the carrier may drift as much as ±3000 Hz.) In AM receivers, no specific circuitry is used to maintain local oscillator stability, yet the results are very satisfactory. On the other hand, with SSB, frequency stability is very important. When the carrier is not transmitted, it must be generated locally in the receiving equipment. For maximum fidelity, this locally generated carrier must be identical to the original transmitter carrier—both in frequency and in phase. In voice communications, phase shifts have negligible effects, but a frequency change of only 10 Hz will be noticeable, while frequency errors beyond

Notice of Proposed Rule-making, FCC-55-987.

†Docket No. 12221.

50 Hz* can destroy intelligibility. However, when transmitting digital information, the requirements are more stringent, and phase shifts cannot be neglected. Consequently, transmitter stabilities in the order of 0.5 ppm (or better) are highly desirable for SSB operation. In receivers, the same high stability is necessary. Quality receivers use either precision frequency-control circuitry or complex AFC circuitry. Lacking these features, continual operator monitoring is required to maintain communication.

Basic Principles

When an AM transmitter is modulated with a single audio frequency, the output varies in amplitude at the audio rate. This familiar waveshape was seen in Fig. 6–2(b). Such a waveshape was analyzed as containing a carrier, an upper side frequency and a lower side frequency. Each of these components is a pure sine wave.

In SSB transmitters, balanced modulators (as described for the Armstrong FM system) are used to generate sidebands while eliminating the carrier component. Then, either band-pass filters are used to pass only the desired sideband, or phase-shift circuitry is used to cancel the unwanted sideband. Although either sideband can be transmitted, standardization favors transmission of the upper sideband.† Obviously, the output for *single-frequency* modulation will be a pure sine wave (constant amplitude) at a frequency of $f_c + f_m$. With complex modulating signals, the output wave will vary in amplitude, but the envelope will bear no resemblance to the modulating signal.

SSB Generation—the Filter Method

This type of transmitter is shown in block-diagram form in Fig. 12–5. The carrier (crystal oscillator output) and the modulating signal are fed to a balanced modulator to suppress the carrier. The double-sideband output is then fed to a band-pass filter to pass the desired sideband and reject the other sideband. It should be realized that the two sidebands might be only a few *hertz* apart. For example, at an audio frequency of 40 Hz, the side frequencies would be $f_c \pm 40$ or only 80 Hz apart. To reduce the burden on the filter circuit requirements, voice frequencies below 100 Hz are suppressed. (This does not affect intelligibility.)

* The amount of tolerable frequency shift will vary, depending on the signal-to-noise ratio.

†A station can simultaneously transmit two messages independently by using the upper sideband for one message and the lower sideband for the second message.

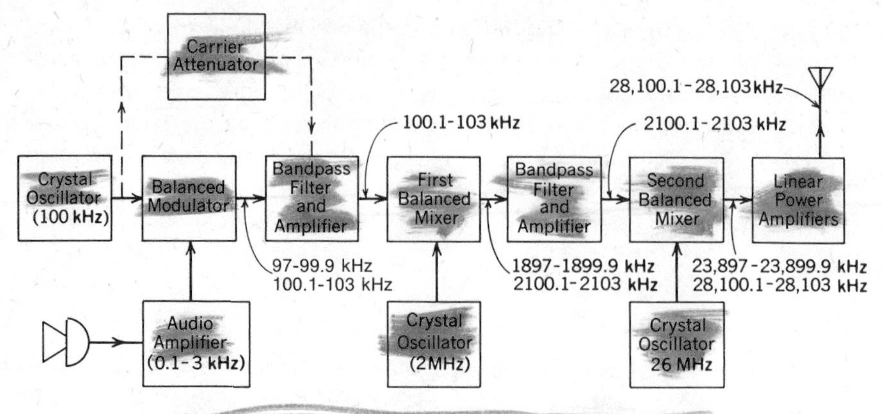

Figure 12-5 Block diagram of a filter-type SSB transmitter.

The degree of separation of the sidebands is also a function of the carrier frequency, as can be shown by a numerical illustration. Assuming a carrier frequency of 2.0 MHz and a modulating frequency of 100 Hz, the two side frequencies would be 2000.1 kHz and 1999.9 kHz. The filter must pass the upper side frequency and reject the other, which is 0.2 kHz away. The percent difference in frequency is only 0.2/2000.1 × 100 or 0.01%. On the other hand, if we use a carrier frequency of 50 kHz, the two side frequencies are still only 0.2 kHz apart, but the percent difference from the desired signal is now 0.2/50.1 × 100 or 0.4%. This is definitely a less stringent requirement. Therefore, when the filter method is used to produce an SSB signal, low-frequency oscillators are preferable as carrier generators. A commonly used frequency is 100 kHz. Excellent crystal oscillator stability can also be obtained at this frequency.

Now, however, there is the problem of raising the sideband to the desired RF value. Obviously, frequency multipliers cannot be used. Such circuits would both alter the intelligence signal and increase the bandwidth. (For example, if a 100-kHz oscillator is modulated with voice frequencies of 100–3000 Hz, the upper sideband would range from 100.1 to 103 kHz. The output from a frequency tripler would then be 300.3–309 kHz. Notice that the intelligence, or audio, has been raised to 300–9000 kHz. This is not the original sound.) To preserve the intelligence signal and bandwidth, frequency translation must be used to change the carrier frequency. This is shown in the block diagram of Fig. 12–5. This transmitter uses two frequency-conversion stages to obtain its final output frequency. Notice that the two sidebands at the output of the balanced modulator are very close in frequency. Yet, at the output of the second mixer, these sidebands are a full 4000 kHz apart, and it is now very easy to reject one without affecting the other. This excellent separation is obtained because of dual frequency conversion, and it results in a high degree of unwanted-sideband (and carrier) suppression. The same final frequency could be obtained with only one step of frequency conversion, but the separation between sidebands would not be as great, and suppression of the unwanted sideband would not be as effective. (See Problem 8.)

Balanced-Modulator Circuits

A balanced-modulator circuit was shown in Fig. 9–28 in connection with the Armstrong FM transmitter. Such a circuit is also used in SSB systems.* (Similarly, any of the circuit variations mentioned in the earlier discussion could also be used.) However, this circuit will not provide good carrier suppression unless the transistors (Q_1 and Q_2 of Fig. 9–28) are perfectly matched. This situation is remedied by providing a *carrier-balance* control. In the absence of a modulating signal, the control is adjusted for equal RF components of collector currents (zero RF output). This balancing action can be readily achieved by controlling any element potential. Figure 12–6 shows a balanced modulator with an *R-C* coupled carrier input and a balance control in the source circuit. The basic circuit action was described previously and needs no further discussion. However, any imbalance in the respective drain currents now can be equalized

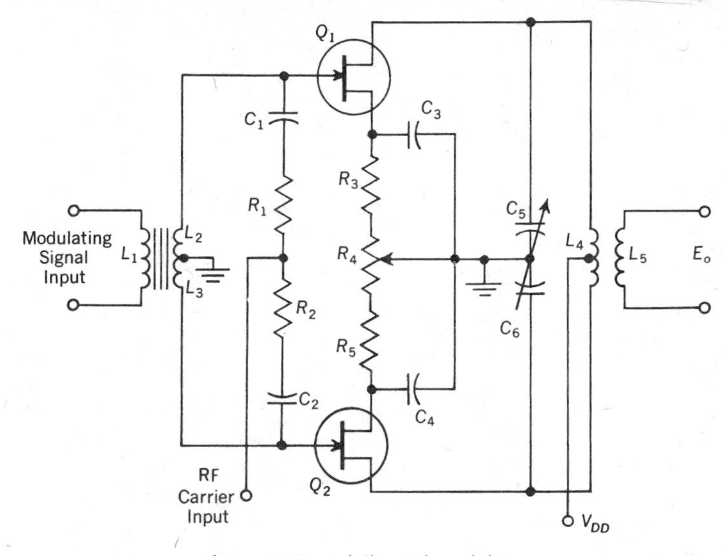

Figure 12-6 A balanced modulator.

* Obviously, winding L_2 of the carrier coupling network would not be necessary now.

by adjusting the potentiometer in the source circuit. Notice, also, that the RF signal is shunt-fed to the two gates—still in phase.

Again, as mentioned in the discussion on FM transmitters, ICs are also being used for balanced-modulator applications. The RF carrier is fed to the base of both transistors in the differential pair (Q_1 and Q_2 of Fig. 3–14). With equal inputs, the output from this pair is zero. Obviously, the carrier frequency is effectively eliminated from the output. The modulating signal (the audio) is fed to what was called "the constant-current source," Q_3. Now, the bias of this unit varies at the audio rate and produces an RF output that is modulated by the audio, i.e., upper and lower sidebands. Many manufacturers use the dual-independent combination of the basic IC in a doubly balanced modulator circuit. An advantage of the IC modulator, as compared to discrete-component circuits, is that far better balance can be obtained. Because the transistors are on the same chip, the characteristics of the differential pair are very closely matched.

Another commonly used circuit is the ring- or lattice-type *modu-*

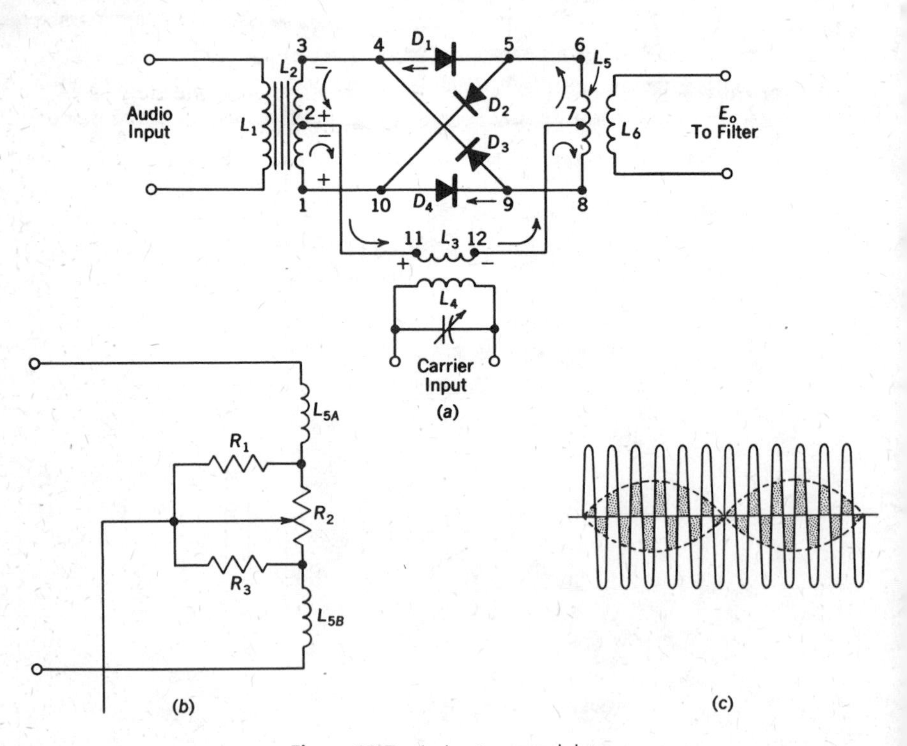

Figure 12-7 A ring-type modulator.

lator shown in Fig. 12–7. Consider first the circuit action when only the carrier is applied. With the polarity shown, current will flow as indicated by the arrows. The current flow through both halves of winding L_5 is equal and opposite. The output voltage, at the carrier frequency, is zero. During the next half-cycle of RF, diodes D_2 and D_3 conduct equally, and again the output is zero. With regard to the audio signal—by itself—current will flow from winding L_2 either through diodes D_3 and D_4 or through D_2 and D_1, but not through L_5. Again, there is no output. For balanced-modulator action, both signals must be applied. Also the RF signal should be appreciably stronger (6 to 8 times) than the modulating signal.* Therefore, conduction is determined by the polarity of the RF carrier. With the RF polarity as shown in Fig. 12–7, current will again flow as indicated by the arrows. However, now the audio voltage will aid or oppose the conduction. For example, with the instantaneous audio polarities shown, diode D_4 will conduct heavier than diode D_1, the current balance in winding L_5 is upset, and an output at the sideband frequencies will be developed in winding L_6. With matched diodes, carrier suppression of 40–60 dB is readily obtained. Carrier balance can also be obtained by splitting the output winding L_5 and inserting the resistive balance network R_1-R_2-R_3. This modification is shown in Fig. 12–7(b). The waveshape in Fig. 12–7(c) shows the switching action as each pair of diodes conducts on alternate half-cycles of carrier input. Notice how the degree of unbalance varies with the amplitude of the audio signal.

The block diagram (Fig. 12–5) shows balanced mixers in addition to the balanced modulator. The circuit action in both cases is identical. The only difference is the nature (and value) of the input signals. In a mixer, the audio signal is replaced by the SSB output from the preceding filter and amplifier. Because of this similarity, any of the circuits already discussed can be used (with minor modifications) as balanced mixers.† For example, audio coupling transformers must be replaced by suitable RF inductive or impedance coupling. However, as a second mixer—since the output is not fed to a band-pass filter—a tank circuit must be used in the output.

Figure 12–8 shows a balanced mixer circuit using bipolar transistors. Tuned circuits are used for both inputs (SSB and Osc.) and for the output. The transistors are biased for nonlinear action by the voltage divider in the base circuit. The oscillator signal is suppressed from the output because it is fed in phase to both emitters, and cancellation is improved by the balance control, R_4.

* The RF level should be 3–6 V for minimum distortion.

†Diode circuits such as the ring modulator would not normally be used because of their lack of gain.

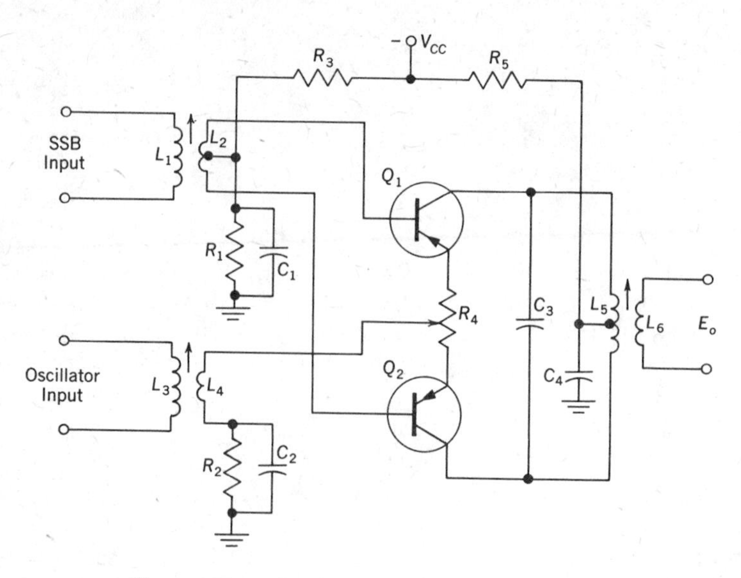

Figure 12-8 A bipolar transistor balanced mixer.

Sideband Filters

Because of the closeness of the two sidebands at the output of the balanced modulator, high-Q multisection filters are necessary. One commonly used filter circuit takes advantage of the high Q of a typical quartz crystal. (The equivalent circuit and impedance characteristics of such a crystal were shown in Fig. 5–16. Note the two resonant frequencies, the sharp change in impedance between the series- and parallel-resonant frequencies, and the relative closeness of these two frequencies.) Excellent band-pass action can be obtained by using two matched pairs of crystals in a bridge circuit called a *lattice network*. Such a *crystal lattice* filter is shown in Fig. 12–9. Crystals X_2 and X_3 are in series with the signal path. They are selected so that their series-resonant frequency falls within the desired pass band, and will have relatively low impedance for a small band

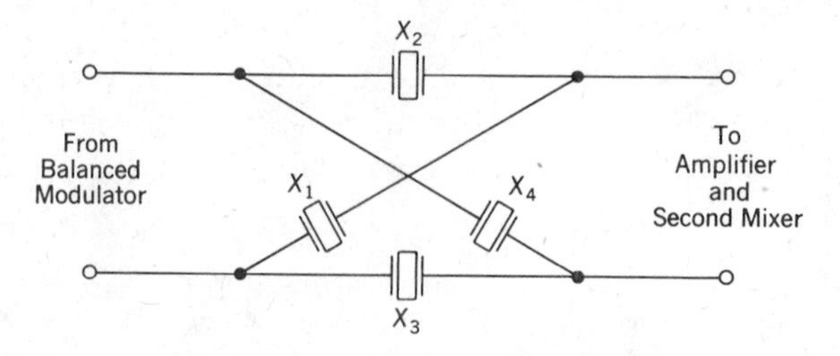

Figure 12-9 A crystal lattice filter.

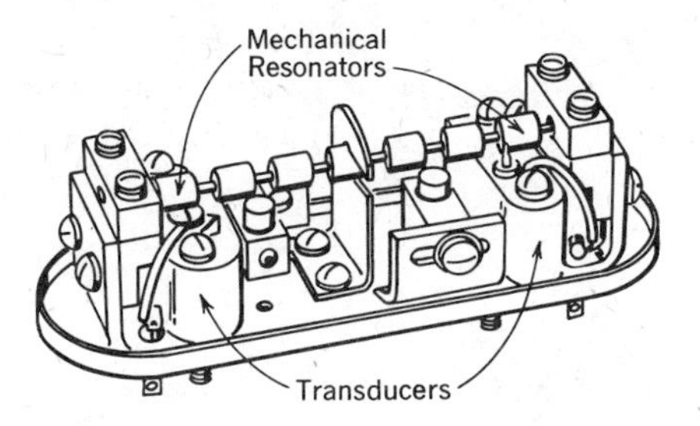

Figure 12-10 An electromechanical filter.

of frequencies to either side. Crystals X_1 and X_4 will complement this action. Their parallel-resonant frequency (high impedance) will correspond to the series-resonant frequency of crystals X_2 and X_3. Therefore, they will not bypass the desired sideband. On the other hand, the series-resonant frequency of crystals X_1 and X_4 will bypass the unwanted sideband to ground, while the high impedance of crystals X_2 and X_3 at these frequencies will block passage of this signal. By proper choice of crystals (series- and parallel-resonant frequencies and Q), such a network can give an almost ideal response curve, with a flat top over the small pass band required, and very steep skirts to either side.

Crystal filters give excellent results, particularly at frequencies around 100 kHz. However, they are costly, and compared to other circuit components, they are relatively bulky. Furthermore, because they cannot withstand appreciable shock or vibration, they are not suitable for mobile applications.

Another type of filter that has won wide acceptance is the *electromechanical filter** (often called a mechanical filter). The input circuit is resonant at the center of the pass band. RF fed to the filter is converted to mechanical energy by a magnetostrictive transducer located within the input circuit. The energy is then passed through a series of mechanical resonators. These could be in the form of plates, rods, or disks. Figure 12-10 shows a filter of the rod type, with seven resonators. (Resonance depends on the dimensions and mode of vibration of the filter rod.) A second transducer at the output converts the mechanical energy back into electrical energy. In the process of passing through the mechanical filters, the energy takes on the response characteristics dictated by the geometry of the mechanical elements.

Electromechanical filters are available to cover the frequency range of 50–600 kHz, with bandwidths up to approximately 10%. These filters

* D. L. Lundgren, "Electromechanical Filters for SSB Applications," *Proceedings IRE*, **44** (December, 1956).

have a low insertion loss, good stability, and can be built to withstand shock and vibration. Optimum results are obtained at frequencies around 250 kHz. Below this value, physical size increases and becomes relatively large, whereas at higher frequencies physical tolerances become a problem.

Linear Power Amplifiers

Since modulation occurs at low levels, it is necessary that the amplifiers following the balanced mixers be linear amplifiers. At low signal levels, maximum linearity and high voltage gain are more important than efficiency. Class-A amplifiers are used in such cases. These stages create no distortion problems. At higher power levels, it would be desirable to use class B for improved efficiency. Unfortunately, the distortion level, even with push-pull, is high.* Not only does this result in a poor signal at the receiver, but in addition, the energy of the transmitted signal may extend one bandwidth to either side of its normal spectrum, causing interference with adjacent channels. Since linearity is of prime importance, power amplifiers are generally operated in class AB, with the no-signal operating point selected for the lowest plate dissipation consistent with an acceptable intermodulation distortion level.† To further reduce distortion, inverse feedback is often added. Such a circuit is shown in the partial schematic of Fig. 12–11. (It should be realized that vacuum tubes are used at the higher power levels.) Since the feedback is taken over two

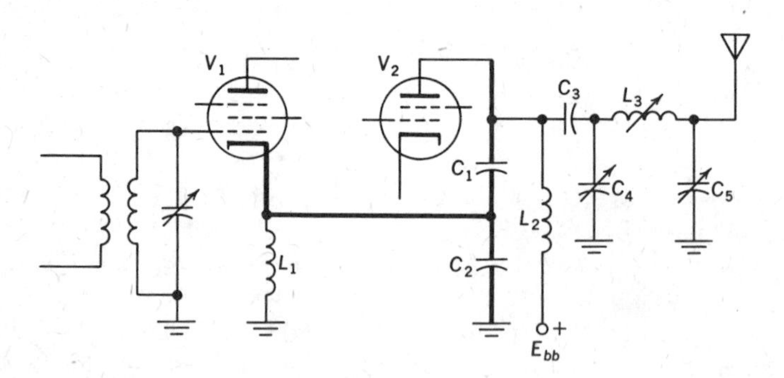

Figure 12-11 Partial schematic of a two-stage Class-AB linear power amplifier with feedback.

*This distortion is an odd-order (third and fifth) intermodulation product. For example, if the SSB signal contains two frequencies f_1 and f_2, nonlinearity produces outputs at $2f_1$-f_2, $2f_2$-f_1, $3f_1$-$2f_2$, $3f_2$-$2f_1$, etc. These intermodulation products are so close to the desired frequency that they cannot be separated by the tank circuit.

†A common point is at a bias halfway between the start of curvature of the dynamic characteristic and cutoff.

stages, the feedback voltage is in phase with the grid input to V_1. It is therefore fed to the cathode to produce inverse action.

Depending on the operating frequency and the type of tube used, tetrodes and beam power tubes may not require neutralization. Yet, high-mu, "zero-bias" triodes are also used because of the simpler power supply requirements. Not only is the screen supply eliminated, but the grid circuitry is simplified because these tubes are designed to operate at zero control-grid bias. Neutralization problems are solved by operating the tubes in grounded-grid circuitry.

Carrier Attenuators

As mentioned earlier in this discussion, SSB *suppressed-carrier* transmission requires excellent oscillator accuracy and stability both in the transmitter and the receiver. A few hertz difference between the transmitter's suppressed carrier and the receiver's reinserted carrier can make speech unintelligible. An error in phasing can destroy coded information. One solution to this problem is to use a precision master oscillator,* not just for the original carrier signal, but for all three oscillators in Fig. 12–5. The higher frequencies required for mixer action are then obtained with properly designed *frequency synthesizers*.† A similar system would also be needed in the receiver.

Another technique used to maintain synchronism between the transmitter and receiver is to transmit a *pilot carrier*. SSB transmitters often have provision for this type of operation. This is the specific function of the *carrier-attenuator* block in Fig. 12–5. The transmitter is designed to achieve maximum suppression of the carrier signal. Then, if a pilot carrier is desired, a controlled amount of carrier signal is added. The level of such pilot carriers is generally in the order of 10–30 dB below the normal AM carrier value.

Figure 12–12 shows a commercial ground-station unit (transmitter and receiver) in which the transmitter circuit closely follows the block diagram of Fig. 12–5, except for the specific frequencies chosen. The main crystal-oscillator frequency in this unit is 457 kHz. (This makes it possible to use the same crystal oscillator to supply the IF carrier for the receiver operation.) The balanced modulator uses the ring-type diode circuit (and this section also serves as the demodulator for the receiver). The band-pass filters for obtaining SSB are of the crystal type. To ensure good sideband separation, the audio range is restricted to 300–2700 Hz. The transmitter is rated at a power output of 1000 W, and except for the

*Oscillators with frequency accuracy and stability good to one part in 10^8 are available.

†Frequency synthesizers will be discussed later in the chapter under suppressed-carrier receivers.

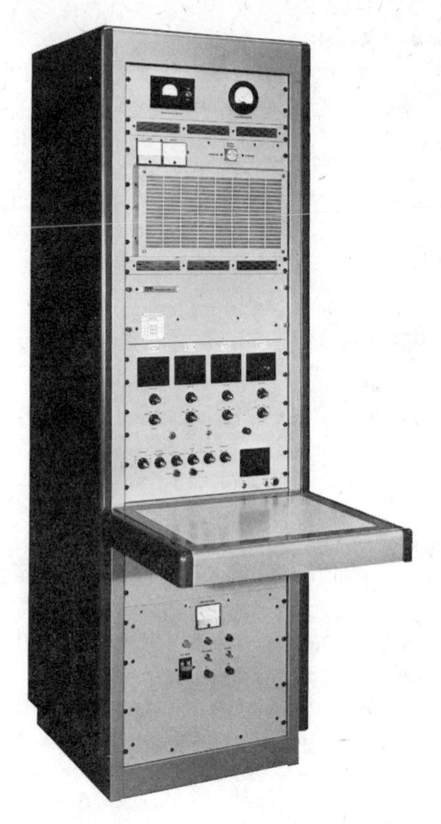

Figure 12-12 A commercial transmitter using the filter method for SSB. (Courtesy Communications Associates, Inc.)

driver and final power-amplifier stages, the rest of the circuitry is solid-state. (A later model uses tubes only in the final power amplifier.) The overall dimensions of the rack are 58 in. × 23 in. × 25 in.

SSB Generation – the Phase-Shift Method

In this type of transmitter, filter circuits are not used to eliminate the unwanted sideband. Instead, this component is suppressed or canceled by producing (and combining) two such components that are 180° out of phase. Because a cancellation process is used, the degree of separation between the upper and lower sideband components is not important. It is therefore possible to start at higher crystal oscillator frequencies. In fact, if multifrequency operation is not required, the modulator can be operated at the final carrier frequency, and frequency translation is not necessary.

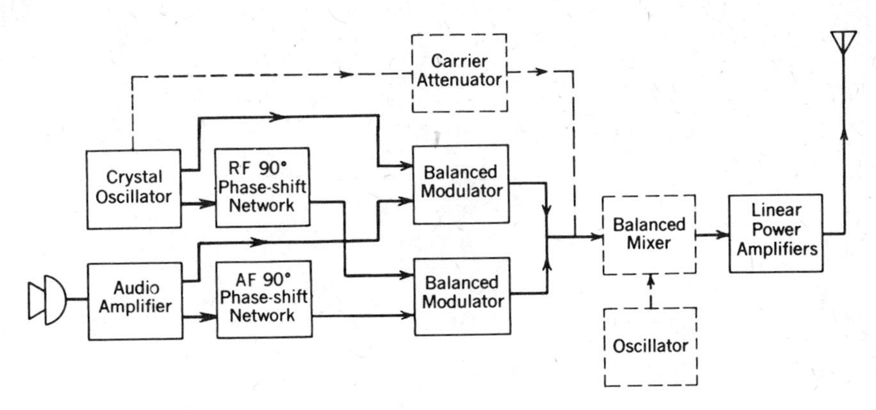

Figure 12-13 SSB phase-shift transmitter.

A block diagram of a phase-shift SSB transmitter is shown in Fig. 12–13. Notice that two balanced modulators are used. The carrier and modulating signals are fed to both modulators. In each case the carrier is suppressed, and the output contains only the upper and lower sidebands. These outputs are then fed to a common network. However, because of the 90° phase shifts in the input signals applied to these stages, one of their sidebands will be in phase-opposition and will cancel. The other sideband is additive, producing a single-sideband output. If this signal is at the desired final transmitting frequency, it is fed directly to the linear amplifiers. Conversely, if a low-frequency crystal oscillator is used in the first block, frequency translation (balanced mixer and oscillator*) is now necessary to raise the SSB signal to the desired final channel frequency. Then it is fed to the linear amplifiers to obtain the desired power output. As in the previous block diagram, a carrier attenuator can be used to provide a pilot carrier.

The action of the phase-shift circuits in causing cancellation of one sideband can be seen from the phasor diagrams in Fig. 12–14. Diagram (a) shows the phase relations that exist between the carrier and its side frequencies *at some particular instant* at balanced modulator A.† Compared to the carrier frequency f_A, the upper side frequency, f_u, is rotating counterclockwise, and the lower side frequency, f_l, clockwise. Simultaneously, the carrier fed to balanced modulator B is shifted 90° with respect to carrier A. It is shown as f_B in Fig. 12–14(b). In addition, because of the 90° shift in the audio components, its side frequencies are shifted 90° behind the 30° positions *they would otherwise have occupied with*

* For multifrequency operation, this second oscillator could be a variable-frequency oscillator. For high stability, its frequency would be obtained from a precision master oscillator with frequency synthesizer.

†This is the same as would be obtained in any modulator. (See Fig. 6–4.)

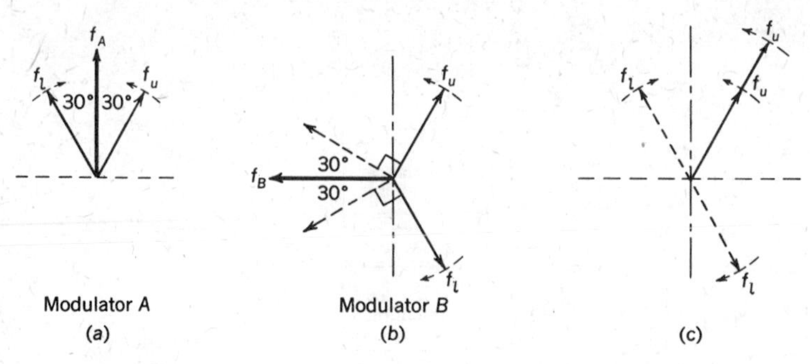

Figure 12-14 Cancellation of the lower side-frequency components.

respect to f_B at this instant. This effect is also shown in Fig. 12–14(b). Notice that the upper side-frequency components of each modulator are in phase, while the lower side-frequency components are in phase-opposition. This is more clearly seen in Fig. 12–14(c), which represents the combined output. (The carriers are not shown, since each carrier is suppressed in the balanced modulator.)

At any instant later, as the side frequencies rotate with respect to their carriers, the same effect will still prevail. This is so because the side-frequency phasors rotate in the same direction (and at the same speeds) and will maintain their relative phasing. If it is desired to transmit the lower side frequency and suppress the upper, all that is necessary is to reverse the phasing of the carriers so that f_A leads f_B by 90°.

In the above analysis, we have been speaking of side frequencies, obviously implying that the modulating signal is a single frequency. However, in a practical situation, the modulation (speech) is a complex wave producing side*bands.* In order to suppress one sideband, all the audio frequencies must be shifted by exactly 90°, and their relative amplitudes must remain unchanged. Any imbalance in the amplitudes fed to the modulators, or any deviation from the quadrature relation between the signals (carriers and modulating signals), will impair the sideband suppression. Because of the exacting component tolerances required, it is difficult to match the degree of suppression obtainable with filters. The problem is further complicated by the difficulty of designing the audio network to maintain a constant (90°) phase shift over a relatively wide band of frequencies (100–3000 Hz).* Quite often such networks are designed so that one output will lead the input by 45°, and the second output will lag the input by 45°. These networks may be passive (*R-L-C*) networks, or they may include tubes or transistors. The latter, active networks, generally have a wider bandwidth.

*D. E. Norgard, "The Phase-Shift Method of Single-Sideband Generation," *Proceedings IRE,* **44** (December, 1956). *See also* R. B. Dome, "Wideband Phase Shift Networks," *Electronics,* **19** (December, 1946).

Controlled Carrier Transmission

Another technique used to maintain proper sideband-to-carrier frequency relations is known as *controlled carrier transmission.* The carrier is transmitted at close to full amplitude (3–6 dB down), but only in bursts. The carrier bursts coincide with the lulls or pauses between modulation levels so that the average power output from the transmitter remains essentially constant. These carrier bursts are then used to activate the receiver AFC and AGC circuits.

SSB RECEIVERS

Receivers used for SSB service are in many respects identical to any other communications receivers. For example, the basic circuit is the superheterodyne; RF amplifiers are generally used to improve the signal-to-noise ratio, to reduce the possibility of oscillator radiation, and to improve image-frequency rejection; dual conversion (and even triple conversion) is used with a high first IF value to improve image rejection, and a low second IF value to improve adjacent-channel rejection; and finally, the audio systems are alike. However, the demodulation processes must differ. Conventional AM receivers use diode detectors, whereas with SSB receivers, since the RF signal does not have a carrier, the simple diode detector is no longer adequate. A locally generated carrier must be supplied before demodulation can take place. The simplest solution is to use a beat frequency oscillator as is done in the reception of CW (A1) transmissions. (Such a circuit was shown in Fig. 7–16.) Obviously, then, any communications receiver with BFO circuitry could be used for reception of SSB signals. The *Radio Amateur's Handbook** gives the following technique:

> When such a signal (SSB) is encountered, it should first be peaked with the main tuning dial. (This centers the signal in the IF pass band.) After this operation, do not touch the main tuning dial. Then set the RF gain control at a very low level and switch off AVC. Increase the audio volume control to maximum, and bring up the RF gain control until the signal can be heard weakly. Switch on the beat oscillator and carefully adjust its frequency until proper speech is heard. If there is a slight amount of carrier present, it is only necessary to zero-beat the beat oscillator with this weak carrier. It will be noticed that with incorrect tuning of an SSB signal, the speech will sound high- or low-pitched or even inverted (garbled) . . . . The use of minimum RF gain and maximum audio gain will insure that no distortion (overload) occurs in the receiver. It may require a readjustment of your tuning habits to tune the receiver slowly enough during the first few trials.

* *Radio Amateur's Handbook, 37th Ed.,* "Receiving Suppressed-Carrier Signals," p. 321.

It may also be added that because of frequency drift of the oscillators in the receiver's frequency conversion stages or of the beat frequency oscillators (or of the transmitter signal), it may be necessary to "ride" the main dial or the BFO tuning control, in order to preserve speech intelligence.

SSB Demodulators

When a diode detector with BFO is primarily intended for use with SSB signals, an amplifier is often inserted between the output of the beat oscillator and the detector. Therefore, the oscillator level* will be appreciably stronger than the peak RF signal level, and drastic reduction of the RF gain is not necessary. The resulting high carrier-to-sideband level is equivalent to a low percent modulation, and it minimizes the distortion level in the detector. Carrier-to-sideband ratios in the order of 10:1 are recommended.

Although the above diode-detector–BFO combination will produce the desired audio output, receivers designed for SSB service generally employ some type of *product detector*. Ideally, the output developed by such circuits is proportional to the product of the reinserted carrier and the SSB signal. For example, if two audio tones of amplitude E_s and frequencies f_a and f_b are used to modulate a carrier E_c, the SSB output will be

$$e_{\text{SSB}} = E_s \cos(\omega_c + \omega_a)t + E_s \cos(\omega_c + \omega_b)t \tag{12-1}$$

Then, if the local carrier is $e_c = E_c \cos \omega_c t$ and these signals are applied to a product detector, the output *current* will contain components corresponding to the product of the input signals or

$$i_s \propto E_c \cos \omega_c t[E_s \cos(\omega_c + \omega_a)t + E_s \cos(\omega_c + \omega_b)t]$$

This expands to

$$i_s \propto \frac{E_s E_c}{2} \cos \omega_a t + \frac{E_s E_c}{2} \cos \omega_b t + \frac{E_s E_c}{2} \cos(2\omega_c + \omega_a)t + \frac{E_s E_c}{2} \cos(2\omega_c + \omega_b)t \tag{12-2}$$

If the product detector has a linear transfer characteristic, there are no distortion terms. By using a suitable output network, an output voltage is developed across the load by the audio components of current (the first two terms), while the RF components are bypassed to ground. An advantage of the product detector is very low intermodulation distortion, even with a low value of inserted carrier voltage.

*A ratio of at least 10:1 (oscillator to RF signal level) is recommended for minimum distortion.

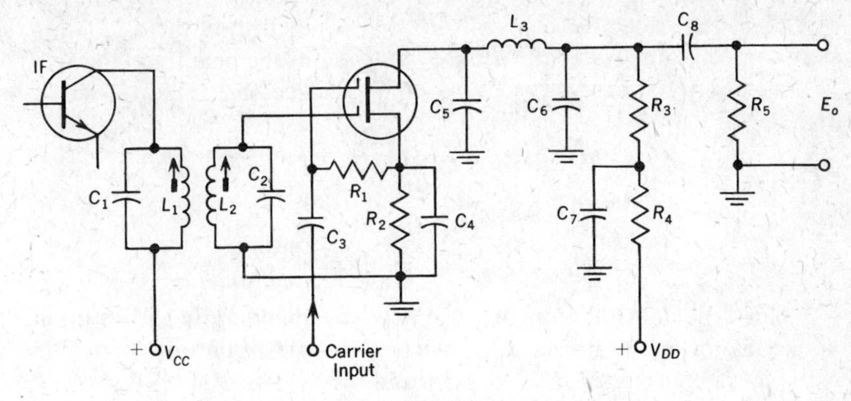

Figure 12-15 Product detector using dual-gate MOSFET.

Figure 12–15 shows a dual-gate MOSFET used as a product detector. The SSB signal (the IF) is fed through the last IF transformer to gate 1, and the carrier (from a local oscillator circuit) is R-C coupled to gate 2. The drain current will therefore contain the product terms as shown in Eq. (12–2). The RF choke, L_3, together with capacitors C_5 and C_6 form a low-pass filter to block RF from the output. The audio voltage is developed across R_3 and coupled to R_5 through C_8.

Since modulation and demodulation are essentially similar processes, it is not surprising to find demodulator circuits similar to the balanced modulators of Figs. 12–6 and 12–7. Figure 12–16 shows a balanced demodulator using JFETs. The SSB signal is inductively coupled from the IF amplifier. Notice that it is converted to a push-pull signal. The carrier input is fed in phase to both gates, so that carrier output is canceled in the output transformer T_1. Adjustment of R_1 will improve the balancing action. Capacitors C_6 and C_7 will filter out any other RF components. The audio output is developed across T_1.

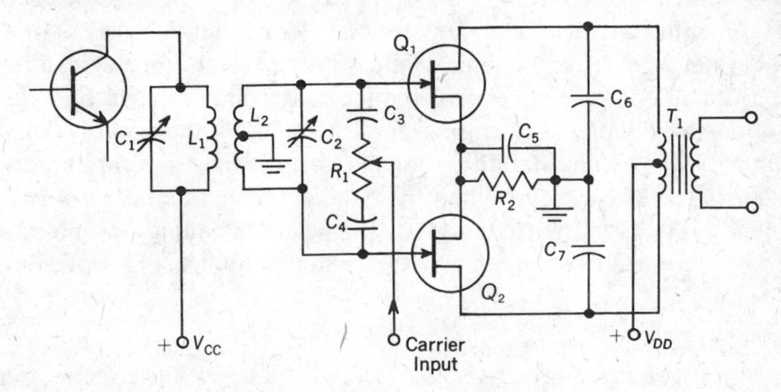

Figure 12-16 A balanced demodulator.

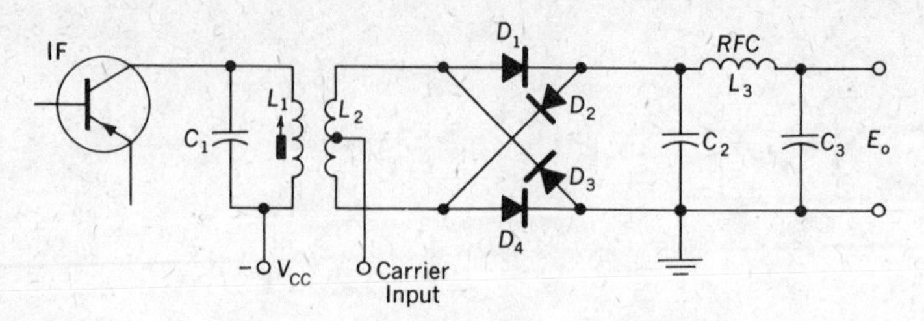

Figure 12-17 A ring-type demodulator.

Ring-type demodulators are also used for obtaining an audio output. The modulator circuit in Fig. 12–7 can be used for this purpose merely by reversing transformers L_1L_2 and L_5L_6, and then feeding the SSB signal to the L_6 input. A variation of this circuit is shown in Fig. 12–17. Notice that both the carrier input signal and the audio output signal are single-ended. Capacitor C_2 should have a very low reactance at the carrier frequency so as to equalize the signal fed to D_1D_2 and D_3 and D_4.

Again showing the similarity between modulation and demodulation, ICs are used as product detectors too. The IC is the familiar basic differential amplifier unit of Fig. 3–14. The IF signal is fed to the base of transistor Q_1; the carrier is applied to the base of Q_3; and the audio output is obtained from the collector of Q_2 through a low-pass filter to remove any RF components. Product detectors using ICs have higher gain and better carrier isolation than discrete-component circuits.

Amateur and General-Purpose Receivers

SSB receivers designed for general monitoring purposes or for amateur use are patterned along the lines of conventional AM communications receivers, but they incorporate one or more modifications to enhance performance. One such change is the reduced bandwidth of the tuned circuits—particularly in the IF system. For an SSB voice channel, 3 kHz is sufficient, whereas conventional AM requires 6 kHz. In addition, crystal or electromechanical filters are used for steep skirt attenuation. Both of these features will reduce noise and improve rejection of the opposite sideband.* Another improvement is the use of oscillator circuitry with better stability. For fixed-frequency operation, crystal oscillators are often used. When the oscillator must be tuned over a full frequency band (and particularly with single-dial tuning), a fine-tune or "clarifier" control may be added to assist in compensating for frequency drift.

* Some stations will transmit suppressed-carrier double-sideband signals—with *separate* intelligence channels on each side. This can cause strong interference if the receiver selectivity is inadequate.

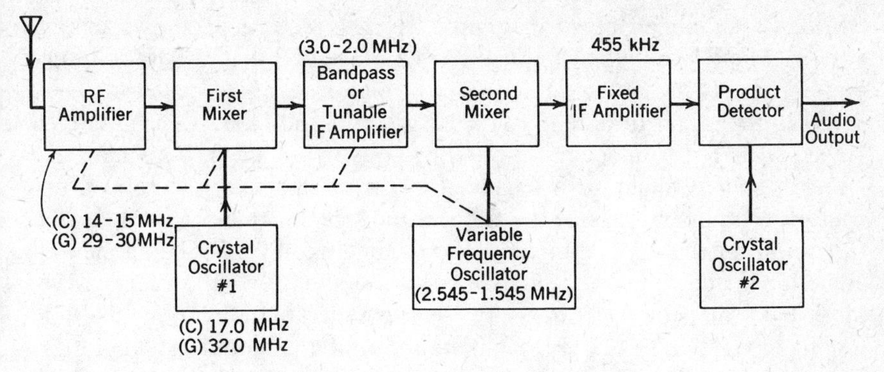

Figure 12-18 An SSB amateur-band receiver.

One technique used to improve stability is shown in Fig. 12–18. Notice that the first conversion uses a crystal oscillator. Obviously, the mixer output will vary in frequency, depending on the RF frequency selected by the RF and mixer stages. For example, when the RF is tuned to 14.5 MHz, the IF output is 2.5 MHz. For full coverage on this tuning band (14–15 MHz), the IF will vary 3.0–2.0 MHz. Consequently, the first IF section must either have a bandpass of 2.0–3.0 MHz, or it must be tunable over this frequency range. When the RF tuning band is switched, a new crystal is inserted, so as to maintain the IF band at the same 2.0–3.0 MHz value. Now notice that a variable-frequency oscillator is used in the second conversion. Its frequency is correlated with the first IF value so that the output of the second mixer will be a fixed frequency (455 kHz). The major gain and selectivity in the receiver is a function of the following IF amplifier. This system has several advantages. At the higher frequency where oscillator stability is more critical, a crystal oscillator is used.* Since the VFO operates at a lower frequency —and tunes the same low-frequency range for all bands—adequate stability is more easily maintained. However, such a receiver requires manual control to prevent excessive drift and to maintain the proper volume level.

Pilot-Carrier Receivers

For commercial point-to-point transmissions, particularly with automatic operation, higher stability is required. Most of these transmissions use a pilot carrier and then employ elaborate circuitry in the receiver to obtain automatic frequency and gain control from the pilot

*By using several fixed crystal frequencies, the full 3–30 MHz range can be covered.

carrier. A simplified block diagram of such a receiver is shown in Fig. 12-19. The RF amplifier tuning covers a range of 3.0–27.5 MHz (in several bands). Its output, together with the high-frequency oscillator signal, is fed to the mixer to obtain a first IF value of 1600 kHz. The amplified IF signal is fed to a second frequency-conversion stage. This may use either a simple mixer or a balanced mixer to improve rejection of unwanted components. Since the transmitted signal includes a pilot carrier, the output of this second mixer will contain a weak carrier component at 100 kHz in addition to the sideband component (100.1–103 kHz). Crystal or electromechanical filters are used to separate the two signals. The intelligence is fed to the IF amplifier, and to the demodulator where it is reduced to audio. The carrier signal is filtered, amplified, and limited. The "reconditioned" carrier is then used to provide AGC, AFC, and demodulation. A high-stability 100-kHz crystal oscillator is used as a reference oscillator. The reconditioned carrier is compared with this reference in the AFC comparator circuit. Any deviation produces a control voltage for the AFC motor, which, in turn, varies a capacitor in the oscillator circuit to correct the error. The AFC action normally holds the carrier to within 1 Hz of the 100-kHz reference oscillator frequency. Notice that the AGC circuit has two outputs. These represent a short time-constant output (variable from 0.1 to 2 s) and a main AGC output with a time constant of about 8 s. By incorporating the additional fast-

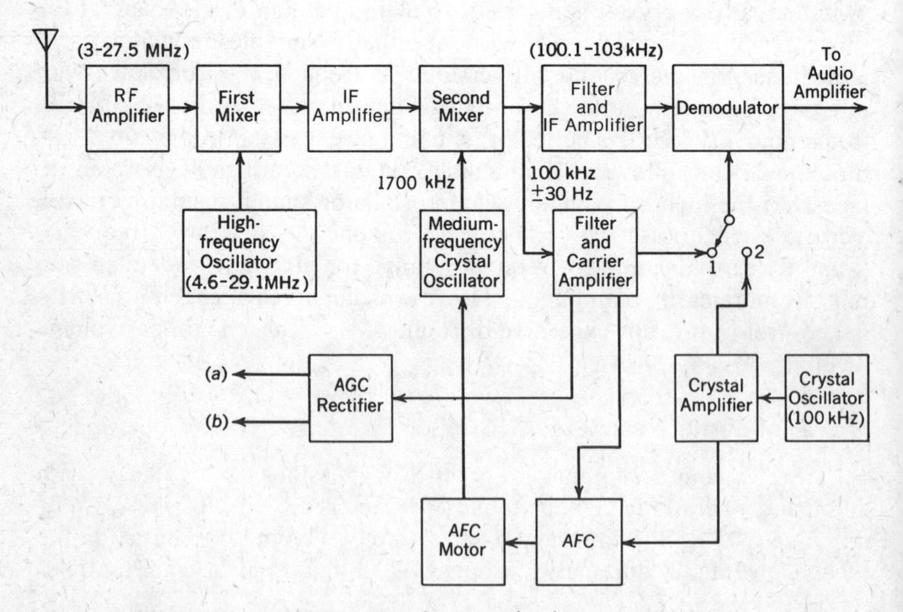

Figure 12-19 An SSB receiver for use with pilot-carrier transmissions.

acting control, it is often possible to correct for flutter variations of the input signal that could be caused by fading or interference. AGC is applied to the RF, IF, and carrier-amplifier stages.

Suppressed-Carrier Receivers

The block diagram (Fig. 12–19) for a pilot-carrier type of receiver seems rather simple. Only three blocks are shown for the AFC circuitry. However, the actual circuitry is quite complex and requires many transistors or tubes. Even so—no matter how complex the circuitry—if selective fading causes the pilot carrier to fall below some minimum amplitude value, AFC action will be lost, and performance can be seriously affected. In addition, it is possible for an interfering carrier—or a deliberate jamming signal—to "capture" the AFC action and introduce large frequency errors that destroy intelligence. For military use, this is a serious weakness. Consequently, suppressed-carrier transmissions (SSSC) are used by the military.

When communication is to be limited to only a few stations, the receiver will use separate (precision) crystals for each channel. Such a receiver block diagram is shown in Fig. 12–20. The RF stage is a wide-band amplifier accepting all signals in the 3–27 MHz band. The fixed-tuned trap ahead of it is needed to prevent direct pickup of interfering signals by the IF system. For a six-channel receiver, the *channel crystal oscillator* block has six crystals (selectable by switching) so as to produce an IF of 1600 kHz for each of the desired RF channels. This receiver is capable of processing either the lower or upper sideband of each channel. Selection is done by switching the frequency of the *sideband-selection crystal oscillator* to either 1143 or 2057 kHz. In either case, the second IF carrier value is 457 kHz. The missing carrier is reinserted by the *carrier crystal oscillator.* The rest of the block diagram needs no further discussion.

The six-channel receiver (Fig. 12–20) requires a total of nine crystals, six of which are in the channel crystal oscillator block. It should be obvious that, if full coverage of any SSB station in the 3–27 MHz band is desired, this separate-crystal-per-channel technique would be too cumbersome. This led to the development of *frequency synthesizers.* At first, these were quite bulky and costly and were used only in military applications. Improvements—in particular, integrated circuitry—have reduced their size and cost, making them practical for commercial applications. Modern frequency synthesizers, using digital techniques, are capable of producing any desired frequency (within a given range) in steps as close as 100 Hz. Although the design of these units is complex, selection of a desired channel requires only the setting of a series of decade switches

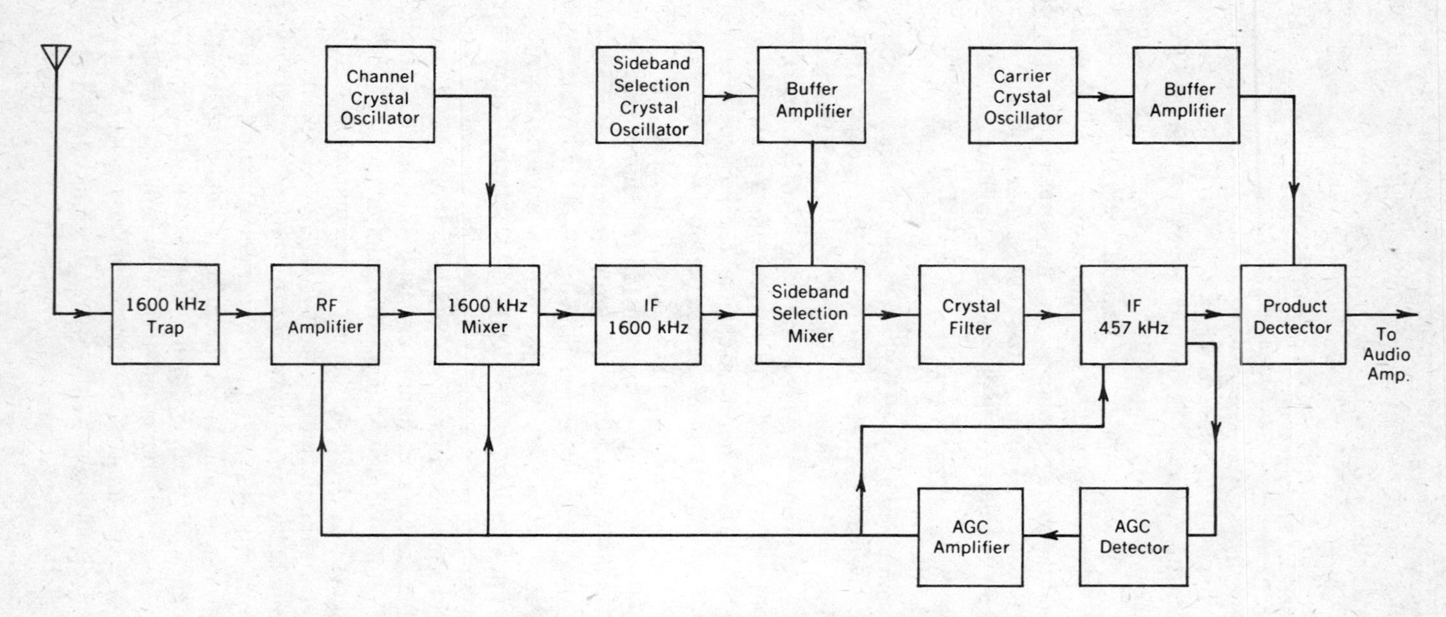

Figure 12-20 A six-channel crystal-controlled SSB receiver. (Courtesy Communications Associates, Inc.)

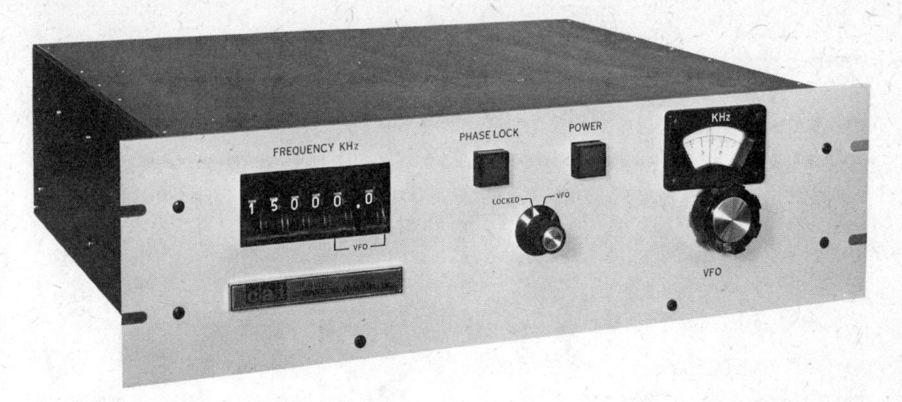

Figure 12-21 A frequency synthesizer. (Courtesy Communications Associates, Inc.)

to the desired frequency. The frequency selected is read directly on the windows or dials of the register. A frequency synthesizer is shown in Fig. 12–21.

When a frequency synthesizer is used in an SSSC receiver, it takes the place of the three separate crystal oscillator blocks of Fig. 12–20. A typical receiver with synthesizer is shown in block-diagram form in Fig. 12–22. (Notice that, except for the choice of second IF value, this diagram is essentially the same as the previous receiver diagram.) Now, however, since the oscillator input to the first mixer is in steps of 100 Hz, it is obvious that associated transmitters cannot operate at any frequency, but must be channelled at even 100-Hz values. This should present no problem. Another variation is in the AGC circuit takeoff. Since there is no carrier, the AGC voltage is generally developed from the audio output.* However, the time constant of such AGC circuits is very important.

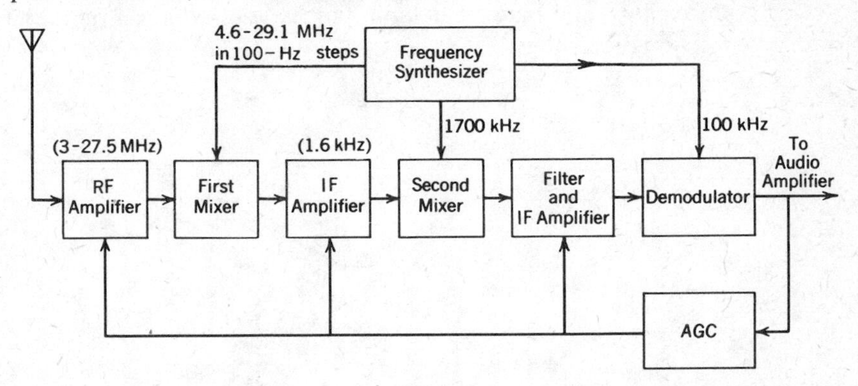

Figure 12-22 An SSB receiver using frequency synthesizer for SSSC transmissions.

*Some receivers obtain this voltage from the modulation envelope at the IF output. Special care must then be taken to isolate the AGC system from the reinserted carrier, since this voltage is at (almost) the same frequency and is much stronger in amplitude.

Elimination of the carrier makes an SSB voice signal intermittent or pulse-like, corresponding to syllables, words, and phrases. The AGC circuit must have a very short "attack" or rise time and the AGC voltage must build up rapidly, to avoid a "thump" or "popping" at the start of each word or syllable. Conversely, to avoid high gain and high background noise between syllables or words, the AGC voltage should maintain at a level corresponding to the average, for several syllables. Such an AGC circuit is said to have a fast-attack, slow-release characteristic.

PULSE MODULATION

Heretofore, we have transmitted the intelligence signal by raising it (translation) to some high radio frequency and transmitting it in its full form. However, it has been found that the signal to be transmitted can be sampled at discrete intervals, and the resulting pulses used to modulate the carrier. If the sampling rate (pulses per second) is at least twice the highest frequency component in the intelligence signal, the original signal can be reconstructed with adequate fidelity.

When pulse modulation is used, energy is transmitted only during the pulse sampling intervals. Therefore, the transmitter duty cycle is quite low. For example, in voice communication, assuming the highest speech frequency is 4000 Hz, a suitable sampling rate would be 8000 pulses per second (2.0×4000). The timing for each pulse is 1/PRR, or one every 125 μs. Since the pulse width is in the order of 1 μs, it is obvious that the transmitter is off more than it is on. This allows either for "sandwiching" several other intelligence channels onto the one carrier (multiplexing), or for very high peak-signal power with only a low average-power requirement. The latter results in a high signal-to-noise ratio and a greater transmission range.

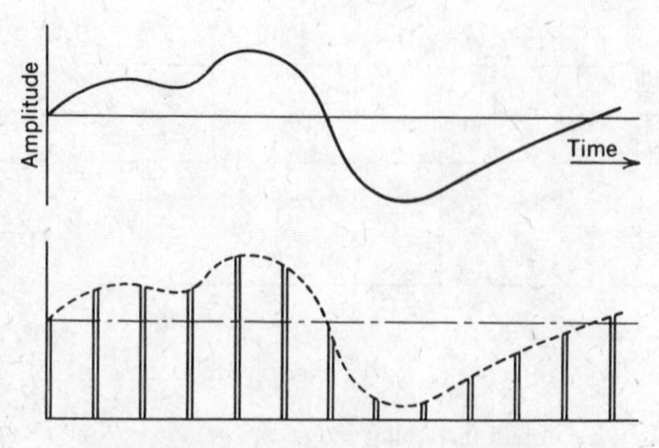

Figure 12-23 Pulse amplitude modulation.

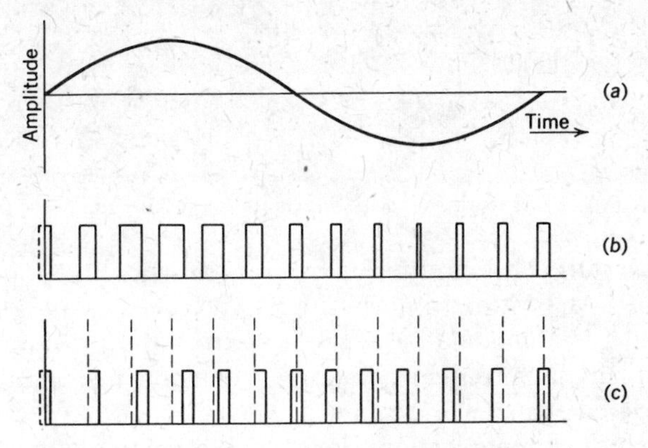

Figure 12-24 Pulse-duration and pulse-position modulation.

Types of Pulse Modulation

A number of techniques have been used to convert the sampled amplitude of the intelligence signal into suitable pulse form. The more important systems are discussed below.

PULSE-AMPLITUDE MODULATION (PAM). The simplest technique is to let the amplitude of the pulse represent the amplitude of the intelligence at that instant. This is shown in Fig. 12–23. Notice that the minimum pulse amplitude corresponds to the negative maximum of the intelligence signal.

PULSE-DURATION MODULATION (PDM).* In this technique, the amplitude of the pulses is kept constant. Instead, the amplitude of the intelligence signal (at the instant of sampling) determines the width of the pulse. This is shown in Fig. 12–24(b). Notice that the pulse width increases to a maximum, corresponding to the positive peak of the intelligence signal, and it decreases to a minimum width when the intelligence signal is at its most negative value.

PULSE-POSITION MODULATION (PPM). This is shown in Fig. 12–24(c). The dashed vertical lines correspond to the center position of the pulse when the intelligence amplitude is zero. An increase in intelligence amplitude (positive) causes the pulse to be displaced to the right (lag), while negative amplitudes displace the pulse to the left (lead). In either case, the amount of displacement is proportional to the intelligence amplitude.

Any of the above pulse signals can then be used to amplitude-modulate, frequency-modulate, or phase-modulate the RF carrier. The

*Also known as pulse-width modulation (PWM).

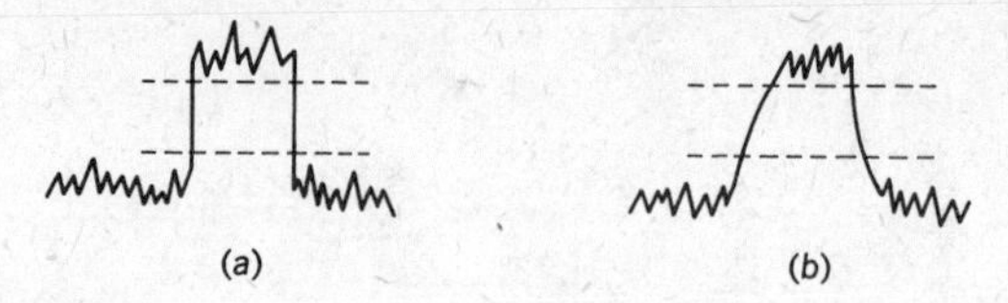

Figure 12-25 Noise suppression with pulse-position modulation.

final transmission can therefore be one of many possible forms, such as PAM-AM, PAM-FM, and PDM-AM.

Since the final modulating signal is a series of short-duration pulses, it contains many high-frequency harmonic components, and the resulting (RF) modulated wave will occupy a wide frequency spectrum. (Consequently, pulse communications are generally restricted to ultrahigh frequencies.) In this respect, PAM is least demanding. Since intelligence is a function of pulse amplitude, distortion of the pulse *shape* due to limited bandwidth will not impair reconstruction of the original intelligence. Unfortunately, PAM—as with any AM signal—can be seriously affected by noise interference. On the other hand, with PPM or PDM, the amplitude of the pulses does not convey any intelligence. Background noise (noise between pulses or "grass" that obscures the lower portion of the pulses), as well as noise that rides on top of the pulses, can be removed by slicing (clipper-limiter action), and retaining only the center portion of the pulse. This is shown in Fig. 12–25(a). When the center portion is further amplified, the reconstituted pulse will show no trace of noise. However, such excellent noise immunity will be obtained only if the transmitter and receiver bandwidths are wide enough to produce the steep pulse rise shown in Fig. 12–25(a).* If the system bandwidth is limited, the pulse shape is distorted as shown in diagram (b). Notice that slicing will now alter the pulse position and pulse width. Consequently, noise can affect the output signal to some degree.

Pulse-Code Modulation (PCM)

In the above systems, a *change* in the pulse (amplitude, width, or position) was used to convey the amount of change in the amplitude of the intelligence. As we just saw, noise or other interference can alter the pulse and produce some error in the reconstructed signal amplitude. In addition, any nonlinearity in the transmitter modulators and amplifiers, or in the receiver translators and amplifiers, will also produce errors. In voice communications, such errors may not be too serious. However, when transmitting coded information, such as that for remote control of a

* The bandwidth required is given by

$$\text{BW} \cong \frac{2}{\text{pulse width}} \tag{12–3}$$

Table 12–1
PULSE-CODE MODULATION (BINARY CODE)

Binary Code	Equivalent Amplitude	Waveform	Binary Code	Equivalent Amplitude	Waveform
0000	0		1000	8	
0001	1		1001	9	
0010	2		1010	10	
0011	3		1011	11	
0100	4		1100	12	
0101	5		1101	13	
0110	6		1110	14	
0111	7		1111	15	

missile, or telemetered data from an orbiting satellite, more accuracy or better noise immunity is desired. Reliability can be greatly increased by converting the analog intelligence or data into digital form and then transmitting in digital units. This system of encoding is known as *pulse-code modulation.* The intelligence signal is sampled as before, but the amplitude of each sample is converted into a digital pulse train.

Because of the convenience and accuracy of bistable multivibrators, pulse-code modulation uses a binary digital code.* The peak-to-peak value of the quantity (or signal) to be transmitted is divided into a number of discrete steps or values. Each step is assigned a digital code number. The number of digits or *bits* used in the binary code determines the total number of steps available and the accuracy of the transmitted values. For example, a 4-bit binary system provides 15 steps, as shown in Table 12–1. If the quantity to be transmitted had a total swing of 30 V, each amplitude step would correspond to 2.0 V. Now assume that at three sampled instants the measured quantities were 4.0, 8.5, and 13.4 V, respectively. These quantities could be transmitted only as discrete binary code steps of 2, 4, and 7 (corresponding to 4.0, 8.0, and 14.0 V), because of the code limitation of Table 12–1. This process is called *quantizing.*

Notice from Table 12–1 that all pulses (bits) have identical amplitudes and shapes. Actually, neither their amplitude nor shape has any significance. For example, an analog amplitude of 7 is shown by a digital

*J. J. DeFrance, *General Electronic Circuits,* Chaps. 19 and 21.

pulse train of 0 1 1 1, while an analog 11 has a pulse train of 1 0 1 1. The only important aspect is the presence or absence of each bit. Any noise picked up in transmission will ride on top of the pulse and/or produce grass between the pulses. But as long as a pulse can be distinguished from the noise, there is no problem. Clean, perfectly shaped pulses can be reconstructed by another multivibrator at the receiving end (or at in-between repeater points if the overall distance is too great). This gives the PCM system its excellent reliability.

Unfortunately, the quantizing process itself creates another form of noise. Notice in the above analog-to-digital conversion using a 4-bit code that the measured sample values of 8.5 V and 13.4 V would be converted to binary 4 and 7, respectively, which at the receiving end would be decoded as 8.0 V and 14.0 V. Obviously, some error is introduced in the quantizing process. This discrepancy is called *quantizing noise* and can be reduced (if the required accuracy warrants it) by using a binary code with more digits. A 5-bit code would allow 32 steps (2^5), while a 7-bit code would result in 128 steps, or better than 1.0% accuracy.

Now we run into another problem: the more bits in the code group, the greater the bandwidth required to transmit these signals. As a compromise between quantizing noise and bandwidth requirements, commercial systems generally use a 6-bit code.

Both quantizing noise and bandwidth can be reduced by using more sophisticated systems. The drawback, of course, is the need for more-complex and more-costly equipment. One technique takes advantage of the fact that most of the speech information is concentrated at low amplitudes, and also that, with equal-increment steps, the percent error is greatest at the lower amplitude levels. The quantizing error can therefore be appreciably reduced by using quantizing steps with logarithmic increments. Another noise reducing technique is to first compress the signal amplitudes at the transmitting location and then expand them back to normal at the receiving end. This technique is called *companding*. By reducing the maximum swing of the voice samples, each digital step can have a smaller incremental value.

Further improvement is obtained by using *adaptive data systems*. Such systems employ redundancy reduction techniques. Any sampled data which would not add significant value to the received information is considered redundant and need not be transmitted. Therefore, instead of taking samples at every fixed (and equal) time interval point, the transmission system scans the intelligence and then selects and transmits only those points in the data waveform that cannot be predicted at the receiver within a given percentage of error. By this technique, it has been found possible to reduce the number of sampled points by at least a 5 to 1 ratio—and in some cases by as much as 160 to 1. By reducing the number of samples taken, the pulses per second of modulation is reduced, and the bandwidth required is reduced. Obviously, as part of a nonredundant transmission, it is necessary to include a *time tag* to tell the receiver how many sample-intervals have elapsed between any two actual samples selected.

MULTIPLEX TRANSMISSIONS

If the highest useful voice frequency is considered as 4000 Hz, the minimum sampling rate should be twice this value or 8000 samples per second. With a 6-bit code, the result is a bit rate of 6×8000 or 48 kb/s (kilobits per second). Since each bit is ideally a rectangular pulse with zero rise time, the bandwidth required to transmit these pulses should be theoretically infinite. Practically, a bandwidth of ten times the bit rate, or 480 kHz, would give satisfactory waveshapes in SSB (or 960 kHz bandwidth for double-sideband transmission). Had we transmitted the voice signal directly, without encoding, the bandwidth requirement would have been only 4 kHz for SSB, or 8 kHz for DSB. Obviously, pulse modulation systems are very wasteful. We have obtained good reliability and noise immunity at a sacrifice of increased bandwidth.

It has been found that more efficient operation can be obtained if a station transmits more than one message on the same carrier. This process (the simultaneous transmission of several messages over the same "pair of wires") is known as *multiplexing* and has been used for many years in *carrier telephony.* The technique should be fairly obvious. Each conversation is raised (translated) to a different band of frequencies. They can then be transmitted over a common line. At the receiving station, each message is recovered by using band-pass filters and then demodulated to its original audio-frequency spectrum. Similar techniques have also been used with wireless (radio) transmissions. Multiplex transmissions have been used in commercial communications, not only for voice channels, but also for teletype and facsimile. SSB transmissions gave further impetus to multiplexing. (Additional messages could now be transmitted on the other sideband.) Multiplex has also been used for many years by broadcast FM transmitters. Many stations get additional revenue by storecasting,* in addition to broadcasting. More recently, the color added to TV transmissions and the stereo added to FM transmissions are further examples of multiplex operations. Finally, remote data transmissions (telemetry) would not be practicable were it not for multiplexing.

* Special transmissions such as music for restaurants or factories.

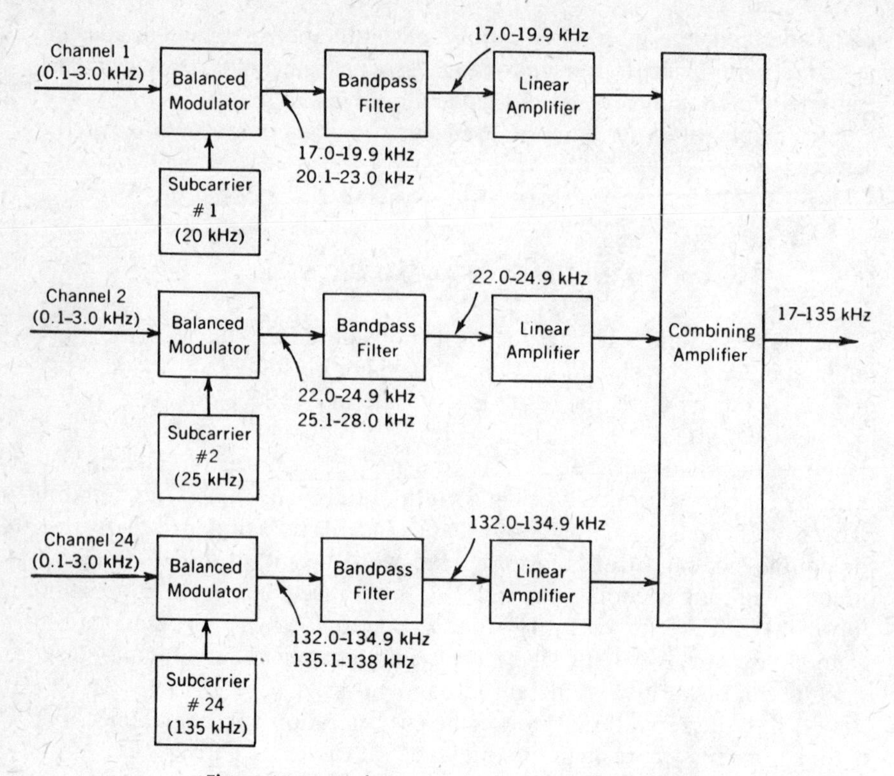

Figure 12-26 A frequency-division multiplex system.

Frequency-Division Multiplex

Two basic systems are used for multiplex transmission—*frequency-division* multiplex and *time-division* multiplex. A typical frequency-division system is shown in Fig. 12–26. Each message is fed to separate balanced modulators, together with a subcarrier signal. The double-sideband output is filtered, and the SSB signal is fed to respective linear amplifiers. Their outputs are combined to form a basic group-modulation band. One commercial system* uses 24 channels in the group, for a total modulation-signal bandwidth of 17–135 kHz. Nine more identical groups are formed, each with its own 17–135 kHz modulation-signal bandwidth. Then, the ten groups are again frequency-multiplexed into a master-group at its own subcarrier and bandwidth. By the same techniques, additional channels are combined into other sets of groups and master-groups, and

* RCA radio relay system, with a final output at 1800 MHz, frequency-modulated ±1.5 MHz.

finally several master-groups are multiplexed into one composite modulating signal. This, in turn, is used to modulate the main transmitter carrier. This final modulation process could be AM or FM.

In the above illustration, the subcarriers were amplitude-modulated (and then converted to SSB). However, in some applications, it may be easier (or more desirable) to frequency-modulate the subcarrier. For example, in telemetry, it may be necessary to measure strain, acceleration, pressure, or temperature, and the transducers used to convert these quantities to electrical equivalents may be variable-capacitance or variable-inductance gauges. Such transducers are ideally suited for frequency-modulating the subcarrier oscillator. Obviously, the subcarrier frequencies must be sufficiently far apart to allow for the frequency deviation (plus a guard band). As an illustration, one subcarrier could be at 100 kHz, and swing ±25 kHz, with the next subcarrier at 175 kHz, and so on. The FM subcarriers would then be used to amplitude- or frequency-modulate the main carrier.

Depending on the type of modulation used for the subcarrier and main carrier, frequency-division multiplex systems can be classified as AM-AM, AM-FM, FM-AM, or FM-FM.

Time-Division Multiplex

As the number of messages to be transmitted increases, the frequency-division technique presents problems. The number of subcarriers needed increases, and stability problems can arise. Additional circuitry is required (both at the transmitter and receiver) to handle each added channel. The bandwidth requirements increase directly with the number of channels. These problems are eliminated to a great extent by using *time-division* multiplex together with pulse modulation.

In time-division multiplex, each intelligence signal to be transmitted (voice or telemetry data) is sampled *sequentially,* and the resulting pulse code is used to modulate the transmitter. Since only one signal modulates the transmitter at any time, no added equipment and no increase in bandwidth is needed when multiplexing. The number of sequential channels that can be handled is limited by the time span required by any one channel pulse and the interval between samples.

A commonly used voice communication system uses PPM, with a 0.5-μs pulse and a pulse displacement of ±4.5 μs. The total time allocation per pulse is approximately 10 μs. A sampling rate of 10,000 pulses per second is used, so that the time between pulses (of any one channel) is 100 μs. Eight separate message channels are multiplexed within this 100-μs time span. This allows for the 10 μs for each pulse, an additional 2-μs *guard band* between channels, and up to 4 μs for a *synchronizing signal.* Each of the eight channel signals modulates the transmitter in sequential fashion, once every 100 μs. At the receiving station, the RF wave is demodulated, and a series of pulses is obtained, the pulses representing the synchronizing signal and the individual channel coded intelligence. This output is fed simultaneously to eight individual channel decoders and amplifiers. However, each channel is held in cutoff by a gating circuit. A timing circuit brings each channel out of cutoff for a 10–12 μs period in a sequence corresponding to the transmitter modulating sequence. The transmitted synchronizing signal locks the timing sequence in step, so that each channel comes out of cutoff only when the pulses for that channel are transmitted.

In telemetry, the rate of signal variation is much slower than in voice communications. Frequency components of such signals are generally below 100 Hz. Therefore, a sampling rate of 250 pulses per second would be more than adequate. The time between the coded signals for any one channel is therefore 1/250 or 4000 μs—or longer. Even allowing 40 μs for a PPM pulse or for a multidigit PCM train, this still makes possible the transmission of 100 separate types of data on the one transmitting carrier. Of course, a synchronizing signal must be used to keep transmitter and receiver in step.

FM STEREOPHONIC BROADCASTING

The aim in stereophonic reproduction is to give sounds a spatial dimension, or directivity. For example, at a concert hall, string instruments may be heard at one side of the orchestra, and the brass section on the other side. When an actor walks across a stage as he speaks, his voice can be heard to move with him across the stage. Stereophonic sound can be reproduced electronically by using two loudspeakers spaced at least 6 feet apart. Of course, the audio signals fed to the two speakers must differ in the same manner as the sound reaching the left and right ears of the audience "on location." This is readily done by using a "left" and a "right" microphone to pick up the original sounds. In monophonic reproduction, the outputs from the two (or more) microphones would be combined. In stereophonic reproduction, each pickup must be processed as a separate message channel all the way through to its own loudspeaker.

In stereophonic recordings (records and tapes), two separate sound tracks are recorded and two pickup heads * are used; the output from each

* In stereo records, the two sound tracks are cut into opposite walls of the same groove, at 45° angles. This makes possible the use of a pickup cartridge with only one stylus (needle)—but two transducer mechanisms. Because of the 90° space relationship of the sound tracks, this one needle will excite each transducer independently, producing "left" and "right" outputs, with negligible interaction.

head is fed to its own amplifier and loudspeaker. Stereophonic sound via radio also requires two separate audio transmissions. For a number of years, some studios achieved this effect by transmitting "right" microphone pickup on the station's AM facilities and "left" microphone on their FM equipment. Stereo reception could then be had by simultaneous use of an AM and an FM receiver.* Unfortunately, such a system impaired monophonic reception. Anyone listening to only the AM station (or only the FM station) during stereo broadcasts would hear only (approximately) half of the studio sound.

In 1961, the Federal Communications Commission authorized true stereophonic broadcasting for commercial FM broadcasts, using frequency-division multiplex transmission to provide the two separate audio signals from the one carrier. At first thought it would seem logical to use the "left" and "right" audio signals as the two separate channels. However, such a system would produce the same objection as the AM-FM receiver combination. A monophonic receiver would reproduce only half the broadcast program.

The approved FCC system not only produces high-fidelity stereophonic sound, but also provides for monophonic compatibility and storecasting.† In order for a standard (monophonic) receiver to develop a normal output, it is obvious that one of the audio channels must be the combined "left" and "right" (L + R) audio signal. Then to produce the stereo effect, the second channel is the difference between the two audio signals (L − R).

Stereo Transmitter Principles

A block diagram of the modulation section of an FM stereo transmitter is shown in Fig. 12-27. Two microphones placed 10–50 ft apart are used to pick up the program material. Their outputs are fed through pre-emphasis circuits to the *matrixing* network. Here the two signals are combined in an *adder* circuit to produce the (L + R) output. In addition, the "right" signal is also inverted and then combined with the "left" signal to produce the (L − R) output. Because of the different treatment of the two signals within the matrix network, the (L + R) output is fed through a delay network to maintain the proper phase and amplitude correlation between the (L + R) and (L − R) signals.

To provide for monophonic reception from a stereophonic transmission, the (L + R) audio signal is used—as is—to modulate the reactance stage (Crosby system) or balanced modulator (Armstrong system) of the FM exciter. The stereo information is contained in the (L − R)

*This will not produce a true stereo effect because of phase shifts between the two "channels" due to differences in the transmission paths and/or circuitry delay.

†FCC's Subsidiary Communications Authorization (SCA).

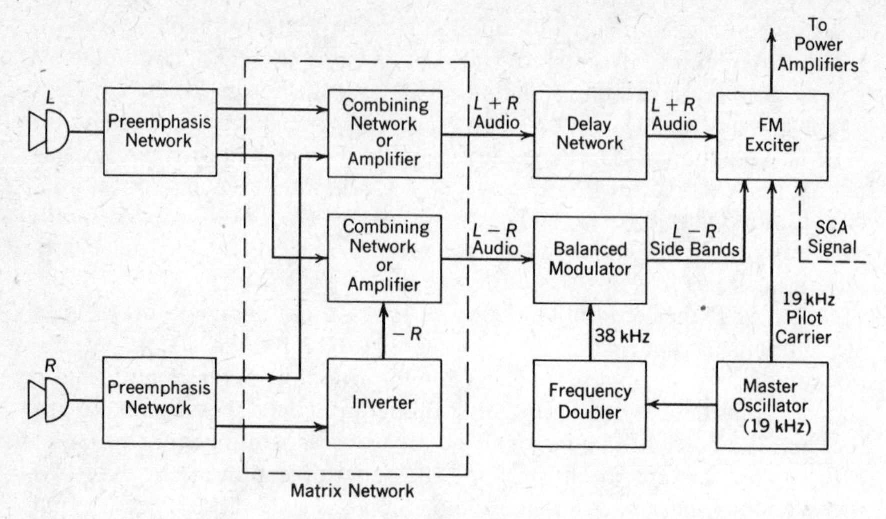

Figure 12-27 A stereo signal generator.

signal. However, this cannot be fed directly to the FM exciter because it would cancel the "right" audio signal. This interference problem is solved by translating the (L − R) intelligence to a spectrum location just above the (L + R) audio signal. For this purpose, a subcarrier signal—38 kHz—is used. The (L − R) and the subcarrier signals are mixed in a *balanced* modulator. The output is the (L − R) sidebands.* For a full-fidelity audio signal, the sideband spectrum extends 23–53 kHz. (See Fig. 12–28). This translated audio signal is also fed to the FM exciter.

Suppression of the 38-kHz subcarrier is very important for good monophonic reception of the stereophonic program. For noise-free reception, the (L + R) signal should fully modulate the FM exciter and produce the maximum allowable deviation (±75 kHz) when this audio signal is at its peak amplitude. This peak value will obviously occur when the (L − R) signal is at its minimum amplitude. By limiting the (L + R) channel to only 90% of maximum deviation, overmodulation cannot occur, and noise immunity is only slightly impaired (1 dB). Conversely, when the (L + R) signal is at minimum amplitude, the (L − R) is at maximum amplitude. Therefore the (L − R) signal can now produce 90% of maximum deviation. This process, wherein each modulating signal (L + R) and (L − R) can independently produce almost full deviation without overmodulation, is called *interleaving*. If the 38-kHz carrier were transmitted with its sidebands, each channel would be limited to only 50% of the full deviation.

* This is *amplitude* modulation of the 38 kHz subcarrier by the (L − R) audio signal to produce a double-sideband suppressed-carrier output.

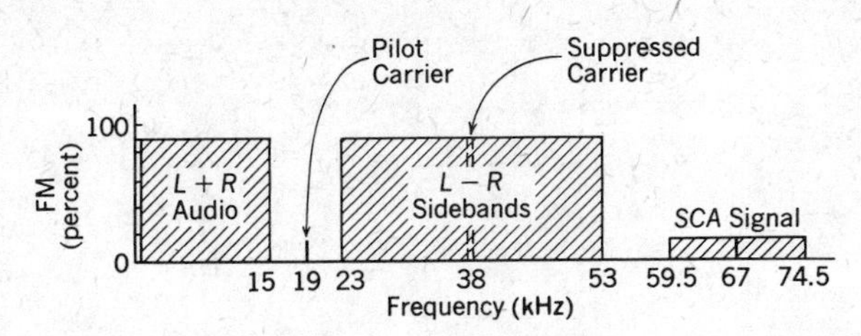

Figure 12-28 Composite modulating signal.

As in SSB operation, at some point in the *receiver,* the (L − R) intelligence must be recovered from the 23–53 kHz double-sideband spectrum. It is then necessary to reinsert the 38-kHz subcarrier, and this signal must be accurate—both in frequency and in phase. This accuracy is obtained by the transmission of a pilot carrier. However, if the 38-kHz subcarrier is transmitted, it would be extremely difficult to separate it from its sidebands at the receiver. Instead, the transmitted pilot signal is at half the subcarrier frequency, or 19 kHz. Notice that this signal is 4 kHz above the (L + R) highest audio frequency and 4 kHz below the lowest (L − R) sideband frequency. (See Fig. 12–28.) Separation at the receiver is relatively simple.

Returning to the block diagram (Fig. 12–27), notice that the 19-kHz signal is fed to the FM exciter, as the pilot carrier, as well as to the frequency doubler to produce the 38-kHz subcarrier. Notice also the SCA signal (shown by broken line) fed to the FM exciter. This is used when the transmitter is engaged in simultaneous storecast transmission. To prevent interference, the storecast program material must be raised to a frequency above the (L − R) spectrum. A second subcarrier is needed. The frequency chosen is 67 kHz. The sidebands produced by the storecast intelligence extend from 59.5 to 74.5 kHz. The total spectrum for the composite modulating signal is shown in Fig. 12–28.

Stereo Receiver Principles

The FM output from a stereo transmitter is in the 88–108 MHz range and has a deviation of ±75 kHz. Since this is the same as for monophonic transmissions, it is obvious that any standard broadcast FM receiver can also tune in a stereo program. The output from its discriminator (or ratio detector) will consist of the composite signal. The (L + R) audio component will produce the normal monophonic output. Meanwhile, the other demodulation components will be bypassed to ground or heavily attenuated by the standard audio circuitry. Furthermore, any signal that might come through would be above the normal hearing range and would not affect monophonic reception.

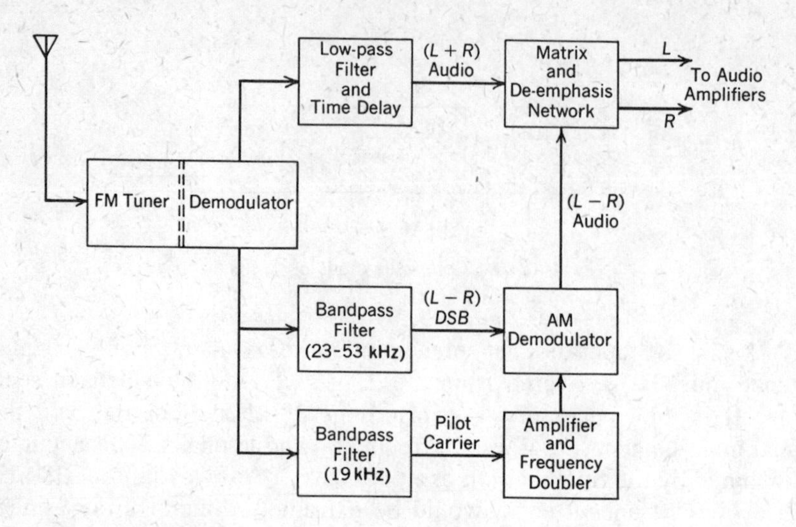

Figure 12-29 The multiplex section of a stereo receiver.

A receiver for stereophonic reception will be identical to the above monophonic receiver up to and including the FM demodulator. Now, additional circuitry must be added to process the (L − R) sidebands and produce separate "left" and "right" outputs. This is shown in block-diagram form in Fig. 12–29. One output from the discriminator (or ratio detector) is fed to a low-pass filter. The output is the (L + R) audio signal. The filter must cut off sharply above 15 kHz to prevent any 19-kHz pilot signal from reaching the matrix. A second output from the FM demodulator is fed to a band-pass filter (23–53 kHz). This isolates the (L − R) double-sideband signal. The 19-kHz pilot carrier is isolated by feeding the composite output from the FM demodulator through a narrow-band filter tuned to that frequency. The pilot carrier is amplified and then converted to 38 kHz by a frequency doubler. The double-sideband signal and the 38-kHz carrier are then fed to the AM demodulator, and the (L − R) audio intelligence is extracted. The (L + R) and the (L − R) signals are now combined in a matrix network to produce separate "left" and "right" signals. At the same time, each of these signals is de-emphasized to correct for the transmitter's pre-emphasis. Finally, the "left" and "right" outputs are fed to separate audio amplifiers and loudspeakers to produce a stereo output.

Receiver Circuits

As mentioned before, the circuits in a stereo receiver are, for the most part, identical to those found in any other FM receiver. Only two of the blocks in Fig. 12–29 need any further discussion—the AM demodulator and the matrix network. Again, many variations are possible.

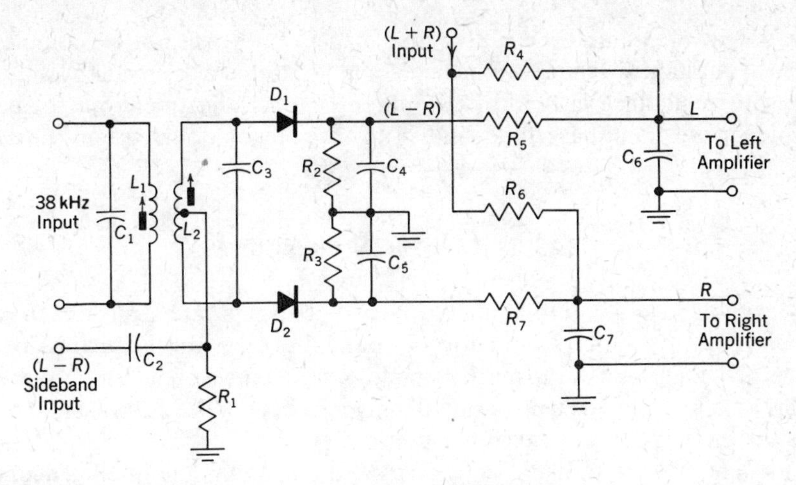

Figure 12-30 An AM-detector and matrix network.

For demodulation of the double-sideband signal, product detectors and synchronous detectors are common. The matrix network could use active elements (transistors) or merely passive (resistive) networks to combine the signals. Figure 12–30 shows a synchronous detector used with a passive matrix network. Since this detector produces both an (L – R) and a –(L – R) output, the simple resistive combining network is sufficient. Conduction of diodes D_1 and D_2 depends on the combined effect of the 38-kHz carrier and the (L – R) sideband signal. Because of nonlinear action, sum and difference frequencies appear in the rectified current waveshape. The output voltages across R_2 and R_3 are the demodulated audio signal (L – R). (Any other components are bypassed to ground.) However, since the diode connections are reversed, the two voltages are 180° out of phase. In other words, one is (L – R) and the other is –(L – R). These signals together with the (L + R) output of the low-pass filter are fed to the matrix and de-emphasis networks. The combination of an (L + R) and an (L – R) signal across C_6 causes cancellation of the R signal. This output will correspond to the "left" audio pickup at the studio. Meanwhile, the frequency discrimination action (R_4C_6 and R_5C_6) will provide the necessary de-emphasis. In similar fashion, the (L + R) and the –(L – R) signals applied to C_7 will produce an output corresponding to the audio pickup by the "right" studio microphone.

Stereophonic sound will result when these signals are amplified and fed to properly phased, properly located loudspeakers. However, to get complete separation of the "left" and "right" outputs, it is necessary that the (L + R) and the (L – R) channels have identical frequency and phase responses. All frequencies contained in *both* channels must arrive at the matrix at the same time. Otherwise, cancellation will not be complete, and separation will suffer. Phasing compensation can be made by adjusting the inductance in the low-pass filter. Similar adjustment may be possible in the band-pass filter and in the frequency-doubler tank circuit.

BASIC TELEVISION PRINCIPLES

Every television station actually transmits two separate signals from two individual transmitters – one for the sound, the other for the picture information. The sound transmitter employs frequency modulation, and except that it is allowed a deviation of only ±25 kHz, it is otherwise identical to the circuitry discussed in Chapter 9.

The picture transmitter is an AM transmitter. The intelligence, instead of being audio voltages from the sounds picked up by a microphone, are now video voltages from the light changes picked up by a "camera" tube (an image orthicon or a vidicon). Since light intensities vary much more rapidly than sound wave pressures, the frequency range of video signals extends out to 4.0 MHz (as compared to only 15 kHz for sound waves). Obviously, video amplifiers should have a frequency response flat to 4.0 MHz, and the RF output from the picture transmitter would require a bandwidth of 8.0 MHz. To reduce this bandwidth, most of the lower sideband is suppressed,* reducing the total picture bandwidth to just over 5.0 MHz. This is called *vestigial-sideband transmission.*

Both RF signals, from the picture and sound transmitters, are fed to a common antenna. For efficient radiation from one antenna, the frequencies from the two transmitters must be fairly close. The separation used between the sound and picture carrier frequencies is 4.5 MHz. The RF spectrum for a television station is shown in Fig. 12–31 using channel 4 as an illustration. Notice that the picture carrier is 1.25 MHz from the low-frequency end of its channel. (This is the vestigial sideband.)

Television transmissions are separated into three frequency groupings:

1. Channels 2 through 6 – from 54 to 88 MHz.
2. Channels 7 through 13 – from 174 to 216 MHz.
3. Channels 14 through 83 – from 470 to 890 MHz.

The first two groups are in the VHF range; the third group is in the UHF band.

*Complete elimination of one sideband would require much more elaborate filtering, at greater cost.

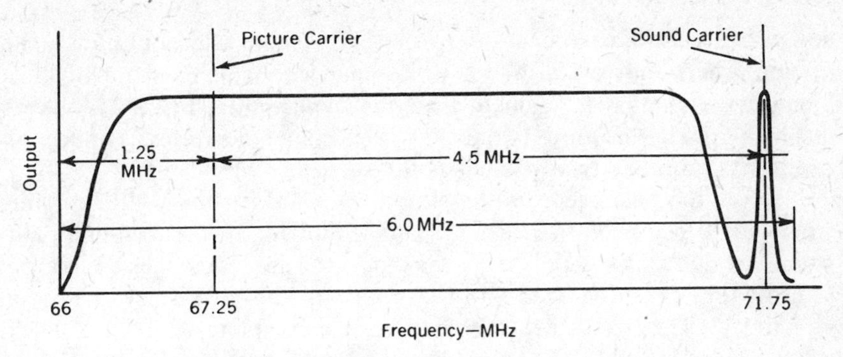

Figure 12-31 Energy distribution in a TV channel.

Scanning

A television picture is not flashed all at once on the screen of the cathode ray tube, in the manner of a film slide projected onto its screen. Instead it is "painted," spot by spot, until the picture is completed. This process, called *scanning,* is similar to the way we read a page—word by word from left to right, and line by line from top to bottom.* Of course, the scanning process is so rapid that we do not notice this buildup. The actual time for the complete scan of a picture is one-thirtieth of a second. But at this speed, some flicker will be noticeable. This is eliminated by the use of *interlaced scanning.* The scanning starts at the upper left, point *a* in Fig. 12–32, and proceeds from *a* to *b*. The beam then retraces (horizontally) very rapidly to *c*, and then scans to *d*. Notice the slight

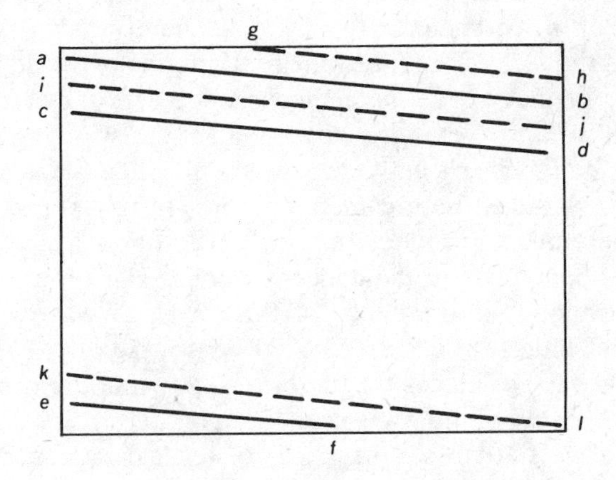

Figure 12-32 Interlaced scanning.

* Please excuse this elementary technique if you are a speed reader.

downward slant of each line. The process is repeated to the bottom of the screen, where only a half-line (*e-f*) is scanned. At this point, a total of 262.5 lines have been scanned. This is called a *field.* From *f* the beam retraces rapidly upward (while still zigzagging horizontally), until it reaches the top center at *g*. This is the *vertical retrace.*

Now the beam repeats the slow downward motion while scanning from left to right. Notice that it scans (following the broken-line path) from *g* to *h*, retraces rapidly to *i*, scans to *j*, etc., until it reaches the bottom right of the screen on scan *k* to *l*. Another field of 262.5 lines has been completed. The two fields, combined, form a complete scan of 525 lines; this is called a *frame.* The time for one frame is still one-thirtieth of a second, but since each field is twice as fast, flicker is not evident. Notice how the half-line in each field produces interlacing of the scanning lines. These lines are called *raster lines,* and a full screen of these lines is referred to as the *raster.* (The raster is all that is seen when no picture information is being transmitted.) The overall size of any raster will obviously depend on the size of the picture tube. However, it is necessary that the relative dimensions of width to height be the same for all transmitters and all receivers. This ratio of width to height is known as the *aspect ratio* and is standardized at 4:3.

At the television studio or field location, the optical system of the camera tube focuses an image of the scene on the target or signal plate.* The optical image is then scanned by an electron beam, as described before, producing a voltage that varies in magnitude, depending on the brightness of the scene at each particular spot. This video voltage is the modulating signal for the picture transmitter. At the receiver, the recovered video signal (after demodulation and amplification) is applied to the control grid of the cathode ray tube. Consequently, as the electron beam scans across the face of the tube (due to the action of the deflection coils), the intensity of the spot varies in accordance with the brightness of the scene at that instant. Obviously, the timing of the scanning at the camera tube and at the receiver's cathode ray tube must be identical.

Television tubes (both camera and picture tubes) use magnetic deflection. Therefore, to produce the scanning motion, a sawtooth wave of current must be applied to the deflection coils. For proper timing of the horizontal motion, the frequency of the sawtooth wave applied to the horizontal deflection coils is 262.5 × 60, or 15,750 Hz. The slow rise of the sawtooth wave produces the uniform steady motion from left to right, and the quick decay of the wave causes the rapid retrace. To scan a complete field in one-sixtieth of a second, the vertical sawtooth frequency is 60 Hz. Notice that this is the same as the power-line frequency. Now,

*In the now obsolete *iconoscope* camera tube, this was a *mosaic* plate made of a thin mica sheet and covered on one side with many globules of light-sensitive material.

should any power line "hum" be picked up in the vertical sawtooth circuit, the ripple produced on the screen will be stationary and not too objectionable. It would be seen only at the edges—if the horizontal width is too narrow.

There is still one more problem. In order for the picture at the receiver to be stable, the timing at the receiver must correspond exactly to the timing at the camera at the televised scene—not only in frequency, but also in phase. Synchronization is necessary. This is accomplished by using triggering pulses at the camera. These same pulses are also transmitted as part of the modulating signal and, after demodulation at the receiver, they are then used as *synchronizing pulses* to trigger the sweep action of the cathode ray tube. One final step—since we do not wish to *see* these synchronizing pulses (nor the regular sawtooth horizontal and vertical retrace lines), a *blanking* pulse is also transmitted. This pulse drives the cathode ray tube into cutoff during retrace and synchronizing-pulse timing intervals. The composite modulating signal consists of:

1. The video signal corresponding to the televised scene.
2. Blanking pulses.
3. Horizontal synchronizing pulses.
4. Vertical synchronizing pulses.

A composite modulating signal for one horizontal line is shown in Fig. 12–33(a). This is standard for all television stations in the United States. It is a negative transmission system—the higher the amplitude, the darker the picture detail. At 75% of maximum amplitude, the screen is blanked out. The corresponding modulated RF wave is shown in diagram (b).

Toward the end of a field, the screen is again blanked in preparation for the vertical retrace. (Approximately twenty horizontal lines are "lost" during the vertical blanking period.) The vertical blanking period

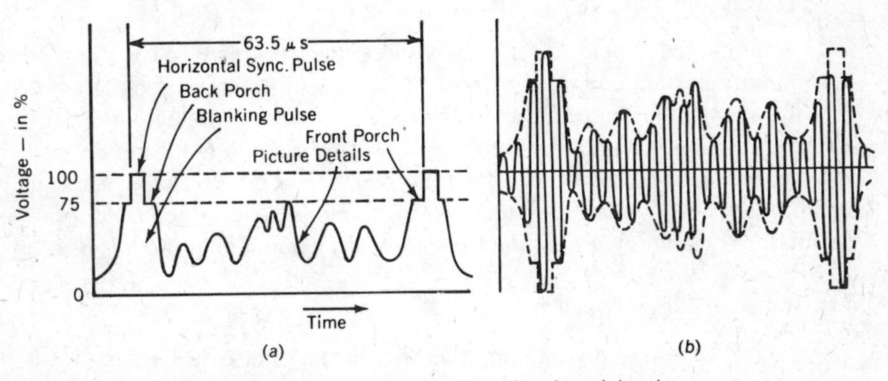

Figure 12-33 Composite video signal and modulated wave.

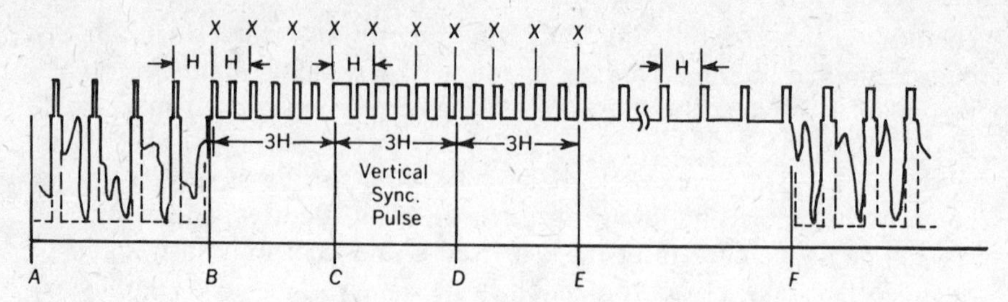

Figure 12-34 Pulse sequence during vertical retrace.

begins with six *equalizing pulses* spaced at one-half the horizontal-line time interval. These are followed by a *vertical synchronizing pulse* with six serrations, also at half-line intervals; then by six more equalizing pulses; and, finally, by (approximately) eleven horizontal synchronizing pulses. This sequence is shown in Fig. 12–34 for each *odd* field. Notice the full H time-interval between the last *active* horizontal line and the first equalizing pulse. Proper horizontal-line triggering is produced by the first, third, and fifth equalizing pulses, and then by the marked edges of the serrated vertical pulse and by the marked equalizing pulses.

On the even fields, which end at the center of the screen, the last active horizontal line is only half an H time-interval from the first equalizer pulse. Therefore, correct horizontal-line triggering is maintained by the second, fourth, and sixth equalizer pulses, and by the alternate serrations of the vertical synchronizing pulses. Obviously, proper synchronization could not be obtained without the half-line timing intervals. The second set of equalizing pulses also allows proper discharge time for capacitors in the vertical sweep integrator circuits, regardless of whether they operate from an odd or even field. This is important for good interlace action.

Transmitter Block Diagram

A simplified block diagram of a video transmitter is shown in Fig. 12–35. Notice the similarity with the AM transmitter of Fig. 6–21. The only apparent difference is the addition of the synchronizing generators to provide blanking and synchronizing pulses. Another difference (not obvious from the block diagram) is that, because of difficulty in obtaining high amounts of video modulating power, the final stage is generally grid-modulated (or low-level modulation followed by linear amplifiers is used).

Receiver Block Diagram

Figure 12–36 shows a simplified block diagram of a television receiver. The circuit is typically that of a superheterodyne receiver. The

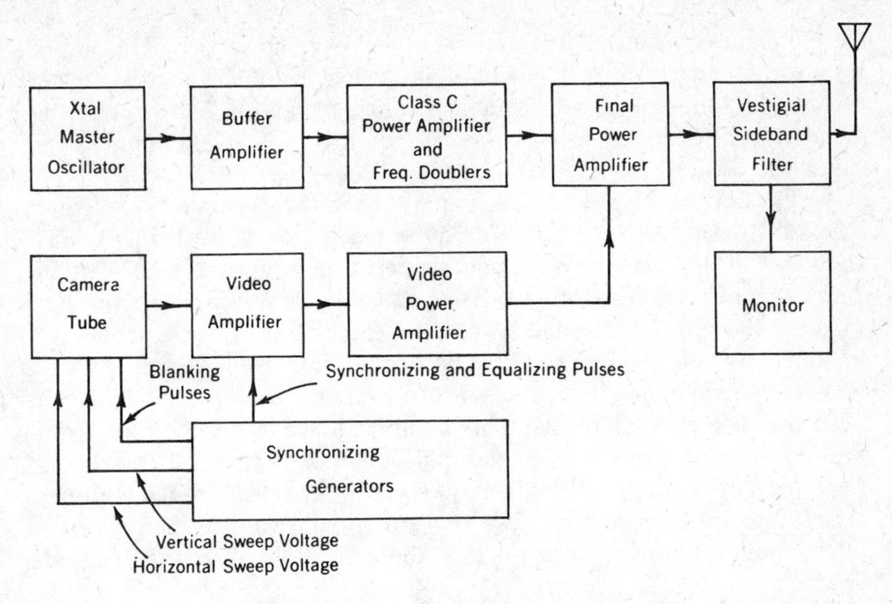

Figure 12-35 Simplified transmitter block diagram.

complete signal (sound and picture RF) are fed through the tuner (RF, mixer, and oscillator) to the IF amplifier. Each of these sections should have a 6-MHz bandwidth. The standard IF values used are 45.75 MHz

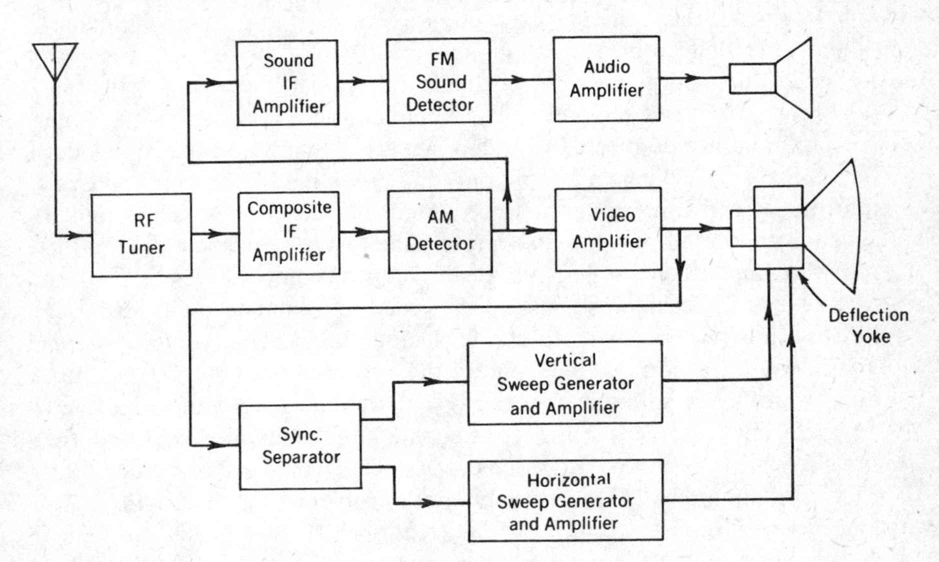

Figure 12-36 Simplified receiver block diagram.

for the picture carrier and 41.25 MHz for the sound carrier. The output from the IF amplifier is then fed to an AM detector. The video output (obtained from the video carrier and its sidebands) is fed through the video amplifiers to the grid of the cathode ray tube. The circuit action should be obvious from previous theory discussions.

The FM sound signal is also amplified in the common IF amplifier and fed to the AM detector. Here, the video carrier, acting as a local oscillator, beats against the sound carrier, translating the sound intelligence to a new IF value of 4.5 MHz.* (The detector circuit is acting as a mixer.) The rest of the sound system (IF amplifier, FM detector, and AF amplifier) is standard and requires no further discussion.

In the video amplifier, the synchronizing pulses are amplified along with the picture intelligence. This combined signal is fed to the sync-separator where the synchronizing pulses are extracted. Next, the vertical and horizontal synchronizing pulses are separated from each other and fed to their respective sawtooth sweep generators. Their outputs, after amplification, are in turn fed to the deflection coils of the cathode ray tube.

Introduction to Color TV

A color TV camera is essentially a combination of three basic cameras. Light from the televised scene passes through the standard camera lens system and strikes a blue-reflecting dichroic mirror. The red and green components of the incident light will pass through, but the blue-light components are deflected to the "blue camera," where the blue light is converted to a video signal. Next, the main light strikes a red-deflecting dichroic mirror, and the red-light components are directed to the "red camera" and are converted to video signals. Meanwhile, the remainder of the main beam, which is now only the green-light component, passes into the "green camera" and is converted to video waves. The original scene is converted into three separate video waves corresponding to the red, blue, and green components of the original color scene.

These three video signals are not used to modulate the transmitter directly for two reasons. First, a black-and-white (*monochrome*) receiver would respond to only one of these signals (if at all), and picture quality would be seriously impaired. This impairment would render all monochrome receivers obsolete. Second, the bandwidth required for separate transmission of the three color signals would greatly exceed the FCC limitation of 6 MHz. To solve both problems, the modulation system proposed in 1953 by the NTSC (National Television System Committee) was adopted. In this system, the three outputs from the color

* Recall that this is the frequency difference between the two carriers.

camera are fed to a matrixing network to produce a *luminance signal, Y*, and two *color-difference signals, I* and *Q*. The luminance signal is obtained by adding the three camera components—in the same proportion as the eye's response to the individual color frequencies, or

$$E_Y = 0.59E_G + 0.30E_R + 0.11E_B$$

All the information necessary for proper functioning of a black-and-white receiver is contained in this luminance signal. This makes color TV transmission compatible with monochrome requirements. The I and Q signals, since they supply the color information (hue and saturation), are called *chrominance* signals. From experience, best color results are obtained by making

$$E_I = 0.74(E_R - E_Y) - 0.27(E_B - E_Y)$$

$$E_Q = 0.48(E_R - E_Y) + 0.41(E_B - E_Y)$$

The video intelligence contained in the Y, I, and Q channels is used to modulate the RF carrier. However, these three signals cannot be used as is, or interference between signals would result. Instead, only the luminance or Y channel modulates the carrier directly. As in a monochrome receiver, this will produce an energy spectrum ranging from 1.25 MHz below the RF carrier to 4.2 MHz above, as shown in Fig. 12–31. However, this energy is not uniformly distributed in this bandwidth, as implied earlier. It is actually bunched in clusters spaced 15,750 Hz apart (the horizontal line frequency). Therefore a frequency-division type of multiplexing can be used to interlace the color information, without increasing bandwidth, and without interference.

The first step is to select a subcarrier frequency. It is found that, if the subcarrier is an odd multiple of half the line frequency, the sideband-energy clusters will fall halfway between the Y-channel clusters. The subcarrier frequency chosen is (approximately) 3.58 MHz* (the 455th multiple of 15,750/2). The Q signal modulates this subcarrier in a balanced-modulator circuit, eliminating the carrier. The sidebands are fed to a filter, limiting them to a spread of from 3.0 to 4.2 MHz. Similarly, the I signal is fed to another balanced modulator and filter, producing sidebands of from 3.0 to 4.2 MHz. However, to prevent interference between the Q and I sidebands, the 3.58-MHz subcarrier is shifted 90° before modulation by the I channel. This makes the chrominance sidebands 90° apart, so that when they add vectorially they produce a resultant that varies in phase and in amplitude, depending on the color of the spot being scanned at the scene.†

* The exact frequency is 3.579545 MHz.

† The phase of the resultant is a function of the hue, while its amplitude is determined by the strength, or level of saturation, of the color.

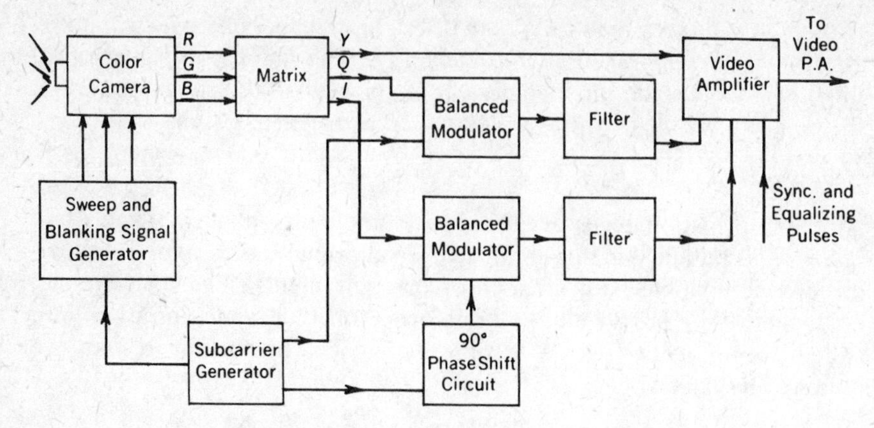

Figure 12-37 Simplified color portion of a TV transmitter.

There is one final step in the process. Notice that the color subcarrier is eliminated by the balanced modulators. Yet, in the receiver, in order to recover the Q and I signals, this carrier must be reproduced with exact frequency and phase. Synchronization is necessary. This is accomplished by transmitting a burst of at least eight cycles of the subcarrier signal. The burst is superimposed on the back porch of each horizontal synchronizing pulse.

A simplified diagram of the synchronizing and modulating-signal portion of a color TV transmitter is shown in Fig. 12–37.

Color TV Receivers

In many respects, a color TV receiver is quite similar to a monochrome receiver. Their tuners, composite IF amplifiers, and AM detectors can be identical—except that, for color reception, flat response over the full 6-MHz bandwidth is more important. Phase shift in the tuned circuits can destroy color fidelity. The entire color TV sound section (IF amplifier, FM detector, and audio amplifier) is identical to that of the monochrome receiver. One change often made to reduce possible interference is to make the sound-channel takeoff at the output of the composite IF amplifier, and to use a separate diode mixer to produce the 4.5-MHz sound IF. The deflection circuits (horizontal and vertical) are also quite similar in the two receivers. The main difference in the color receiver is the expansion of the video amplifier section to include the chroma circuitry.

A block diagram of the video and chroma section of a color receiver is shown in Fig. 12–38. The composite video signal is passed into the luminance block for further amplification. The output is fed to the three cathodes of a three-gun cathode ray tube. This signal, of itself, will provide a complete black-and-white picture. (A 3.58-MHz trap is often included to prevent possible interference from the color subcarrier.)

The output of the first video block is also fed to a band-pass filter and amplifier. Only the subcarrier bursts and the sidebands will pass through this stage. These signals are fed simultaneously to the *Q* demodulator, the *I* demodulator, and the burst gating amplifier. This last stage conducts only during the burst interval. Its output is used (through the AFC and feedback circuits) to maintain correct phase and frequency of the 3.58-MHz subcarrier generator. This color subcarrier is fed to the *Q* demodulator, producing the *Q* video signal. Notice that the subcarrier is shifted 90° before it is applied to the *I* demodulator. This is to compensate for the original 90° shift at the transmitter. Therefore, we now recover the *I* video signal. The matrix network converts these inputs into *R-Y, G-Y,* and *B-Y* signals. These are fed to the separate grids of the three-gun tube. The combination of the *Y* signal on the cathodes and the *R-Y, G-Y,* and *B-Y* signals on the grids produces three electron beams that are dependent on the amplitude of the red, green, and blue components of the televised scene.

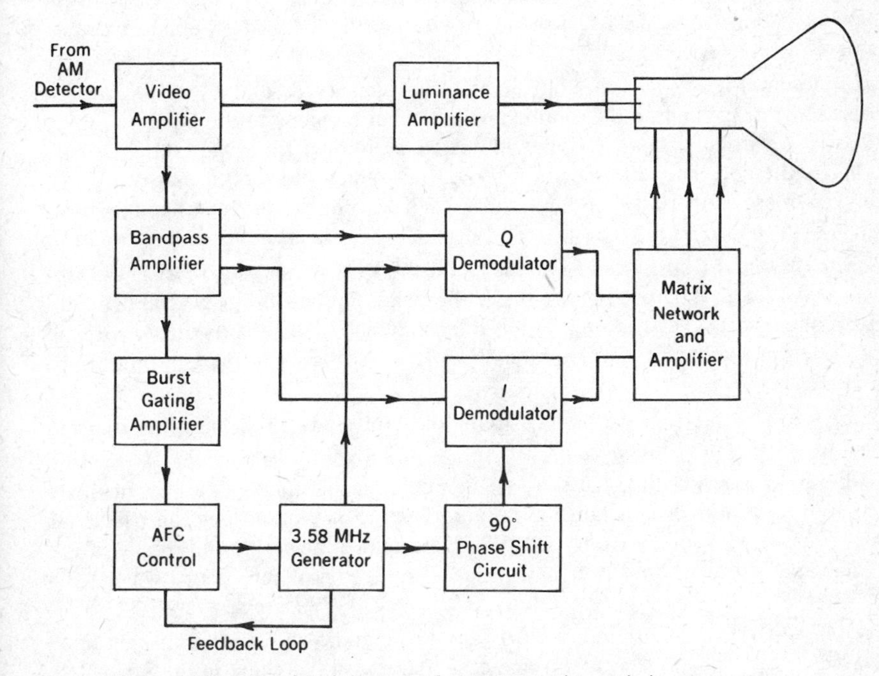

Figure 12-38 Simplified diagram of a receiver video and chroma section.

The color TV tube most commonly used is the shadow-mask tube. As already noted, this tube has three separate electron-gun structures mounted within its neck. The guns are positioned so that their beams cross over, a short distance from the screen, as they pass through holes in a shadow mask. The screen itself is made up of thousands of "elements," each consisting of a trio of phosphor dots. When struck by an electron beam, one dot emits red light, another emits green light, and the third emits blue light. The holes in the mask—one for each element—are so placed that the beam from the red gun will strike only the red-emitting dot, etc. To the eye, the three-dot trio appears as one element. The rest is obvious. The color seen at any spot on the screen will depend on the percentage of red, green, and blue light emitted by the phosphor trio.

REVIEW QUESTIONS

1. (*a*) Is the ampliphase transmitter an AM, PM, or FM transmitter? (*b*) What was the main reason for developing this type of transmitter?

2. (*a*) What is the disadvantage of high-level amplitude modulation? (*b*) What is the disadvantage of low-level modulation? (*c*) How does the ampliphase system "correct" both weaknesses? (*d*) State two other advantages claimed for transmitters of this type.

3. Refer to Fig. 12–1. (*a*) What do the two phasors E_1 and E_2 in diagram (a) represent? (*b*) When modulation is applied, what happens to the position of E_1? (*c*) If at a particular instant E_1 has moved an additional 10° towards the vertical, what is the corresponding position of E_2? (*d*) Which phasors in diagram (b) (if any) correspond to 100% modulation? (*e*) What accounts for the (a) and (b) positions of E_1 and E_2?

4. (*a*) If a final transmitter frequency of 1200 kHz is desired, what is the frequency of the crystal oscillator? (*b*) Within what limits will this crystal maintain its frequency? (*c*) How is such precision maintained? (*d*) What is the waveshape of the output voltage from this block? (*e*) How is this waveshape of output obtained?

5. (*a*) To what block is the crystal oscillator output fed? (*b*) What other signal is also fed to this next block?

6. Refer to Fig. 12–2(a). (*a*) Why is the audio signal shown as a straight line? (*b*) Why is it shown constant at the zero level? (*c*) Where does the triangular wave marked RF come from? (*d*) What determines the width and timing of the rectangular pulse in the lower half of diagram (a)? (*e*) Where is this waveshape obtained?

7. Refer to Fig. 12–2(b) and (c). (*a*) Why is the audio shown as a straight line? (*b*) Why is it a negative value in one case and a positive value in the other? (*c*) What determines the timing and width of the phase-modulator output pulse?

8. Refer to Fig. 12–3. (*a*) Where is the RF sine-wave carrier actually developed? (*b*) In channel *A*, what determines the timing or phasing of this sine wave? (*c*) What two extra blocks are incorporated in channel *B*? (*d*) Why is the inverter block required? (*e*) Why is the delay-multivibrator block required? (*f*) At what modulation level is the delay multivibrator adjusted, and for what effect?

9. Refer to Fig. 12–4. (*a*) To what stage(s) does this phasor diagram apply? (*b*) What do the broken-line phasors, I_1 and I_2, represent? (*c*) Is there any load output at this instant? Explain. (*d*) For what condition do the phasors I_{1a} and I_{2a} apply? (*e*) What (if any) is the output at this instant? (*f*) At what instant do phasors I_{1b} and I_{2b} apply? (*g*) What is the output now? (*h*) State two disadvantages of the constant-current conditions shown in Fig. 12–4. (*i*) What feature shown in the block diagram alleviates these drawbacks?

10. (*a*) In Fig. 12–3, what is applied to the *intermediate power amplifier* block from the *drive regulator* block? (*b*) Specifically, where is it fed and what type of circuitry is used? (*c*) What is the effect of this feature on the final output conditions?

11. (*a*) How does an SSB transmission differ from AM transmission? (*b*) State five advantages of SSB over conventional AM. (*c*) What is the major disadvantage of SSB transmission systems?

12. (*a*) Why is frequency stability important in SSB transmissions? (*b*) How much frequency error is tolerable in voice communications? (*c*) When is phase shift detrimental? Why?

13. (*a*) Is frequency stability important in SSB *receivers*? Explain. (*b*) State two techniques used to maintain frequency.

14. (*a*) In SSB transmitters, how is the carrier suppressed? (*b*) State two methods used to suppress one sideband. (*c*) Which sideband is generally transmitted? (*d*) If the modulating signal is a pure sine wave, describe the output waveshape from an SSB transmitter. (*e*) Why is this so?

15. Refer to Fig. 12–5. (*a*) Why isn't a 100-kHz signal present at the output of the balanced modulator? (*b*) What are the frequencies shown called? (*c*) How are these values obtained? (*d*) Why is only one of these frequency bands shown as the output of the next block? (*e*) What is the general purpose of the balanced mixers? (*f*) Why aren't frequency multipliers used for this purpose? (*g*) Why isn't a band-pass filter used at the output of the second mixer? (*h*) Why are *linear* amplifiers used in the final block?

16. The transmitter shown in Fig. 12–5 is to be used in a commercial airliner, and it is necessary to contact many airfields at different operating frequencies. What simple change in Fig. 12–5 would provide this feature?

17. (*a*) Give two reasons why modulating frequencies below 100 Hz are not transmitted in SSB speech transmitters. (*b*) If the final frequency is in the order of 10 MHz, why is such a low starting frequency as 100 kHz used? (*c*) Why isn't the frequency translation done in one step, by using a higher oscillator frequency?

18. Refer to Fig. 12–6. (*a*) What is the output from this circuit at the modulating-signal frequency? Why? (*b*) What is the phase relation between the RF signals fed to Q_1 and Q_2? (*c*) What is the output from this circuit when only the RF signal is applied? Why? (*d*) What is the function of R_7? (*e*) If for any reason the RF output from Q_1 is less than from Q_2, how is balance achieved? Explain.

19. (*a*) What type of IC can be used as a balanced modulator? (*b*) Referring to Fig. 3–14(a), how is the RF carrier signal applied? (*c*) What is the output at the carrier frequency? Explain. (*d*) Where is the audio signal applied? (*e*) What is the effect of this signal on the IC conduction? (*f*) What is the output when both RF and audio inputs are applied? (*g*) Which would have better balance action, the IC or a discrete-component circuit? Explain.

20. Refer to Fig. 12–7. (*a*) Trace the current flow due to carrier input only, when point 11 is negative compared to point 12. (*b*) Why is the output zero? (*c*) Trace the current flow due to audio input only, when point 3 is negative compared to point 1. (*d*) Why is the output zero? (*e*) Trace the current flow due to both inputs when the instantaneous polarities are as shown. (*f*) Is the output zero? Explain. (*g*) Why is a strong carrier input required (compared to the modulating-signal level) to reduce distortion?

21. What do the dark areas in Fig. 12–7(c) represent?

22. (*a*) Compare the input signals to a balanced modulator and a balanced mixer. (*b*) What modification(s) must be made to the circuit in Fig. 12–6 for use as a balanced mixer?

23. Refer to Fig. 12–8. (*a*) What identifies this circuit as a mixer rather than a modulator? (*b*) What components determine the emitter-base bias of each transistor? (*c*) Why is the R_4 value made variable? (*d*) What change(s) would convert this to a balanced modulator?

24. Name two types of filters used in SSB transmitters.

25. Refer to Fig. 12–9. (*a*) What frequency characteristics are desired for crystals X_2 and X_3? (*b*) for crystals X_1 and X_4? (*c*) At what frequency will they give optimum results? (*d*) State three disadvantages of crystal filters.

26. (*a*) What are the three basic actions in an electromechanical filter? (*b*) At what frequency do these give optimum results? (*c*) Over what range of frequencies can they be used? (*d*) What determines the specific frequency for a given filter?

27. (*a*) Why are linear amplifiers used as power amplifiers in SSB transmitters? (*b*) What class of operation is generally used? Why?

28. Refer to Fig. 12–11. (*a*) Why is an inductor used in the cathode circuit of V_1? (*b*) What is the function of C_1 and C_2? (*c*) What is the function of L_2? (*d*) What type of output circuit is used in the final stage?

29. (*a*) What is the function of the *carrier-attenuator* block in Fig. 12–5? (*b*) Why is it connected to the other stages by dashed lines? (*c*) How does such operation differ from conventional AM?

30. Compare the filter type and phase-shift type of SSB transmitters with regard to (*a*) method of carrier suppression, (*b*) method of suppressing the unwanted sideband, (*c*) need for linear power amplifiers, (*d*) use of pilot carrier, (*e*) starting "carrier" frequency value, and (*f*) need for balanced mixers.

31. Refer to Fig. 12–13. (*a*) At what RF frequency does the RF phase-shift network produce a 90° phase shift? (*b*) At what particular audio frequency does the audio network produce a 90° phase shift? (*c*) How does the frequency output from each balanced modulator compare? (*d*) Why are two balanced modulators needed? (*e*) Why is the balanced mixer and its oscillator shown in dashed-line form?

32. Refer to Figure 12–14(a). (*a*) What do these three phasors represent? (*b*) What do the dashed-line arrows over two of these phasors represent? (*c*) Why are these arrows in opposite directions? (*d*) Is the angle between the center and side phasors always 30°? Explain.

33. Refer to Fig. 12–14(b). (*a*) Why is carrier f_B drawn horizontally instead of vertically upward as for f_A? (*b*) What is the instantaneous phase angle between f_u and f_B? (*c*) Why is it drawn so? (*d*) What is the angle between f_l and f_B? (*e*) Why are the 90° increments taken in opposite directions?

34. Compare Figs. 12–13(a) and (b). (*a*) What is the relative phasing of the upper side frequencies from modulators *A* and *B*? (*b*) What is the magnitude of the resultant f_u? (*c*) What is the relative phasing of the lower side frequencies from modulators *A* and *B*? (*d*) What is their resultant magnitude? (*e*) At some instant later when f_u in modulator *A* leads its carrier f_A by 90°, what happens to their resultant magnitude? Explain. (*f*) At this same instant, what is the resultant magnitude of the lower side frequencies?

35. (*a*) State two factors that determine the degree of suppression of the unwanted sidebands. (*b*) Which of these factors is more difficult to obtain? Why?

36. In controlled-carrier transmission: (*a*) At what amplitude is the carrier transmitted? (*b*) Is it maintained at this level throughout the transmission? Explain. (*c*) How does this affect the peak power requirement of the equipment?

37. (*a*) What basic type of receiver is used for reception of SSB signals? (*b*) Is a conventional AM diode detector suitable for SSB use? Explain. (*c*) Under what conditions can an AM communications receiver be used with SSB signals?

38. (*a*) When a beat frequency oscillator is used with a diode detector for demodulation of SSB signals, what oscillator voltage level is desired? (*b*) Why is this so?

39. (*a*) What input signals are fed to a product detector? (*b*) Why is it called a product detector? (*c*) Give two advantages of this circuit over a diode detector with BFO.

40. Refer to Fig. 12–15. (*a*) What type of active device is used as the detector? (*b*) Where does the carrier input signal come from? (*c*) How is it coupled to this circuit? (*d*) What is the function of L_3? (*e*) What other components (if any) aid in this function?

41. Refer to Fig. 12–16. (*a*) How much output is produced when the only input is the signal voltage coupled into L_2? Explain. (*b*) How much output is produced if the only input is the carrier signal? Explain. (*c*) What determines the capacitance values required for C_6 and C_7? (*d*) What is the function of R_1? (*e*) Explain how it accomplishes this function.

42. Refer to Fig. 12–17. (*a*) Which diodes conduct when the carrier signal is positive? (*b*) Due to carrier alone, how much output voltage is developed? Why?

(*c*) Due to the SSB signal alone, which diodes conduct when the upper terminal of L_2 is positive? (*d*) How much output will this produce? Why? (*e*) What two factors determine the value of C_2?

43. When using an IC as a balanced demodulator: (*a*) To what point is each input fed? (*b*) From what point is the output taken?

44. (*a*) What bandwidth is generally used for the IF system of an SSB receiver? (*b*) Is this bandwidth adequate? Explain. (*c*) How is this bandwidth generally obtained? (*d*) Would a wider bandwidth be an advantage? Explain.

45. In an SSB receiver using a clarifier control: (*a*) What is the technical function of this control? (*b*) Why may it be necessary? (*c*) What effect would be noticed if this compensation were not made?

46. Refer to Fig. 12–18. (*a*) What does the notation "14–15 MHz" alongside the *RF-amplifier* block mean? (*b*) What does the "29–30 MHz" mean? (*c*) How is this second band obtained? (*d*) Is the mixer stage tunable? (*e*) When the RF amplifier is on band C, over what frequency range does the mixer stage tune? (*f*) Over what frequency range does oscillator #1 tune? (*g*) Why are two frequency values shown for the oscillator? (*h*) If the RF signal is at 14.0 MHz, what is the IF value? (*i*) What is the IF value for an RF of 15.0 MHz? (*j*) for an RF of 29.0 MHz? Explain. (*k*) State two ways by which this IF coverage can be obtained. (*l*) When the first IF value is 2.0 MHz, what is the second IF value? (*m*) How is this obtained? (*n*) What is the second IF value corresponding to a first IF value of 3.0 MHz? (*o*) How is this obtained? (*p*) What is the frequency coverage of the variable-frequency oscillator when the RF is on band C? on band G? (*a*) What is the proper frequency for crystal oscillator #2?

47. In Fig. 12–18, what is the advantage of using a variable-frequency oscillator in the second conversion rather than in the first conversion?

48. Is the receiver in Fig. 12–18 suitable for commercial automatic point-to-point transmission? Explain.

49. State two advantages obtained by transmitting a pilot carrier.

50. Refer to the block diagram of Fig. 12–19. (*a*) Which blocks represent tunable stages? (*b*) When the RF amplifier is tuned to 10 MHz, what are the frequencies at the mixer, high-frequency oscillator, and first IF amplifier? (*c*) What does the frequency span 100.1–103 kHz at the output of the second mixer represent? (*d*) What does the 100-kHz output represent? (*e*) What type of circuit could be used in the *demodulator* block? (*f*) What frequencies might its output contain? (*g*) What functions are accomplished in the *filter and carrier amplifier* block?

51. Refer again to Fig. 12–19. (*a*) What is the bandwidth of the first IF amplifier? (*b*) If the high-frequency oscillator should drift off frequency by +2 kHz (no AFC action), would the pilot-carrier signal get through? Explain. (*c*) Why isn't the bandwidth of the IF increased to allow for such drift? (*d*) How does the AFC circuit prevent oscillator drift? (*e*) Why are two AGC outputs shown? (*f*) To what stages is the AGC signal fed?

52. Explain a disadvantage of pilot-carrier transmissions.

53. Refer to Fig. 12–20. It is desired to tune in a channel frequency of 10.245 MHz. (*a*) To what frequency is the RF amplifier tuned? (*b*) What should the channel crystal oscillator frequency be? (*c*) If the intelligence is in the upper sideband, to what frequency should the sideband-selector oscillator be set? (*d*) What is the frequency of the carrier crystal oscillator?

54. What is a frequency synthesizer?

55. Refer to Fig. 12–22. (*a*) Why isn't AFC circuitry shown? (*b*) Give the pass band of the first IF section. (*c*) What frequencies are present in the output of the second mixer? (*d*) Why is the AGC excitation taken from the audio end? (*e*) Why should the AGC system in an SSSC receiver have a short attack time? (*f*) What other characteristic must it have? (*g*) Why?

56. (*a*) In pulse modulation, is the intelligence a pulse? (*b*) Is the complete intelligence signal transmitted? Explain. (*c*) How often should the intelligence signal be sampled? (*d*) What is the transmitted power between samples? (*e*) Explain one way of utilizing this low duty cycle. (*f*) Give another advantage of such operation.

57. Name three types of pulse modulation with respect to the effect of the modulation on the pulse.

58. Refer to Fig. 12–23. Which is the pulse amplitude corresponding to (*a*) zero level of the intelligence signal? (*b*) positive maximum value of the intelligence signal? (*c*) maximum negative value of the intelligence signal?

59. Refer to Fig. 12–24. (*a*) What effect does the intelligence signal have on the amplitude of these pulses? (*b*) Which of these diagrams represents PDM? (*c*) What intelligence level corresponds to maximum pulse width? (*d*) What condition results in minimum pulse width?

60. In Fig. 12–24(c): (*a*) What do the dashed vertical lines represent? (*b*) What is the effect of positive intelligence-signal values on the pulses? (*c*) What is the effect of negative intelligence-signal values? (*d*) What determines the amount of shift?

61. After the intelligence signal is converted to a pulse form, how are these pulses transmitted?

62. (*a*) Is pulse modulation used with low-frequency carriers? Explain. (*b*) Which type of pulse modulation requires the least bandwidth? Why? (*c*) What is a disadvantage of this technique?

63. Refer to Fig. 12–25. (*a*) What do the irregular jagged waveshapes represent? (*b*) How is noise removed from the pulses? (*c*) What causes the curvature in the pulse of diagram (b)? (*d*) How could this curvature be reduced? (*e*) What is the effect of this curvature?

64. (*a*) In PAM systems, how does component tolerance or nonlinearity in the receiver affect the reconstructed intelligence signal? (*b*) Why is this so? (*c*) Will *transmitter* tolerance or nonlinearity affect the reconstructed signal? (*d*) Will this effect also exist in PDM systems? (*e*) in PPM systems? (*f*) Give an example of a type of intelligence signal that could be seriously affected by such distortion. (*g*) Name a pulse modulation system that is not subject to such distortion.

65. (*a*) What basic coding system is used for pulse-code modulation? (*b*) When converting a given intelligence amplitude into pulse code, does the coded value generally represent the exact amplitude? (*c*) Explain why this is so. (*d*) What is this process called? (*e*) What does the accuracy of the converted values depend on? (*f*) How many discrete amplitude values can be encoded using a 3-bit binary code? (*g*) a 4-bit code? (*h*) a 6-bit code?

66. Refer to Table 12–1. (*a*) How many bits does this code have? (*b*) How many discrete steps are available? (*c*) Amplitude values ranging from zero to 60 V are to be transmitted using uniform incremental steps. What binary code value would be used for an amplitude of 60 V? (*d*) 24 V? (*e*) 23 V? (*f*) 22.0 V? (*g*) What is the error introduced in the last two cases called? (*h*) How can this error be decreased? (Give three methods.)

67. (*a*) What is multiplex transmission? (*b*) Explain briefly how it could be used to transmit two messages. (*c*) Give two examples of multiplex transmission as used in broadcasting.

68. Name two general techniques used for multiplexing.

69. Refer to Fig. 12–26. (*a*) Are the channel-1 and channel-2 inputs identical signals? Explain. (*b*) What do the two frequency bands at the output of the channel-1 modulator represent? (*c*) What is the function of the band-pass filter? (*d*) Why is a linear amplifier used after this filter? (*e*) What frequencies are present at the output of the channel-1 linear amplifier? at the output of the channel-2 amplifier? (*f*) Why do these two outputs differ? (*g*) When the 24 channel signals are combined, what is the output (frequency band) from the combining amplifier? (*h*) How is this output used? (*i*) If the final output of this multiplex transmitter is conventional AM, what is the bandwidth of the modulated wave?

70. Explain how the first two channels' audio messages are recovered from the above modulated wave at the receiving location.

71. With reference to multiplex operation, what do each of the following abbreviations designate: (*a*) AM-FM? (*b*) FM-FM? (*c*) AM-AM? (*d*) FM-AM?

72. In frequency-division multiplexing, how does increasing the number of messages affect (*a*) transmitter and receiver bandwidth requirements? Explain. (*b*) subcarrier requirements? Explain.

73. In time-division multiplexing, how does the number of simultaneous transmissions affect (*a*) the number of subcarriers needed? Explain. (*b*) the transmitter and receiver bandwidth requirements? Explain.

74. A PAM system transmits 2-μs pulses every 60 μs. How many simultaneous channels could be transmitted by time-division multiplexing? Explain.

75. (*a*) Why are synchronizing pulses needed with time-division multiplex transmissions? (*b*) How is a pulse for channel 3 kept out of the other channels?

76. Why is it possible to send many more messages in time-division telemetry applications than with voice communications?

77. (*a*) Assuming excellent audio fidelity, why does the sound heard from a "live" performance differ from that reproduced from a single-loudspeaker system? (*b*) What is this latter type of sound called? (*c*) Will using two (or more) spaced loudspeakers fed from the one amplifier produce the true "live-sound" effect? Explain. (*d*) What other requirement is needed? (*e*) What is this "live-sound" effect called?

78. How is stereophonic sound obtained with tape recordings?

79. (*a*) To obtain stereo sound from radio transmissions, how many "messages" must be transmitted? (*b*) What are these "messages" called? (*c*) What system is used to send these messages on one transmitted carrier?

80. What does the term "compatibility" mean, with reference to stereo transmission systems?

81. (*a*) Can stereophonic sound be obtained if the "left" sound is transmitted as one message and the "right" sound as the other? (*b*) Is this system compatible? Explain. (*c*) What technique is used for compatible stereo transmission?

82. Refer to Fig. 12–27. (*a*) Why are pre-emphasis circuits used? (*b*) If the frequencies entering each microphone range 30–15,000 Hz, what frequency range does the (L + R) output span? (*c*) What frequency range does the (L − R) output span? (*d*) How is an (L − R) output obtained? (*e*) How does the output from the balanced modulator differ from its input? (*f*) What frequency span is covered by the (L − R) sidebands? (*g*) Why aren't the (L + R) and (L − R) signals fed directly to the FM exciter? (*i*) What additional functions are accomplished in the main transmitter?

83. In FM stereo transmission: (*a*) What is the total (maximum) deviation allowed to the FM output? (*b*) How much deviation (maximum) will the (L + R) modulating component produce? (*c*) How much will the (L − R) sideband signal produce? (*d*) Will this cause overmodulation beyond the FCC allowance? Explain. (*e*) What is this process called? (*f*) Is the 38-kHz subcarrier transmitted? (*g*) What effect would this have on the maximum allowable modulation level to the (L + R) and (L − R) signals? (*h*) Why is this so?

84. (*a*) Why is a pilot carrier transmitted? (*b*) What is the frequency of this signal? (*c*) Why isn't a 38-kHz pilot carrier transmitted?

85. Refer to Figure 12–28. (*a*) What is this composite signal used for? (*b*) What is the highest audio intelligence frequency contained in the (L + R) signal? the (L − R) sidebands? the SCA signal? (*c*) How much FM modulation (deviation) is caused by the (L + R) signal? by the (L − R) sidebands? by the pilot carrier? by the SCA signal? (*d*) How is overmodulation prevented when a transmitter is storecasting in addition to broadcasting?

86. (*a*) Can any FM broadcast receiver pick up a stereo program? Explain. (*b*) What components are present at the demodulation output of an ordinary FM receiver when tuned to a stereo program? (*c*) Will the output be stereo? Explain. (*d*) Which demodulator signal will produce the monophonic output? (*e*) What happens to the other components at the demodulator output?

87. How does an FM *tuner* for stereo differ from a monophonic tuner? Explain.

88. Refer to Fig. 12–29. (*a*) What is the function of the low-pass filter? (*b*) At what frequency should it cut off? Why? (*c*) What is the function of the 23–53 kHz filter? (*d*) Could a high-pass filter be used for this function? Explain. (*e*) Why is it necessary to isolate the pilot carrier? (*f*) What is the output of the amplifier and frequency doubler? (*g*) Since this is an FM receiver, why is an AM demodulator used here? (*h*) What is the function of the matrix network? (*i*) How does it accomplish this effect?

89. Refer to Fig. 12–30. (*a*) What constitutes the load for the diode D_1 circuit? (*b*) What is the load for diode D_2? (*c*) Will any RF voltage appear across these loads? Explain. (*d*) How does the phasing of the output voltages across these loads compare? Why? (*e*) Explain why an L output is obtained across capacitor C_6. (*f*) Why is an R output obtained across C_7? (*g*) Is any de-emphasis action obtained? Explain.

90. (*a*) What effect results if the (L + R) and (L − R) channels do not have correct phasing when fed to the matrix? (*b*) What could cause improper phasing? (*c*) How can this be remedied?

91. (*a*) How many transmitters are used for a complete TV program? (*b*) What are they? (*c*) What kind of modulation is used in each? (*d*) What is the bandwidth of each?

92. Refer to Fig. 12–31. (*a*) To what TV channel does this diagram apply? (*b*) What is the separation between carriers? (*c*) What type of transmission is used for the picture information?

93. Refer to Fig. 12–32. (*a*) What is the significance of the solid lines and broken lines? (*b*) What is each scan from top to bottom called? (*c*) How much time is allowed to scan all the solid lines? (*d*) What are two full consecutive scans called? (*e*) How many lines are there in a full picture scan? (*f*) What is the time for scanning one line? (*g*) What are the scanning lines called?

94. (*a*) How does the width-to-height ratio vary from a 12-inch screen to a 25-inch screen? Explain. (*b*) What is this ratio called?

95. (*a*) In television camera and picture tubes, how is the scanning motion obtained? (*b*) What frequency is used for the vertical scan? (*c*) for the horizontal scan? (*d*) How are camera and picture-tube scans made to coincide?

96. Name four components of the composite modulating signal.

97. Refer to Fig. 12–33. (*a*) What is the distinction between diagrams (a) and (b)? (*b*) In diagram (a), what corresponds to the brightest part of the scene? (*c*) When the video signal level is higher than 75% of maximum, is the screen black or very bright? (*d*) Is this desirable? Explain.

98. Refer to Fig. 12–34. (*a*) What does section *A-B* represent? (*b*) What are the pulses in section *B-C* called? (*c*) Why are these pulses used? (*d*) How many pulses are there in section *C-D*? Explain. (*e*) Why is it serrated? (*f*) What are the pulses in section *D-E*? (*g*) Why are they used? (*h*) How long—in horizontal lines—does the retrace take? (*i*) Is the screen dark or bright during this interval? (*j*) How does this affect what is seen by the eye? Explain.

99. Refer to Fig. 12–35. (*a*) Does this represent the sound transmitter, picture transmitter, or both? (*b*) What basic type of transmitter is this? (*c*) How does it differ from an AM radio transmitter?

100. Refer to Fig. 12–36. (*a*) What circuits are contained in the RF tuner? (*b*) What basic type of receiver is this? (*c*) What are the frequencies of the picture and sound carriers in the IF amplifier? (*d*) What is the frequency fed from the AM detector to the sound IF amplifier? (*e*) How is this frequency obtained? (*f*) What is the function of the sync generator? (*g*) What type of waveform does the horizontal sweep generator produce?

101. Refer to Fig. 12–37. (*a*) What is the nature of the input to the camera? (*b*) What is the nature of the output from the camera? (*c*) An all-yellow scene is scanned by the camera. Is there any output? Explain. (*d*) Why aren't the red, green, and blue video voltages from the camera output used directly (after suitable amplification) to modulate the transmitter RF power amplifier? (*e*) What is the frequency range of the *Y* output from the matrix? (*f*) Why are the *Q* and *I* signals fed to a balanced modulator before being used to modulate the transmitter? (*g*) Why is the subcarrier shifted 90° in the *I* modulator? (*h*) What is the standard frequency used for this subcarrier? (*i*) Why are *balanced* modulators used? (*j*) How is subcarrier synchronization maintained between transmitter and receiver?

102. Refer to Fig. 12–38. (*a*) Is this a complete receiver diagram? (*b*) How does the rest of the color TV receiver compare with a monochrome receiver? (*c*) What is the frequency spread of the signal fed to the luminance amplifier? (*d*) To what (specifically) is the output of the luminance amplifier fed? (*e*) What is the frequency spread of the signal fed to the band-pass amplifier? (*f*) What signals are present in its output? (*g*) What is the function of the *Q* and *I* demodulators? (*h*) Why is a 90° phase-shift circuit used between the 3.58-MHz generator and the *I* demodulator? (*i*) What is the purpose of the burst gating amplifier, the AFC control, and the feedback loop? (*j*) What are the three outputs from the matrix network and amplifier? (*k*) To what (specifically) are these outputs fed?

PROBLEMS AND DIAGRAMS

1. Draw the block diagram of an ampliphase AM transmitter.

2. Show, with the aid of a phasor diagram, how an output at 80% modulation is obtained. Give specific values.

3. An AM transmitter has a carrier output of 80 W. (*a*) At 100% modulation, what is the power in one sideband? (*b*) What is the total power output? (*c*) What percent of the total power is contained in one sideband?

4. An AM transmitter, when fully modulated, has a power output (average) of 36 kW. (*a*) How much of this is carrier power, and how much is sideband power? (*b*) What is the peak power output? (*c*) If only one sideband is transmitted, how much average power would be required; what is the peak envelope power; and what is the saving (in dB) using SSB?

5. A particular AM transmitter operating at 22 MHz has a frequency stability of 70 ppm. What is the maximum carrier drift?

6. An SSB transmitter has an operating (suppressed carrier) frequency of 22 MHz. If the frequency drift should not exceed 50 Hz, what stability tolerance should it have?

7. In Fig. 12–5, the output of the first band-pass filter (100.1–103 kHz) is fed to a frequency-multiplier chain consisting of a doubler and two triplers. What is the output? Is this satisfactory?

8. In the transmitter of Fig. 12–5, we wish to eliminate the second-frequency translation stages, and yet retain the same final frequency. (*a*) What oscillator frequency should be used at the first mixer? (*b*) What frequencies are developed at the output of this mixer? (*c*) Compare this technique to the method shown in Fig. 12–5. Which is preferable? Why?

9. Draw the circuit diagram of a two-transistor balanced modulator utilizing a push-push output circuit. Include a carrier-balance control.

10. Draw the circuit diagram for a ring-type balanced modulator. Show signal polarities and indicate direction of current flow.

11. Draw the circuit diagram for a balanced mixer using FETs.

12. Draw the circuit diagram for a balanced mixer using NPN bipolar transistors.

13. Convert the circuit diagram of Fig. 12–8 for use as a balanced modulator.

14. Draw the complete circuit diagram for a two-stage tetrode linear power amplifier with feedback.

15. An AM transmitter has a carrier power of 50 kW. How much saving in carrier power would result if operation is converted to SSB using a pilot carrier 20-dB down?

16. Repeat Problem 15 for a pilot carrier 25-dB down.

17. Draw phasor diagrams, similar to Fig. 12–14, to show the instantaneous phase relations (carriers and side frequencies) corresponding to the instant when the upper side frequency of modulator A is in phase with its carrier.

18. Repeat Problem 17, corresponding to the instant when the lower side frequency lags by 30°.

19. Repeat Problem 17, corresponding to the instant when the upper side frequency lags by 30°, but with carrier f_A leading f_B by 90°.

20. Draw the circuit diagram for a product detector using a dual-gate MOSFET.

21. Repeat Problem 20 using JFETs.

22. Draw the circuit diagram for a balanced demodulator circuit using transistors and incorporating a carrier-balance control.

23. In the receiver shown in Fig. 12–18, the first crystal oscillator has a stability of 10 ppm and the variable-frequency oscillator only 100 ppm. Calculate the possible frequency drift when the receiver is tuned to an input frequency of 30 MHz.

24. Repeat Problem 23 using the variable-frequency oscillator in the first conversion stage; a first IF value at 2.0 MHz; and the crystal oscillator (1.545 MHz) in the second conversion stage. Which circuitry results in better stability?

25. When the SSB receiver of Fig. 12–19 is tuned to a nominal frequency of 20 MHz, the input SSB signal covers a range of 20.0001–20.0030 MHz. If the high-frequency oscillator is tuned to 21.6000 MHz, find the frequencies present (*a*) in the first IF amplifier, (*b*) in the second IF amplifier, and (*c*) in the demodulator output. (Use the 100-kHz crystal oscillator for demodulation.)

26. Repeat Problem 25, for an uncompensated high-frequency oscillator drift of +0.001%. (Assume that the IF stages will allow these signals to pass through.) How will this drift affect the audio output?

27. Draw a block diagram for an SSB receiver for use with pilot-carrier transmission.

28. Repeat Problem 27 for SSSC transmission.

29. A PAM system is to be used for voice frequencies (100–3000 Hz). (*a*) Calculate the transmitter duty cycle for a pulse width of 1 μs and a sampling rate of 2.0 times the highest audio component frequency. (*b*) Show by diagram three pulses, and mark the "off time" between pulses. (*c*) Allowing a 5-μs guard band between signals, how many other intelligence channels can be sandwiched between two pulses of one channel?

30. The above timing is to be used for a PPM system. The pulses will be shifted ±4.5 μs at the peak intelligence-signal level. (*a*) What is the total time span that any one pulse may occupy? (*b*) Allowing a 5-μs guard band between signals, how many other intelligence channels can be sandwiched between two pulses of one channel?

31. A 5-bit binary code is to be used to telemeter temperature information ranging −30 to +60°C. (*a*) Using uniform increments, what is the incremental value for each step? (*b*) What is the maximum possible error (in degrees)? (*c*) What is the maximum percent error at a temperature near 60°C? (*d*) What is the maximum percent error at a temperature near 3°C? (*e*) How could this variation be reduced?

32. Draw a block diagram for a three-channel frequency-division multiplex transmitter. Include the main transmitter stages. Use amplitude modulation of the subcarriers and main carrier.

33. Repeat Problem 32 using frequency-modulated subcarriers.

34. Draw a block diagram for an eight-channel time-division multiplex transmitter. Include the main transmitter stages.

35. A PDM system uses an unmodulated pulse width of 5 μs and a pulse repetition rate of 7500. Modulation causes the pulse width to vary ±4 μs. Allowing a 2-μs guard band, how many simultaneous messages can this system handle?

36. A PCM telemetry system uses a 4-bit pulse train 17 μs long. Pulse sampling is repeated at a frequency of 200 per second. Allowing approximately 5 μs between pulse trains, and 4 μs for synchronizing signals, how many separate quantities can be encoded on one carrier?

37. Draw a block diagram for a complete stereo FM transmitter. Include exciter and main transmitter details.

38. In modulating the exciter of Fig. 12–27, 20 volts of combined (L + R) audio and (L − R) sidebands will produce 100% (±75 kHz) deviation. Assuming that the (L − R) sideband crest amplitude is equal to the (L − R) audio amplitude, and that the L and R channels each have an amplitude of 10 V, then for L at positive maximum and R at negative maximum, find (*a*) the peak (L + R) signal level, (*b*) the (L − R) signal level, (*c*) the (L − R) sideband level, and (*d*) the percent modulation of the exciter.

39. Repeat Problem 38 for L = +10 and R = +10.

40. Repeat Problem 38 for L = +10 and R = +5.

41. With the L and R channel amplitudes as in Problem 38, it is desired to transmit the 38-kHz subcarrier at a level such that it is 100% amplitude-modulated. (*a*) What subcarrier amplitude is necessary? (*b*) What is the crest value of the (L − R) amplitude-modulated wave? (*c*) What is the percent modulation (deviation) of the FM exciter output? (*d*) What amplitude level of (L − R) audio and 38-kHz subcarrier would restore 100% FM modulation. (*e*) What is the maximum (L + R) modulation obtainable now?

42. Draw a block diagram showing all stages of an FM stereo receiver.

43. Draw the block diagram of a monochrome TV transmitter.

44. Draw the block diagram of a TV monochrome receiver.

45. Draw the block diagram of a TV color transmitter.

46. Draw the block diagram of a TV color receiver.

TYPICAL FCC QUESTIONS

4.109 What is the device called which is used to derive a standard frequency of 10 kilohertz from a standard-frequency oscillator operating on 100 kilohertz?

4.251 Why is a scanning technique known as "interlacing" used in television broadcasting?

4.252 Does the video transmitter at a television broadcast station employ frequency or amplitude modulation?

4.253 Does the sound transmitter at a television broadcast station employ frequency or amplitude modulation?

4.254 What is a monitor picture tube at a television broadcast station?

4.255 Describe scanning as used by television broadcast stations. Describe the manner in which the scanning beam moves across the picture in the receiver.

4.256 What is a mosaic plate in a television camera?

4.257 What is the purpose of synchronizing pulses in a television broadcast signal?

4.259 Besides the camera signal, what other signals and pulses are included in a complete television broadcast signal?

4.260 What are synchronizing pulses in a television broadcasting and receiving system?

4.261 What are blanking pulses in a television broadcasting and receiving system?

4.262 For what purpose is a voltage of sawtooth waveform used in a television broadcast receiver?

4.263 In television broadcasting, what is the meaning of the term "aspect ratio"?

4.264 How many frames per second do television stations transmit?

4.265 In television broadcasting, why is the field frequency made equal to the frequency of the commercial power supply?

4.266 If the cathode ray tube in a television receiver is replaced by a larger tube such that the size of the picture is changed from 8 by 6 inches to 16 by 12 inches, what change if any is made in the number of scanning lines per frame?

4.267 If a television broadcast station transmits the video signal in channel 6 (82 to 88 MHz), what is the center frequency of the aural transmitter?

4.270 Numerically, what is the aspect ratio of a picture as transmitted by a television broadcast station?

4.271 What is meant by vestigial-sideband transmission of a television broadcast station?

4.274 How wide is a television broadcast channel?

4.276 What is the range of audio frequencies that the aural transmitter of a television broadcast station is required to be capable of transmitting?

4.277 What is meant by 100% modulation of the aural transmitter at a television broadcast station?

4.285 Describe the composition of the chrominance subcarrier used in the authorized system of color television.

4.287 Draw a block diagram of a typical monochrome television transmitter, indicating the function of each part.

4.288 Describe the scanning process employed in connection with color TV broadcast transmission.

4.290 In a transmitted monochrome television signal, what is the relationship between peak carrier level and the blanking level?

S-3.153 Draw a block diagram of an SSSC transmitter (filter type) with a 20-kilohertz oscillator and emission frequencies in the range of 6 MHz. Explain the function of each stage.

S-3.154 Explain briefly how an SSSC emission is detected.

S-4.098 Draw a block diagram of a multiplex FM broadcast transmitter complete from the microphone (and/or camera) input to the antenna output. State the purpose of each stage and explain briefly the overall operation of the transmitter.

Appendix

CURRENT AND VOLTAGE LETTER SYMBOLS

Current and voltage values are often represented by letter symbols. This is especially true in equations and diagrams. Much confusion can arise if these symbols are interpreted incorrectly. The symbols used in this text are in accordance with IEEE standards. An explanation follows.

General

1. All *fixed* values—such as dc (or average), rms, and maximum values—are represented by capital letters (E or V for voltages, and I for currents).
2. All *instantaneous* values are represented by lowercase letters (e or v, and i).

To further distinguish current and voltage values, suitable subscripts are added. In general, it is necessary to distinguish between dc components, ac (or varying) components, and *total* values (dc + ac). These subscripts also indicate the element or elements to which the currents or voltages apply.

Application to Transistors

1. Supply voltages (dc)—use *doubled capital-letter subscript* for the element being supplied. Examples:
   (*a*) V_{EE} emitter supply voltage
   (*b*) V_{DD} drain supply voltage
2. All other dc values—use appropriate *capital-letter subscript(s)* for the elements involved. Examples:
   (*a*) I_B base current
   (*b*) V_{BE} voltage between base and emitter
3. Values of the ac (varying) component only—use appropriate *lowercase subscript(s)*. Examples:
   (*a*) I_e rms value of emitter current
   (*b*) I_{em} maximum value of emitter current
   (*c*) i_b instantaneous value of base current
   (*d*) v_{ce} instantaneous value of collector-to-emitter voltage

4. Total values (ac + dc)—use appropriate *capital-letter subscripts*.
   Examples:
   (*a*) i_D instantaneous value of the *total* drain current
   (*b*) I_{BM} maximum value of the *total* base current
   (*c*) v_{BE} instantaneous value of the total base-to-emitter voltage.

It should be noted that total values (ac + dc) are measured with a zero (absolute) value as reference, whereas ac components are measured with their own average values as references. The above designations are illustrated in Fig. A-1 for the drain current in a field-effect transistor.

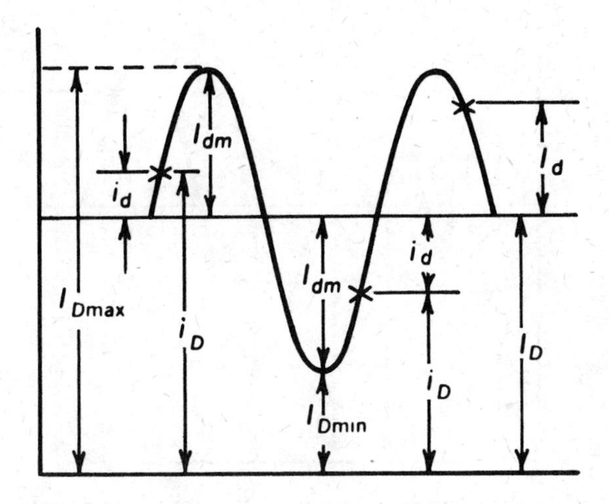

I_{Dmax} — maximum value of the total current.

i_d — instantaneous value of the ac component.

i_D — instantaneous value of the total current.

I_{dm} — maximum value of the ac component.

I_{Dmin} — minimum value of the total current.

I_D — average value, or dc component.

I_d — rms value of the ac component.

Figure A-1 Letter symbols for FET drain currents.

Application to Electron Tubes

Letter symbols are used in much the same manner with electron tubes. However, instead of using lowercase and capital-letter subscripts to distinguish between ac and total or dc components, we now use *different* letters, and we always use lowercase subscripts. In the grid circuit, g denotes an ac component, while c denotes a total or a dc component. In the plate circuit, p is used for the ac component, and b is used for the dc or for the total value. This nomenclature is illustrated in Fig. A–2 to distinguish voltage values in the plate circuit of a vacuum tube.

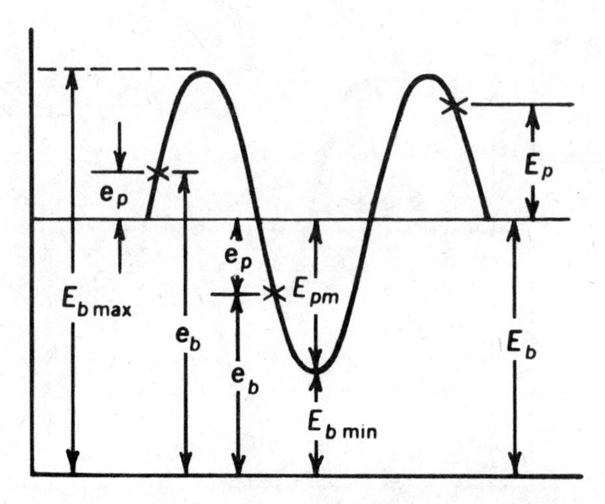

$E_{b\,max}$ — maximum value of the total plate voltage.

e_p — instantaneous value of the ac component.

e_b — instantaneous value of the total plate voltage.

E_{pm} — maximum value of the ac component.

$E_{b\,min}$ — minimum value of the total plate voltage.

E_p — rms value of the ac component.

E_b — average value, or dc component.

Figure A-2 Letter symbols for vacuum-tube plate voltages.

Index

E

F

T

U